Data Assimilation and its Applications

Edited by
Maithili Sharan
Jean Pierre Issartel

Previously published in *Pure and Applied Geophysics*
(PAGEOPH), Volume 169, No. 3, 2012

 Birkhäuser

Editors
Maithili Sharan
Indian Institute of Technology
Centre for Atmospheric Sciences
Hauz Khas
110016 New Delhi
India

Jean Pierre Issartel
Centre d'Etudes du Bouchet
rue Lavoisier 5
91710 Vert le Petit Cedex
France

ISBN 978-3-0348-0441-7   ISBN 978-3-0348-0442-4 (eBook)
DOI 10.1007/978-3-0348-0442-4

Library of Congress Control Number: 2012935550

Cover illustration: Based on Fig. 7 from "Identification of a Point of Release by Use of Optimally Weighted Least Squares" by J.-P. Issartel, M. Sharan and S. K. Singh.

Cover design: deblik, Berlin.

Printed on acid-free paper

Springer Basel AG is part of Springer Science+Business Media

www.birkhauser-science.com

Contents

Pure Appl. Geophys. 169 (2012), 309–310
© 2011 Springer Basel AG
DOI 10.1007/s00024-011-0388-x

Data Assimilation and its Applications

Maithili Sharan[1] and Jean Pierre Issartel[2]

1. Preface

Significant advances have been made in the meteorological observations through direct and indirect measurements. New techniques have been emerged to provide the improved observations. In addition, satellites are providing the valuable data over the globe. Specific local and regional field programs have been launched, in the recent years, in various parts of the world, to understand the atmospheric and oceanic processes. In view of public awareness and importance associated with natural hazards, such as tropical cyclones, hurricanes, drought, flood, the extensive, better resolved observations, both over atmosphere and ocean, are being collected. Simultaneously, with the availability of computational power, high resolution models have been developed for the improved prediction of meteorological variables governing the weather and climate.

Data assimilation has been used operationally for meteorological modeling and prediction. It is also useful as an inverse modeling technique for diagnosing pollutant emission source locations and strengths as a parametric estimation problem. In this context, assimilating the data in the models is a key issue. For this purpose, various data assimilation techniques within Bayesian and non-Bayesian

framework ranging from Least-Square, nudging, 3-D variational (3DVAR), 4-D variational (4DVAR), Kalman filter, etc., have evolved over the years.

In view of the above, this topical issue on "Data Assimilation and Its Applications" has been planned. This issue covers both theoretical and application aspects of data assimilation to atmospheric sciences including atmospheric pollution and ocean.

The problem of variational data assimilation for a nonlinear evolution model is formulated by Shutyaev and Le Dimet in their paper as an optimal control problem to find the initial condition function. The numerical algorithm is developed to compute the sensitivity coefficients for the analysis error using the fundamental control functions. A local ensemble transform Kalman filter (LETKF) is implemented with the Weather Research and Forecasting (WRF) model by Miyoshi and Kunii and real observations are assimilated to assess the newly developed WRF-LETKF system. The paper by Xie and MacDonald addresses the problem of how to choose the momentum variables for 3DVAR analysis, because there are three options, and all three have been used in certain popular and even operation 3DVAR systems. This paper uses theoretical analysis, an idealized case, and a real case to demonstrate that eastward and northward velocity, or vorticity and divergence, are preferable control variables than stream function and velocity potential for variational systems.

The field of meteorology has been a breeding ground for data assimilation techniques because of the pressing need to provide accurate forecasts to the public. The paper by Mishra and Krishnamurti summarizes the results obtained by them when they

[1] Centre for Atmospheric Sciences, Indian Institute of Technology Delhi, Hauz Khas, New Delhi 110016, India. E-mail: mathilis@cas.iitd.ac.in
[2] Centre d'Etudes du Bouchet, 5, rue Lavoisier, BP 3, 91710 Vert le Petit Cedex, France.

conducted an Observing System Simulation Experiment to quantify the effect of assimilating Global Precipitation Measurement mission data, in the form of precipitation estimates, into the Florida State University global spectral model. The paper by Zagar reviews basics of the 4-D variational data assimilation in the tropics based on a simplified tropical assimilation model. This work provides a better understanding of the relative impact of different equatorially trapped wave solutions in the variational analysis in the tropics. MOHANTY et al. in their paper made an attempt to evaluate the impact of the 3DVAR data assimilation within WRF modeling system to simulate two heavy rainfall events along west coast of India. The next three papers are concerned with the assimilation of satellite data. The paper by MACKARO et al. examines physical and computational issues encountered in land surface data assimilation of satellite skin temperatures to improve surface flux specifications in weather forecast and air quality models. The paper by FORE et al. deals with the rain-corrections applied to the scatterometer data for wind retrieval specifically for the hurricane cases with ultimate objective of assimilation of IR and microwave data in numerical models. In the paper by SINGH et al. assimilation experiments are performed with the WRF models' 3D-Var scheme to evaluate the impact of directly assimilating the Advanced Television and Infrared Observation Satellite Operational Vertical Sounder (ATOVS) radiance, including AMSU-A, AMSU-B and HIRS, on the analysis and forecasts of a mesoscale model over the Indian region during the 2008 summer monsoon.

Data assimilation has important applications to air quality studies. The paper by PENENKO et al. presents a concept of environmental modeling and forecasting based on the variational technique along with an example for the Siberian region. The problem of data assimilation is considered part of methodology for assessment of the atmospheric state and for solution of direct and inverse problems. Studies by ISSARTEL et al., SHARAN et al., and YEE touch on a research area of atmospheric pollution monitoring and modeling—identification of a point release of air contaminants. ISSARTEL et al. have used, in their paper, the inverse modeling approach by optimally weighted least squares, and suggest using the covariance matrix of the measurement errors and alternative weighing which enters in a more unified approach. SHARAN et al. present a least-square data assimilation approach for identification of single and simultaneous multiple-point releases from the atmospheric concentration measurements. As a part of the methodology, an algorithm for optimization, free from initialization and taking care of singularities in a natural way, is proposed. YEE, in his paper, proposes a Bayesian framework to solve an inverse problem of estimating tracer emissions to the atmosphere from a finite set of concentration measurements and the methodology is validated against a full-scale atmospheric dispersion experiment involving a multiple point source release. The paper by SCHMEHL et al. presents a Genetic algorithm variational approach to data assimilation and its use in atmospheric transport and dispersion problems. The Genetic algorithm is applied to determine source information for a volcanic eruption from satellite data. The ensemble based Kalman filter, an alternative data assimilation approach, is used in the paper by BEI et al. to explore the targeted meteorological observation locations for improving ozone prediction in Houston and the surrounding area.

In their paper, AGOSHKOV and ZALESNY formulated the problems of the variational assimilation of observed data of sea level height, as well as the temperature and salinity measurements from the Argo system of buoys, with the use of the global 3-D model of ocean dynamics developed at the authors Institute. In the paper by FUJII et al., a full-fledged data assimilation system has been set up simulating successfully the structure of the barrier layer in the equatorial Pacific Ocean.

We would like to place on record our gratitude to all the authors who have contributed to this special issue. Their associations to this special issue are very much appreciated. We wish to thank all the reviewers for their valuable comments/suggestions. We also thank Dr. Renata Dmowska, Editor-in-Chief for Topical Issues of PAGEOPH, for her assistance in bringing out this special issue.

Finally, we express our sincere thanks to the editorial staff of PAGEOPH for the meticulous care they have taken in preparing this special edition.

Pure Appl. Geophys. 169 (2012), 311–320
© 2011 Springer Basel AG
DOI 10.1007/s00024-011-0371-6

Fundamental Control Functions and Error Analysis in Variational Data Assimilation

V. Shutyaev[1] and F.-X. Le Dimet[2]

Abstract—The problem of variational data assimilation for a nonlinear evolution model is formulated as an optimal control problem to find the initial condition function. The equation for the error of the optimal solution (analysis) is derived through the errors of the input data (background and observation errors). The numerical algorithm is developed to compute the sensitivity coefficients for the analysis error using the fundamental control functions. Application to the variational data assimilation problem for a model of ocean thermodynamics is considered.

Key words: Data assimilation, error analysis, fundamental control functions.

1. Statement of the Problem

Consider the mathematical model of a physical process that is described by the evolution problem

$$\begin{cases} \frac{\partial \varphi}{\partial t} = F(\varphi) + f, & t \in (0, T) \\ \varphi|_{t=0} = u, \end{cases} \quad (1)$$

where $\varphi = \varphi(t)$ is the unknown function belonging for any t to the Hilbert space X, $u \in X$, F is a nonlinear operator mapping X into X. Let $Y = L_2(0, T; X), \|\cdot\|_Y = (\cdot, \cdot)_Y^{1/2}, f \in Y$. Let us introduce the functional

$$J(u) = \frac{\alpha}{2} \|u - u_0\|_X^2 + \frac{1}{2} \|C\varphi - \varphi_{\text{obs}}\|_{Y_{\text{obs}}}^2, \quad (2)$$

where $\alpha = \text{const} \geq 0$, $u_0 \in X$, $\varphi_{\text{obs}} \in Y_{\text{obs}}$ are prescribed functions (observational data), Y_{obs} is a Hilbert space (observational space), $C: Y \to Y_{\text{obs}}$ a linear continuous operator.

---
[1] Institute of Numerical Mathematics, Russian Academy of Sciences, Gubkina 8, 119333 Moscow, Russia.
[2] MOISE project (CNRS, INRIA, UJF, INPG); LJK, Université de Grenoble, BP 51, 38051 Grenoble Cedex 9, France. E-mail: Francois-Xavier.Le-Dimet@imag.fr

Consider the following data assimilation problem with the aim to identify the initial condition: find u and φ such that they satisfy (1), and on the set of solutions to Eq. 1, the functional (2) takes the minimum value, i.e.

$$\begin{cases} \frac{\partial \varphi}{\partial t} = F(\varphi) + f, & t \in (0, T) \\ \varphi|_{t=0} = u \\ J(u) = \inf_v J(v). \end{cases} \quad (3)$$

Problems in the form (3) were studied by Pontryagin *et al.* (1962), Lions (1968, 1988) (see also Agoshkov and Marchuk, 1993; Blayo *et al.*, 1998; Glowinski and Lions, 1994; Kurzhanskii and Khapalov, 1991; Le Dimet and Talagrand, 1986; Marchuk *et al.*, 1996; Navon, 1995, and others).

The necessary optimality condition reduces the problem (3) to the following optimality system:

$$\begin{cases} \frac{\partial \varphi}{\partial t} = F(\varphi) + f, & t \in (0, T) \\ \varphi|_{t=0} = u, \end{cases} \quad (4)$$

$$\begin{cases} -\frac{\partial \varphi^*}{\partial t} - (F'(\varphi))^* \varphi^* = -C^*(C\varphi - \varphi_{\text{obs}}), \\ \varphi^*|_{t=T} = 0, \end{cases} \quad (5)$$

$$\alpha(u - u_0) - \varphi^*|_{t=0} = 0 \quad (6)$$

with the unknowns φ, φ^*, and u, where $(F'(\varphi))^*$ is the adjoint to the Frechet derivative of F, and C^* is the adjoint to C defined by $(C\varphi, \psi)_{Y_{\text{obs}}} = (\varphi, C^*\psi)_Y$, $\varphi \in Y, \psi \in Y_{\text{obs}}$.

Suppose that $u_0 = \bar{u} + \xi_1$, $\varphi_{\text{obs}} = C\bar{\varphi} + \xi_2$, $f = \bar{f} + \xi_3$, where $\xi_1 \in X$, $\xi_2 \in Y_{\text{obs}}$, $\xi_3 \in Y$, and $\bar{\varphi}$ is the solution to the problem (1) with $u = \bar{u}$, $f = \bar{f}$:

$$\begin{cases} \frac{\partial \bar{\varphi}}{\partial t} = F(\bar{\varphi}) + \bar{f}, & t \in (0, T) \\ \bar{\varphi}|_{t=0} = \bar{u}. \end{cases} \quad (7)$$

The solution $\bar{\varphi}$ may be treated as "exact", and the functions ξ_1, ξ_2, ξ_3 may be treated as the errors of the input data u_0, φ_{obs}, f, respectively, ξ_2 being the observational error, and ξ_3 the model error.

Having supposed that the solution of the problem (4)–(6) exists, we study the influence of the errors ξ_1, ξ_2, ξ_3 on the optimal solution u.

2. Error Equation via Hessian

The system (4)–(6) with the three unknowns φ, φ^*, u may be treated as an operator equation of the form

$$\mathcal{F}(U, U_d) = 0, \qquad (8)$$

where $U = (\varphi, \varphi^*, u)$, $U_d = (u_0, \varphi_{\text{obs}}, f)$.

The following equality holds for the exact solution:

$$\mathcal{F}(\bar{U}, \bar{U}_d) = 0, \qquad (9)$$

with $\bar{U} = (\bar{\varphi}, \bar{\varphi}^*, \bar{u})$, $\bar{U}_d = (\bar{u}, C\bar{\varphi}, \bar{f})$, $\bar{\varphi}^* = 0$. From (8)–(9), we get

$$\mathcal{F}(U, U_d) - \mathcal{F}(\bar{U}, \bar{U}_d) = 0. \qquad (10)$$

Let $\delta U = U - \bar{U}$, $\delta U_d = U_d - \bar{U}_d$. Then (10) gives

$$\mathcal{F}(\bar{U} + \delta U, \bar{U}_d + \delta U_d) - \mathcal{F}(\bar{U}, \bar{U}_d) = 0. \qquad (11)$$

From (11), with an accuracy of the second order in δU, δU_d, we obtain

$$\mathcal{F}'_U(\bar{U}, \bar{U}_d)\delta U + \mathcal{F}'_{U_d}(\bar{U}, \bar{U}_d)\delta U_d = 0, \qquad (12)$$

where $\mathcal{F}'_U, \mathcal{F}'_{U_d}$ are the Gateâux derivatives with respect to U and U_d.

Let $\delta\varphi = \varphi - \bar{\varphi}$, $\delta u = u - \bar{u}$; then $\delta U = (\delta\varphi, \varphi^*, \delta u)$, $\delta U_d = (\xi_1, \xi_2, \xi_3)$. By calculating the derivatives $\mathcal{F}'_U, \mathcal{F}'_{U_d}$, it is easily seen that Eq. 12 is equivalent to the system:

$$\begin{cases} \frac{\partial \delta\varphi}{\partial t} - F'(\bar{\varphi})\delta\varphi &= \xi_3, \ t \in (0, T), \\ \delta\varphi|_{t=0} &= \delta u, \end{cases} \qquad (13)$$

$$\begin{cases} -\frac{\partial \varphi^*}{\partial t} - (F'(\bar{\varphi}))^*\varphi^* &= (F''(\bar{\varphi})\delta\varphi)^*\bar{\varphi}^* - C^*(C\delta\varphi - \xi_2), \\ \varphi^*|_{t=T} &= 0, \end{cases}$$

$$\qquad (14)$$

$$\alpha(\delta u - \xi_1) - \varphi^*|_{t=0} = 0, \qquad (15)$$

where $\bar{\varphi}$ is the solution of the original problem (7).

The problem (13)–(15) is a linear data assimilation problem; for $\bar{\varphi}^* = 0$ it is equivalent to the following minimization problem: find u and φ such that

$$\begin{cases} \frac{\partial\varphi}{\partial t} - F'(\bar{\varphi})\varphi &= \xi_3, \quad t \in (0, T) \\ \varphi|_{t=0} &= u \\ J_1(u) &= \inf_v J_1(v), \end{cases} \qquad (16)$$

where

$$J_1(u) = \frac{\alpha}{2}\|u - \xi_1\|_X^2 + \frac{1}{2}\|C\varphi - \xi_2\|_{Y_{\text{obs}}}^2. \qquad (17)$$

Consider the Hessian $H(u)$ of the functional (2) which is defined by the successive solution of the following problems:

$$\begin{cases} \frac{\partial\varphi}{\partial t} &= F(\varphi) + f, \quad t \in (0, T) \\ \varphi|_{t=0} &= u, \end{cases}$$

$$\begin{cases} -\frac{\partial\varphi^*}{\partial t} - (F'(\varphi))^*\varphi^* &= -C^*(C\varphi - \varphi_{\text{obs}}), \\ \varphi^*|_{t=T} &= 0, \end{cases}$$

$$\begin{cases} \frac{\partial\psi}{\partial t} - F'(\varphi)\psi &= 0, \ t \in (0, T), \\ \psi|_{t=0} &= v, \end{cases}$$

$$\begin{cases} -\frac{\partial\psi^*}{\partial t} - (F'(\varphi))^*\psi^* &= (F''(\varphi)\psi)^*\varphi^* - C^*C\psi, \\ \psi^*|_{t=T} &= 0, \end{cases}$$

$$H(u)v = \alpha v - \psi^*|_{t=0}.$$

For $u = \bar{u}, \xi_2 = \xi_3 = 0$, the Hessian $H(\bar{u})$ coincides with the Hessian H of the functional (17); it is defined by the successive solutions of the following problems:

$$\begin{cases} \frac{\partial\psi}{\partial t} - F'(\bar{\varphi})\psi &= 0, \ t \in (0, T), \\ \psi|_{t=0} &= v, \end{cases} \qquad (18)$$

$$\begin{cases} -\frac{\partial\psi^*}{\partial t} - (F'(\bar{\varphi}))^*\psi^* &= -C^*C\psi, \ t \in (0, T) \\ \psi^*|_{t=T} &= 0, \end{cases}$$

$$\qquad (19)$$

$$Hv = \alpha v - \psi^*|_{t=0}. \qquad (20)$$

Let us introduce the operator R_2 acting on the functions $g \in Y_{\text{obs}}$ according to the formula

$$R_2 g = \theta^*|_{t=0}, \qquad (21)$$

where θ^* is the solution to the adjoint problem

$$\begin{cases} -\frac{\partial\theta^*}{\partial t} - (F'(\bar{\varphi}))^*\theta^* &= C^*g, \ t \in (0, T) \\ \theta^*|_{t=T} &= 0. \end{cases} \qquad (22)$$

We introduce also the operator $R_3 : Y \to X$ defined successively by the formulas:

$$\begin{cases} \frac{\partial \theta_1}{\partial t} - F'(\bar{\varphi})\theta_1 & = h, \ h \in Y, \\ \theta_1|_{t=0} & = 0, \end{cases} \quad (23)$$

$$\begin{cases} -\frac{\partial \theta_1^*}{\partial t} - (F'(\bar{\varphi}))^*\theta_1^* & = -C^*C\theta_1, \ t \in (0,T) \\ \theta_1^*|_{t=T} & = 0, \end{cases}$$
$$(24)$$

$$R_3 h = \theta_1^*|_{t=0}. \quad (25)$$

From (18)–(25) we conclude that the system (13)–(15) is equivalent to the control equation for δu:

$$H\delta u = R_1 \xi_1 + R_2 \xi_2 + R_3 \xi_3, \quad (26)$$

where $R_1 = \alpha E$, and E is the identity operator in X.

For $\alpha > 0$, the Hessian H is coercive, H^{-1} exists, and the error δu of the optimal solution depends on the errors ξ_1, ξ_2, ξ_3 linearly and continuously. The influence of the errors ξ_1, ξ_2, ξ_3 on the value of δu is determined by the operators $H^{-1}R_1$, $H^{-1}R_2$, $H^{-1}R_3$, respectively. The values of the norms of these operators may be considered as sensitivity criteria: the less is the norm of the operator $H^{-1}R_i$, the less impact on δu is given by the corresponding error ξ_i. This criteria may be used also to choose the regularization parameter α (TIKHONOV et al., 1995; MOROZOV et al., 1987; ALIFANOV, 1988).

3. Fundamental Control Functions for Error Analysis

Consider first the case that $Y_{obs} = Y$, $C = E$ (the identity operator). We assume that the operator H defined by (18)–(20) is positive (i.e. the inequality holds: $(Hv, v)_X > 0$, $v \neq 0$) and has a complete orthonormal system in X of eigenfunctions v_k corresponding to the eigenvalues μ_k:

$$Hv_k = \mu_k v_k, \quad (27)$$

where $(v_k, v_l)_X = \delta_{kl}$, $k, l = 1, 2, \ldots$, and δ_{kl} the Kronecker delta.

It is easily seen that the eigenvalue problem (27) is equivalent to the system:

$$\begin{cases} \frac{\partial \varphi_k}{\partial t} - F'(\bar{\varphi})\varphi_k & = 0, \ t \in (0,T), \\ \varphi_k|_{t=0} & = v_k, \end{cases} \quad (28)$$

$$\begin{cases} -\frac{\partial \varphi_k^*}{\partial t} - (F'(\bar{\varphi}))^*\varphi_k^* & = -\varphi_k, \ t \in (0,T) \\ \varphi_k^*|_{t=T} & = 0, \end{cases} \quad (29)$$

$$\alpha v_k - \varphi_k^*|_{t=0} = \mu_k v_k. \quad (30)$$

We say that the system of functions $\{\varphi_k, \varphi_k^*, v_k\}$ satisfying (28)–(30) is the system of *fundamental control functions*. It can be easily proved that the functions $\{\varphi_k\}$, $\{\varphi_k^*\}$, $\{v_k\}$ defined by formulas (28)–(30) form complete orthonormal systems in the corresponding spaces.

Using the fundamental control functions, we can obtain the solution of the error Eq. 26 in the explicit form. The Eq. 26 is equivalent to the system (13)–(15) and may be written as the following system:

$$\begin{cases} \frac{\partial \psi}{\partial t} - F'(\bar{\varphi})\psi & = 0, \ t \in (0,T), \\ \psi|_{t=0} & = \delta u, \end{cases} \quad (31)$$

$$\begin{cases} -\frac{\partial \psi^*}{\partial t} - (F'(\bar{\varphi}))^*\psi^* & = -\psi, \ t \in (0,T) \\ \psi^*|_{t=T} & = 0, \end{cases} \quad (32)$$

$$\alpha \delta u - \psi^*|_{t=0} = P, \quad (33)$$

where $P = R_1 \xi_1 + R_2 \xi_2 + R_3 \xi_3$ is the right-hand side of (26).

The solution ψ, ψ^*, δu of the system (31)–(33) may be represented in the form (SHUTYAEV, 1991):

$$\psi = \sum_k a_k \varphi_k, \quad \psi^* = \sum_k a_k \varphi_k^*, \quad \delta u = \sum_k a_k v_k,$$
$$(34)$$

where $\varphi_k, \varphi_k^*, v_k$ are the fundamental control functions defined by (28)–(30), $a_k = (P, v_k)_X/\mu_k$.

From (34), we have the representation for the Fourier coefficients $(\delta u)_k$ of the error δu:

$$(\delta u)_k = (\delta u, v_k)_X = a_k = \frac{1}{\mu_k}(R_1\xi_1 + R_2\xi_2 + R_3\xi_3, v_k)_X.$$
$$(35)$$

Note that

$$(R_1\xi_1, v_k)_X = \alpha(\xi_1, v_k)_X. \quad (36)$$

By definition of R_2, R_3,

$$(R_2\xi_2, v_k)_X = (\theta^*|_{t=0}, v_k)_X,$$

$$(R_3\xi_3, v_k)_X = (\theta_1^*|_{t=0}, v_k)_X,$$

where θ^* is the solution of (22) for $g = \xi_2$, and θ_1, θ_1^* are the solutions of (23)–(24) for $h = \xi_3$. From (22) and (28) we get

$$(\theta^*|_{t=0}, v_k)_X = (\xi_2, \varphi_k)_Y.$$

Hence,

$$(R_2\xi_2, v_k)_X = (\xi_2, \varphi_k)_Y. \qquad (37)$$

Analogously, from (24) and (28),

$$(\theta_1^*|_{t=0}, v_k)_X = (-\theta_1, \varphi_k)_Y.$$

Further, (23) and (29) give

$$(\theta_1, -\varphi_k)_Y = (\theta_1|_{t=0}, \varphi_k^*|_{t=0})_X + (\xi_3, \varphi_k^*)_Y = (\xi_3, \varphi_k^*)_Y.$$

Hence,

$$(R_3\xi_3, v_k)_X = (\xi_3, \varphi_k^*)_Y. \qquad (38)$$

From (35)–(38) we obtain the expression for the Fourier coefficients $(\delta u)_k$ of the error δu of the optimal solution through the errors ξ_1, ξ_2, ξ_3:

$$(\delta u)_k = \frac{\alpha}{\mu_k}(\xi_1, v_k)_X + \frac{1}{\mu_k}(\xi_2, \varphi_k)_Y + \frac{1}{\mu_k}(\xi_3, \varphi_k^*)_Y, \qquad (39)$$

where $\{\varphi_k, \varphi_k^*, v_k\}$ are the fundamental control functions defined by (28)–(30).

In more general case, when $Y_{obs} \neq Y$, $C \neq E$, we may also define the control functions $\varphi_k, \varphi_k^*, v_k$ as the solutions to the system:

$$\begin{cases} \frac{\partial \varphi_k}{\partial t} - F'(\bar\varphi)\varphi_k = 0, \ t \in (0, T), \\ \varphi_k|_{t=0} = v_k, \end{cases} \qquad (40)$$

$$\begin{cases} -\frac{\partial \varphi_k^*}{\partial t} - (F'(\bar\varphi))^* \varphi_k^* = -C^*C\varphi_k, \\ \varphi_k^*|_{t=T} = 0, \end{cases} \qquad (41)$$

$$\alpha v_k - \varphi_k^*|_{t=0} = \mu_k v_k. \qquad (42)$$

It may be easily verified that in this case the error relationship (39) changes to

$$(\delta u)_k = \frac{\alpha}{\mu_k}(\xi_1, v_k)_X + \frac{1}{\mu_k}(\xi_2, C\varphi_k)_{Y_{obs}} + \frac{1}{\mu_k}(\xi_3, \varphi_k^*)_Y. \qquad (43)$$

From (39), (43), it is seen that the fundamental control functions play a role of "sensitivity functions"; they are the weight-functions for the corresponding errors ξ_1, ξ_2, ξ_3 in the representations (39), (43). Note that the fundamental control functions $\{\varphi_k, \varphi_k^*, v_k\}$ do not depend on the structure of the errors ξ_1, ξ_2, ξ_3 and may be calculated beforehand for each k in need.

4. Singular Vectors and Sensitivity Coefficients

Consider the Error Eq. 26. Under the hypotheses of Sect. 3, we may rewrite (26) as

$$\delta u = H^{-1}R_1\xi_1 + H^{-1}R_2\xi_2 + H^{-1}R_3\xi_3. \qquad (44)$$

Suppose that the errors ξ_1, ξ_2, ξ_3 do not correlate and the following relation is satisfied:

$$\|\delta u\|_X^2 = \|T_1\xi_1\|_X^2 + \|T_2\xi_2\|_X^2 + \|T_3\xi_3\|_X^2, \qquad (45)$$

where $T_i = H^{-1}R_i$. From (45),

$$\|\delta u\|_X^2 = (T_1^*T_1\xi_1, \xi_1)_X + (T_2^*T_2\xi_2, \xi_2)_{Y_{obs}} + (T_3^*T_3\xi_3, \xi_3)_Y, \qquad (46)$$

where $T_1^* : X \rightarrow X$, $T_2^* : X \rightarrow Y_{obs}$, $T_3^* : X \rightarrow Y$ are the adjoints to T_i, $i = 1, 2, 3$.

Each summand in (46) determines the impact given by the corresponding error ξ_i. We have

$$(T_1^*T_1\xi_1, \xi_1)_X \leq \|T_1^*T_1\|\|\xi_1\|_X^2,$$

$$(T_2^*T_2\xi_2, \xi_2)_{Y_{obs}} \leq \|T_2^*T_2\|\|\xi_2\|_{Y_{obs}}^2, \qquad (47)$$

$$(T_3^*T_3\xi_3, \xi_3)_Y \leq \|T_3^*T_3\|\|\xi_3\|_Y^2,$$

and the i-th inequality becomes an equality when ξ_i is the singular vector of T_i corresponding to the largest singular value $\sigma_{max}^2 = \|T_i^*T_i\|$. The values $r_i = \sqrt{\|T_i^*T_i\|}$ may be considered as *sensitivity coefficients* which clearly demonstrate the measure of influence of the corresponding error upon the optimal solution. The higher the relative sensitivity coefficient, the more effectual is the error in question.

As above, we assume that the operator H defined by (18)–(20) is positive and has a complete orthonormal system in X of eigenfunctions v_k corresponding to the eigenvalues $\mu_k : Hv_k = \mu_k v_k$, $(v_k, v_l)_X = \delta_{kl}$.

Consider the operator T_1. Since $T_1 = H^{-1}R_1 = \alpha H^{-1} = T_1^*$, the singular vectors of T_1 are the eigenvectors v_i of H, and the corresponding sensitivity coefficient is equal to

$$r_1 = \sqrt{\|T_1^*T_1\|} = \frac{\alpha}{\mu_{min}}, \qquad (48)$$

where μ_{min} is the lower spectrum bound of H.

For the operator $T_2: Y_{obs} \to X$ the following statement is valid (LE DIMET and SHUTYAEV, 2005).

The singular values σ_k^2 and the corresponding orthonormal (right) singular vectors $w_k \in Y_{obs}$ of the operator T_2 are defined by the formulas:

$$\sigma_k^2 = \frac{\mu_k - \alpha}{\mu_k^2}, \quad w_k = \frac{1}{\sqrt{\mu_k - \alpha}} C\varphi_k, \quad (49)$$

where μ_k are the eigenvalues of the control operator H, and φ_k are the fundamental control functions defined by (28). The left singular vectors of T_2 coincide with the eigenvectors v_k of H:

$$T_2 T_2^* v_k = \sigma_k^2 v_k, \quad k = 1, 2, \ldots$$

The sensitivity coefficient $r_2 = \sqrt{\|T_2^* T_2\|}$ is defined by the formula:

$$r_2 = \max_k \frac{\sqrt{\mu_k - \alpha}}{\mu_k}. \quad (50)$$

The equality $(T_2^* T_2 \xi_2, \xi_2)_{Y_{obs}} = r_2^2 \|\xi_2\|_{Y_{obs}}^2$ holds if $\xi_2 = w_{k_0}$, where w_{k_0} is the singular vector of T_2 corresponding to the largest singular value $\sigma_{k_0}^2$.

Consider now the operator $T_3 = H^{-1} R_3$. To determine the sensitivity coefficient $r_3 = \sqrt{\|T_3^* T_3\|}$, we need to derive R_3^*. For $h \in Y, p \in X$, we have from (23)–(25):

$$(R_3 h, p)_X = (\theta_1^*|_{t=0}, p)_X = -(C^* C\theta_1, \phi)_Y$$
$$= (C\theta_1, C\phi)_{Y_{obs}},$$

where θ_1, θ_1^* are the solutions to (23)–(25), and ϕ is the solution to (18) for $v = p$. Further,

$$(R_3 h, p)_X = -(\theta_1, C^* C\phi)_Y = (h, \phi^*)_Y$$

and $R_3^* p = \phi^*$, where ϕ^* is the solution to the adjoint problem:

$$\begin{cases} -\frac{\partial \phi^*}{\partial t} - (F'(\bar{\varphi}))^* \phi^* = -C^* C\phi, \ t \in (0, T) \\ \phi^*|_{t=T} = 0. \end{cases}$$
$$(51)$$

The operator $R_3 R_3^*: X \to X$ may be defined as follows: for given $p \in X$ find ϕ as the solution of (18) for $v = p$, find ϕ^* as the solution of (51), and for $h = \phi^*$ find θ_1, θ_1^* as the solutions of (23)–(25); then, put $R_3 R_3^* = \theta_1^*|_{t=0}$.

Therefore, the operator $T_3 T_3^* = H^{-1} R_3 R_3^* H^{-1}$ is defined by the successive solutions of the following problems (for given $v \in X$):

$$Hp = v, \quad (52)$$

$$\begin{cases} \frac{\partial \phi}{\partial t} - F'(\bar{\varphi})\phi = 0, \ t \in (0, T), \\ \phi|_{t=0} = p, \end{cases} \quad (53)$$

$$\begin{cases} -\frac{\partial \phi^*}{\partial t} - (F'(\bar{\varphi}))^* \phi^* = -C^* C\phi, \ t \in (0, T) \\ \phi^*|_{t=T} = 0. \end{cases}$$
$$(54)$$

$$\begin{cases} \frac{\partial \theta_1}{\partial t} - F'(\bar{\varphi})\theta_1 = \phi^*, \ t \in (0, T) \\ \theta_1|_{t=0} = 0, \end{cases} \quad (55)$$

$$\begin{cases} -\frac{\partial \theta_1^*}{\partial t} - (F'(\bar{\varphi}))^* \theta_1^* = -C^* C\theta_1, \ t \in (0, T) \\ \theta_1^*|_{t=T} = 0, \end{cases}$$
$$(56)$$

$$Hw = \theta_1^*|_{t=0}, \quad (57)$$

then

$$T_3 T_3^* v = w, \quad (58)$$

and for the sensitivity coefficient r_3 we have

$$r_3 = \sqrt{\|T_3 T_3^*\|}. \quad (59)$$

From the formulas (48), (50), (59), one can derive the typical behaviour of the sensitivity coefficients r_1, r_2, r_3, depending on the parameter α. The sensitivity coefficient r_1 is small and r_2, r_3 are large if α goes to zero and μ_{min} is close zero. If α goes to 1, the coefficient r_1 also is close to 1, and r_2, r_3 are decreasing, being less than r_1. Thus, with α increasing, the output regularization error increases, whereas the sensitivity to the observation and model errors decreases. If α goes to zero, the output regularization error vanishes; however, the sensitivity to the observation and model errors is increasing (it is usually due to the fact that the Hessian of the cost functional becomes ill-conditioned with μ_{min} close to zero). In most cases, $r_3 < r_2$, that is, the optimal solution is more sensitive to variations of the observation errors than to the model errors.

5. Variational Data Assimilation Problem for a Model of Ocean Thermodynamics

Consider the ocean thermodynamics problem in the following form:

$$T_t + (\bar{U}, \mathrm{Grad})T - \mathrm{Div}(\hat{a}_T \cdot \mathrm{Grad}\, T)$$
$$= f_T \text{ in } D \times (t_0, t_1),$$

$$T = T_0 \text{ for } t = t_0 \text{ in } D,$$

$$-v_T \frac{\partial T}{\partial z} = Q \text{ on } \Gamma_S \times (t_0, t_1),$$

$$\frac{\partial T}{\partial N_T} = 0 \text{ on } \Gamma_{w,c} \times (t_0, t_1),$$

$$\bar{U}_n^{(-)}T + \frac{\partial T}{\partial N_T} = \bar{U}_n^{(-)}d_T + Q_T \text{ on } \Gamma_{w,op} \times (t_0, t_1),$$

$$\frac{\partial T}{\partial N_T} = 0 \text{ on } \Gamma_H \times (t_0, t_1), \qquad (60)$$

where $T = T(x, y, z, t)$ is an unknown temperature function, $t \in (t_0, t_1), (x, y, z) \in D = \Omega \times (0, H), \Omega \subset R^2, H = H(x, y)$ is the function of the bottom relief, $Q = Q(x, y, t)$ is the total heat flux, $\bar{U} = (u, v, w), \hat{a}_T = \mathrm{diag}((a_T)_{ii}), (a_T)_{11} = (a_T)_{22} = \mu_T, (a_T)_{33} = v_T, f_T = f_T(x, y, z, t)$ are given functions. The boundary of the domain $\Gamma \equiv \partial D$ is represented as a union of four disjoint parts $\Gamma_S, \Gamma_{w,op}, \Gamma_{w,c}, \Gamma_H$, where $\Gamma_S = \Omega$ (the unperturbed ocean surface), $\Gamma_{w,op}$ is the liquid (open) part of vertical lateral boundary, $\Gamma_{w,c}$ is the solid part of the vertical lateral boundary, Γ_H is the ocean bottom. The other notations and a detailed description of the problem statement can be found in (AGOSHKOV et al., 2008).

Problem (60) can be written in the form of the following operator equation:

$$T_t + LT = \mathcal{F} + BQ, \quad t \in (t_0, t_1),$$
$$T = T_0, \quad t = t_0, \qquad (61)$$

where the equality is understood in the weak sense, namely,

$$(T_t, \hat{T}) + (LT, \hat{T}) = \mathcal{F}(\hat{T}) + (BQ, \hat{T}) \ \forall \hat{T} \in W_2^1(D), \qquad (62)$$

in this case $L, \mathcal{F}, B$ are defined by the following relations:

$$(LT, \hat{T}) \equiv \int_D (-T\mathrm{Div}(\bar{U}\hat{T}))\mathrm{d}D + \int_{\Gamma_{w,op}} \bar{U}_n^{(+)}T\hat{T}\mathrm{d}\Gamma$$
$$+ \int_D \hat{a}_T\mathrm{Grad}(T) \cdot \mathrm{Grad}(\hat{T})\mathrm{d}D,$$

$$\mathcal{F}(\hat{T}) = \int_{\Gamma_{w,op}} (Q_T + \bar{U}_n^{(-)}d_T)\hat{T}\mathrm{d}T + \int_D f_T\hat{T}\mathrm{d}D,$$

$$(T_t, \hat{T}) = \int_D T_t\hat{T}\mathrm{d}D, \quad (BQ, \hat{T}) = \int_\Omega Q\hat{T}|_{z=0}\mathrm{d}\Omega,$$

and the functions $\hat{a}_T, Q_T, f_T, Q$ are such that equality (62) makes sense. The properties of the operator L were studied in (AGOSHKOV et al., 2008).

Consider the data assimilation problem for the ocean surface temperature (see AGOSHKOV et al., 2008). Suppose the function $Q \in L_2(\Omega \times (t_0, t_1))$ is unknown in problem (60). Let also $T_{\mathrm{obs}}(x, y, t)$ be the only function on $\bar{\Omega} \equiv \Omega \cup \partial\Omega$ obtained for $t \in (t_0, t_1)$ by processing the observation data, and this function in its physical sense be an approximation to the surface temperature function on Ω, i.e. to $T|_{z=0}$. We suppose that $T_{\mathrm{obs}} \in L_2(\Omega \times (t_0, t_1))$, but the function T_{obs} may not possess greater smoothness, and hence, it cannot be used for the boundary condition on Γ_S. We admit the case when T_{obs} is present only on some subset of $\Omega \times (t_0, t_1)$, denote the characteristic function of this set by m_0. For definiteness sake, we assume that T_{obs} is trivial outside this subset.

Consider the data assimilation problem for the surface temperature in the following form: find T and Q such that

$$\begin{cases} T_t + LT = F + BQ, & \text{in } D \times (t_0, t_1), \\ T = T_0, & t = t_0 \\ J(Q) = \inf_Q J(Q), \end{cases} \qquad (63)$$

where

$$J(Q) = \frac{\alpha}{2}\int_{t_0}^{t_1}\int_\Omega |Q - Q^{(0)}|^2\mathrm{d}\Omega \mathrm{d}t$$
$$+ \frac{1}{2}\int_{t_0}^{t_1}\int_\Omega m_0|T|_{z=0} - T_{\mathrm{obs}}|^2\mathrm{d}\Omega \mathrm{d}t, \qquad (64)$$

and $Q^{(0)} = Q^{(0)}(x, y, t)$ is a given function, $\alpha = \mathrm{const} > 0$.

For $\alpha > 0$ this variational data assimilation problem has a unique solution. The existence of the optimal solution follows from the classic results of the theory of extremal problems, because it is not difficult to show that the solution to problem (60) continuously depends on the flux Q (a priori estimates are valid in the corresponding functional spaces), the functional J is weakly semicontinuous from below, and the space of admissible controls $L_2(\Omega \times (t_0, t_1))$ is weakly compact.

For $\alpha = 0$ the problem does not always have a solution, but, as was shown in AGOSHKOV *et al.* (2008), there is unique and dense solvability, which allows one to construct a sequence of regularized solutions minimizing the functional.

The optimality system determining the solution of the formulated variational data assimilation problem according to the necessary condition grad $J = 0$ has the form:

$$T_t + LT = \mathcal{F} + BQ \quad \text{in } D \times (t_0, t_1),$$
$$T = T_0, \quad t = t_0, \tag{65}$$

$$-(T^*)_t + L^*T^* = B^*m_0(T - T_{\text{obs}}) \quad \text{in } D \times (t_0, t_1),$$
$$T^* = 0, \quad t = t_1, \tag{66}$$

$$\alpha(Q - Q^{(0)}) + T^* = 0 \quad \text{on } \Omega \times (t_0, t_1), \tag{67}$$

where L^*, B^* are the operators adjoint to L and B, respectively.

In what follows we assume that $\text{supp}(m_0) = \bar{\Omega} \times [t_0, t_1]$ and study the problem of the sensitivity of the optimal solution Q to input data, namely, to errors in the functions T_{obs} and $Q^{(0)}$. For this application, we suppose that the model is perfect, so the model errors are not considered.

Let

$$Q^{(0)} = \bar{Q} + \xi_1, \quad T_{\text{obs}} = \bar{T}|_{z=0} + \xi_2, \tag{68}$$

where $\xi_1, \xi_2 \in L_2(\Omega \times (t_0, t_1))$, and $\bar{T}$ is the exact solution to the original problem for $Q = \bar{Q}$, :

$$\bar{T}_t + L\bar{T} = \mathcal{F} + B\bar{Q} \quad \text{in } D \times (t_0, t_1),$$
$$\bar{T} = T_0, \quad t = t_0. \tag{69}$$

The functions ξ_1, ξ_2 may be considered as errors in the input data $Q^{(0)}$ and T_{obs}, respectively. We are interested in the influence of these errors on the optimal solution Q obtained from the optimality system (65)–(67).

6. *Error Analysis via Hessian*

Let $\delta T = T - \bar{T}$, $\delta Q = Q - \bar{Q}$. Then from (65)–(67) we get the system for δT, δQ:

$$\delta T_t + L\delta T = B\delta Q \quad \text{in } D \times (t_0, t_1),$$
$$\delta T = 0, \quad t = t_0, \tag{70}$$

$$-(T^*)_t + L^*T^* = B^*m_0(\delta T - \xi_2) \quad \text{in } D \times (t_0, t_1),$$
$$T^* = 0, \quad t = t_1, \tag{71}$$

$$\alpha(\delta Q - \xi_1) + T^* = 0 \quad \text{on } \Omega \times (t_0, t_1). \tag{72}$$

This system is equivalent to the auxiliary data assimilation problem for determination of δT, δQ such that

$$\begin{cases} \delta T_t + L\delta T &= B\delta Q \quad \text{in } D \times (t_0, t_1), \\ \delta T &= 0, \quad t = t_0, \\ J_1(\delta Q) &= \inf_Q J_1(Q), \end{cases} \tag{73}$$

where

$$J_1(\delta Q) = \frac{\alpha}{2} \int_{t_0}^{t_1} \int_{\Omega} |\delta Q - \xi_1|^2 d\Omega dt$$
$$+ \frac{1}{2} \int_{t_0}^{t_1} \int_{\Omega} m_0 |\delta T|_{z=0} - \xi_2|^2 d\Omega dt. \tag{74}$$

Introduce the Hessian $\mathcal{H}$ of functional (74). It is easily seen that

$$\mathcal{H} = \mathcal{H}_0 + \alpha E,$$

where E is the identity operator and $\mathcal{H}_0$ is the operator defined on $v \in L_2(\Omega \times (t_0, t_1))$ by a successive solution of the problems:

$$\psi_t + L\psi = Bv \quad \text{in } D \times (t_0, t_1),$$
$$\psi = 0, \quad t = t_0, \tag{75}$$

$$-(\psi^*)_t + L^*\psi^* = B^*m_0\psi \quad \text{in } D \times (t_0, t_1),$$
$$\psi^* = 0, \quad t = t_1, \tag{76}$$

$$\mathcal{H}_0 v = \psi^* \quad \text{on } \Omega \times (t_0, t_1). \tag{77}$$

Introduce the auxiliary operator $R : L_2(\Omega \times (t_0, t_1)) \rightarrow L_2(\Omega \times (t_0, t_1))$ acting on the functions $g \in L_2(\Omega \times (t_0, t_1))$ by the formula

$$Rg = \theta^*|_{z=0},$$

where θ^* is the solution to the adjoint problem

$$-(\theta^*)_t + L^*\theta^* = B^*m_0g \quad \text{in } D \times (t_0, t_1),$$
$$\theta^* = 0, \quad t = t_1. \tag{78}$$

From (75)–(78) we conclude that the system (70)–(72) is equivalent to the control equation for δQ:

$$\mathcal{H}\delta Q = \alpha\xi_1 + R\xi_2. \tag{79}$$

It is not difficult to verify that the operator $\mathcal{H}$ acts in $L_2(\Omega \times (t_0, t_1))$ with the domain of definition $D(\mathcal{H}) = L_2(\Omega \times (t_0, t_1))$, it is bounded, self-adjoint and positive. If $\alpha > 0$, then $\mathcal{H}$ is positive definite. The operator R is bounded.

For $\alpha > 0$ Eq. 79 has the unique solution

$$\delta Q = \alpha\mathcal{H}^{-1}\xi_1 + \mathcal{H}^{-1}R\xi_2. \tag{80}$$

This gives

$$\|\delta Q\| \le \|T_1\xi_1\| + \|T_2\xi_2\|, \tag{81}$$

where $T_1 = \alpha\mathcal{H}^{-1}, T_2 = \mathcal{H}^{-1}R$, $T_i : L_2(\Omega \times (t_0, t_1)) \to L_2(\Omega \times (t_0, t_1))$, $i = 1, 2$, and the norm means the norm in $L_2(\Omega \times (t_0, t_1))$.

Each summand in (80) determines the effect of the corresponding error ξ_i on the error of the optimal solution δQ. We have

$$\|T_1\xi_1\| \le \sqrt{\|T_1^*T_1\|}\|\xi_1\|, \quad \|T_2\xi_2\| \le \sqrt{\|T_2^*T_2\|}\|\xi_2\|, \tag{82}$$

where $T_i^* : L_2(\Omega \times (t_0, t_1)) \to L_2(\Omega \times (t_0, t_1))$ are the operators adjoint to T_i, $i = 1, 2$, and in this case the ith inequality turns into an equality if ξ_i is a singular vector of the operator T_i corresponding to the maximal singular value $\sigma_{\max}^2 = \|T_i^*T_i\|$.

The values

$$r_1 = \sqrt{\|T_1^*T_1\|}, \quad r_2 = \sqrt{\|T_2^*T_2\|} \tag{83}$$

are considered as sensitivity coefficients demonstrating the influence of the corresponding error ξ_i on the optimal solution. The greater the coefficient r_i (the norm of the response operator T_i), the greater can be the influence of the error ξ_i on the optimal solution.

Formulas (80)–(82) imply

$$\|\delta Q\| \le \sqrt{\|T_1^*T_1\|}\|\xi_1\| + \sqrt{\|T_2^*T_2\|}\|\xi_2\|$$
$$= r_1\|\xi_1\| + r_2\|\xi_2\|.$$

Consider the operator T_1. Since $T_1 = \alpha\mathcal{H}^{-1} = T_1^*$, the $T_1^*T_1 = \alpha^2\mathcal{H}^{-2}$ and

$$r_1 = \sqrt{\|T_1^*T_1\|} = \frac{\alpha}{\alpha + \mu_{\min}}, \tag{84}$$

where $\mu_{\min}$ is the lower spectrum bound of the operator $\mathcal{H}_0$. Note that $r_1 \le 1$ (because $\mu_{\min} \ge 0$).

The following representation holds for T_2:

$$T_2T_2^* = \mathcal{H}_0\mathcal{H}^{-2},$$

where $\mathcal{H}$ is the Hessian of the functional (74), and $\mathcal{H}_0$ is the operator determined by the formulas (75)–(77).

In fact, $T_2 = \mathcal{H}^{-1}R$, $T_2T_2^* = \mathcal{H}^{-1}RR^*\mathcal{H}^{-1}$, and R^* is determined for $p \in L_2(\Omega \times (t_0, t_1))$ by the formula

$$R^*p = \phi|_{z=0},$$

where ϕ is the solution to the problem

$$\phi_t + L\phi = Bp \quad \text{in } D \times (t_0, t_1),$$
$$\phi = 0, \quad t = t_0. \tag{85}$$

Using the definitions of the operators R and $\mathcal{H}$, we see that $RR^* = \mathcal{H}_0$ and $T_2T_2^* = \mathcal{H}_0\mathcal{H}^{-2}$.

Hence, the sensitivity coefficient r_2 is defined by the formula:

$$r_2 = \sqrt{\|\mathcal{H}_0\mathcal{H}^{-2}\|}, \tag{86}$$

where for the norm $\|\mathcal{H}_0\mathcal{H}^{-2}\|$ one can consider the upper spectrum bound of the operator $\mathcal{H}_0\mathcal{H}^{-2}$.

In the finite-dimensional case, if $\alpha = 0, T_2T_2^* = \mathcal{H}_0^{-1}$ and the coefficient r_2 is defined by the formula:

$$r_2 = \frac{1}{\sqrt{\mu_{\min}}}, \tag{87}$$

where $\mu_{\min}$ is the lower spectrum bound of the control operator $\mathcal{H}_0$. Note that under the finite-dimensional approximation the operator $\mathcal{H}_0$ becomes an ill-conditioned matrix with the least eigenvalue close to zero. In this case the coefficient r_2 may be sufficiently large, i.e., the optimal solution may be very sensitive to observation errors.

The typical behaviour of the sensitivity coefficients r_1 and r_2, demonstrated by numerical experiments, is given in Fig. 1. Here the values of the sensitivity coefficients are presented versus the regularization parameter α.

Figure 1 displays that with $\alpha \ge \alpha_0 > 0$, the sensitivity coefficients r_1, r_2 are bounded. As α tends to

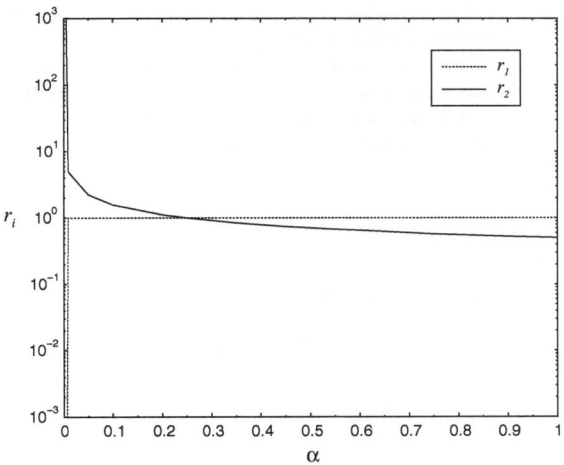

Figure 1
Sensitivity coefficients versus the parameter α

1, the coefficient r_1 also tends to 1, and r_2 decreases and is less than r_1. The situation is different when α goes to zero. If $\alpha = 0$, then $r_1 = 0$, but r_2 is large, and it is rapidly increasing with $\alpha \to 0$. This means that it is necessary to introduce a regularizer with the parameter $\alpha > 0$, which makes the problem of finding the optimal solution stable with respect to observation errors.

7. Conclusions

The sensitivity of the optimal solution to the input errors is determined by the value of the sensitivity coefficients which are the norms of the specific response operators relating the error of the input to the error of the optimal solution. The maximum error growth for the output is given by the singular vectors of the corresponding response operator. The singular vectors are the fundamental control functions which form complete orthonormal systems in specific functional spaces and may be used for error analysis.

Acknowledgments

This work was carried out within the MOISE project (CNRS, INRIA, UJF, INPG) and the project 09-01-00284 of the Russian Foundation for the Basic Research, within the project ADAMS (INRIA) and under the Federal Purpose Program "Kadry".

REFERENCES

AGOSHKOV V.I., MARCHUK G.I. *On solvability and numerical solution of data assimilation problems.* Russ. J. Numer. Analys. Math. Modelling, 1993, *8*, no.1, 1–16.

AGOSHKOV V.I., PARMUZIN E.I., SHUTYAEV V.P. *Numerical algorithm of variational assimilation of the ocean surface temperature data.* J. Comp. Math. Math. Phys., 2008, *48*, 1371–1391.

ALIFANOV O.M. *Inverse problems of heat exchange,* Mashinostroenie, Moscow, 1988.

E.BLAYO, J.BLUM, J.VERRON. Assimilation variationnelle de données en océanographie et réduction de la dimension de l'espace de contrôle, *Équations aux Dérivées Partielles et Applications,* Paris, Elsevier, 1998, pp. 205–219.

CHAVENT G. *Local stability of the output least square parameter estimation technique.* Math. Appl. Comp., 1983, *2*, 3–22.

DONTCHEV A.L. *Perturbations, Approximations and Sensitivity Analysis of Optimal Control Systems.* (Lecture Notes in Control and Information Sciences; 52), Berlin, Springer, 1983.

GEJADZE, I., LE DIMET, F.-X., SHUTYAEV, V.P. *On analysis error covariances in variational data assimilation.* SIAM J. Sci. Comput. (2008), *30*, no. 4, 1847–1874.

GEJADZE, I., LE DIMET, F.-X., SHUTYAEV, V.P. *On optimal solution error covariances in variational data assimilation problems.* J. Comp. Phys. (2010), *229*, 2159–2178.

GEJADZE I.YU. and SHUTYAEV V.P. *An optimal control problem of initial data restoration.* Comput. Math. Math. Phys., 1999, *39*, no. 9, 1416–1425.

Glowinski R., LIONS J.L. *Exact and approximate controllability for distributed parameter systems.* Acta Numerica, 1994, *1*, 269–378.

KRAVARIS, C. and SEINFELD, J.H. (1985) *Identification of parameters in distributed parameter systems by regularization.* SIAM J. Control and Optimization, 1985, *23*, 217–241.

KURZHANSKII, A.B. and KHAPALOV, A.YU. *An observation theory for distributed-parameter systems.* J. Math. Syst. Estimat. Control, 1991, *1*, 389–440.

LE DIMET F.-X., NAVON I.M., DAESCU D.N. *Second-order information in data assimilation.* Monthly Weather Review, 2002, *130*, no. 3, 629–648.

LE DIMET F.-X., NGODOCK, LUONG B., VERRON J. *Sensitivity analysis in variational data assimilation.* J. Meteorol. Soc. Japan, 1997, *75*(1B), 245–255.

LE DIMET F.X., TALAGRAND O. *Variational algorithms for analysis and assimilation of meteorological observations: theoretical aspects.* it Tellus, 1986, *38A*, 97–110.

LE DIMET F.-X., SHUTYAEV V. *On deterministic error analysis in variational data assimilation.* Nonlinear Processes in Geophysics, 2005, *12*, 481–490.

LIONS J.L. *Contrôle optimal des systèmes gouvernés par des équations aux dérivées partielles,* Paris, Dunod, 1968.

LIONS J.L. *Contrôlabilité Exacte Perturbations et Stabilisation de Systèmes Distribués,* Paris, Masson, 1988.

MARCHUK G.I., AGOSHKOV V.I., SHUTYAEV V.P. *Adjoint and Perturbation Algorithms in Nonlinear Problems,* New York, CRC Press Inc., 1996.

11

MOROZOV, V.A. *Regular Methods for Solving the Ill-Posed Problems*, Nauka, Moscow, 1987.

NAVON I.M. Variational data assimilation, optimal parameter estimation and sensitivity analysis for environmental problems. *Computational Mechanics'95*, New York, Springer, 1995, v. 1, pp. 740–746.

PONTRYAGIN L.S., BOLTYANSKII V.G., GAMKRELIDZE R.V., MISCHENKO E.F. *The Mathematical Theory of Optimal Processes*, New York, John Wiley, 1962.

SHUTYAEV, V.P. *On a class of insensitive control problems.* Control and Cybernetics, 1994, *23*, 257–266.

SHUTYAEV, V.P. *Control Operators and Iterative Algorithms for Variational Data Assimilation Problems*, Nauka, Moscow, 1991.

TIKHONOV, A.N., LEONOV, A.S., YAGOLA, A.G. *Nonlinear Inverse Problems*, Nauka, Moscow, 1995.

(Received November 3, 2010, accepted April 19, 2011, Published online July 29, 2011)

Pure Appl. Geophys. 169 (2012), 321–333
© 2011 Springer Basel AG
DOI 10.1007/s00024-011-0373-4

The Local Ensemble Transform Kalman Filter with the Weather Research and Forecasting Model: Experiments with Real Observations

TAKEMASA MIYOSHI[1] and MASARU KUNII[1]

Abstract—The local ensemble transform Kalman filter (LET-KF) is implemented with the Weather Research and Forecasting (WRF) model, and real observations are assimilated to assess the newly-developed WRF-LETKF system. The WRF model is a widely-used mesoscale numerical weather prediction model, and the LETKF is an ensemble Kalman filter (EnKF) algorithm particularly efficient in parallel computer architecture. This study aims to provide the basis of future research on mesoscale data assimilation using the WRF-LETKF system, an additional testbed to the existing EnKF systems with the WRF model used in the previous studies. The particular LETKF system adopted in this study is based on the system initially developed in 2004 and has been continuously improved through theoretical studies and wide applications to many kinds of dynamical models including realistic geophysical models. Most recent and important improvements include an adaptive covariance inflation scheme which considers the spatial and temporal inhomogeneity of inflation parameters. Experiments show that the LETKF successfully assimilates real observations and that adaptive inflation is advantageous. Additional experiments with various ensemble sizes show that using more ensemble members improves the analyses consistently.

Key words: Data assimilation, numerical weather prediction, ensemble Kalman filter.

1. Introduction

Along with the recent improvement of computational capability, high-resolution numerical weather prediction (NWP) is becoming capable of direct quantitative representation of extreme mesoscale weather. Moreover, the high resolving capability improves the predictability of larger-scale extreme weather such as tropical cyclones and organized convective systems, in which multi-scale interactions play an important role. For better prediction of such extreme weather events, data assimilation is a key component to make most use of available observations and to provide better initial conditions.

A number of data assimilation studies have been published using the Weather Research and Forecasting (WRF) model (SKAMAROCK *et al.* 2005), a widely-used community mesoscale NWP model. Some used three-dimensional and four-dimensional variational methods from a community package known as the WRF-VAR (HUANG *et al.* 2009); others used ensemble Kalman filter (EnKF) methods including the one within the Data Assimilation Research Testbed (DART; ANDERSON 2009) framework and another using the serial ensemble square root filter (EnSRF; WHITAKER and HAMILL 2002) approach (e.g., ZHANG *et al.* 2004). Previous studies using these data assimilation systems helped improve our knowledge on the mechanisms of mesoscale weather and contributed to further improvements of the mesoscale NWP capabilities.

In this study, the local ensemble transform Kalman filter (LETKF; HUNT *et al.* 2007) is developed with the WRF model and is assessed with real observations. The LETKF is a kind of EnSRF based on the ensemble transform Kalman filter (ETKF; BISHOP *et al.* 2001), but the algorithm is designed to be particularly efficient in parallel computer architecture by taking an advantage of independent local analyses of the local ensemble Kalman filter (LEKF; OTT *et al.* 2004). A number of studies have shown promise of the LETKF with wide applications including global and regional atmosphere and ocean models (e.g., SZUNYOGH *et al.* 2005, 2008; MIYOSHI and ARANAMI 2006; MIYOSHI and YAMANE 2007; MIYOSHI *et al.* 2010) and even a Martian atmosphere model (HOFFMAN *et al.* 2010).

[1] Department of Atmospheric and Oceanic Science,
University of Maryland, College Park, Maryland, MD 20742, USA.
E-mail: miyoshi@atmos.umd.edu

This paper focuses on the description of the newly developed WRF-LETKF system and presents the performance of assimilating real observations. This provides the basis of future research on mesoscale data assimilation using the WRF-LETKF system, an additional testbed to the existing EnKF systems with the WRF model used in the previous studies. Section 2 describes the WRF-LETKF system and related software. Then, experiments are presented in Sect. 3. Finally, summary and discussion are provided in Sect. 4.

2. The WRF-LETKF System

HUNT et al. (2007) proposed the LETKF algorithm as a kind of EnSRF, in which the analysis ensemble members are constructed by a linear combination of the forecast ensemble members. The ensemble transform matrix, composed of the weights of the linear combination, is computed for each local subset of the state vector independently, which allows essentially perfectly parallel computations. The local subset depends on the covariance localization (ANDERSON and ANDERSON 1999). Without localization, a single ensemble transform matrix is shared for all components of the state vector to apply the same linear combination globally. With a limited ensemble size in realistic applications, it is known that covariance localization is indispensable, which results in different ensemble transform matrices for each local subset of the state vector. Typically a local subset of the state vector contains all variables at a grid point, which is the default setting in the WRF-LETKF. However, inter-variable localization (KANG et al. 2011) may be considered according to the dynamical nature, in which case a local subset contains only a part of the variables at a grid point. There is no theoretical limitation in the choice of the local subset.

The Massage Passing Interface (MPI)-parallelized Fortran90 code of the LETKF in this study is based on the system initially developed by MIYOSHI (2005) and has been continuously improved thereafter. The features of the current version include spatial covariance localization with physical distance (MIYOSHI et al. 2007), four-dimensional EnKF (4D-EnKF) for

appropriate treatment of asynchronous observations (HUNT et al. 2004) and temporal covariance localization. Among many improvements (e.g., MIYOSHI and YAMANE 2007; MIYOSHI et al. 2007, 2010), the most recent updates include the adaptive inflation scheme (MIYOSHI 2011), which finds the covariance inflation factors adaptively for each local subset. The approach is essentially equivalent to the adaptive inflation scheme of ANDERSON (2009) which is implemented and available in the DART system. The LETKF code has been applied to and assessed with the Lorenz 40-dimensional model (LORENZ 1996; LORENZ and EMANUEL 1998), a low-dimensional atmospheric general circulation model (AGCM) known as the SPEEDY model (MOLTENI 2003; MIYOSHI 2005) and its extension with the carbon dioxide variable (KANG 2009), realistic atmospheric models such as the AGCM for the Earth Simulator (AFES; OHFUCHI et al. 2004; MIYOSHI and YAMANE 2007) and the Japan Meteorological Agency (JMA) operational global and mesoscale models (MIYOSHI and ARANAMI 2006; MIYOSHI and SATO 2007; MIYOSHI et al. 2010), Regional Ocean Modeling System (ROMS; SHCHEPETKIN and MCWILLIAMS 2005), a global ocean model known as the Geophysical Fluid Dynamics Laboratory (GFDL) Modular Ocean Model 2.b (MOM2, PENNY 2011), and GFDL Mars AGCM (WILSON and HAMILTON 1996; GREYBUSH 2011). All applications showed successful data assimilation using the LETKF code, some of which showed promising results with real observations. Among the wide applications, the core part of the LETKF code is shared, so that findings and improvements from each application can benefit other applications directly. This is an important advantage of using this system.

Figure 1 shows the flowchart of the WRF-LETKF system, where the shaded boxes indicate the processes newly developed in this study. The white boxes are the processes provided by the WRF model package. In this study, the Advanced Research WRF (ARW) version 3.2 is used. The National Centers for Environmental Prediction (NCEP) operational global data assimilation system (GDAS) uses the observation data of the PREPBUFR format (KEYSER 2010). The NCEP PREPBUFR data are available through the University Corporation for Atmospheric Research (UCAR) data server for registered users. In this study, the "PREPBUFR Decoder" is developed, in which the PREPBUFR data are decoded for the input to the LETKF. Then, the LETKF reads in the observation data and m-member ensemble forecasts, and writes out the m-member analyses and ensemble mean states. The WRF model requires inputs of not only the atmospheric initial condition but also boundary conditions such as sea-surface temperature (SST), soil temperature and moisture, and surface topography. These input files are generated by the PRE-WRF process, which includes converting the NCEP global atmosphere, land, and ocean analysis data to the suitable formats for running the WRF forecasts. The atmospheric initial condition generated by PRE-WRF is replaced by the LETKF analysis in the MERGE process, whereas the other files generated by PRE-WRF, e.g., boundary conditions, land and ocean states, and topography, remain untouched. MERGE contains the computation of the column dry air mass variable (mu), which is a diagnostic variable but the WRF model requires in the initial condition. Then the WRF model evolves each member, and the forecast

outputs are post-processed in the POST-WRF process. POST-WRF creates pressure-level data, which are saved in a storage disk space. The pressure-level data include the initial time, i.e., the analysis, and forecast times according to the WRF model settings. Finally, the CONV process converts the forecast data files to the simple 4-byte binary format of the LETKF. A particular attention was made to design the input/output (i/o) interface of the LETKF to be simple, so that it can be adapted relatively easily to other mesoscale NWP models.

Since the forecast/data assimilation cycle interval of the NCEP GDAS is 6 h, the PREPBUFR data exist every 6 h for the analyses at 0000, 0600, 1200, and 1800 UTC and include observation data within a 6-h period centered at the analysis time. The WRF-LETKF cycle interval is chosen to be the same 6 h in this study, so that 9-h WRF forecasts create the first guess, and the PREPBUFR Decoder makes the observing times rounded to the hour for the hourly input to the 4D-LETKF (Fig. 2). Here, no observation

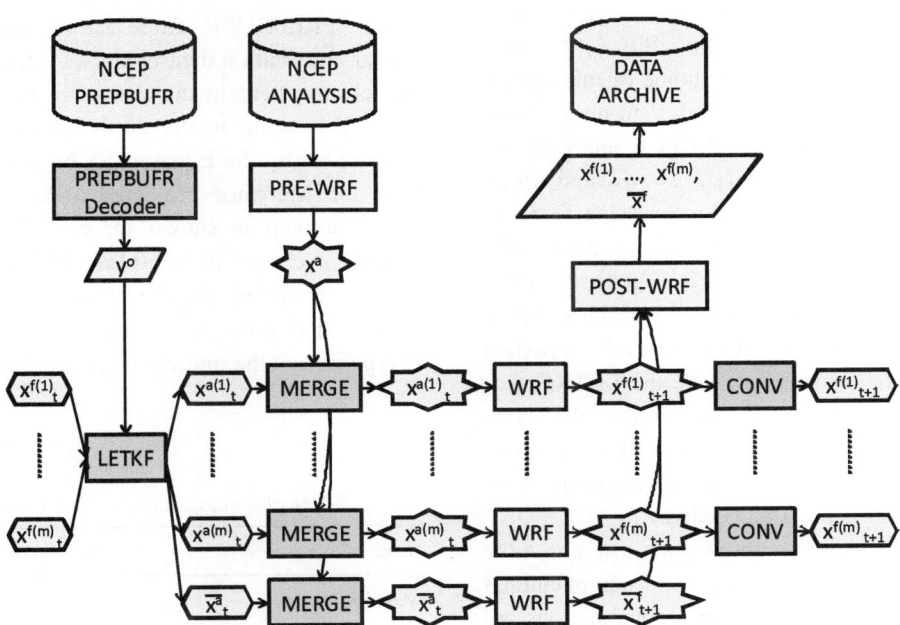

Figure 1

The flowchart of the WRF-LETKF system. $x_t^{a(k)}$ and $x_t^{f(k)}$ denote the analysis and forecast state vectors of the k-th ensemble member at time t, respectively. m denotes the ensemble size, and *overbar* denotes the ensemble mean. The data box of y^o is the observation data for the input to the LETKF. The *white* and *grey rectangular boxes* are the processes provided by the WRF model package and by this study, respectively. The *hexagonal boxes* indicate simple binary formatted data files of the state vector for the LETKF i/o. The *star-shaped boxes* are netCDF formatted data files of the state vector for the WRF model i/o

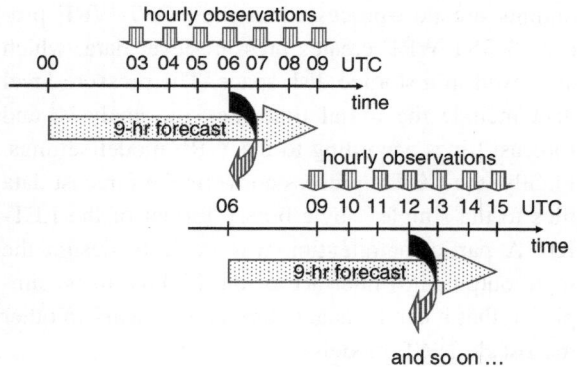

Figure 2

Schematic showing the forecast-analysis cycle (adapted from Fig. 1 of MIYOSHI *et al.* 2010)

is used twice; e.g., in the case of Fig. 2, there is no intersection between the observation sets at 0900 UTC for the analyses of 0600 and 1200 UTC.

Since the LETKF requires only a subset of the output variables from the WRF model, the CONV process generates the LETKF input composed of three-dimensional wind components (u, v, w), temperature (T), three-dimensional pressure (p), three-dimensional geopotential (ph), humidity (qv), water/ice variables depending on the choice of microphysics schemes, surface pressure (ps), 2-m temperature ($t2$), and 2-m humidity ($q2$). LETKF analyzes the prognostic variables of u, v, w, T, ph, qv, and water/ice variables, which affects the successive forecast. The other variables (p, ps, $t2$, and $q2$) are used in the observation operators. Although it is generally difficult to analyze w, ph, and water/ice variables mainly due to the limited observations and imperfect knowledge of the error covariance, the LETKF analyzes them by default by considering the error covariance from ensemble forecasts. Inter-variable localization allows the covariance between these variables to be turned off as needed.

The WRF model executions for each ensemble member are independent. With a parallel computational environment, the ensemble forecast computations are parallelized ideally perfectly with the entire parallel processors. In order to minimize the inter-node communication, the analysis and forecast data are temporarily stored in the local scratch disk spaces. Efficient parallel i/o is implemented in the LETKF, so

that the LETKF accesses simultaneously the ensemble forecast data spread out in each local disk space. Once the LETKF reads in the ensemble forecasts, they are scattered and gathered among the nodes at the beginning and end of the LETKF. These are the only two inter-node communications during the entire procedure of Fig. 1. For most parallel computational environments, the inter-node communication is ignorable due to perfectly parallel ensemble forecasts and parallel-efficient LETKF analyses.

3. Experiments with Real Observations

3.1. Experimental Setup

The WRF-LETKF system is tested with the real observations from the NCEP PREPBUFR data which include upper sounding data from radiosondes and dropsondes (ADPUPA), surface stations (ADPSFC), ships and buoys (SFCSHP), aircrafts (AIRCFT), wind profilers (PROFLR), and satellite-based winds (SAT-WND, VADWND, SPSSMI, QKSWND). The six-character codes of the observation types are defined in the PREPBUFR Table 1.a (KEYSER 2010). The global observation dataset allows setting up the WRF model anywhere in the world. In this study, WRF-LETKF is setup in the northwestern Pacific region surrounded by the Equator, 55°N, 100°E, and 180°E with the Mercator projection; the entire computational domain is shown in, e.g., Fig. 5. The grid spacing is chosen to be 60 km at 22.5°N, giving 136 by 108 grid points for the variables other than u and v. The horizontal staggered grid of the Arakawa C type makes the number of grid points to be 137 by

Table 1

The list of experiments with 20 members

Name of experiment	Description
NOOBS	Pure ensemble forecasts without inflation or any other ensemble transform (without observation data)
ADAPT	LETKF assimilation with adaptive inflation
FIXED	LETKF assimilation with a fixed inflation parameter

108 for u, and 136 by 109 for v. The vertical resolution is chosen to be 39 levels for the variables other than w and ph, which have 40 levels due to the vertical staggered grid of the Lorenz type.

Throughout this study, the localization parameters of the LETKF are chosen to be 400 km in the horizontal, 0.4 ln p in the vertical, and 3 h in time. The localization parameters correspond to the one-sigma length at which the Gaussian localization function becomes $e^{-0.5}$. Fitting the Gaussian function to the GASPARI and COHN (1999) fifth order function (HAMILL et al. 2001) gives the localization radius of influence to be the one-sigma length multiplied by a factor of about 3.65; i.e., the 400-km localization setting corresponds to about the 1,460-km radius of influence. The localization parameter values are chosen from those with which MIYOSHI et al. (2007, 2010) assimilated real observations successfully with the LETKF using the T159 AFES and JMA global models at an approximately 80 km resolution. The sensitivities to the localization parameters are out of scope of this study.

The initial time of the experiments is chosen to be 1200 UTC 3 September 2008. The main focus of the period of experiments is Typhoon Sinlaku which was generated on 0000 UTC 8 September at 1,004 hPa and was rapidly intensified after 1200 UTC 8 September to reach a devastating 935 hPa on 1200 UTC 10 September, southeast off the coast of Taiwan. Sinlaku started weakening slowly as it approached Taiwan, but it kept its intensity at 950 hPa when it struck Taiwan around 1200 UTC 13 September. The Observing System Research and Predictability Experiment (THORPEX) Pacific Asian Regional Campaign (T-PARC) made special observations for Sinlaku including dropsondes from aircrafts (e.g., ELSBERRY and HARR 2008; PARSONS et al. 2008; CHOU et al. 2011). Other foci during the experimental period are the synoptic high pressure system over the eastern offshore of northern Japan, near the northeastern boundary of the model domain, and the related precipitation system on the western side of the high pressure system (Fig. 4a).

The initial conditions to start the data assimilation cycle experiments are generated by spinning-up for 5 days the states from the NCEP analyses at 1200

UTC on randomly chosen dates in August 2007 and 2008. The boundary conditions for the 5-day forecasts from 1200 UTC 29 August to 1200 UTC 3 September are given by the NCEP analyses, so that all ensemble members use the same boundary conditions. Inevitably all ensemble members are very similar to each other near the boundaries, but they substantially differ in the inner domain. The single boundary conditions from the NCEP analyses are used for all ensemble members throughout this study, although previous studies (e.g., ZHANG et al. 2004; SAITO et al. 2009, 2010) showed important advantages of perturbing boundary conditions. The impact of unperturbed boundary conditions is not essential in this study as shown in the following results.

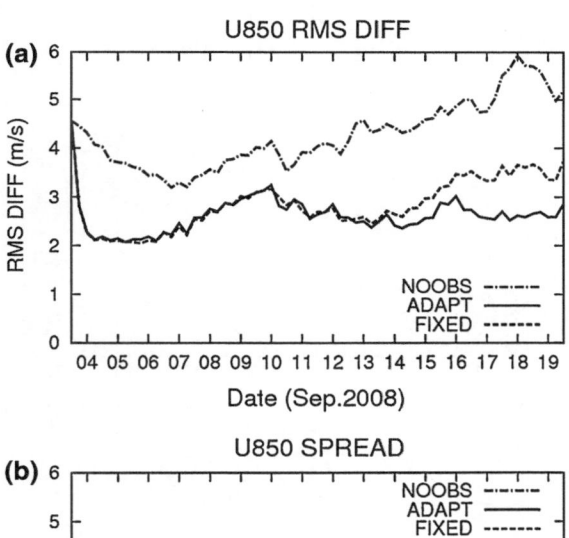

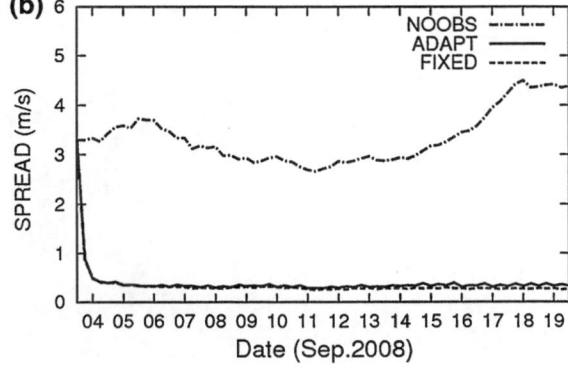

Figure 3
Time series of **a** analysis root mean square differences from the NCEP analysis and **b** analysis ensemble spreads for zonal wind component (m s^{-1}) at 850 hPa, averaged over the entire model domain

3.2. *Experiments with and without Adaptive Inflation*

First, the ensemble size is fixed at 20, and three experiments are performed as shown in Table 1. The NOOBS experiment does not assimilate observations, i.e., it is pure forecasts from the randomly chosen initial conditions. NOOBS goes through the flows shown in Fig. 1, but PREPBUFR data are replaced by empty files. With the empty observation input, the LETKF computes only the ensemble averages, and the analysis fields are simply the first guess fields without inflation or any other ensemble transform. NOOBS is useful to ensure the proper computational flows as well as to find the impact of the unperturbed

boundary conditions. The ADAPT and FIXED experiments assimilate PREPBUFR observation data with and without adaptive inflation, respectively. The comparison reveals the impact of adaptive inflation. The fixed inflation parameter is chosen to be 1.20 (20% covariance inflation), which gives the best performance among 3, 10, 15, and 20% inflation settings.

The effects of the fixed boundary conditions are found to be limited only near the boundaries. Although the boundary conditions are from the NCEP analysis, Fig. 3a shows that forecast fields of NOOBS do not approach to the NCEP analysis during the 16-day period of the experiment. In

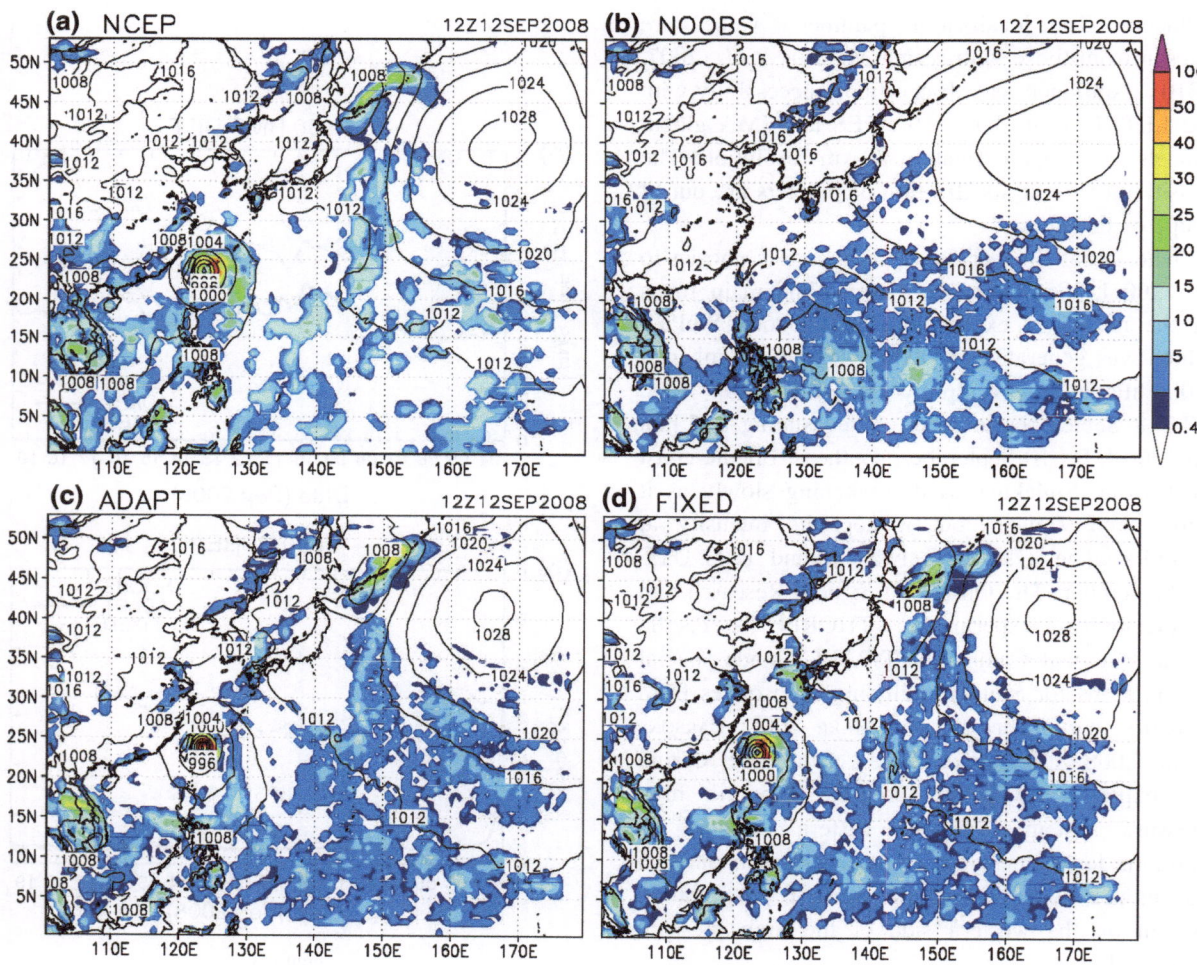

Figure 4

Horizontal maps of 6-h accumulated precipitation (mm, *shades*) and mean sea-level pressure (hPa, *contours*) at 850 hPa on 1200 UTC 12 September 2008 **a** NCEP analysis **b** NOOBS, **c** ADAPT, **d** FIXED. **b**, **c**, and **d** show 6-h deterministic forecasts initiated by the ensemble mean states

addition, Fig. 3b shows essentially no decay of the ensemble spread of NOOBS, and Fig. 5a shows that perturbations grow within the domain even though no perturbation is given at the boundaries. Moreover, Fig. 4b shows that NOOBS does not generate Typhoon Sinlaku; namely, the randomly chosen initial conditions and the boundary conditions do not have the source of the generation of Sinlaku.

Figures 3 and 4 also indicate that the LETKF assimilates observations successfully. Figure 3 shows that both ensemble spread and root mean square (RMS) differences from the NCEP analyses drop rapidly in the first 12 h. Smaller RMS differences indicate assimilating real observations makes the analyses closer to the NCEP analyses. In addition,

Fig. 4 shows that both ADAPT and FIXED generate Sinlaku similarly to the NCEP analysis. It also shows that the synoptic high near the northeastern boundary of the domain is significantly better represented by ADAPT and FIXED than NOOBS. The related precipitation system is missing in NOOBS, but it is well represented in ADAPT and FIXED. Generally, the three WRF experiments show larger areas of weak precipitation than the NCEP analysis, probably due to the characteristics of the WRF model.

The ensemble spread corresponds to the uncertainty of the state estimates. The ensemble spreads of both ADAPT and FIXED tend to be small; e.g., Fig. 3 shows about 0.3 m s^{-1} wind spread at 850 hPa which is probably smaller than the actual uncertainty.

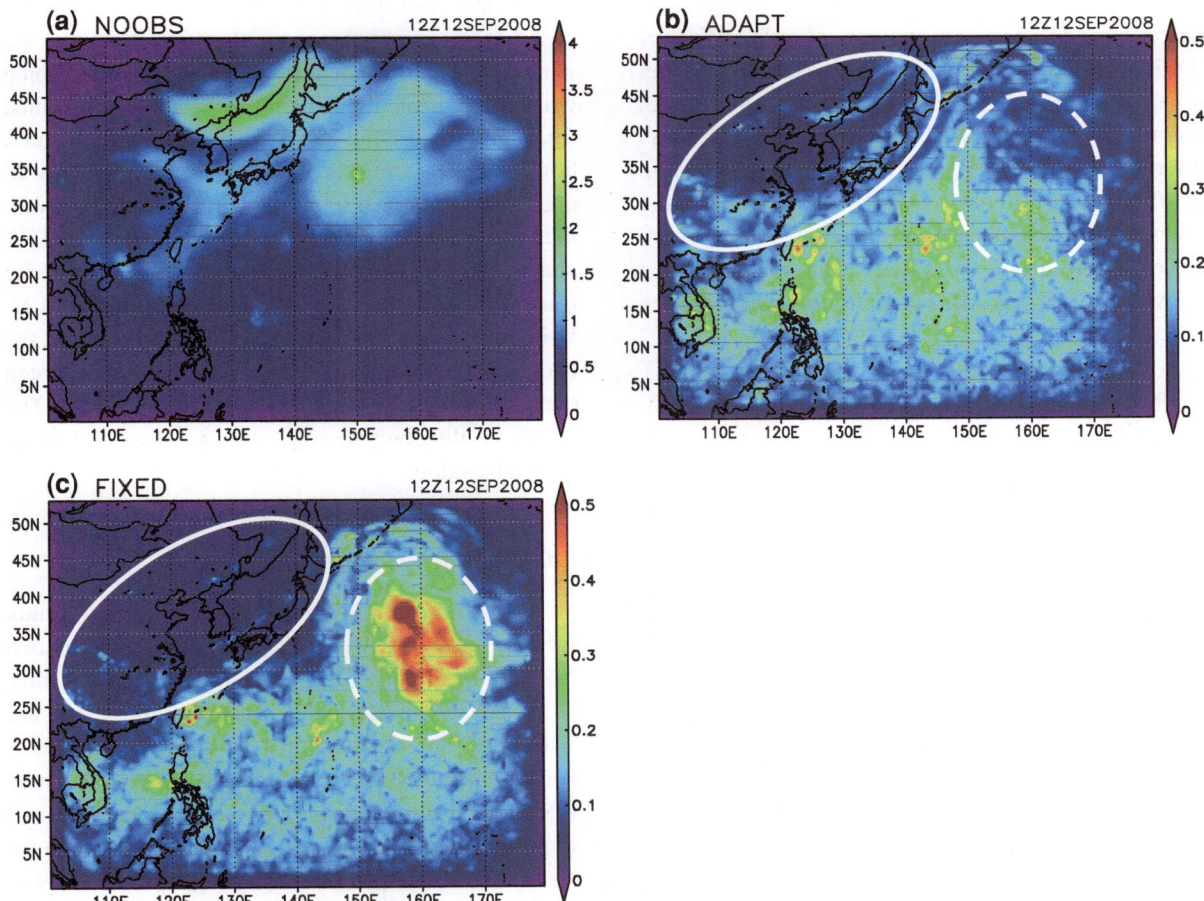

Figure 5

Horizontal maps of the temperature ensemble spread (K) at 500 hPa on 1200 UTC 12 September 2008 for **a** NOOBS, **b** ADAPT, and **c** FIXED. The *solid (dashed) circles* indicate typical areas with dense (sparse) observations. Note that NOOBS has an order of magnitude different color scale

19

The ensemble spreads of the 6-h forecasts are only slightly larger than the analysis ensemble spreads. The adaptive inflation technique is designed to meet the statistics between the ensemble spread and observation-minus-forecast innovations, but it is still spinning up. Yet, Fig. 5 indicates the general tendency of smaller ensemble spreads over densely observed areas where the uncertainty is expected to be smaller.

Adaptive inflation is advantageous over fixed inflation. Figure 3a indicates that ADAPT shows significantly smaller RMS differences from the NCEP analyses than FIXED after about 10-day cycle. Probably due to the spin-up of the adaptive inflation parameters, the advantage is clearer near the end of the experiments. Figure 6 shows the verification of 6-h forecasts relative to radiosonde observations, also indicating consistent advantages of ADAPT over FIXED. Moreover, the ensemble spread is also better in ADAPT. Figure 5c shows overly small (large) ensemble spread of FIXED in the densely (sparsely) observed areas, which agrees with what MIYOSHI et al. (2010) showed in their Fig. 7. The adaptive inflation scheme captures under-dispersion (over-dispersion) of the ensemble from the observation-minus-forecast innovations and tries to compensate it by adapting

the inflation factors at each grid point. Figure 7 shows that the large (small) adaptive inflation values are estimated over the densely (sparsely) observed areas, which make the ensemble spread significantly larger (smaller) in the densely (sparsely) observed areas, as highlighted by solid (dashed) circles in Fig. 5.

3.3. Sensitivity to the Ensemble Size

In order to investigate the sensitivity to the ensemble size, three additional experiments are performed with 27, 34, and 41 ensemble members. Here, adaptive inflation is adopted. With more members, Fig. 8 shows that the analysis RMS difference from the NCEP analyses becomes smaller, and the ensemble spread is increased consistently. In addition, Fig. 9 shows that 6-h forecasts fit better to radiosonde observations with more members, particularly for winds (Fig. 9a). Although Fig. 9 shows only zonal wind component, almost identical results are obtained for meridional wind component. For temperature, however, increasing the ensemble size more than 27 does not show much improvement (Fig. 9b). Overall, increasing the ensemble size improves the LETKF analysis consistently, and biggest improvements are found when increasing the ensemble size from 20 to 27. Using more members requires more computations; the choice of the ensemble size is a tradeoff between the analysis accuracy and the computational cost. The details on the computational time are provided in the next subsection.

3.4. Computational Efficiency

This study employed a cluster of seven Linux servers, each of which was equipped with two 2.6 GHz quad-core CPUs (Central Processing Unit), i.e., total 56 cores. The ensemble forecasts are distributed among the cores as shown in Table 2. For example, in the 20-member experiments, three WRF model forecasts using two cores were executed simultaneously in each node. The wall-clock time for the 20-member forecasts using total 42 cores was approximately 264 s. The LETKF was executed with the number of cores shown in the right-most column of Table 2. Namely, the 20-member LETKF was executed using 42 cores, i.e., six processes per node;

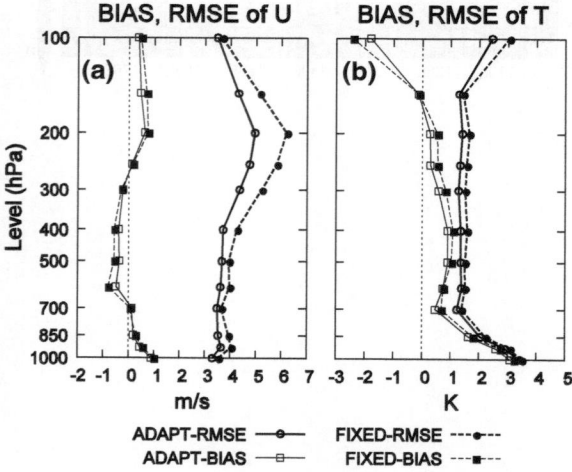

Figure 6
Verifications of 6-h forecasts relative to radiosonde observations for **a** zonal wind component (m s^{-1}) and **b** temperature (K), averaged over 14 days from 1200 UTC 5 September 2008 to 1200 UTC 19 September 2008. *Solid* and *dashed lines* correspond to ADAPT and FIXED, and *thick* and *thin lines* indicate root mean square errors (RMSE) and biases, respectively

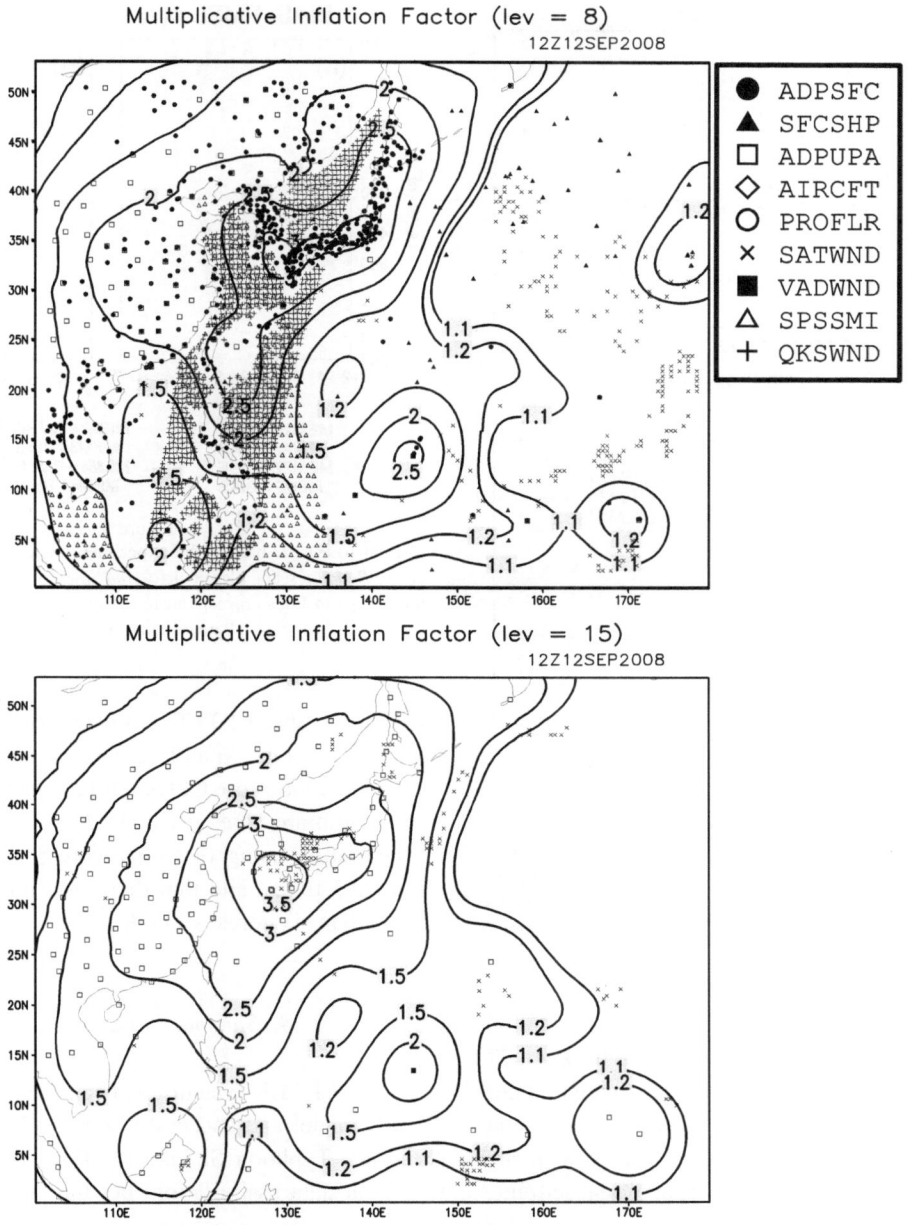

Figure 7

Horizontal maps of the adaptive inflation estimates (*contours*) on 1200 UTC 12 September 2008 at **a** the 8th model level ($\sim$850 hPa) and **b** the 15th model level ($\sim$500 hPa). The values indicate the inflation factors multiplied to the background error covariance (e.g., 1.0 is no inflation, 1.1 is 10% covariance inflation). *Marks* as in the legend indicate assimilated observations near the corresponding model levels, where the six-character codes of the observation types are as in PREPBUFR Table 1.a (Keyser 2010)

the wall-clock time was approximately 142 s on average. Here, the number of observations per 6-h cycle ranges from about 14,000–28,000, and the average number is about 22,000; the computational time of the LETKF scales nearly linearly to the number of observations. Since it took about 1 min for PRE-WRF, POST-WRF, and file copy processes, it took roughly 8 min per 6-h cycle with 20 members.

A larger ensemble size requires more computational time; the wall-clock time per 6-h cycle for each

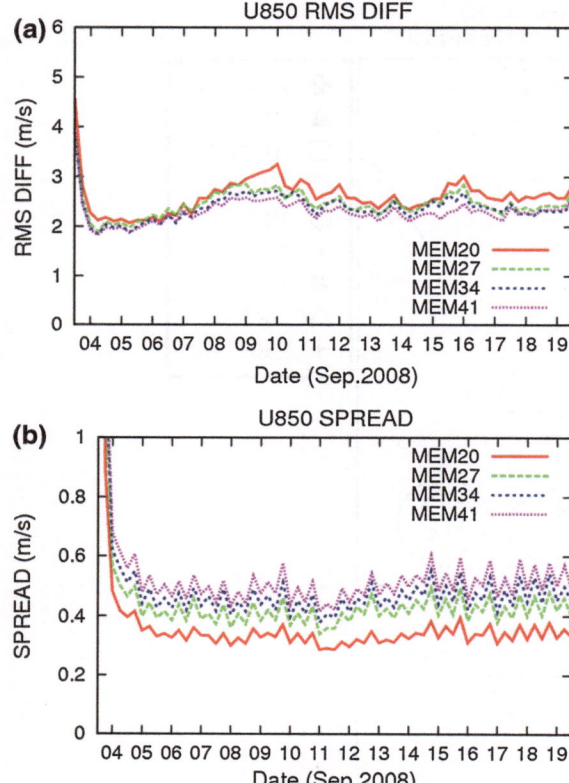

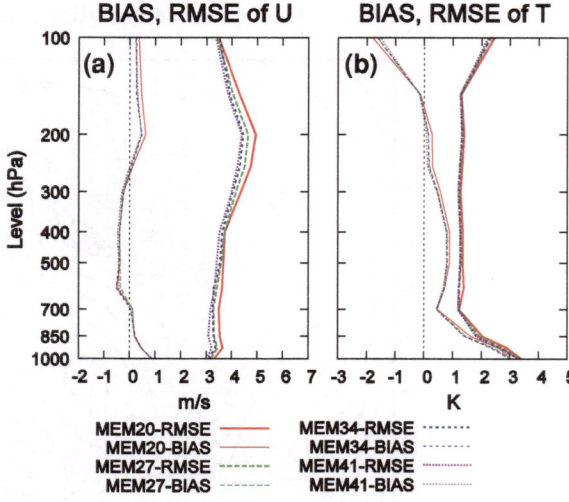

Figure 9

Similar to Fig. 7, but for the experiments with different ensemble sizes. *Red lines* are identical to *sold lines* of Fig. 7. *Green, blue,* and *magenta* lines correspond to the experiments with 27, 34, and 41 members, respectively

Figure 8

Similar to Fig. 3, but for the experiments with different ensemble sizes. *Red curves* correspond to the 20-member experiment and are identical to *solid curves* of Fig. 3. *Green, blue,* and *magenta* curves correspond to the experiments with 27, 34, and 41 members, respectively

the 20-member LETKF at a single step using 6 and 7 nodes, i.e., total 36 and 42 cores, respectively. However, the multi-core CPUs limit the parallel efficiency of using more cores per node; Fig. 11 shows that the acceleration using two quad-core processors was at most 2.5 times, and that there was no gain by using more than six cores per node.

4. Summary and Discussion

The LETKF system has been developed with the WRF model and successfully assimilated real observations. Typhoon Sinlaku (2008) was not generated without data assimilation, but the LETKF generated Sinlaku with the proper position and strength. The adaptive inflation scheme was advantageous, providing better analyses and ensemble spread than fixed inflation. It was also shown that using more ensemble members improved the analysis consistently, although the biggest improvements were found when the ensemble size was increased from 20 to 27. The computational time of 20-member WRF-LETKF was about 8 min per 6-h cycle using seven Linux servers with total 14 quad-core processors. The WRF-LETKF system would be useful for future studies on mesoscale

ensemble size is shown in Fig. 10. The computational time for ensemble forecasts shows a jump between 27 and 34 members; the jump corresponds to the number of CPU cores per forecast member. The computational time for the LETKF scales nearly linearly with the ensemble size with the current computational settings.

Although eight cores exist in each node, using more than six cores per node did not accelerate the computations significantly due to the generally inefficient nature of multi-core CPUs; namely, multi-core CPUs do not scale ideally. The LETKF algorithm is designed to be particularly efficient with parallel architecture, and a parallelization ratio higher than 99.9% was achieved by MIYOSHI and YAMANE (2007) with the Japanese Earth Simulator supercomputer (their Fig. 11). This study also obtained an approximately 99.4% parallelization ratio from the wall-clock time of

Table 2

Parallel computation settings of ensemble forecasts

Ensemble size	CPU cores per member	Members per node	Total CPU cores for ensemble forecasts	CPU cores for LETKF
20	2	3	42	42
27	2	4	56	42
34	1	5	35	42
41	1	6	42	42

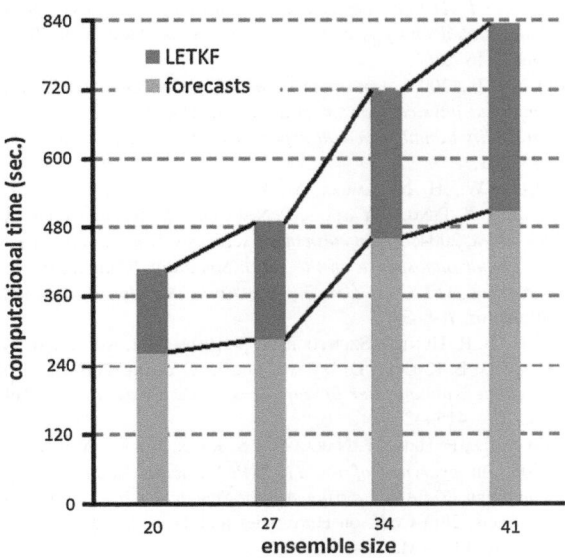

Figure 10

Computational time (seconds) per 6-h forecast-analysis cycle of the WRF-LETKF using four different ensemble sizes

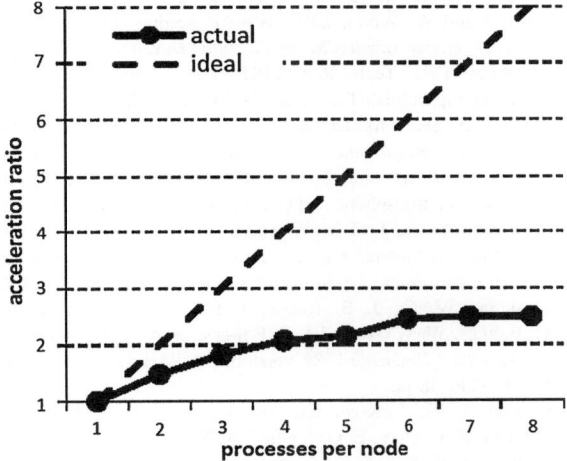

Figure 11

Acceleration ratios of the LETKF computational time using multiple processes per node, limited mainly by parallel efficiency of multi-core CPUs. Each node is equipped with two quad-core processors. The *dashed line* corresponds to the ideal parallelization

data assimilation. The LETKF code and related software are made available to public through the internet at http://www.code.google.com/p/miyoshi/.

Although this study provided a successful first step, there were some limitations. The spatial resolution was only 60 km grid spacing, which is considered to be too coarse as a mesoscale data assimilation system. Higher resolution applications will be an important future issue. Moreover, no inter-variable localization was applied. The sensitivity to the spatial, temporal, and inter-variable localization is another important issue.

Acknowledgments

The authors thank the members of the UMD Weather-Chaos Group for fruitful discussions. The NCEP PREPBUFR observation data were obtained from the UCAR data server, while several missing files were kindly provided by Daryl Kleist of NCEP. This study was supported by the Office of Naval Research (ONR) Grant N000141010149 under the National Oceanographic Partnership Program (NOPP).

REFERENCES

ANDERSON, J. L., 2009: *Spatially and temporally varying adaptive covariance inflation for ensemble filters.* Tellus, *61A*, 72–83.

ANDERSON, J. L., T. HOAR, K. RAEDER, H. LIU, N. COLLINS, R. TORN, and A. AVELLANO, 2009: *The Data Assimilation Research Testbed A Community Facility.* Bull. Amer. Meteor. Soc., *90*, 1283–1296.

ANDERSON, J. L. and S. L. ANDERSON, 1999: *A Monte Carlo implementation of the nonlinear filtering problem to produce ensemble assimilations and forecasts.* Mon. Wea. Rev., *127*, 2741–2758.

BISHOP, C. H., B. J. ETHERTON, and S. J. MAJUMDAR, 2001: *Adaptive sampling with the ensemble transform Kalman filter. Part I: Theoretical aspects.* Mon. Wea. Rev., *129*, 420–436.

CHOU, K.-H., C.-C. WU, P.-H. LIN, S. D. ABERSON, M. WEISSMANN, F. HARNISCH, T. NAKAZAWA, 2011: *The Impact of Dropwindsonde Observations on Typhoon Track Forecasts in DOTSTAR and T-PARC*, Mon. Wea. Rev., *139*, 1728–1743. doi:10.1175/2010MWR3582.1.

ELSBERRY, R. L. and P. A. HARR, 2008: *Tropical Cyclone Structure (TCS08) field experiment science basis, observational platforms, and strategy*. Asia-Pacific J. Atmos. Sci., *44*, 209–231.

GASPARI, G. and S. E. COHN, 1999: *Construction of correlation functions in two and three dimensions*. Quart. J. Roy. Meteor. Soc., *125*, 723–757.

GREYBUSH, S., 2011: *Mars Weather and Predictability: Modeling and Ensemble Data Assimilation of Spacecraft Observations*. Ph.D. Dissertation, University of Maryland, College Park.

HAMILL, T. M., J. S. WHITAKER, and C. SNYDER, 2001: *Distance-dependent filtering of background error covariance estimates in an ensemble Kalman filter*. Mon. Wea. Rev., *129*, 2776–2790.

HOFFMAN, M. J., S. J. GREYBUSH, R. J. WILSON, G. GYARMATI, R. N. HOFFMAN, E. KALNAY, K. IDE, E. J. KOSTELICH, T. MIYOSHI, and I. SZUNYOGH, 2010: *An Ensemble Kalman Filter Data Assimilation System for the Martian Atmosphere: Implementation and Simulation Experiments*. Icarus, *209*, 470–481.

HUANG, X.-Y., and COAUTHORS, 2009: *Four-Dimensional Variational Data Assimilation for WRF: Formulation and Preliminary Results*. Mon. Wea. Rev., *137*, 299–314.

HUNT, B. R., E. KALNAY, E. J. KOSTELICH, E. OTT, D. J. PATIL, T. SAUER, I. SZUNYOGH, J. A. YORKE, and A. V. ZIMIN, 2004: *Four-dimensional ensemble Kalman filtering*. Tellus, *56A*, 273–277.

HUNT, B. R., E. J. KOSTELICH and I. SZUNYOGH, 2007: *Efficient Data Assimilation for Spatiotemporal Chaos: A Local Ensemble Transform Kalman Filter*. Physica D, *230*, 112–126.

KANG, J.-S., 2009: *Carbon cycle data assimilation using a coupled atmosphere-vegetation model and the local ensemble transform Kalman filter*. Ph.D. dissertation, University of Maryland, College Park, 164 pp.

KANG, J.-S., E. KALNAY, J. LIU, I. FUNG, T. MIYOSHI, and K. IDE, 2011: *"Variable localization" to improve carbon cycle data assimilation in an Ensemble Kalman Filter*. J. Geophys. Res., *116*, D09110. doi:10.1029/2010JD014673

KEYSER, D., 2010: *PREPBUFR PROCESSING AT NCEP*, http://www.emc.ncep.noaa.gov/mmb/data_processing/prepbufr.doc/document.htm.

LORENZ, E., 1996: *Predictability: a problem partly solved*. Proceeding of the ECMWF Seminar on Predictability, vol. 1, Reading, UK.

LORENZ, E. and K. EMANUEL, 1998: *Optimal Sites for Supplementary Weather Observations: Simulation with a Small Model*. J. Atmos. Sci., *55*, 399–414.

MIYOSHI, T., 2005: *Ensemble Kalman filter experiments with a primitive-equation global model*. Ph.D. dissertation, University of Maryland, College Park, 197 pp.

MIYOSHI, T., 2011: *The Gaussian Approach to Adaptive Covariance Inflation and its Implementation with the Local Ensemble Transform Kalman Filter*. Mon. Wea. Rev., *139*, 1519–1535. doi:10.1175/2010MWR3570.1.

MIYOSHI, T. and K. ARANAMI 2006: *Applying a Four-dimensional Local Ensemble Transform Kalman Filter (4D-LETKF) to the JMA Nonhydrostatic Model (NHM)*. SOLA, *2*, 128–131.

MIYOSHI, T. and Y. SATO, 2007: *Assimilating Satellite Radiances with a Local Ensemble Transform Kalman Filter (LETKF) Applied to the JMA Global Model (GSM)*. SOLA, *3*, 37–40.

MIYOSHI, T. and S. YAMANE, 2007: *Local ensemble transform Kalman filtering with an AGCM at a T159/L48 resolution*. Mon. Wea. Rev., *135*, 3841–3861.

MIYOSHI, T., S. YAMANE, and T. ENOMOTO, 2007: *Localizing the Error Covariance by Physical Distances within a Local Ensemble Transform Kalman Filter (LETKF)*. SOLA, *3*, 89–92. doi:10.2151/sola.2007-023.

MIYOSHI, T., Y. SATO, and T. KADOWAKI, 2010: *Ensemble Kalman filter and 4D-Var inter-comparison with the Japanese operational global analysis and prediction system*. Mon. Wea. Rev., *138*, 2846–2866.

MOLTENI, F., 2003: *Atmospheric simulations using a GCM with simplified physical parametrizations. I: Model climatology and variability in multi-decadal experiments*. Climate Dynamics, *20*, 175–191.

OHFUCHI, W., H. NAKAMURA, M. K. YOSHIOKA, T. ENOMOTO, K. TAKAYA, X. PENG, S. YAMANE, T. NISHIMURA, Y. KURIHARA, and K. NINOMIYA, 2004: *10-km mesh meso-scale resolving simulations of the global atmosphere on the Earth Simulator-Preliminary outcomes of AFES (AGCM for the Earth Simulator)*. J. Earth Simulator, *1*, 8–34.

OTT, E., B. R. HUNT, I. SZUNYOGH, A. V. ZIMIN, E. J. KOSTELICH, M. CORAZZA, E. KALNAY, D. J. PATIL, and J. A. YORKE, 2004: *A local ensemble Kalman filter for atmospheric data assimilation*. Tellus, 56A, 415–428.

PARSONS, D., P. HARR, T. NAKAZAWA, S. JONES, and M. WEISSMANN, 2008: *An overview of the THORPEX-Pacific Asian Regional Campaign (T-PARC) during August-September 2008*. Extended Abstracts, 28th Conf. on Hurricanes and Tropical Meteorology, Orlando, FL, AMS, 1–6.

PENNY, S., 2011: *Data Assimilation of the Global Ocean Using the 4D Local Ensemble Transform Kalman Filter (4D-LETKF) and theModular Ocean Model (MOM2)*. Ph.D. Dissertation, University of Maryland, College Park.

SAITO, K., M. KUNII, M. HARA, H. SEKO, T. HARA, M. YAMAGUCHI, T. MIYOSHI and W. WONG, 2010: WWRP Beijing 2008 *Olympics Forecast Demonstration/Research and Development Project (B08FDP/RDP)*. Tech. Rep. MRI, *62*, 210 pp. (http://www.mri-jma.go.jp/Publish/Technical/DATA/VOL_62/62_en.html).

SAITO, K., H. SEKO, M. KUNII, M. HARA and T. MIYOSHI, 2009: *Influence of lateral boundary perturbations on the mesoscale EPS using BGM and LETKF*. CAS/JSC WGNE Research Activities in Atmospheric and Oceanic Modelling, *39*, 5.21–5.22.

SHCHEPETKIN, A. and J. C. MCWILLIAMS, 2005: *The Regional Oceanic Modeling System: A split-explicit, free-surface, topography-following-coordinate ocean model*. Ocean Modell., *9*, 347–404.

SKAMAROCK, W. C., J. B. KLEMP, J. DUDHIA, D. O. GILL, D. M. BARKER, W. WANG, and J. G. POWERS, 2005: *A description of the Advanced Research WRF version 2*. NCAR Tech. Note TN-468 + STR, 88 pp.

SZUNYOGH, I., E. J. KOSTELICH, G. GYARMATI, D. J. PATIL, B. R. HUNT, E. KALNAY, E. OTT, and J. A. YORKE, 2005: *Assesing a local ensemble Kalman filter: perfect model experiments with the National Centers for Environmental Prediction global model*. Tellus, *57A*, 528–545.

SZUNYOGH, I., E. J. KOSTELICH, G. GYARMATI, E. KALNAY, B. R. HUNT, E. OTT, E. SATTERFIELD, and J. A. YORKE, 2008: *A local ensemble transform Kalman filter data assimilation system for the NCEP global model*. Tellus, *60A*, 113–130.

WHITAKER, J. S. and T. M. HAMILL, 2002: *Ensemble data assimilation without perturbed observations*. Mon. Wea. Rev., 130, 1913–1924.

WILSON, R. J. and K. P. HAMILTON, 1996: *Comprehensive model simulation of thermal tides in the martian atmosphere*. J. Atmos. Sci., *53*, 1290–1326.

ZHANG, F., C. SNYDER, and J. SUN: 2004: *Impacts of Initial Estimate and Observation Availability on Convective-Scale Data Assimilation with Ensemble Kalman Filter*. Mon. Wea. Rev., *132*, 1238-1253.

(Received October 21, 2010, accepted January 15, 2011, Published online July 17, 2011)

Pure Appl. Geophys. 169 (2012), 335–351
© 2011 AOCS (outside the USA)
DOI 10.1007/s00024-011-0374-3

Selection of Momentum Variables for a Three-Dimensional Variational Analysis

YUANFU XIE[1] and ALEXANDER E. MACDONALD[1]

Abstract—Three choices of control variables for meteorological variational analysis (3DVAR or 4DVAR) are associated with horizontal wind: (1) streamfunction and velocity potential, (2) eastward and northward velocity, and (3) vorticity and divergence. This study shows theoretical and numerical differences of these variables in practical 3DVAR data assimilation through statistical analysis and numerical experiments. This paper demonstrates that (a) streamfunction and velocity potential could potentially introduce analysis errors; (b) A 3DVAR using velocity or vorticity and divergence provides a natural scale dependent influence radius in addition to the covariance; (c) for a regional analysis, streamfunction and velocity potential are retrieved from the background velocity field with Neumann boundary condition. Improper boundary conditions could result in further analysis errors; (d) a variational data assimilation or an inverse problem using derivatives as control variables yields smoother analyses, for example, a 3DVAR using vorticity and divergence as controls yields smoother wind analyses than those analyses obtained by a 3DVAR using either velocity or streamfunction/velocity potential as control variables; and (e) statistical errors of higher order derivatives of variables are more independent, e.g., the statistical correlation between U and V is smaller than the one between streamfunction and velocity potential, and thus the variables in higher derivatives are more appropriate for a variational system when a cross-correlation between variables is neglected for efficiency or other reasons. In summary, eastward and northward velocity, or vorticity and divergence are preferable control variables for variational systems and the former is more attractive because of its numerical efficiency. Numerical experiments are presented using analytic functions and real atmospheric observations.

Key words: Control variables, variational data assimilation, error covariance, background, observations, streamfunction and velocity potential, vorticity and divergence, Bayesian theorem, long wave and short wave.

1. Introduction

A variational data assimilation analysis (three- or four-dimensional variational analysis, 3DVAR or 4DVAR) is derived from the Bayesian theorem (LORENC, 1986). This theorem allows the use of any form of control variables in the background terms as long as the background error covariance accurately represents the error distribution. For meteorological data assimilation, three different sets of control variables can be used as background for horizontal wind: (1) streamfunction, ψ and velocity potential, χ (2) eastward, U, and northward, V, velocity, and (3) vorticity, ζ and divergence, δ. Based upon the Bayesian theorem, any of these three sets of variables can be used in the background term to form a variational cost function and used as the control variables. The theorem gives no preference to any of these variables. For simplicity, a 3DVAR is discussed in this paper and the conclusion is also applicable to any 4DVAR system. For convenience, a 3DVAR system using streamfunction and velocity potential is referred to as a ψ–χ 3DVAR system, one using velocity components of U and V as a u–v 3DVAR, and one using vorticity and divergence as a ζ–δ 3DVAR, hereafter. Although ψ–χ and u–v are widely used in many 3DVAR systems [National Centers of Environmental Prediction's Grid-point Statistical Interpolation, Rapid Update Cycle (RUC) and National Center of Atmospheric Research's WRF-3DVAR, for examples], the National Meteorological Center's Spectral Statistical-Interpolation Analysis system (PARRISH AND DERBER 1992), a global analysis system, is a ζ–δ 3DVAR in spectral space.

Consideration of selecting control variables for wind analysis has been given on the cross-correlation between control variables (e.g., ψ and χ), on computation efficiency (e.g., a u–v 3DVAR does not require a Poisson solver), and on meteorological decomposition

¹ NOAA Earth System Research Laboratory (ESRL), NOAA/OAR/ESRL/GSD, 325 Broadway, Boulder, CO 80305, USA. E-mail: Yuanfu.Xie@noaa.gov

of the wind fields (e.g., rotational and un-rotational winds). In this study, several aspects of selecting control variables for a variational analysis will be examined and their differences are revealed.

In a meteorological data assimilation system, an "optimal" correction to the background can be viewed as coming from two sources: the deterministic information resolved by observations, and statistical estimation of the uncertainty from both background and observations. This does not contradict to the general data assimilation theory, where an accurate error covariance is assumed known. If a data assimilation analysis is decomposed in spectral space, the deterministic information is the component whose wavelengths can be resolved by a given observation network within the observation error ranges and the statistical information is other component whose wavelengths cannot be resolved or seen by this network. For observational resolvable wavelength scales, say, $2\Delta n_r$ (where Δn_r is the average observation data spacing), the observations can deliver deterministic information within the ranges of estimated errors. Hence, the uncertainty is mainly over the unresolvable wavelength scales. It is found that over these shorter wavelength scales, 3DVAR systems perform quite differently. A $\psi-\chi$ 3DVAR system tends to produce analyses that possess large-scale motions of the background field and retrieve small-scale motions from the observations (XIE et al., 2002). This 3DVAR may ignore the deterministic information of observation networks and treat it as an uncertainty unless a perfect covariance is known. If there are differences between backgrounds and observations at observation sites on the scales that can be resolved by an observation network (i.e., longer wavelength scales), a $\psi-\chi$ 3DVAR may treat these differences as smaller scales. It not only misses the opportunity to correct the background fields on these scales, but also damages the analysis on the shorter waves.

In addition, if model state variables for momentum are velocity, a $\psi-\chi$ 3DVAR has to deal with a complicated boundary value problem for converting velocity to streamfunction and velocity potential when it is applied over a limited area (CHEN and KUO, 1992a, b). In this study, it is pointed out that this conversion cannot uniquely determine the streamfunction and velocity potential. The differences between these multiple solutions of streamfunction and velocity potential are

not only simple constants but may be linear functions. That means that the converted streamfunction and velocity potential might derive a completely different velocity field. Thus, a $\psi-\chi$ 3DVAR may use an incorrect background to derive the increments and result in serious analysis errors. Some simplified implementations may result in even more analysis error to the 3DVAR analysis; for example, assume that streamfunction and velocity potential have zero gradients on the boundary in order to use a simple sine FFT transform for the conversion. However, as demonstrated, a $u-v$ 3DVAR does not have the conversion, and a novel implementation is proposed in Sect. 4 for conversion of velocity to vorticity and divergence in a $\zeta-\delta$ 3DVAR. A common practice is to enlarge the analysis domain for the conversion so that the boundary problems can be reduced, for examples, WRF 3DVAR and GSI.

In this paper, a further investigation of the selection of control variables for a variational analysis is performed in the hope of providing some recommendation in improving the analysis simply by selecting better control variables. It does not conclude the analyses from a $\psi-\chi$ system are wrong or no good but they have required to have more additional effort, for example, using a large domain for converting wind velocity to $\psi-\chi$ as mentioned above. Other control variables could further improve these analyses as shown in this paper.

A theoretical and statistical analysis of these three forms of 3DVAR is illustrated in Sect. 2 explaining why a $\psi-\chi$ system tends to introduce analysis errors. A Fourier analysis is given in Sect. 3 showing that a $u-v$ or $\zeta-\delta$ 3DVAR naturally imposts a wave dependent influence radius according the analysis wavelengths. Boundary conditions are discussed in Sect. 4 when applying a $\psi-\chi$ and $\zeta-\delta$ 3DVAR for a regional analysis. Numerical experiments using one-dimensional and two-dimensional data are demonstrated in Sect. 5. Real observation data analyses are presented in Sect. 6. Conclusions are given in Sect. 7.

2. Error Distributions and Additional Analysis Error from a $\psi-\chi$ 3DVAR

In meteorological data assimilation, covariance matrices characterize the uncertainty from both background and observations. SCHLATTER (1975) demonstrated forecast spatial error correlation for height field as a function of separation distance between every pair of 50 US radiosonde stations and found that the distribution is quite close to a Gaussian. This conclusion supports the Gaussian error distribution assumption in data assimilation. In this study, the Eta and RUC (Rapid Update Cycle) models have been used to demonstrate a long-term statistical error behavior of the momentum fields. RUC background error statistics is extracted from 20 km RUC, and Eta background error statistics is reported by using Eta 48-km Advanced Weather Interactive Processing System (AWIPS) CONUS forecasts and analyses. A statistical analysis was set up for 2 months in 2004 and the time period is selected randomly. The model background errors of streamfunction, velocity potential, velocity, vorticity and divergence have been approximated by the differences between an analysis and its 6-h forecast produced 6-h earlier. The spatial correlations of all the variables show Gaussian type distributions with different de-correlation distances (Fig. 1). Higher derivatives show relatively smaller influence radius, which could be relatively easy to approximate. From numerical computation point of view, the smaller influence radius is, the better condition of the background error covariance, for example, if the influence radius tends to very large, the covariance tends to be singular; if the influence radius tends to zero, the covariance tends to be an identity matrix. A smaller influence radius means less statistical correlation estimation, that is, it would be easier to estimate the error covariance for vorticity and divergence than those for velocity and streamfunction and velocity potential.

Now most data assimilation techniques assume that model background errors follow Gaussian distribution in order to estimate the observation and background uncertainty while some other non-Gaussian spatial correlations are considered (BUEHNER, 2004). Among the three forms of 3DVAR, if one set of momentum variables is assumed to follow a Gaussian spatial error distribution, it indirectly implies that the background errors of the others may not follow a Gaussian. For example, assume background spatial errors of streamfunction and velocity potential are Gaussian,

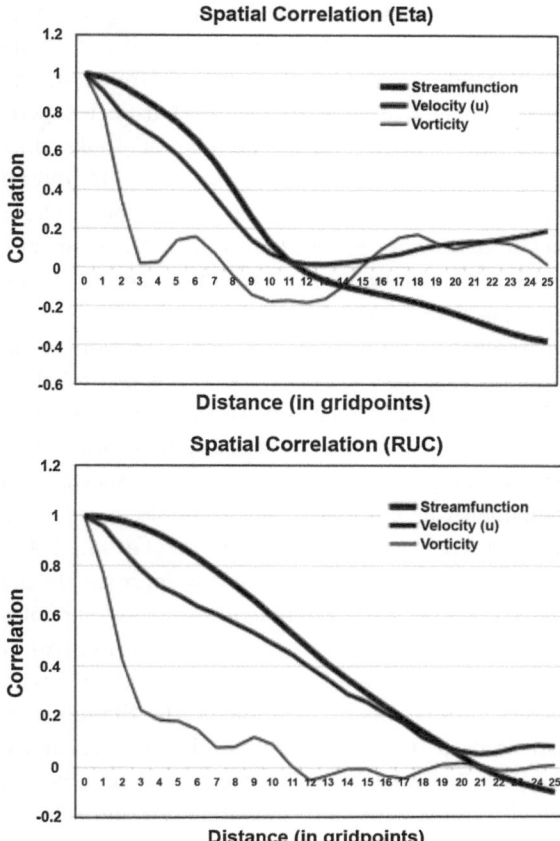

Figure 1

Spatial error correlation of the variables: ψ, u and ζ, as functions of grid points ($\Delta x = 48$ km for Eta model and $\Delta x = 20$ km for RUC model)

$$\text{err}_\psi \propto e^{-[(x-x_c)^2+(y-y_c)^2]/\sigma_\psi^2},$$
$$\text{err}_\chi \propto e^{-[(x-x_c)^2+(y-y_c)^2]/\sigma_\chi^2}.$$

Velocity error distribution should follow

$$\text{err}_u \propto A(y-y_c)e^{-[(x-x_c)^2+(y-y_c)^2]/\sigma_\psi^2}$$
$$+ B(x-x_c)e^{-[(x-x_c)^2+(y-y_c)^2]/\sigma_\chi^2},$$
$$\text{err}_v \propto A(x-x_c)e^{-[(x-x_c)^2+(y-y_c)^2]/\sigma_\psi^2}$$
$$+ B(y-y_c)e^{-[(x-x_c)^2+(y-y_c)^2]/\sigma_\chi^2},$$

for some constant coefficients of A and B, which are no longer Gaussian (neither are the vorticity and divergence). Similarly, if the spatial error correlation of U and V follows a Gaussian distribution, the others may not. A similar analysis had been shown (SCHLATTER, 1975) when the velocity error correlation was derived from error correlation of heights under the geostrophic balance or vice versa; that is, if the height field has a Gaussian error correlation, the velocity field does not. From our statistical experiment, the velocity error correlation looks like Gaussian as well as the other variables. Therefore, one needs to determine a set of variables whose background errors more likely follow Gaussian distributions so that the analysis results less errors by the Gaussian distribution assumption.

Another statistical aspect of selecting momentum variables as a 3DVAR background is the correlation between variables. Some 3DVAR systems assume that the correlations are zero between different variables. A typical approach is to assume the unbalanced wind components un-correlated. For example, a recursive filter (HAYDEN and PURSER, 1995; WU et al., 2002; PURSER et al., 2003a, b) is used to approximate the spatial covariance of each individual control variables only leaving the cross-variables correlation as zeros. It is desirable to choose a set of momentum variables with smaller correlations between them. The Eta model is also used to investigate cross-variable correlations. The error correlations of vorticity and divergence, velocity components and streamfunction and velocity potential are shown as a function of pressure in Fig. 2. To reflect the correlation more accurately, the average of absolute error correlations at all vertical levels is shown in Table 1. Even though the cross-variable correlations are generally weaker than their spatial correlations, two-month statistics show that vorticity and divergence have the smallest correlation, and streamfunction and velocity potential have the largest. It could be intuitive that the higher the derivative is, the noisier the field becomes, which means higher order derivatives are less correlated not only in space but also between variables. Based on the statistics, we believe that either the u–v 3DVAR or ζ–δ 3DVAR analysis could improve the ψ–χ 3DVAR from the cross-variable correlation consideration if the cross-variable correlation is assumed to be zero in a 3DVAR system.

Beyond the statistical consideration, it would be interesting to compare the analysis differences of these three forms of 3DVAR in data assimilation. Consider a single U observation case. An interesting phenomenon of a ψ–χ 3DVAR analysis usually has been seen. While the increment field of U tends to

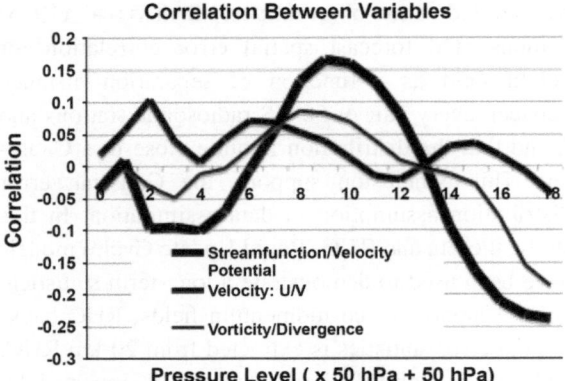

Correlation Between Variables

Figure 2

Correlations between the variables: streamfunction and velocity potential; velocity, u and v; and vorticity and divergence. It shows that the correlation between streamfunction and velocity potential is the strongest; the one between vorticity and divergence is the weakest

Table 1

Two-month error correlation averages between variables over all heights

Between	Vorticity and divergence	U and V	Streamfunction/ velocity potential
Correlations	0.0185	0.0473	0.0856

adjust the analysis approximating the single observation value of U, the increment field shows adjustment in the opposite direction in its near neighborhood around the center of the observation site (see Fig. 3) if there is no spatial correlation. Since there is only one observation in such a domain, the negative values must be analysis errors. These errors always exist but vary in sizes and distances away from the observation site, with different influence radius of background error covariance. This phenomenon motivates a further study of the ψ–χ 3DVAR formulation. A perfect or accurate background error covariance could be thought of handling these opposite increment errors. However, a practical statistical approach is to estimate the errors from the forecast of streamfunction and velocity potential. In forecasts or transformed from forecast status, streamfunction and velocity potential are smoother fields and there is usually no extreme oscillation locally. Thus, these statistics derived from these smooth fields could not have had correlation

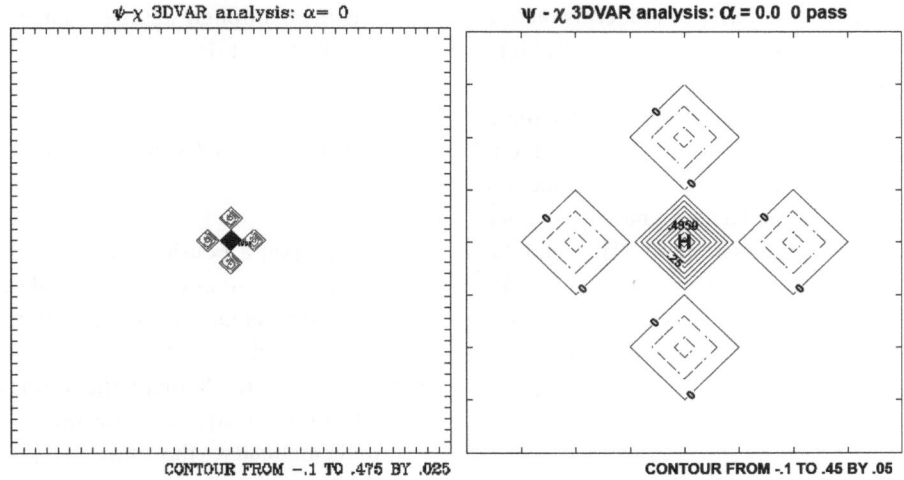

$\psi-\chi$ 3DVAR analysis: $\alpha=0$

$\psi-\chi$ 3DVAR analysis: $\alpha=0.0$ 0 pass

CONTOUR FROM −.1 TO .475 BY .025

CONTOUR FROM −.1 TO .45 BY .05

Figure 3

A single-observation analysis of U from a $\psi-\chi$ 3DVAR: (*Left*) whole domain; (*right*) the central region only. *Solid curves* indicate positive values; *dashed curves* indicate negative values. Each *tick mark* on the perimeter indicates a grid point over a dimensionless domain

information to overcome the opposite increment issues as pointed out here.

It is found that these errors are actually caused by the formulation of a $\psi-\chi$ 3DVAR system. Consider an idealized case where an observation of U is at the grid point (i, j, k) for simplicity. A $\psi-\chi$ 3DVAR using a second order finite difference scheme has the following term in the observation part J_o of its cost function,

$$\ldots \underbrace{O_{ijk}^{-1} \underbrace{\left(-\frac{\psi_{i,j+1,k} - \psi_{i,j-1,k}}{2\Delta y} + \frac{\chi_{i+1,j,k} - \chi_{i-1,j,k}}{2\Delta x} - U_{i,j,k}^o \right)}_{\gamma}^2}_{\tau} + \ldots$$

where $O_{ijk} > 0$ is an observation error variance at this site. Denote this term as τ and the expression inside the parenthesis as γ, i.e., $\tau = O_{ijk}^{-1} \gamma^2$. The derivatives of this term are

$$\frac{\partial \tau}{\partial \psi_{i,j+1,k}} = -\frac{O_{ijk}^{-1}\gamma}{\Delta y} \qquad \frac{\partial \tau}{\partial \psi_{i,j-1,k}} = \frac{O_{ijk}^{-1}\gamma}{\Delta y}$$

$$\frac{\partial \tau}{\partial \chi_{i+1,j,k}} = \frac{O_{ijk}^{-1}\gamma}{\Delta x} \qquad \frac{\partial \tau}{\partial \chi_{i-1,j,k}} = -\frac{O_{ijk}^{-1}\gamma}{\Delta x}$$

If γ is positive, for instance,

$$\frac{\partial \tau}{\partial \psi_{i,j+1,k}} < 0 \qquad \frac{\partial \tau}{\partial \psi_{i,j-1,k}} > 0.$$

Then a gradient-dependent minimization algorithm will increase the value of $\psi_{i,j+1,k}$ and reduce $\psi_{i,j-1,k}$, that is, correcting the streamfunction along a negative gradient direction of the cost function. If rotational wind dominates the velocity, this results in negative increments on $U_{i,j+2,k}$ and $U_{i,j-2,k}$ as, for example,

$$U_{i,j+2,k} \approx -\frac{\psi_{i,j+3,k} - \psi_{i,j+1,k}}{2\Delta y}.$$

Without grid correlations, two-delta x and y waves can be formed (Fig. 3). In this study, we use a simple first order recursive filter (HAYDEN and PURSER, 1995) to approximate the background error covariance of numerical experiments as shown in Sect. 4. Other filters would do similar but maybe a little less serious. Increasing the smoothing parameter of a recursive filter causes these two-delta waves to become longer, but the negative increments remain and propagate away from the observation site (i, j, k), as shown in Sect. 4. If the smoothed wavelengths are longer than the distance between observations, the negative values may not be seen, but the tendency toward the negative direction casts doubt on its accurate representation of the true wind field. If the smoothed wavelengths are shorter, these negative increments become nonphysical errors to the increment field. Thus a $\psi-\chi$ 3DVAR could introduce the nonphysical errors into its

31

analysis. Even though small background terms reduce the errors or make the error less obvious, the problem remains for any of the $\psi-\chi$ 3DVARs.

The errors are not caused by this particular finite difference scheme used in the above demonstration. Rather, they are caused by the use of streamfunction and velocity potential in the background terms under the Gaussian distribution assumption. This can be explained in a continuous form of a $\psi-\chi$ 3DVAR, independent of any discretization scheme. For simplicity, consider the rotational wind

$$u = -\frac{\partial \psi}{\partial y}, \quad v = \frac{\partial \psi}{\partial x} \quad or$$

$$\psi = \psi_0 - \int_0^y u\,d\eta, \quad \psi = \psi_0 + \int_0^x v\,d\eta$$

where η is the integration variable along either x or y axis. That is, streamfunction and velocity potential are integrals of the corresponding velocity field. Thus a background using streamfunction and velocity potential implies that the minimization of a $\psi-\chi$ 3DVAR cost function tends to restrict the analyzed velocity field to maintain their integrals of wind velocity as the same values as the background. Therefore, if the analyzed U component of the velocity field approaches an observation that is different from the background, it will change the integrals. The U values in the neighborhood must move in the opposite direction so that the integrals remain the same values as much as possible. The less correlated the Gaussian distribution, the more these U values move toward the opposite direction. Numerical experiments in Sect. 4 demonstrate this clearly in one-dimensional space.

These 3DVAR systems preserve different quantities in their background fields. A $\psi-\chi$ system preserves integrals of velocity, a $u-v$ system preserves the velocity itself (i.e., making minimal velocity increment), and a $\zeta-\delta$ system preserves derivatives of velocity. Although the Bayesian theorem allows $\psi-\chi$ as control variables in a variational analysis, their analyses could be quite different. The preservation of integrals seems to be causing problems for a $\psi-\chi$ system, as illustrated above because the observations are usually in velocity. In the numerical experiments of this paper, the problem is demonstrated through analytic functions and operational observation data.

3. Advantage of Using Derivatives as Control Variables

Based upon a Fourier analysis, it is more natural to use a $u-v$ or $\zeta-\delta$ as control variables. They provide natural wavelength dependent influence radius. However, if a $\psi-\chi$ analysis is used as shown in this section it would alternate the error correlation of velocity in an unexpected direction.

Mathematically, the three sets of momentum variables can be linearly converted into any other form based upon the following relations of these variables:

$$\nabla^2 \psi = -u_y + v_x = \zeta$$
$$\nabla^2 \chi = u_x + v_y = \delta \tag{1}$$

or

$$\nabla^2 u = -\zeta_y + \delta_x \quad \left(u = -\psi_y + \chi_x\right)$$
$$\nabla^2 v = \zeta_x + \delta_y \quad \left(v = \psi_x + \chi_y\right) \tag{2}$$

where ∇^2 is a Laplacian operator, and the subscript stands for a partial differential operator. That means that a linear operator, for example, can transform any pair of control variables to another, e.g., $(u, v) = M(\psi, \chi)$. By combining the operator into an error covariance matrix, two 3DVAR formulations can be identical,

$$(X_{uv} - X_{uv}^b)^T B_{uv}^{-1}(X_{uv} - X_{uv}^b) + \cdots$$
$$= (X_{\psi\chi} - X_{\psi\chi}^b)^T M^T B_{uv}^{-1} M(X_{\psi\chi} - X_{\psi\chi}^b) + \cdots$$
$$= (X_{\psi\chi} - X_{\psi\chi}^b)^T B_{\psi\chi}^{-1}(X_{\psi\chi} - X_{\psi\chi}^b) + \cdots$$

where $X_{uv} = (u, v)$, $X_{\psi\chi} = (\psi, \chi)$ and $B_{uv} = M^T B_{\psi\chi} M$. Thus, these 3DVAR systems are equivalent if the linear operators of these conversions are included in the background error covariance matrix. However, these operators change the statistical characteristics of the error distributions. For example, the error covariance of $u-v$ will be $B_{uv} = M^T B_{\psi\chi} M$ if background errors of $\psi-\chi$ is assumed to follow a Gaussian distribution. Therefore, under certain assumption of the error covariance, the choice of different control variables would make differences in their analyses. The M matrix contains differential operators and in Fourier space, it has the form of

$$i \begin{pmatrix} -l & k \\ k & l \end{pmatrix}$$

and it is unbounded. As XIE et al., (2002) pointed out, it uses background more on long waves and takes more information from observations on short waves as k and l are small for long waves and large for short waves. This is not ideal for data assimilation as observations observes long wave better but cannot resolve short wave information. In addition to that conclusion, another interesting conclusion can be drawn from this analysis in this paper. If the error distribution of $\psi-\chi$ is Gaussian, the $u-v$ error distribution will have smaller influence radius on longer waves as M becomes smaller but larger one on shorter waves as M becomes larger, comparing to the Gaussian of $\psi-\chi$. It is not natural or physical to use relatively smaller radius for longer waves in velocity and larger ones on short waves. On the opposite, a $u-v$ 3DVAR would have the inverse matrix effect of M. Thus, it implies to larger influence radius on longer waves and smaller ones on short waves on streamfunction and velocity potential. It naturally provides a scaling dependent correlation following the wavelength. Based on the same argument, a $\zeta-\delta$ 3DVAR would be better choices for improving a $\psi-\chi$ analysis.

4. Regional Implementation of These 3DVAR Systems

Most of the 3DVAR applications are regional, that is, they are implemented over a limited area. For a $u-v$ 3DVAR, it is straightforward for prediction models use $u-v$ as prognostic variables, such as the Weather Research and Forecasting model (WRF). However, one has to deal with nonperiodic boundary value problem of Poisson equations for a $\psi-\chi$ and $\zeta-\delta$ system for these prediction models, while for global systems, these are periodic boundary value problems. One has to solve the Poisson equations in (1) or (2), respectively, with proper boundary conditions, because conventional momentum variables of observations and model backgrounds are usually in form of

velocity. Inappropriate implementation may result different background fields and add more errors in the analysis. Thus, this section intends to form correct boundary value problems of Poisson equations for $\psi-\chi$ and $\zeta-\delta$ 3DVAR when those numerical prediction models using $u-v$ as prognostic variables are used for backgrounds.

4.1. $\psi-\chi$ 3DVAR Systems

A system of Poisson equations of (1) is solved for conversion from velocity to streamfunction and velocity potential required by a $\psi-\chi$ 3DVAR system at the beginning of its assimilation with Neumann-type boundary conditions

$$\begin{aligned} -\psi_y + \chi_x &= u \\ \psi_x + \chi_y &= v. \end{aligned} \tag{3}$$

Even though this may not be expensive forming a $\psi-\chi$ 3DVAR, only one conversion at the beginning but it may need to solve for each statistical estimation. In this sense, it is not less expensive comparing to implement a $\zeta-\delta$ 3DVAR. Beside the computational expense, it is a complicated problem as the boundary condition mixes the variables together so that ψ and χ cannot be solved separately. There is always a possibility that the analysis is erroneous due to incorrect boundary treatment. This system of Poisson equations does not have a unique solution, and how to solve it has been discussed for many years (SHUKLA and SAHA, 1974; STEPHENS and JOHNSON, 1978; CHEN and KUO, 1992a, b). However, the most challenging issue is that its solutions of Eqs. (1) and (3) may not differ by just a simple constant. As an example, assume a solution of ψ^* and χ^* satisfying (1) and (3). Consider linear functions

$$\begin{aligned} \psi' &= a_0 + a_{10}x + a_{01}y \\ \chi' &= b_0 + b_{10}x + b_{01}y. \end{aligned}$$

It is obvious that $\psi^* + \psi'$ and $\chi^* + \chi'$ satisfy (1). If

$$a_{10} = -b_{01} \quad \text{and} \quad a_{01} = b_{10}, \tag{4}$$

they also satisfy (3). Thus, its numerical solutions are difficult to compute. CHEN and KUO (1992a, b) discussed a solution using harmonic-sine and harmonic-cosine series expansions and obtain one solution among these multiple ones. All of these multiple solutions represent the same velocity field and would not cause any numerical problem. However, potential errors come from the 3DVAR analysis since the corresponding components to x and y of the 3DVAR increments may not satisfy Eq. (4). Therefore, the boundary value problem for a $\psi-\chi$ 3DVAR opens another error source in general.

It is worth mentioning that decoupling ψ and χ for solving (1) by forcing their gradients to zero over the boundary is not appropriate. Larger errors are introduced even in the background fields. By forcing the gradients to zero, the streamfunction and velocity potential do not represent the model background velocity at all. If such streamfunction and velocity potential are put in a $\psi-\chi$ 3DVAR's background terms, the incorrect background could give incorrect increment fields and seriously damage the analysis. Many operational $\psi-\chi$ 3DVARs use larger domains than their analysis domains for the conversion in the hope of removing or reducing the incorrect boundary condition problem. This can reduce the problem but add additional computations. In addition, the other major problem of the zero gradient boundary conditions is that such $\psi-\chi$ 3DVAR systems fail to make corrections over the boundary. This will not only result in zero horizontal velocity over the boundary but will also impose large errors in the interior of the limited area unless the background fields have correct boundary values, which is generally not the case.

Thus, to convert velocity into streamfunction and velocity potential is an error prone process for implementing a $\psi-\chi$ 3DVAR system. To solve (1) along with (3) in our Eta and RUC model error correlation statistics, a minimization of the residuals of (1) and (3) is solved, which only represents one particular set of streamfunction and velocity potential, and the statistics reflects the error characteristics of this set of variables.

4.2. $\zeta-\delta$ 3DVAR Systems

For a $\zeta-\delta$ 3DVAR system, the background fields are straightforward by directly differentiating model prognostic variables if they are in the form of u and v, and Poisson equations have to be solved at every

optimization iteration converting analyzed vorticity and divergence into new velocity in order to evaluate observation terms of the cost function. For a spectral analysis system, this can be done efficiently (PARRISH and DERBER, 1992). For a finite-difference analysis system over a limited area, this transformation is extra cost. However, the total minimization iterations for solving a 3DVAR depends on the condition number of the Hessian (second order derivatives) matrix of the cost function, where the condition number is defined as the ratio between the largest and smallest eigenvalues of the matrix. In our recent numerical experiments, it is found that a $\zeta-\delta$ 3DVAR system has a better condition number than a $\psi-\chi$ 3DVAR system, and a theoretical investigation is under way to confirm this fact. With a better condition number, a fewer optimization iterations could make the expense of the repeated Poisson solver less significant. Contrary to a $\psi-\chi$ 3DVAR, the boundary conditions of (2) can be well determined for a $\zeta-\delta$ 3DVAR system, a typical Dirichlet boundary value problem and its solution is well determined.

When using $\zeta-\delta$ 3DVAR systems for a regional analysis, boundary conditions are necessary for solving velocity from given vorticity and divergence at each optimization iteration. However, for a regional analysis, these conditions are often unknown unless we also assume they are equal to the background fields, which makes the velocity increments zero, as discussed above regarding a $\psi-\chi$ 3DVAR system. To solve this problem, the boundary conditions of U and V are used as control variables. That is, in the 3DVAR minimization process, the observation terms of the cost function are implicit functions of boundary values of U and V through the Poisson equations. At each optimization iteration it will provide not only corrected vorticity and divergence fields but also corrected boundary values for velocity. These boundary values can be used to form Poisson equations with Dirichlet boundary conditions, which has a unique solution,

$$\nabla^2 u = -\zeta_y + \delta_x \quad (x, y) \in \Omega$$
$$\nabla^2 v = \zeta_x + \delta_y \quad (x, y) \in \Omega \tag{5}$$
$$u = u_{3DVAR}, \quad v = v_{3DVAR}, \quad (x, y) \in \partial\Omega$$

where Ω is the regional domain, $\partial\Omega$ is the boundary of Ω, u_{3DVAR} and v_{3DVAR} are the boundary values from an optimization iteration of the ζ–δ 3DVAR. In our numerical experiments, a ζ–δ 3DVAR system is implemented following this procedure.

Both ψ–χ and ζ–δ 3DVAR systems require solutions of Poisson equations. These solutions could be complicated and expensive to compute. A ζ–δ 3DVAR has deterministic solutions of Poisson equations comparing to a ψ–χ 3DVAR system, but it costs extra computation for the Poisson equations. Based on the complication of Poisson equations with proper boundary treatment for a ψ–χ 3DVAR systems or extra expense for a ζ–δ 3DVAR system, a u–v 3DVAR system seems more attractive in operational applications.

5. Numerical Experiment I: Analytic Test Cases

Based on the discussions in the previous sections, a ψ–χ 3DVAR system is found to preserve the integral values of wind velocity fields, and this preservation results in certain analysis errors in addition to the boundary problems over a limited area. In this section, numerical experiments are constructed to illustrate these errors compared to other forms of 3DVAR systems. These experiments will show that a u–v 3DVAR analysis and a ζ–δ 3DVAR analysis do not contain integral analysis errors like those found in the ψ–χ 3DVAR analysis.

First, we consider a reduced ψ–χ 3DVAR to show the preservation of integrals as well as the differences of three forms of 3DVAR systems. A one-dimensional variational assimilation (1DVAR) is constructed simulating ψ–χ, u–v, and ζ–δ 3DVAR systems. Consider grid functions u_i, u_i^b (background), and u_i^o (observation) over an interval of $[x_a, x_b]$ divided by x_i with equal grid spacing while u_i^o is only used in a subset of all the grid points, K. Assuming a u–v 1DVAR system has the following discrete form:

$$\min \frac{1}{2} \left[\sum_{i,j} \sigma_{i,j}^u (u_i - u_i^b)(u_j - u_j^b) + \sum_{k \in K} o_k^{-1}(H_k u - u_k^o)^2 \right]$$

where $\sigma_{i,j}^u$ is the element of the inverse of background error covariance matrix, K is the index set of all observations, and assume that the observation error covariance matrix, O, is a diagonal matrix with its

diagonal element o_k^{-1}, i.e., $O^{-1} = \mathrm{diag}(o_1^{-1}, o_2^{-1}, \ldots)$. A streamfunction ψ satisfies

$$u_i = \frac{\psi_{i+1} - \psi_{i-1}}{2\Delta x}, \psi_i = \psi_o + 2\Delta x \sum_{j=1}^{i} u_j$$

where the Δx is the uniform grid distance between two adjacent grid points. A corresponding ψ–χ 1DVAR may have the following form:

$$\min \frac{1}{2} \left[\sum \sigma_{i,j}^{\psi}(\psi_i - \psi_i^b)(\psi_j - \psi_j^b) \right.$$
$$\left. + \sum_{k \in K} o_k^{-1}(H_k u - u_k^o)^2 \right].$$

In these experiments, a first order recursive filter is used to approximate the background error covariance of a Gaussian distribution for both u–v 1DVAR and ψ–χ 1DVAR. The recursive filter consists of left and right passes (Hayden and Purser, 1995). The left pass filter is

$$u_k' = \alpha u_{k-1}' + (1 - \alpha)u_k \quad (6a)$$

and similarly, the right pass is

$$u_k'' = \alpha u_{k+1}'' + (1 - \alpha)u_k'. \quad (6b)$$

A ζ–δ 1DVAR can be formed the same as the ψ–χ 1DVAR but no recursive filter is used, that is, $\sigma_{i,i}^{\zeta} = 1$, $\sigma_{ij}^{\zeta} = 0$.

Simulating real data analysis, the following two functions are used to serve as a background and an observed wind field, respectively

$$u^o(x) = \cos(2\pi x) + 0.3\cos(8\pi x)$$
$$u^b(x) = \cos(1.8\pi x - f_0) + 0.2\cos(7.2\pi x - f_0)$$

and thus a "streamfunction" of the background field is

$$\psi^b(x) = \frac{\sin(1.8\pi x - f_0)}{1.8\pi} + \frac{0.2\sin(7.2\pi x - f_0)}{7.2\pi}$$

where $f_0 = 0.3$ for shifting background away from the observation. For a given set of observation sites, two cases have been run with/without a recursive filter to see the differences and recursive filter effects. The analyses without a recursive filter (equivalently no correlation between grid points in the background field) are shown in Fig. 4. The observation values are marked as solid circles on observation curve (thick dashed curve). A thinner dashed curve is the background, and the thinnest dashed curve is the analysis. Without any correlation, only a ζ–δ 1DVAR produces a reasonable (smooth) data analysis. This may be counter-intuitive as a derivative is usually rougher than the function itself. However, for a variational data assimilation or an inverse problem, a 3DVAR using derivatives as controls yields smoother analyses than the one using the functions themselves does. For a u–v 1DVAR, the analysis is the same as the background field but approaches the observations at the observation sites. Interestingly, the analysis of a ψ–χ 1DVAR not only approaches the observations at the observation sites but also produces two surrounding peak values. This is the preservation of the integral effect. The integral of U, the area surrounded by x–y axes and the background curve (e.g., the x–y axes of $x = 0$, $x = 162$, and $y = 0$ and background curve marked with "o"), is changed by the analysis approaching the observation. The integral preservation creates two opposite peaks in order to maintain the same value of the integral, i.e., the same value of the area. This is exactly the preservation of the integral value property of a ψ–χ 3DVAR discussed in Sect. 2.

Since a ζ–δ 1DVAR shows good correlation without a recursive filter in Fig. 4, only ψ–χ and u–v 1DVAR are shown how their analysis varies with the recursive filter's parameter, α. Figure 5 shows the analyses with a recursive filter ($\alpha = 0.3$ with three left and right passes; see HAYDEN and PURSER, 1995). Notice that the recursive filter makes the analyses smoother, but the analysis of a ψ–χ 1DVAR keeps the similar features of area preservation, and the analysis of a u–v 1DVAR shows a Gaussian type of approximation toward the observation values from background. Figure 6 shows the same analysis of a ψ–χ 1DVAR with $\alpha = 0.5$ and 0.8. An interesting observation is that a ψ–χ 1DVAR tends to generate noisier fields. This is also confirms that in a variational analysis, the higher the derivatives used as background, the smoother the analysis.

The 1DVAR analyses confirm the conclusion made in previous sections, and now we present an experiment designed to show the differences between the 3DVAR systems in higher dimensions. Because the operators mapping streamfunction and velocity

potential to velocity and mapping velocity to vorticity and divergence are two-dimensional in x and y space, a two-dimensional variational analysis over x and y space is adequate to illustrate errors of a ψ–χ 3DVAR analysis compared to a u–v 3DVAR analysis or a ζ–δ 3DVAR analysis.

Assume there is a single U observation at the center of the domain and the background wind fields are the same for these analyses. First of all, Fig. 7 shows what the increment fields of these three 3DVAR analyses look like when there is no grid correlations, i.e., $B = I$ (identity matrix). For this single observation of U, the ψ–χ 3DVAR yields some negative values in its increment in fitting the observation while the other two do not. According to the observation information, the negative increment values are erroneous as there is no observation to support the negative values. Thus, a ψ–χ 3DVAR could introduce unnecessary erroneous short waves.

Similar to one-dimensional cases, a ζ–δ 3DVAR provides smooth analysis even without a recursive filter, we only show how the recursive filter parameters affect analyses of the ψ–χ and u–v 3DVAR systems. For a ψ–χ and u–v 3DVAR, we apply a recursive filter Eq. (6a, 6b) to its two-dimensional grid with one direction followed by the other. In a cycle of recursive filtering, the right filter is applied right after the left one. Three or more passes of this cycle of the filtering are applied in the numerical experiments here. With multiple applications of a recursive filter, the ψ–χ and u–v 3DVAR analyses possess similar characteristics. Figure 8 shows the ψ–χ analyses with 1, 3, and 10 passes of a recursive filter with $\alpha = 0.5$, and Fig. 9 shows the u–v analyses. Notice that the u–v analyses are more Gaussian-like than the ψ–χ analyses with the same number of recursive filter passes. More recursive filter passes are needed for a ψ–χ system in order to achieve the similar error distribution as a u–v system, that is, a ψ–χ system is more expensive in order to approximate a similar Gaussian distribution.

An interesting experiment with these analyses is to examine the integral preservation of the ψ–χ 3DVAR concluded in previous theoretical analysis of a ψ–χ 3DVAR system. For the analyses, one could sum the positive grid values and negative grid increment values, which are approximations of the

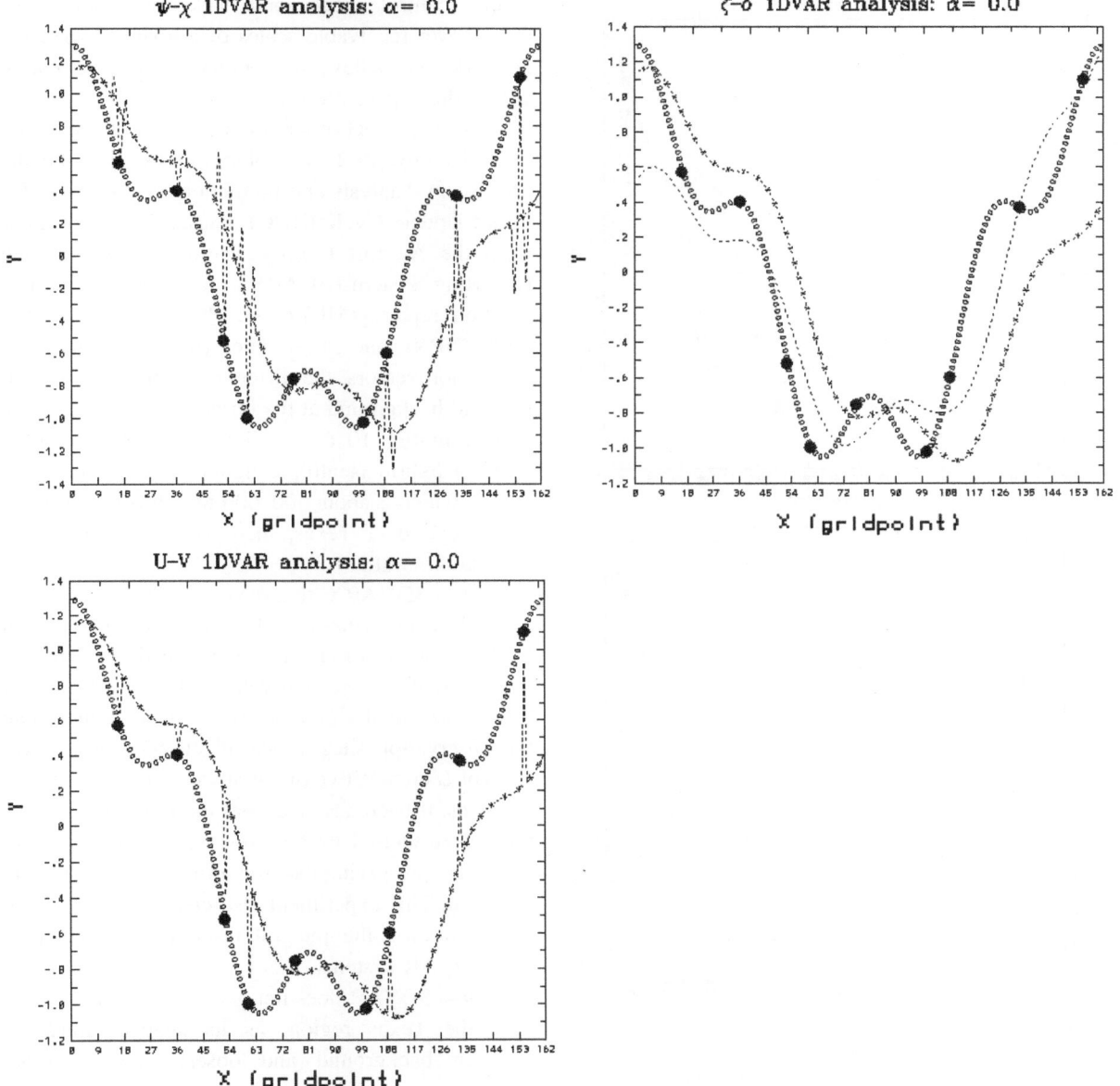

Figure 4

Analyses of a $\psi-\chi$ (*left*), $u-v$ (*middle*), and $\zeta-\delta$ 1DVAR (*right*) with no correlations using analytic functions, where the *thick dashed curve* is the observation; the *finer dashed curve* is the background; and the *finest dashed curve* is the analysis

integral of U increment multiplied by the grid sizes. These summations are shown in Table 2, as well as facts that the recursive filter reduces the preservation and the analyses preserve more over a larger domain. It shows that when analyzing smaller scales, the preservations will be strong and a $\psi-\chi$ analysis may result more negative or erroneous values.

In general, a $\psi-\chi$ 3DVAR system indeed tends to preserve the integral values of the wind field, which results in much shorter waves in fitting the observed wind. A $u-v$ 3DVAR with a recursive filter, or a $\zeta-\delta$ 3DVAR without a filter could provide more reasonable increment fields with less computations. These conclusions are more important to those analyses with

37

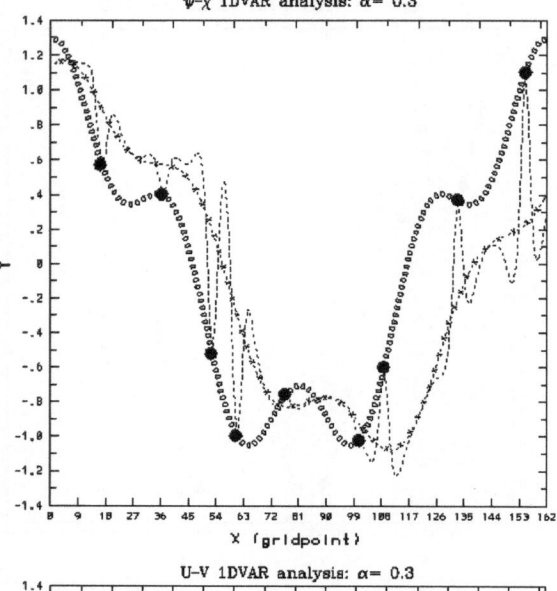

ψ-χ 1DVAR analysis: α = 0.3

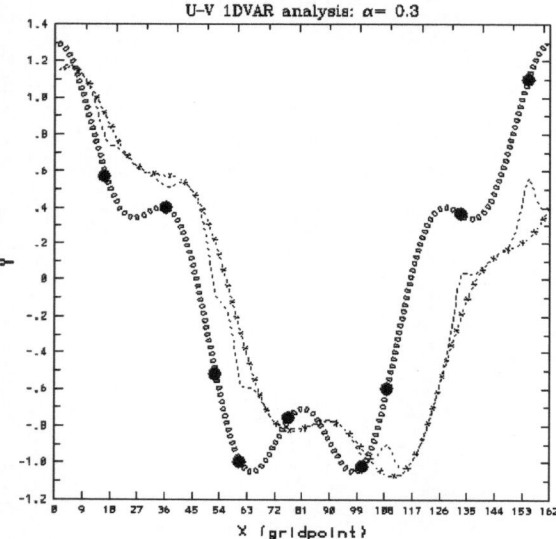

U–V 1DVAR analysis: α = 0.3

Figure 5

Analyses of a ψ-χ (*left*) and u–v 1DVAR (*right*) with a recursive filter ($\alpha = 0.3$)

inhomogeneous observation datasets in practice, as a ψ-χ 3DVAR may likely introduce additional errors.

6. *Numerical Experiment II: a Real Data Experiment*

To analyze real data assimilation results, validating their accuracy is more difficult because the true meteorological wind field is unknown. Therefore, we can rely on the observations that have satisfied quality control checks and background fields. For waves over resolvable scales of a given observation network, we believe the observations are accurate within the expected errors.

Consider a real observation dataset obtained 1200 UTC 13 July 2001. The observations are from the Mesoscale Analysis and Prediction System (MAPS)/ Rapid Update Cycle (RUC) dataset, including those from the Aircraft Communication Addressing and Reporting System (ACARS), routine meteorological aviation reports (METARs), NOAA Profilers Network (NPN), and radiosondes (RAOBs). The wind innovation vectors, the differences between observation and background at the observation locations, are plotted in Fig. 10 at the sixth level of the 40-km RUC, a hybrid isentropic-sigma vertical coordinate model, which is about 500–600 hPa. It is important to notice that over Texas, there is no westerly wind innovation at all.

A ψ-χ 3DVAR with a recursive filter ($\alpha = 0.5$) is applied to this dataset, and its analyzed incremental field is plotted in Fig. 11. It is interesting to observe the negative U increment values (shadow areas) over Texas associated with a positive U increment around the observation sites. Since all observation innovations of U are positive (no westward wind innovation) in Texas, it becomes clear that these negative values are indeed caused by the preservation of wind integrals, i.e., preserving the streamfunction and velocity potential. This experiment also confirms our conclusion regarding the integral reservation property of ψ-χ 3DVAR systems.

A u–v 3DVAR does not generate negative values over the Texas region, as its analysis must lie between background and observations. A more interesting or maybe more extreme experiment is to apply a ψ-χ 3DVAR to the same observation dataset. Figure 12 shows the analyzed U increment component without applying a recursive filter. Even though it results in quite long waves resolved by the observation network, it does not introduce errors.

7. *Conclusions*

A ψ-χ 3DVAR system has been shown to possess a property of preserving the wind integral values, the

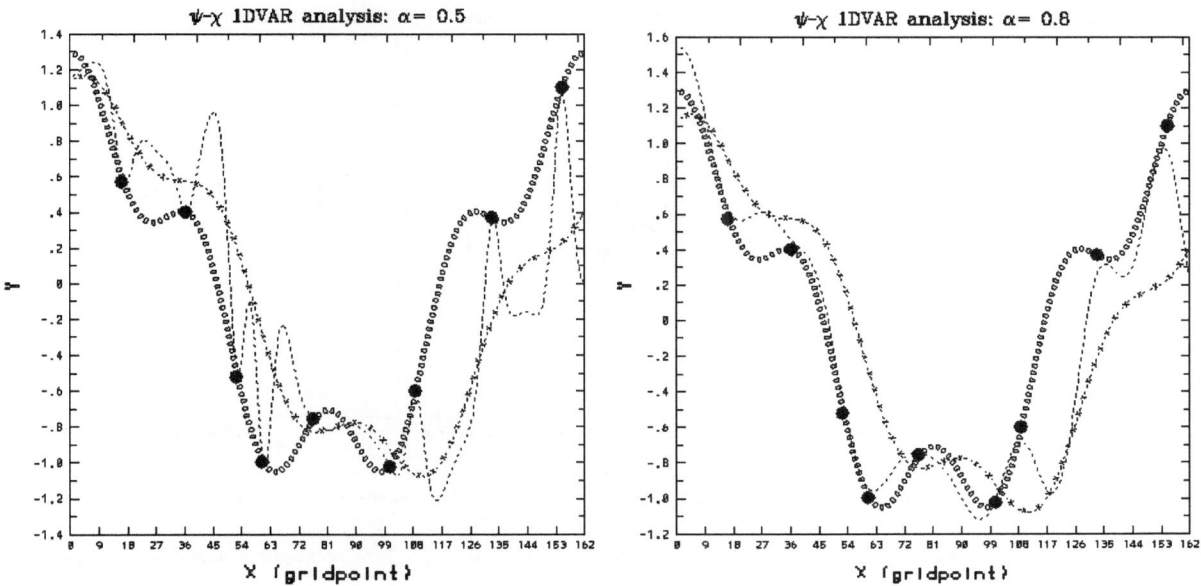

Figure 6
An analysis of a ψ–χ 1DVAR ($\alpha = 0.5$ *left* and 0.8 *right*)

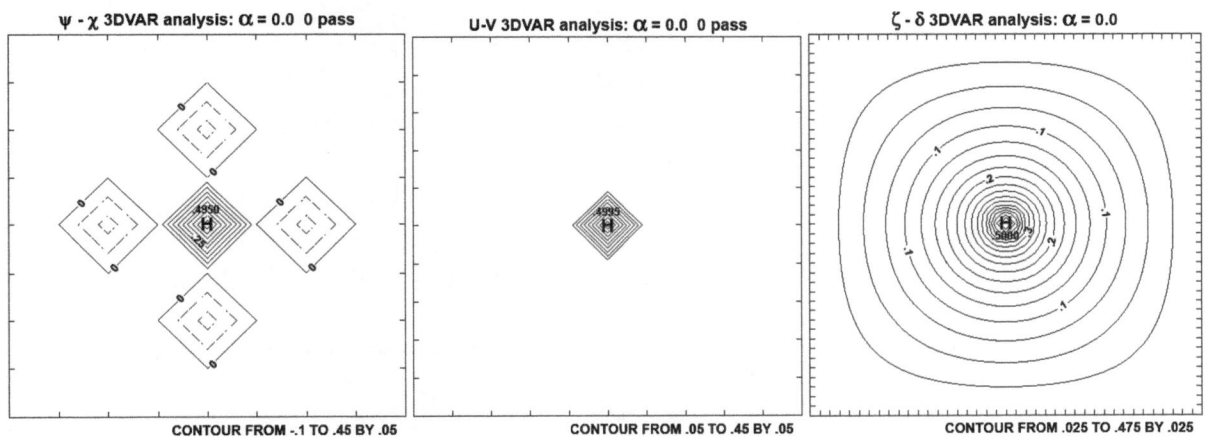

Figure 7
A ψ–χ (*left*), u–v (*middle*) and ζ–δ 3DVAR (*right*) analyses of a single U observation without a filter. Note: for ψ–χ and u–v, only central regions are plotted in order to view the details as the analyses are zero over other regions. Here each *tick* around the perimeter stands for a grid point and the analysis is dimensionless

values of streamfunction, and velocity potential. This property potentially introduces some nonphysical errors into its analysis increment fields even though either larger correlation scales or higher order filters can reduce these errors. These errors may or may not be obvious depending upon the spatial density of an observation network, the influence radius of the background error covariance, and the magnitude of the covariance matrix (accurate background statistics usually produces small magnitude of B, and then strong integral preservation is imposed on the 3DVAR system. It is common to enlarge the magnitudes of B in practice). For large influence radius over a denser spatial dataset, these errors may be small. However, in an operational environment, observations are sparse, particularly in three-dimensional space, and highly inhomogeneous. If there are inaccurate approximations to the statistical background error covariance

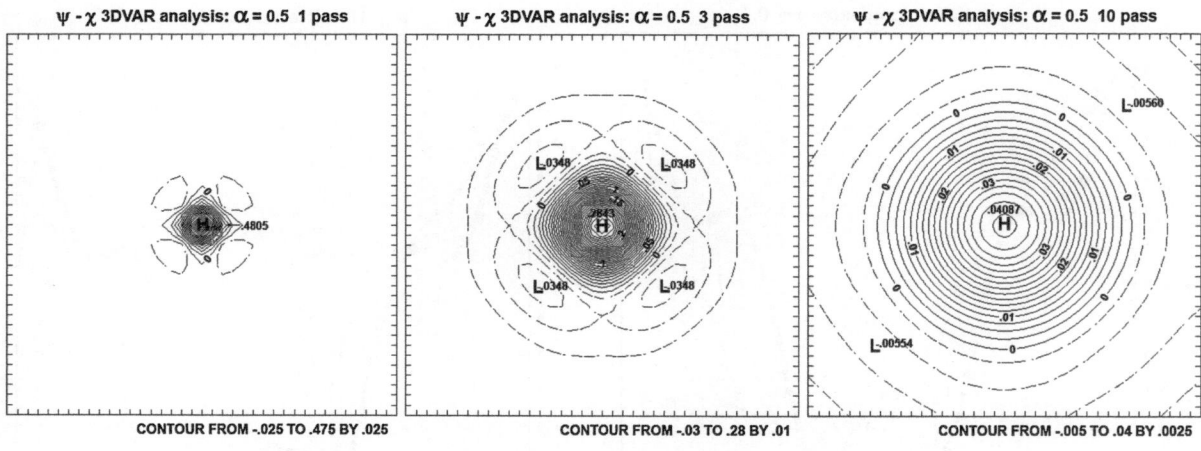

Figure 8

The $\psi-\chi$ analyses of a single U observation with one, three, and ten passes of a recursive filter

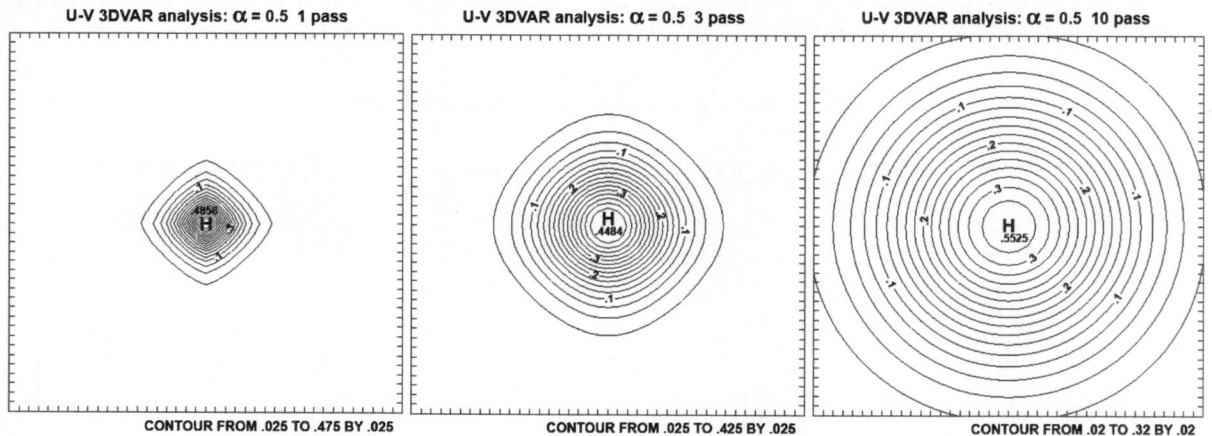

Figure 9

The $u-v$ analyses of a single U observation with one, three, and ten passes of a recursive filter

Table 2

Summations of positive/negative grid values over domains with different grid points

	No RF	1 pass of RF	3 passes of RF	10 passes of RF
41×41 grid	0.4997/$-$0.4997	3.3397/$-$3.3395	11.0589/$-$10.9578	6.8175/$-$4.5225
81×81 grid	0.4999/$-$0.4999	3.4402/$-$3.4402	16.3488/$-$16.3487	21.9957/$-$21.9664

over data-sparse regions, a $\psi-\chi$ 3DVAR could yield large errors in its analysis. In addition, a $\psi-\chi$ 3DVAR reduces the influence radius on large scales of velocity analysis and enlarge the radius for small scales as shown in Sect. 3.

The numerical experiments demonstrated that a 3DVAR using derivatives as control variables produces smoother analyses. For example, a $\zeta-\delta$ 3DVAR produces a smoother wind analysis than a $u-v$ or $\psi-\chi$ 3DVAR. It is counter-intuitive as in calculus, a derivative is usually rougher than the function itself and an integral is smoother. For variational data assimilation or an inverse problem, it is just the opposite. A $\psi-\chi$ 3DVAR uses the smoothest

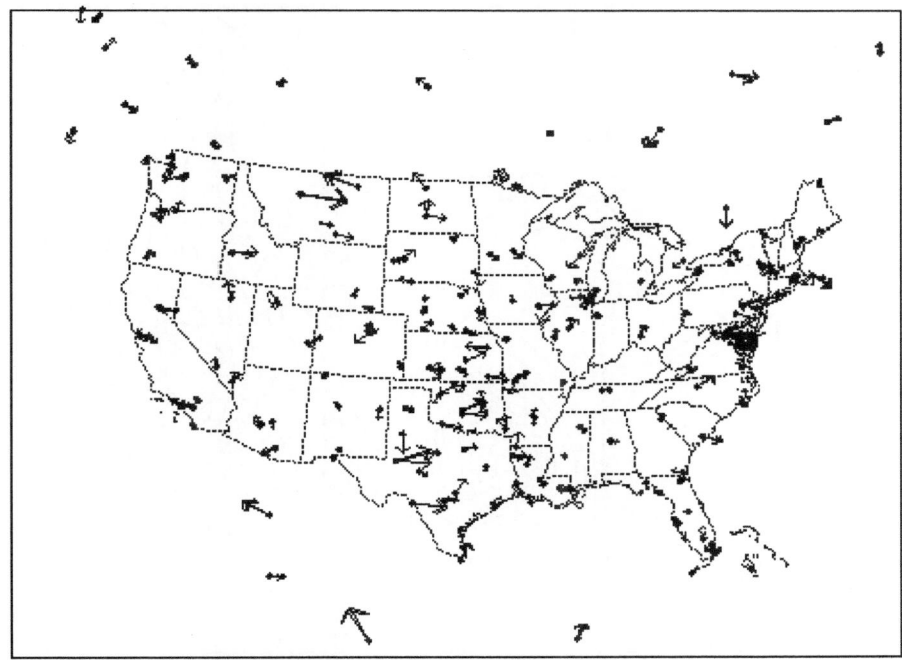

Figure 10
Innovation vectors, the differences between background and observation at the observation sites, where the maximum vector length represents
6 ms^{-1} wind speed

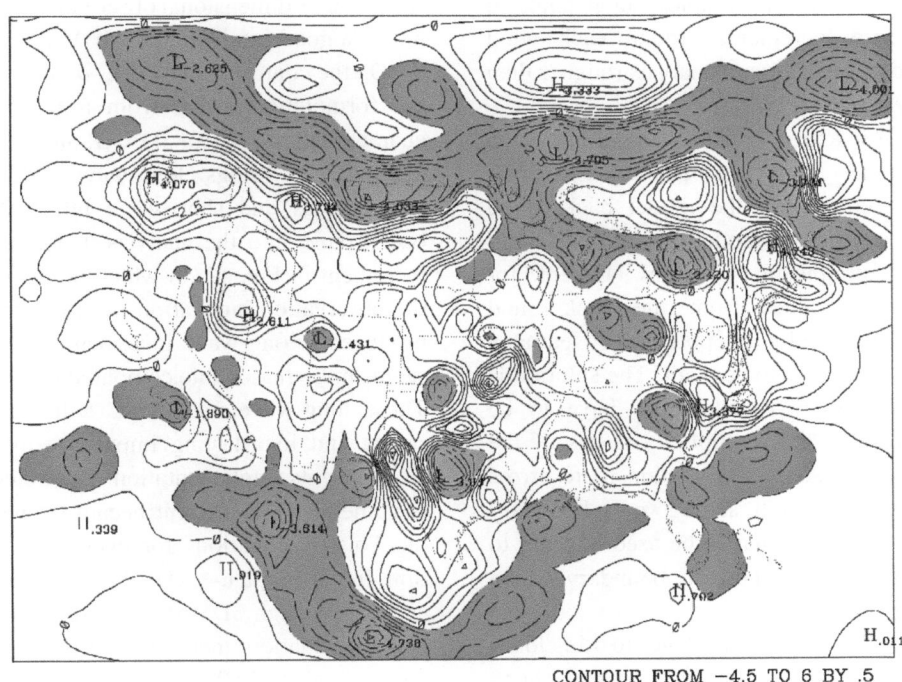

CONTOUR FROM −4.5 TO 6 BY .5

Figure 11
The increment field of a ψ–χ 3DVAR analysis, where *shadowed areas* represent negative values

41

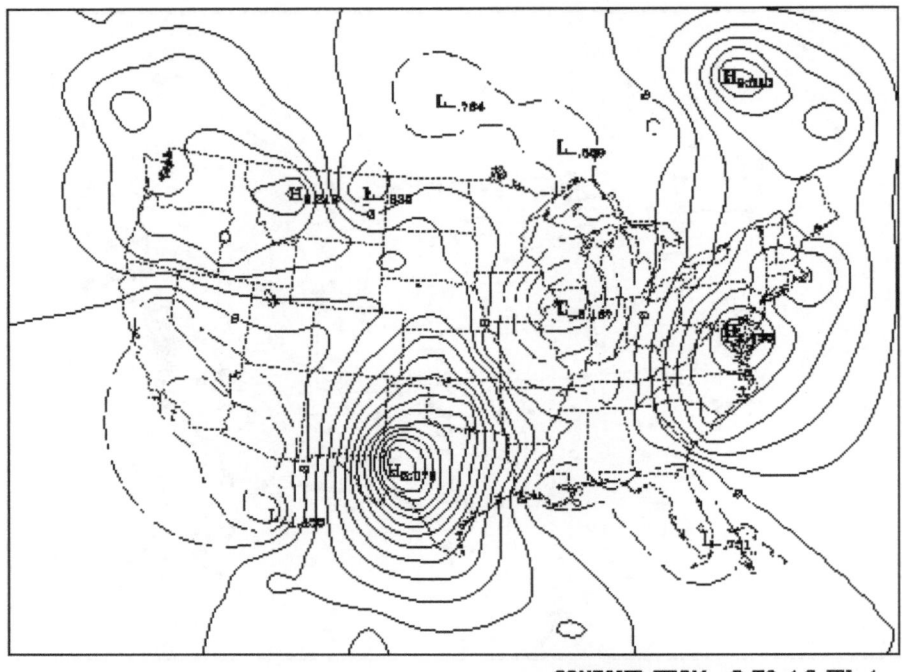

CONTOUR FROM −2 TO 4.8 BY .4

Figure 12
The increment field of a ζ–δ 3DVAR analysis, where *dashed contours* represent negative values

fields among these control variables but it tends to introduce opposite increments around each wind observation and makes its wind analysis rougher than the other 3DVARs.

A difficulty for applying a ψ–χ 3DVAR over a limited area comes from the boundary condition treatment when computing the model background error covariance from a model background with velocity as its prognostic variables. Except for accurate boundary values from the background streamfunction and velocity potential, Neumann boundary conditions have to be dealt. The solution is not unique and numerical techniques have to be considered to resolve the multiplicity issues of the solution. A ψ–χ 3DVAR could contain more errors with a boundary condition setting zero boundary values and a large domain has to be used in order to reduce or remove the incorrect boundary value problems.

In addition, a ψ–χ 3DVAR also tends to bring long-wave information from the background and short-wave information from observations into its analysis (XIE *et al.*, 2002). With highly inhomogeneous and sparse three-dimensional observation networks, it may not be a good candidate in a 3DVAR analysis.

On the contrary, a u–v or ζ–δ 3DVAR do not have the formulation errors from a ψ–χ 3DVAR. The characteristics of a ζ–δ 3DVAR are opposite to a ψ–χ 3DVAR. It brings long-wave information from observations and short-wave information from the background into the analysis. Consider the analysis increment in Fig. 12, in which the ζ–δ 3DVAR used the observation innovation from the background and produced a long-wave increment over Texas, which seems more reasonable compared to the observations. These long-wave increments will definitely have a strong and long-lasting impact on model forecasts. However, its implementation is more complex and it is numerically inefficient because it requires solving the Poisson equations for every optimization iteration. In general, a u–v 3DVAR seems more attractive.

Even though 3DVAR systems are widely used in data assimilation, there are still many issues or challenges in a 3DVAR or even 4DVAR analysis. These variational systems can be further improved. A form of 3DVAR may be carefully selected in designing a

3DVAR system for improvement. This study may further assist in the selection of a 3DVAR or 4DVAR system in meteorological data assimilation.

Acknowledgments

The authors thank Dr. S. Koch at the Global Systems Division (GSD) for many helpful suggestions on the paper, Dr. D. Devenyi of the GSD for providing RUC data assimilation data input routines in the real data analysis and thank Dr. Guo Yong-run of National Center for Atmospheric Research for his constructive discussions on the boundary value problem for a ψ–χ 3DVAR system. The authors also thank Dr. Paul Schultz and Ms. N. Fullerton at GSD for the internal review and technical editing of this paper.

References

BUEHNER, M., (2004) *Ensemble-derived stationary and flow-dependent background error covariances: evaluation in a quasi-operational NWP setting.* Q. J. R. Meteorol. Soc. **128**, 1–31.

CHEN, Q and KUO, Y. (1992a) *A harmonic-sine series expansion and its application to partitioning and reconstruction problem in a limited area.* Mon. Wea. Rev., **120**, 91–112.

CHEN, Q and KUO, Y. (1992b) *A Consistency condition for wind-field reconstruction in a limited area and a harmonic-cosine series expansion.* Mon. Wea. Rev., **120**, 2653–2670.

HAYDEN, C. M. and PURSER, R.J. (1995) *Recursive filter objective analysis of meteorological fields: applications to NESDIS operational processing.* J. Appl. Meteor., **34**, 3–15.

LORENC, A.C., (1986) *Analysis methods for numerical weather prediction.* Quart. J. Roy. Meteor. Soc., **112**, 1177–1194.

PARRISH, D.F. and DERBER, J. (1992) *The National meteorological Center's spectral statistical-interpolation analysis system.* Mon. Wea. Rev., **120**, 1747–1763.

PURSER, R.J., WU, W.-S., PARRISH, D.F., ROBERTS, N.M. (2003a) *Numerical aspects of the application of recursive filters to variational statistical analysis. Part I: Spatially homogeneous and isotropic Gaussian covariances.* Mon. Wea. Rev., **131**, 1524–1535.

PURSER, R.J., WU, W.-S., PARRISH, D.F., ROBERTS, N.M. (2003b) *Numerical aspects of the application of recursive filters to variational statistical analysis. Part II: Spatially inhomogeneous and anisotropic general covariances.* Mon. Wea. Rev., **131**, 1536–1548.

SCHLATTER, T.W., (1975) *Some experiments with a multivariate statistical objective analysis scheme.* Mon. Wea. Rev., **103**, 246–257.

SHUKLA, J., and SAHA, K.R. (1974) *Computation of non-divergent streamfunction and irrotational velocity potential from the observed winds.* Mon. Wea. Rev., **102**, 419–425.

STEPHENS, J.J., and JOHNSON, K.W. (1978) *Rotational and divergent wind potentials.* Mon. Wea. Rev., **106**, 1452–1457.

WU, W.-S., PURSER, R.J. and PARRISH, D.F. (2002) *Three-dimensional variational analysis with spatial inhomogeneous covariance.* Mon. Wea. Rev., **130**, 2905–2916.

XIE, Y., LU, C. and BROWNING, G. (2002) *Impact of formulation of cost function and constraints on three-dimensional variational data assimilation.* Mon. Wea. Rev., **130**, 2433–2447.

(Received December 11, 2010, accepted March 22, 2011, Published online August 7, 2011)

Pure Appl. Geophys. 169 (2012), 353–365
© 2011 Springer Basel AG
DOI 10.1007/s00024-011-0383-2

▌Pure and Applied Geophysics

Observing System Simulation Experiment for Global Precipitation Mission

A. K. Mishra[1] and T. N. Krishnamurti[1]

Abstract—From the suite of future Global Precipitation Mission (GPM) satellites we have selected 11 of the possible contributors to the NASA's International precipitation measurement program. The Observing System Simulation Experiments (OSSE) presented here explores the predictive usefulness of this suite of satellites. In order to carry out such experiments a Nature Run based on results from a state of the model is required. For that purpose we have selected recent past runs from the European Center for Medium Range Forecasts (ECMWF). These were designated as special data sets for OSSEs in partnership between NASA, NCEP/EMC, and NOAA. In order to test the usefulness of these future GPM-based precipitation measurements we first identify the typical orbits of eleven satellites. Along these orbital tracks we generate proxy precipitation data sets from the ECMWF Nature Run. This method of extraction of precipitation data set from a Nature Run is described in this paper. This methodology also requires a fraternal twin model (different from the Nature Run) in which the usefulness of the proposed GPM proxy data sets from the Nature Run are systematically evaluated in a forecast mode. The procedure for incorporation of the rainfall data sets is called the rain rate initialization. Data from one or more satellites are sequentially introduced into the fraternal twin model (which is the Florida State University Global Spectral Model) during the initialization phase for a number of experiments. After the initialization of such precipitation data sets, forecast experiments are carried out with the fraternal twin. The question asked is, as we introduce more and more GPM satellites how close do the forecasts from the fraternal twin approach the Nature Run? The results from this experimentation show that very promising improvements for short-range precipitation forecast skills are attainable from the proposed suite of GPM satellites.

Key words: Observing system simulation experiment, global precipitation mission, nature run, physical initialization, global spectral model.

Abbreviations

AMSR-E	Advanced Microwave Scanning Radiometer for the Earth Observing System
ATMS	Advanced Technology Microwave Sounder
DMSP	Defense Meteorological Satellite Program
DWL	Doppler Wind Lidar
ECMWF	European Centre for Medium-Range Weather Forecasts
EM	Ensemble Mean
EMC	Environmental Modeling Center
ESA	European Space Agency
ETS	Equitable Threat Score
FSU	Florida State University
GARP	Global Atmospheric Research Program
GCOM-W	Global Change Observation Mission-Water
GMI	GPM Microwave Imager
GPM	Global Precipitation Mission
GSM	Global Spectral Model
JAXA	Japanese Aerospace Exploration Agency
JPSS	Joint Polar Satellite System
MADRAS	Microwave Analysis and Detection of Rain and Atmospheric Structures
MetOp	Meteorological Operational satellite
MHS	Microwave Humidity Sounder
NASA	National Aeronautics and Space Administration
NCEP	National Centers for Environmental Prediction
NOAA	National Oceanic and Atmospheric Administration
NPOESS	National Polar-orbiting Operational Environmental Satellite System
NPP	NPOESS Preparatory Project
OLR	Outgoing Longwave Radiation
OSSE	Observing System Simulation Experiment
PBL	Planetary Boundary Layer

[1] Department of Earth, Ocean and Atmospheric Science, Florida State University, Tallahassee, FL 32306, USA. E-mail: mishra.nwp@gmail.com

RAS   Rain-rate Assimilation System
RMS   Root Mean Square
SC   Spatial Correlation
SSMI/S   Special Sensor Microwave Imager/
             Sounder
TRMM   Tropical Rainfall Measuring Mission

1. Introduction

The Observing System Simulation Experiment (OSSE) provides useful information on the use of new observing systems, in the context of their usefulness for the numerical weather prediction problem (ARNOLD and DEY, 1986; Lord *et al.,* 1997; ATLAS *et al.*, 1985, ATLAS 1997; MASUTANI *et al.*, 2006, 2010). The first observing system simulation experiment was carried out for the first Global Atmospheric Research Program (GARP) experiment by NITTA (1975). Various OSSEs studies were carried out in order to improve numerical weather prediction by HALEM AND DLOUHY (1984), ROHALY AND KRISHNAMURTI (1993), ATLAS (1997), STOFFELEN *et al.* (2006). ARNOLD (1986) described the usefulness and limitations of OSSEs in detail. In their study they proposed using fraternal twin instead of conventional identical twin in OSSE studies. Previously the same NWP models were used to simulate the atmosphere and assimilate the proxy data and to run the forecast, these were referred to as "identical twin".

The Global Precipitation Mission (GPM) is planned to be in operation within a few years from now. This is a constellation of satellites that will collectively provide estimates of precipitation. These indirect estimates will largely come from microwave and radar-based measurements of the brightness temperature and the radar reflectivity provided by the atmospheric hydrometeors. During the current era of the Tropical Rainfall Measurement Mission (TRMM), which started in 1997, one satellite provided such estimates of rain rates. Given a diverse constellation of satellites, the OSSEs reported in this study utilize a number of components: a high resolution Nature Run, extraction of precipitation data along the proposed orbits of GPM from the Nature Run, a fraternal twin model, assimilation of Nature Run-based proxy rainfall

observations along the satellite orbits into the initial states of the fraternal twin, carrying out prediction experiments with the fraternal twin using none or one to as many as eleven proposed GPM satellite-based data sets. The Fraternal Twin extracts the proxy precipitation estimates from the Nature Run along the orbits of the GPM suite of satellites. The FSU GSM (KRISHNAMURTI *et al.*, 2006) is chosen as the fraternal twin model. This model is sufficiently different from the Nature Run, and hence can be used as a fraternal twin. The objective of the OSSEs is to start with a somewhat degraded initial state and incorporate pseudo observations from the Nature Run (by introducing errors into the observing system) being tested and carry out assimilation and forecast experiments. Those forecasts are compared with a control run from the fraternal twin model to assess the effects of incorporation of the pseudo observations. This work is guided by the similar OSSEs that were recently published in the context of the Atmospheric Dynamics Mission Aeolus (ADM-Aeolus) of the European Space Agency (ESA) for Doppler Wind Lidar (DWL) following STOFFELEN *et al.* (2006). Their study included the use of pseudo wind observations for evaluation of the ADM-Aeolus mission using the ECMWF database for the Nature Run and a different European model as the fraternal twin.

2. The Observing System Simulation Experiments

2.1. Global Precipitation Mission

GPM is the National Aeronautics and Space Administration's (NASA's) international cooperative program with the objectives of improve understanding of the rainfall process and improving the frequency and accuracy of high-resolution rainfall measurements from space. NASA and the Japanese Aerospace Exploration Agency (JAXA) have together initiated this international mission as an extension and/or replacement of the highly successful TRMM mission, which is scheduled to end somewhere around late 2012 or early 2013 according to current estimates. GPM is designed to have better global coverage as than the TRMM. In order to better understand rainfall processes, cloud structure and dynamics, and microphysics, with accurate global precipitation measurement with high temporal and spatial resolution, GPM will comprise more than 10 satellites. The GPM constellation will carry spacecraft from the Global Change Observation Mission-Water (GCOM-W) series of satellites, the Defense Meteorological Satellite Program (DMSP) F series, Megha-Tropiques, Meteorological Operational satellite (MetOp), NOAA's Joint Polar Satellite System (JPSS), and NOAA-19 series, and NASA's National Polar-orbiting Operational Environmental Satellite System (NPOESS) Preparatory Project (NPP). These satellites will carry radiometers, namely the Advanced Microwave Scanning Radiometer for the Earth Observing System (AMSR-E), GPM Microwave Imager (GMI), Microwave Analysis and Detection of Rain and Atmospheric Structures (MADRAS), Special Sensor Microwave Imager/Sounder (SSMI/S), Microwave Humidity Sounder (MHS), and Advanced Technology Microwave Sounder (ATMS) with almost similar characteristics. In addition GPM core spacecraft will also carry dual frequency precipitation radar to obtain accurate and detailed microphysical information, for example drop size distribution (DSD), identification of liquid, ice, and mixed-phase and particle density, and their size distribution.

In this study spacecraft passage data for only eleven satellites (names are bold and italic in Table 1) are used. In the future more satellites can be added in the GPM mission; if spacecraft passage plans are provided OSSEs can designed for additional satellites. Figure 1 shows the hypothetical coverage of the globe by GPM satellites. Top left panel shows the area covered by the ten satellites of the GPM

Table 1

GPM Constellation of satellites

No	Satellite name	Mission instruments	Inclination (deg.)	Altitude (km)	Local time	Launch date (target)
Low Inclination Observatory						
1	*GPMCore*	DPR GMI	65	407	–	2013.07
2	GPM constellation	GMI	40	600	–	2014
3	GPM Br	GMI or MADRAS	30	600	–	2014
Megha Tropiques						
4	*Megha Tropiques*	MADRAS SAPHIR	20	867	–	2009.04
GCOM series						
5	*GCOM-W1*	AMSR-2	98.19	699.6	13:30	2012.01
DMSP series						
6	DMSP F17	SSMI/S	98.8	848	17:31	2006.11.04
7	*DMSP F18*	SSMI/S	98.7	850	20:00	2008.11
8	*DMSP F19*	SSMI/S	98.7	850	17:30	2010.10
9	DMSP F20	SSMI/S	98.7	850	17:30	2012.10
METOP series						
10	METOP A	MHS	98.8	837	21:30	2006.10.19
11	*METOPB*	MHS	98.8	840	21:30	2011.08
NOAA series						
12	*NOAA 19 (NOAA-N')*	MHS	98.75	870	14:00	2009.02
NPOESS series						
13	*NPP*	CrIMSS(CriS + ATMS)	98.75	824	13:30	2010
14	NPOESS Cl	–	98.75	833	13:30	2013.01
FY-3 series						
15	FY-3 B	MWHS	98.728	836	13:30	2009
16	FY-3 C	MWRI	98.728	836	10:30	2011
17	FY-3D	MWHS MWRI MWHS	98.728	836	13:30	2013

Bold italics indicate the satellites used in this study

Source: GPM data working group (GDaWG)

constellation in just 1 h, the second panel on the top right of Fig. 1 depicts the coverage in 2 h. Current estimates suggest that GPM constellation will cover 99% of the earth in 3 h. The bottom two panels in Fig. 1 showing the global coverage of GPM satellites in 4 and 5 h, respectively, confirm the above statement. Having a global precipitation data-set almost every 3 h would surely help to improve the NWP models and their forecasts.

2.2. Nature Run

This experiment was not possible without EC-MWF's Nature Run (MASUTANI Coauthors, 2006) datasets. The Nature Run is described as long uninter-rupted forecast using the best available atmosphere–ocean coupled model, where the lower boundary condition (e.g. SST and ice cover) is provided. The Nature Run data used in our study was generated using the ECMWF model. We were given access to the NASA's OSSEs web portal and Nature Run dataset.

This dataset is available for the period of May 2005 through May 2006 at very high spatial (T511, approx-imately 0.35° × 0.35°) and temporal (stored every 3 h) resolution. Further details about the Nature Run experiment can be accessed by following the link http://sivo.gsfc.nasa.gov/OSSE/documents/t511. Total precipitation is derived from diagnostics stored at every 3 h by using convective and large-scale precipitation.

2.3. FSU Global Spectral Model: A Fraternal Twin

The FSU GSM (KRISHNAMURTI et al., 2006) is used as the fraternal twin in this study. This model was run at a resolution of T255, which has a transform grid separation of roughly 55 km in the horizontal and carries 28 vertical levels. The FSU model has a simplified Arakawa Schubert (GRELL, 1993) cumulus parameterization scheme whereas ECMWF's Nature Run used TIEDTKE (1989) mass flux cumulus convec-tion. A biosphere–atmosphere transfer scheme is used to address land surface processes in the FSU model

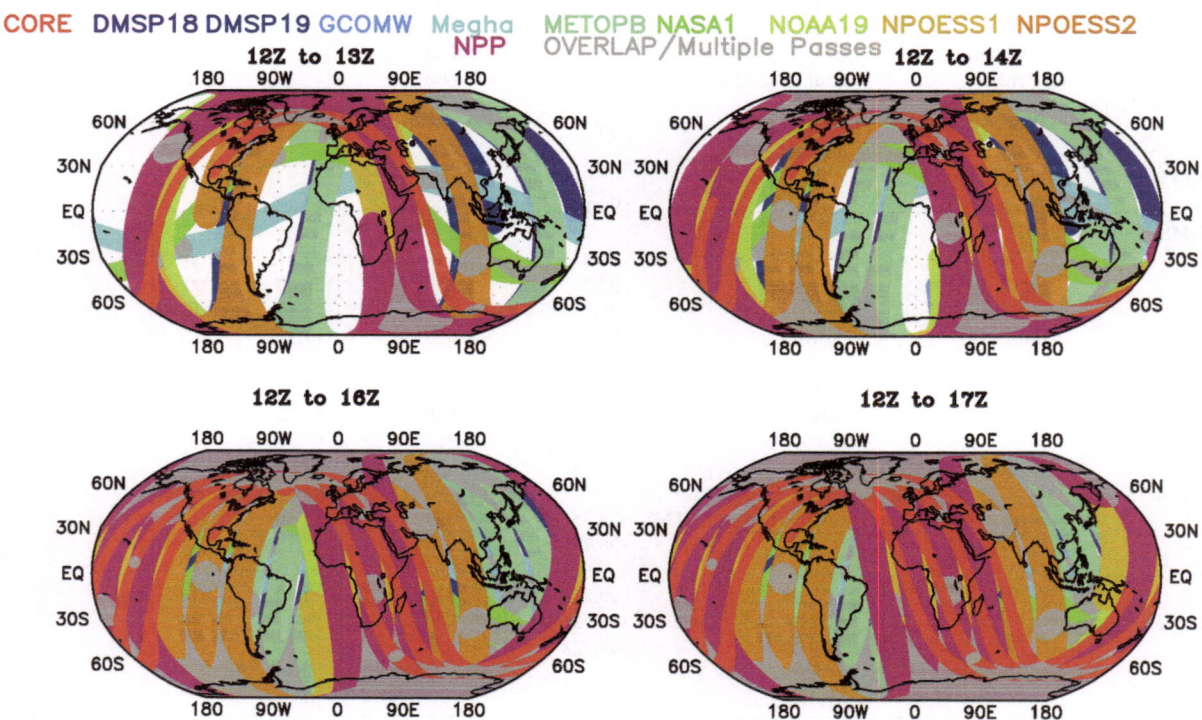

Figure 1
Global coverage by the GPM constellation of satellites

whereas the ECMWF uses the hydrology tilted ECMWF scheme. The planetary boundary layer (PBL) scheme used in the ECMWF is the Martin Köhler PBL scheme (2005) whereas the FSU model uses similarity theory for PBL. For radiative transfer the FSU model uses the BAND model unlike ECMWF's rapid radiative transfer model for long-wave radiation. This clearly shows the FSU model has completely different physics and dynamics and thus can be used as a fraternal twin.

To assimilate the synthetic precipitation data, the physical initialization mode of the FSU GSM is used. KRISHNAMURTI et al. (1991, 1993) developed a procedure of physical initialization that assimilates observed measures of rain rates into an atmospheric model. During this process, the surface fluxes of moisture, the vertical distribution of the humidity variable, the mass divergence, the convective heating, the apparent moisture sink (YANAI et al., 1973), and the surface pressure experience a spin-up consistent with the model physics and the imposed (observed) rain rates. This is accomplished through a number of reverse physical algorithms within the assimilation mode of the FSU model. These include a reverse similarity algorithm, a reverse cumulus parameterization algorithm, and an algorithm that restructures the vertical distribution of relative humidity to provide a match between the model-calculated outgoing longwave radiation (OLR) and its satellite-based observations.

FSU physical initialization is used as one of the very important components of this experiment. The main advantages of using the method of assimilation by physical initialization, compared with variational assimilation is its simple and less complicated approach. In the variational assimilation several background parameters need to be adjusted to reconcile the observed brightness temperatures with the forward calculations from forecast values of the underlying variables. This procedure may appear to be very detailed and rigorous but can in fact be self-defeating because the forward radiative transfer calculations rely on very imperfect descriptions of hydrometeors. Variational data assimilation in this case can deceive the user into thinking they got the right analyzed values for the right reasons, when in fact they can easily make the wrong adjustments to

the wrong variables. This problem is largely avoided by the physical initialization. The FSU physical initialization had improved to the point that it was possible to have consistent correlations for the nowcasting of rain from the model and those derived from TRMM observations at a level close to 0.9 (KRISHNAMURTI et al., 2001). This method of direct rain rate assimilation in the model is very cost effective, requires less computational resource and results in very high skill scores of day 0 rain rate in comparison to other data assimilation methods widely used. Ensemble Kalman filter or variational methods require huge amounts of computer time.

2.4. The Experiment

Figure 2 outlines OSSE plans of this study, which includes the Nature Run, extraction of proxy GPM observations, assimilation via physical initialization within the fraternal twin, and medium range forecasts with and without the GPM observations. The final part of this flow chart includes the evaluation metrics for both deterministic and probabilistic error estimates and the effects from the GPM. After retrieval of the precipitation data from the Nature Run, rainfall-analysis maps are prepared for various satellites every 3 hours. First guess is used from 24-hour precipitation forecast of the FSU control run stored at 12Z daily. Similar to every observing system, the GPM can also have some percentage of errors, this may be instrumentation errors, observation bias errors, or, in this case, representativeness errors. To incorporate these errors in this study we have introduced 0.05 mm/day random errors in every rainfall map. NOAA interpolated OLR data is used to perform physical initialization. Several experiments were carried out using the Nature Run and the FSU fraternal twin. Fraternal twin runs were made between May 1st and 10th 2005. Eleven runs were made for every day to assess the sensitivity of the GPM observations.

3. Results and Discussion

Figure 3 shows a typical example of the performance of the FSU fraternal twin's physical

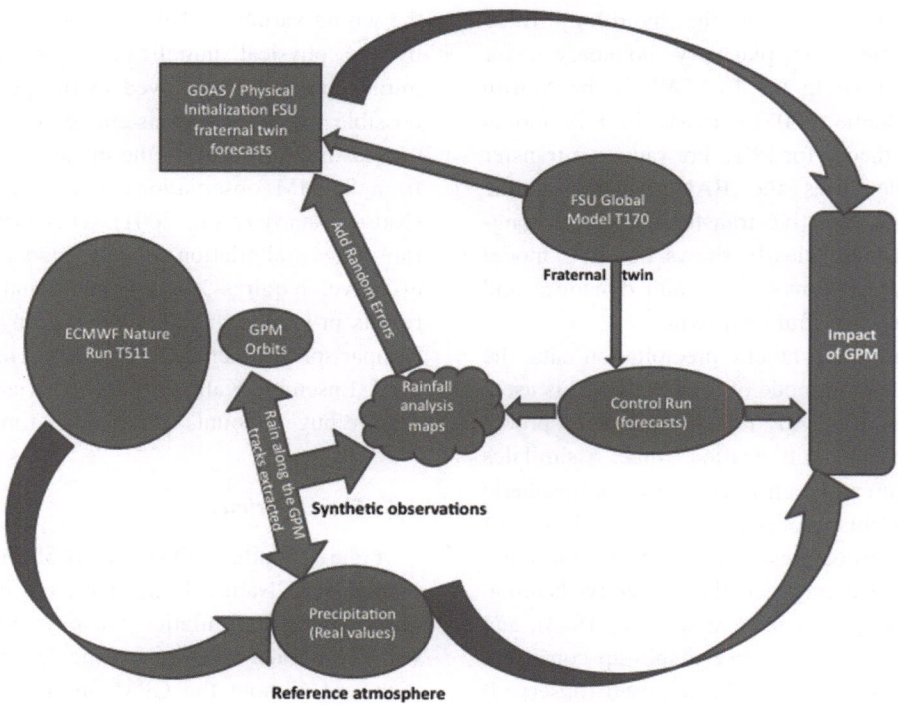

Figure 2
Flow chart of the Observing System Simulation Experiment for the Global Precipitation Mission

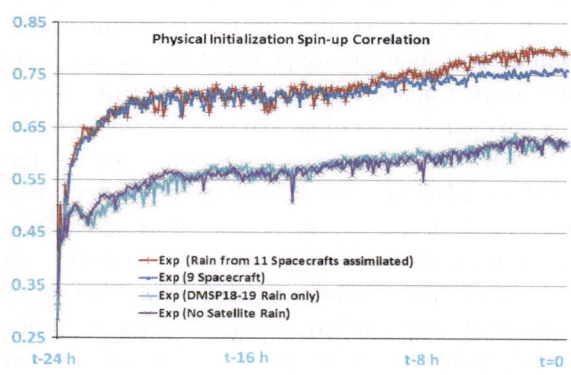

Figure 3
Rainfall correlation coefficients calculated at every model time step
during the assimilation of GPM proxy data

initialization run as a time series of rainfall correlation coefficients during the 24-hour spin-up between day −1 and day 0. As evident from the graph the least correlation is shown by the Rain rate Assimilation System (RAS) when no proxy data is introduced and simply first guess precipitation is passed to the RAS, resulting in very low correlation starting around 0.3 at day −1 and reaching to 0.6 on day 0. However, by

introducing synthetic rain data from the two sensors DMSP18 and 19 a slight improvement is seen in the correlation coefficients, still correlation at day 0 is <0.65. When proxy rainfall data from nine spacecraft are introduced in the RAS a drastic improvement is seen in first 3 h of the assimilation cycle, and day 0 correlation reaches 0.75, it is worth noting here that these nine spacecraft are other than DMSP18 and 19. Injecting the data from two sensors of DMSP (18 and 19) along with nine spacecraft shows further improvement in the skills especially in the last 6 h of the assimilation cycle. It is interesting to see that data from DMSP's two sensors does not have much effect when used alone; however, use with nine other GPM constellation satellites resulted in higher skill. Further study is required to understand this strange behavior of the RAS, but it is safe to say that together GPM satellites have a better effect.

Figure 4 is constructed by averaging the skills of ten fraternal twin RAS runs between May 1st and May 10th (once daily at 12Z). The top panel shows anomaly correlation of the precipitation with and without GPM proxy rain from day 0 through day 5 of the forecast. Similarly, the bottom panel depicts root mean square (RMS) errors (in mm/day) during the same period. A huge increase in the anomaly correlation from 0.35 to approximately 0.8 can be seen on day 0 precipitation by using GPM proxy rain from all the eleven satellites. RMS errors are also reduced on day 0 from 6 mm/day to 4 mm/day. A noticeable 30% reduction of RMS errors and 125% increase in anomaly correlation on day 0 very strongly justifies the need for GPM to improve weather prediction. These results also represent the strength of the FSU RAS. From day 1 through day 5 noticeable improvements in the skill of the forecasts are observed in both the panels of Fig. 4. Although these skill improvements are not as big as on day 0, but fraternal twin runs with GPM rain shows improvement over the control experiment for all days of forecast. Relatively small improvement in the forecast skills beyond day 3 can be because of the model's physics, but addressing these errors is beyond the scope of this study.

An example of day 0 analysis with and without proxy GPM precipitation data is compared with the Nature Run in Fig. 5a. The top panel represents the 24-hour total precipitation (in mm) valid at 12Z on May 5th 2005. The FSU fraternal twin started its spin-up from 12Z of 4th May 2005 and the RAS is used to assimilate the rainfall rate. The middle panel shows the experiment when satellite proxy rain data are not assimilated while the bottom panel depicts the rainfall analysis map resulted after assimilating rain data from 11 sensors of GPM spacecraft for last 24 h. At the top of both middle and bottom panels root mean square (RMS) errors and spatial correlation (SC) coefficients are shown for comparison of the skills from the two experiments. Skills presented in Fig. 4 are the average of the correlation and RMS errors, however these maps of Fig. 5 are just one example of rainfall distribution carrying similar skill matrices. The map shows that high rainfall belts in the equatorial Indian Ocean and equatorial Pacific Ocean were completely absent from the experiment without GPM rain assimilation, assimilation of GPM rain using RAS has almost reproduced the observation (Nature Run) by capturing most high precipitation events and placing them in the right place with almost the same intensity, for example the equatorial Indian Ocean and the equatorial Pacific Ocean. It is also evident that correlation increased from 0.36 to 0.79, and errors decreased from 5.98 to 3.55 mm on assimilating GPM rain from the eleven satellite sensors. Figure 5b shows the similar results except for the day 1 forecast valid at 12Z on 6th May 2005. Using the initial rainfall maps of Fig. 5a's middle and bottom panels, if the FSU fraternal twin is allowed to run in forecast mode, the next 24 h precipitation total would look like the middle and bottom panels of Fig. 5b. Here RMS errors show a decrease from 8.9 to 6.2 mm, while correlation increased from 0.25 to 0.39, because of the rain assimilation initialized forecast. Precipitation forecasts are also calibrated using the equitable threat scores (ETS) and bias scores, where categorical rain rate skills of the forecasts are assessed.

Figure 6a shows the ETS (in the left panels) and bias scores (in the right panels) for day 0 analysis (top two panels), and day 1 and day 2 forecasts (averaged between May 1st and 10th 2005) globally. Each plot

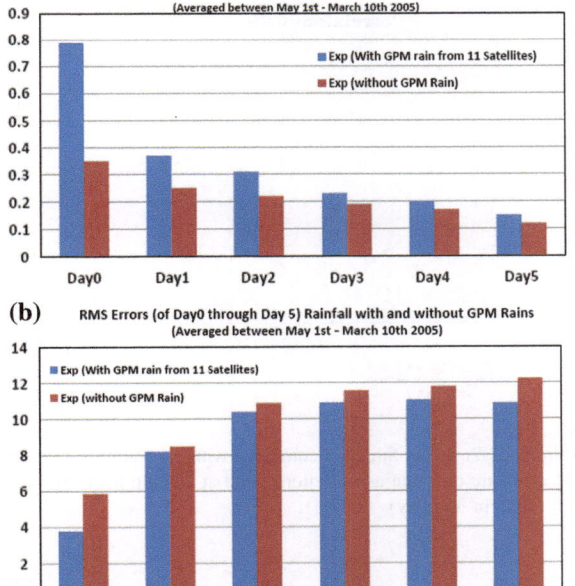

Figure 4
Day 0 through day 5 forecasts (averaged for May 1st to 10th 2005) **a** Anomaly correlations. (*top panel*) and **b** RMS errors (*bottom panel*) with and without GPM rain from 11 different satellites

(a)

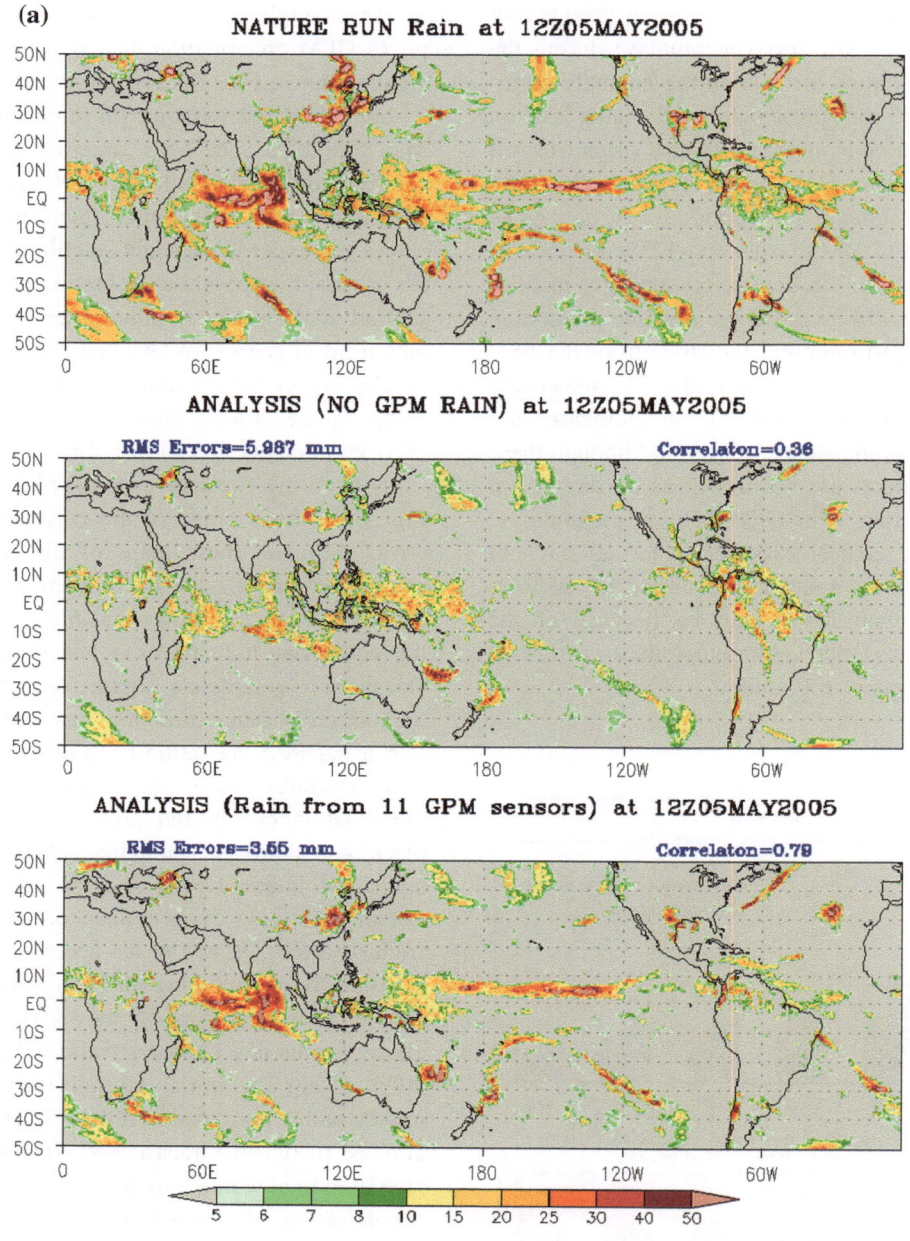

Figure 5

a Day 0 analysis (with and without GPM synthetic rain rate assimilation) valid at 12Z on May 5th 2005 compared with the Nature Run (units in mm/day). **b** Day 1 forecast using fraternal twin (with and without GPM synthetic rain rate assimilation) valid at 12Z on May 6th 2005 compared with the Nature Run (units in mm/day)

shows thresholds for different rainfall (in mm/day) along the abscissa, and the ordinate measures the skills (ETS or bias). Lines with "filled squares" represent skills for the experiment without GPM proxy rain assimilation whereas lines with "filled circles" are for the GPM rain assimilation experiment. For the rainfall threshold 3 mm/day and above, a very big jump is ETS skills is observed with the GPM rain assimilation. ETS for the rainfall rates 10 mm/day has increased from 0.15 to 0.4 because of

(b)

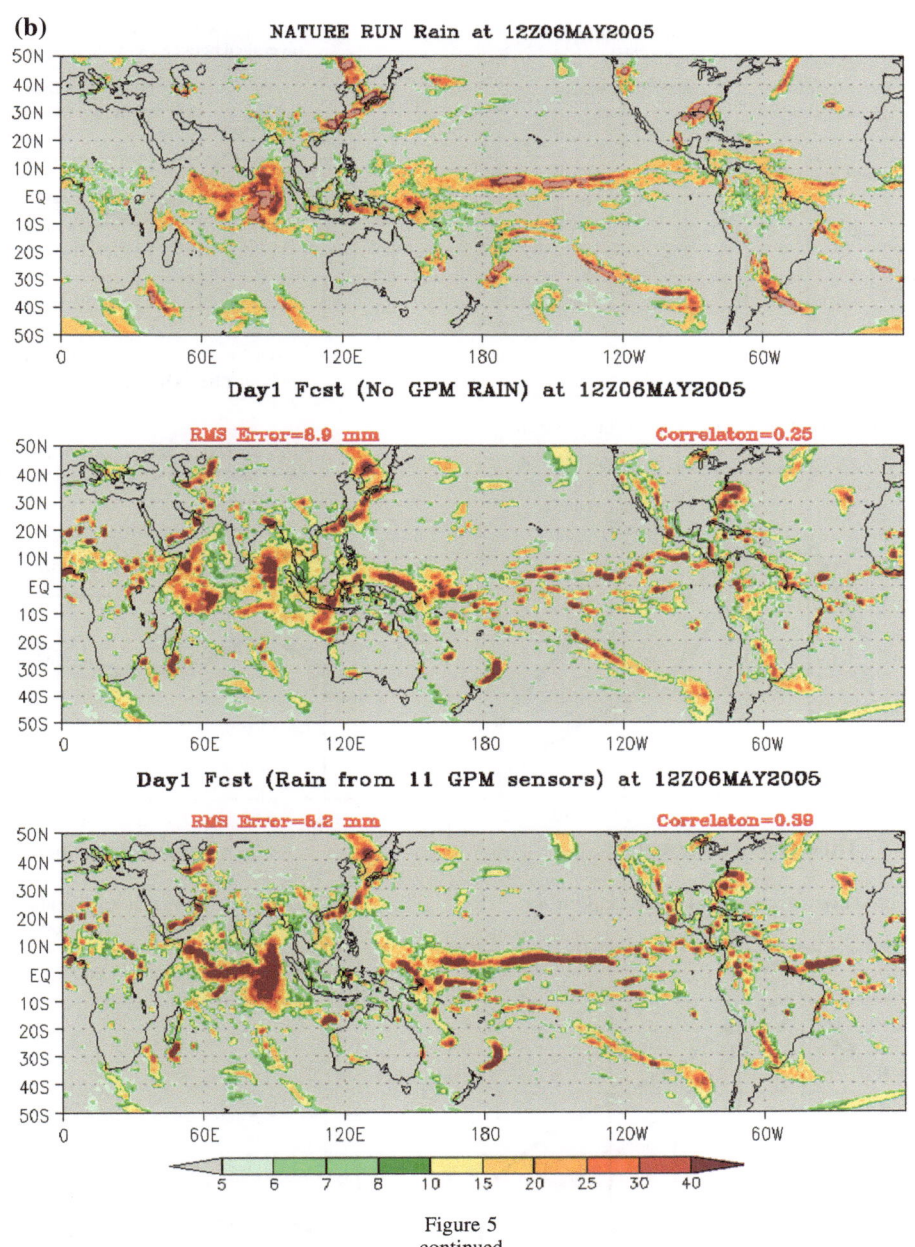

Figure 5
continued

the rainfall assimilation from GPM satellites. Bias score, which was <0.8 before GPM rain assimilation for the same threshold of 10 mm/day, has reached 0.9. Together this is a great improvement in the precipitation skills. Capturing heavy precipitation events or flash floods has always been a problem of numerical weather prediction. It is clear from these skills that by GPM rain assimilation the skills for such a high precipitation events are improved drastically, e.g. ETS for 30 mm/day threshold has increased from 0.02 to 0.38 and bias is improved from 0.2 to 0.95, these are excellent results showing the effect of GPM proxy rainfall assimilation. ETS and bias for day 1 and day 2 forecasts with and without GPM rain assimilation are compared in the middle and bottom plots and skills from the

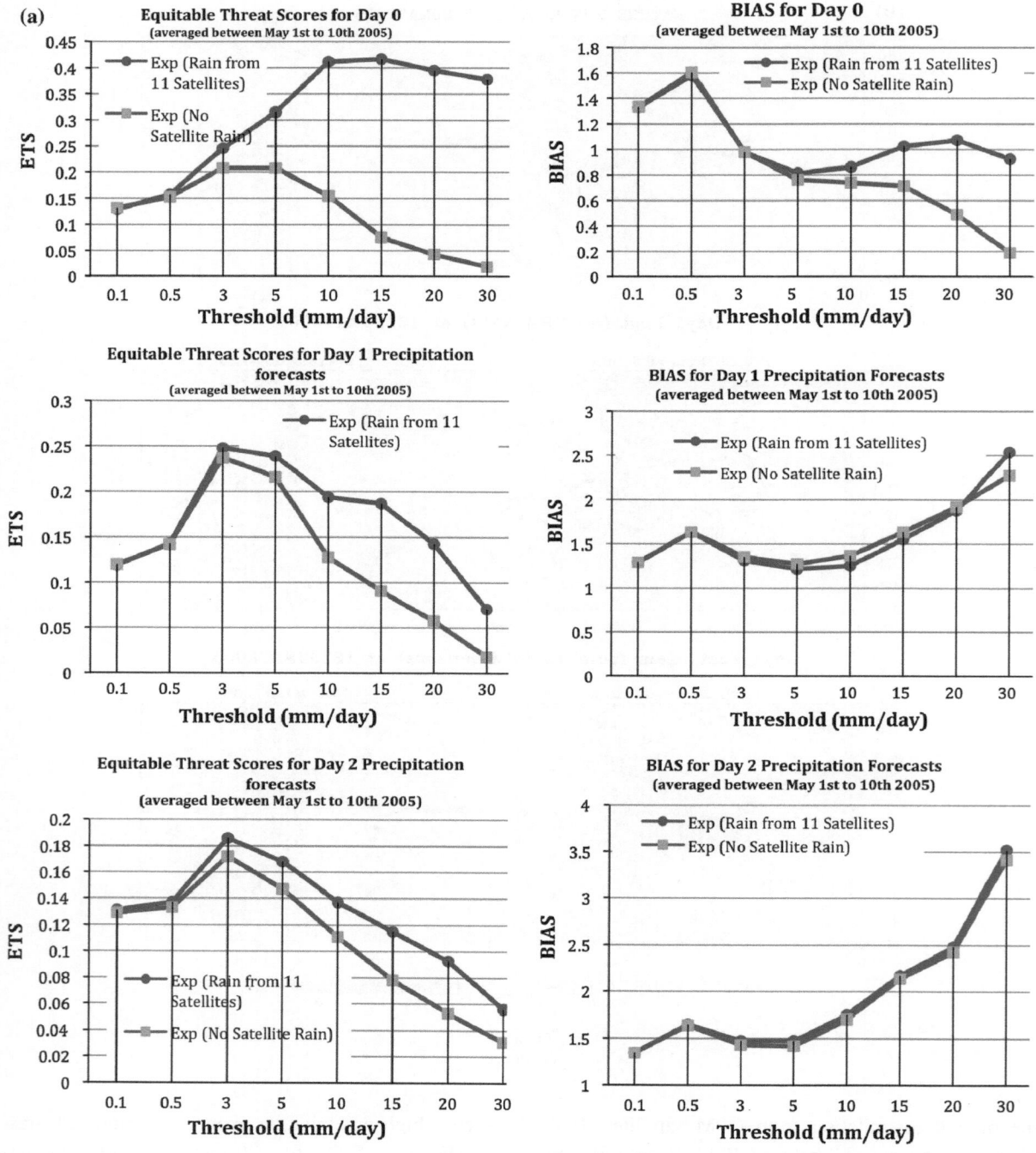

Figure 6

a Averaged (forecast issued on 12Z of May 1st through 10th) equitable threat scores (*left panel*) and bias scores (*right panel*) over the globe for day 0, day 1, and day 2 forecasts. **b** Same as Fig. 6a except for the monsoon domain (60–120E, 25S–35N)

assimilation experiment are consistently better than in the control experiment for all the rainfall thresholds. Figure 6b shows similar results over the

monsoon region defined between the longitudes of 60°E and 120°E, and latitudes of 25°S and 35°N. Almost similar results are seen over the monsoon

(b)

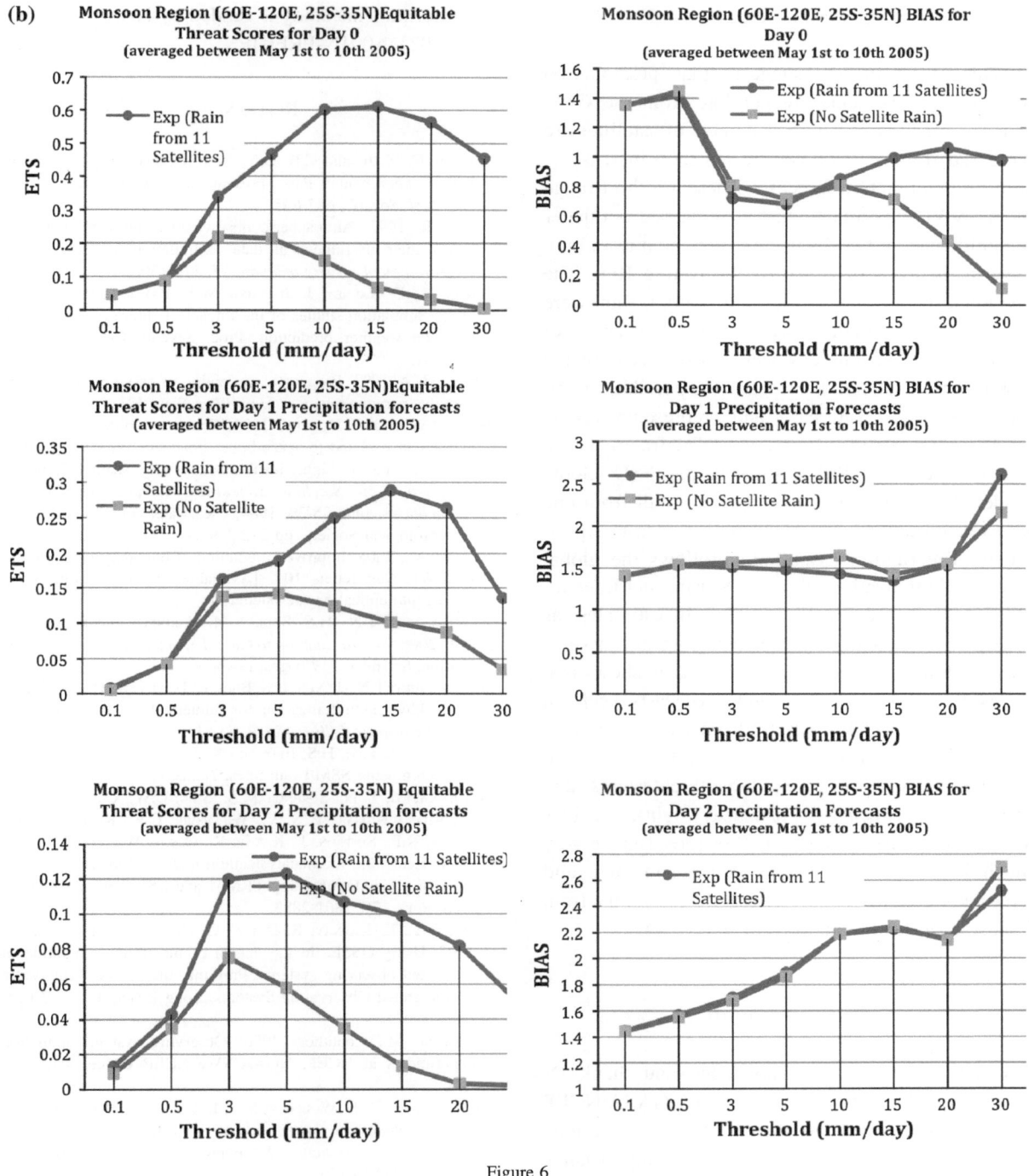

Figure 6
continued

region with even better day 0 analysis showing ETS scores close to 0.6 for 10, 15 and 20 mm/day thresholds. Results from current experiment clearly suggest that because of global coverage and the very high accuracy of the GPM sensors there would be a noticeable effect on numerical weather prediction.

4. Summary and Future Work

This study clearly suggests that precipitation forecast improvements are possible from the GPM suite of satellites. When the numbers of satellites are increased from 1 to 11 we see a steady increase of skill for precipitation forecasts. These are short-range global NWP experiments that were carried out under the guidelines of OSSEs using a very high resolution experiment from ECMWF as the Nature Run. The precipitation estimates from the Nature Run were supplied at the satellite locations. Random errors of the order of 0.05 mm/day were introduced into the Nature Run data sets to account for possible instrumentation, observational, and representativeness errors for the satellite observations. The fraternal twin that carried the GPM OSSEs is the FSU global model. This model assimilated the rainfall from one to eleven satellites, in sequential forecast experiments. The rainfall assimilation follows the global physical initialization of rain (KRISHNAMURTI et al. 1991), using the satellites' rainfall estimates as proxies extracted from the Nature Run and assimilated within the fraternal twin. This study shows that the skills of short-range rainfall forecasts keep on increasing even as we increase from 9 to 11 satellites (Fig. 3). These skills are measured using the spatial correlations, RMS errors, equitable threat scores (that distinguish between light and heavy rain), and bias scores. The skills for days 1 to 3 of forecasts are very much improved over the Asian monsoon belt and over the global tropics for light and heavy rains from inclusion of the 11 satellites for the GPM.

Acknowledgments

The author wishes to acknowledge and thank Dr Masutani and OSSEs group in NOAA/NASA/NCEP for providing access to their Nature Run results, which were a vital component of this study. Thanks are also due to Dr Xin Lin from NASA's Godderd Earth Science and Technology Center (GEST) for providing orbital data of GPM constellations. We would like to thank anonymous reviewers for their valuable suggestions and help in improving this manuscript. This study is funded by NASA grant no. NNH09ZDA001N-PRECIP.

REFERENCES

ARNOLD, C. P., Jr. and C.H. DEY, 1986. Observing-systems simulation experiments: Past, present, and future. *Bull. Amer. Meteorol. Soc.,*67, 687-695.

ATLAS, R. 1997: Atmospheric observations and experiments to assess their usefulness in data assimilation. *Journal of the Meteorological Society ofJapan*, 75, 111-130.

ATLAS, R., E. KALNEY, J. SUSSKIND, W.E. BAKER and M. HALEM, 1985. Simulation studies of the impact of future observing systems on weather prediction. Proc. Seventh Conf. on NWP, 145-151.

GRELL, G. A.,1993. Prognostic evaluation of assumptions used by cumulus parameterization. *Mon. Wea. Rev., 121*, 764–787.

HALEM, M. and R. DLOUHY, 1984. Observing system simulation experiments related to space-borne lidar wind profiling. Part 1: Forecast impact of highly idealized observing systems. Preprints, *Conference on Satellite Meteorology/Remote Sensing and Applications*, June 25-29, 1984, Clearwater, Florida, American Meteorological Society, pp 272-279.

KÖHLER, M., 2005: Improved prediction of boundary layer clouds. ECMWF Newsletter, 104, [Available online at http://www.ecmwf.int/publications/newsletters/].

KRISHNAMURTI, T.N., H.S. BEDI, V.M. HARDIKER and L. RAMASWAMY, 2006. *An Introduction to Global Spectral Modeling*, Second Edition, Springer, 317 pages, textbook.

KRISHNAMURTI T.N., J. XUE, H.S. BEDI, K. INGLES and D. OOSTERHOF, 1991. Physical initialization for numerical weather prediction over the tropics. *Tellus, 43A-B*, 53-81.

KRISHNAMURTI, T.N., H.S. BEDI and K. INGLES, 1993. Physical initialization using SSM/I rain rates. *Tellus, 45A*, 247-269.

KRISHNAMURTI, T. N., SURENDRAN, S., SHIN, D.W., CORREA-TORRES, R., VIJAYA KUMAR, T. S. V., WILLIFORD, C.E., KUMMEROW, C., ADLER, R.F., SIMPSON, J., KAKAR, R., OLSON, W., and TURK, F.J. 2001. Real time multianalysis/multimodel superensemble forecasts of precipitation using TRMM and SSM/I products. *Mon. Wea. Rev., 129*, 2861-2883.

LORD, S. J., E. KALNAY, R. DALEY, G. D. EMMITT, and R. ATLAS 1997: Using OSSEs in the design of the future generation of integrated observing systems. Preprint volume, 1st Symposium on Integrated Observation Systems, Long Beach, CA, 2-7 February 1997.

MASUTANI, M., Coauthors 2006: Observing system simulation experiments at NCEP. NOAA/NWS/NCEP Office Note *451*, 34 pp.

MASUTANI, M., J. S. WOOLLEN, S. J. LORD, et al. 2010, *Observing system simulation experiments at the National Centers for Environmental Prediction*, J. Geophys. Res., *115*, D07101, doi: 10.1029/2009JD012528.

NITTA, T., 1975. Some analyses of observing systems simulation experiments in relation to First GARP Global Experiment. *GARP Working Group on Numerical Experimentation, Report No. 10*, US GARP Plan, pp 1-35. [Available from the National Academy of Sciences, 2101 Constitution Ave. N.W., Washington, D.C. 20418.].

ROHALY, G. D., T. N. KRISHNAMURTI, 1993: An *Observing System Simulation Experiment for the Laser Atmospheric Wind Sounder (LAWS)*. J. Appl. Meteor., *32*, 1453–1471. doi: 10.1175/1520-0450(1993)032<1453:AOSSEF>2.0.CO;2.

STOFFELEN, A., G.J. MARSEILLE, F. BOUTTIER, D. VASILJEVIC, S. DE HAAN and C. CARDINALI, 2006. *ADM-Aeolus Doppler wind lidar Observing System Simulation Experiment*. Q.J. R. Meteorol. Soc., *619*, 1927-1948.

TIEDTKE, M., 1989. *A comprehensive mass flux scheme for cumulus parameterization in large- scale models*. Mon. Wea. Rev., *117*, 1779-1800.

YANAI, M., S. ESBENSEN and J.H. CHU, 1973. Determination of bulk properties of tropical cloud clusters from large-scale heat and moisture budgets. 611-627.

(Received December 11, 2010, accepted February 27, 2011, Published online July 24, 2011)

Pure Appl. Geophys. 169 (2012), 367–379
© 2011 Springer Basel AG
DOI 10.1007/s00024-011-0375-2

Multivariate Data Assimilation in the Tropics by Using Equatorial Waves

NEDJELJKA ŽAGAR[1,2]

Abstract—This paper reports recent advances in understanding of dynamical aspects of the tropical data assimilation. In contrast with the mid-latitudes, there is no a well-defined approach for the tropical data assimilation in numerical weather prediction (NWP) community which has traditionally been concentrated on the mid-latitude analysis problem. In particular, the impact of the equatorial Rossby, inertio-gravity, and mixed Rossby-gravity waves on the tropical forecast-error covariances is difficult to quantify. Various tropical waves are characterized by different couplings between the mass field and the wind field. The average mixture of these waves, built into the background-error covariance matrix for data assimilation provides analysis increments which appear nearly univariate even though they result from the advanced multivariate assimilation methodology. This applies to both dry and moist idealized tropical systems as well as to a 4D-Var NWP assimilation system.

Key words: Tropics, Data assimilation, 4D-Var, Multivariate relationships, Equatorial waves, Moist processes.

1. Introduction

Faithful description of weather and climate requires observations of atmospheric mass-field variables (temperature, humidity, pressure) as well as the field of motion (three wind components). Despite the increasing amount of observations available, these will hardly ever encompass the entire atmosphere. In addition, observations have errors, especially observations derived from satellites. Satellite observations are dominated by the observations of radiances and they make about 95% of all observations actually used for numerical weather prediction (NWP) (this data percentage is for the ECMWF assimilation system).

However, the fact that temperature observations dominate in the global observing system does not necessarily reflect their relative importance. This fact was recognized in the early days of NWP. Thus SMA-GORINSKY (1969) described that fact in Orwellian tones: "First of all, not all data are equal in their information-yielding capacity. Some are more equal than others. This tells us that if there is a choice as to what can be measured, then one variable may be preferable over another." Sparsity of observations applies most to wind field profiles (WMO, 2000; STOFFELEN *et al.*, 2005).

In the data assimilation process, observations are combined with a priori information derived from an NWP model to produce the analysis. Analyses display atmospheric phenomena with an accuracy limited by the amount of available observations and their quality as well as by the physics and the resolution of the model. The models are never going to be error-free and a reliable estimate of their errors poses a real challenge (e.g. RABIER *et al.*, 1998).

In the mid-latitudes, Rossby waves are the most important elements of atmospheric motion, and their accurate analysis is the primary concern in data assimilation for the large-scale NWP. To ensure that the observational information is assimilated primarily in terms of Rossby modes, initialization procedures and methods for generating balanced analysis fields have been developed for early analysis systems (e.g. MACHENHAUER, 1977; WERGEN, 1988). Today, advanced data-assimilation methods such as four-dimensional variational data assimilation (LE DIMET and TALAGRAND, 1986), provide analysis solutions which hardly need any initialization (e.g. GAUTHIER and THÉPAUT, 2001).

In the tropics, on the other hand, important dynamics occur in the vertical direction; i.e. the tropical circulation evolves in response to diabatic

¹ Faculty of Mathematics and Physics, University of Ljubljana, Jadranska 19, 1000 Ljubljana, Slovenia. E-mail: nedjeljka.zagar@fmf.uni-lj.si
² Center of Excellence SPACE-SI, Askerceva cesta 12, 1000 Ljubljana, Slovenia.

heating caused primarily by latent heat release. Of great importance for tropical dynamics is the change in the sign of the Coriolis force at the equator. This permits additional wave motions compared with conditions in the mid-latitudes. Namely, besides the Rossby and inertio-gravity waves (IG) present in the mid-latitudes, the tropical atmosphere supports Kelvin and mixed Rossby-gravity (MRG) waves (e.g. MATSUNO, 1966).

Consequently, data assimilation for the tropical region needs to address the relative importance of a larger variety of wave motions than encountered in the mid-latitudes, as well as the interaction of these dynamical fields with deep convection. A unified dynamical framework, analogous to quasi-geostrophic theory for the mid-latitudes, that would link the organized tropical convection, equatorial wave theory and the observed equatorial waves of the lower stratosphere has yet to fully emerge. It should include the impact of the equatorially trapped Kelvin and MRG waves, other equatorial IG waves as well as the equatorial Rossby waves.

A priori (background or first guess) information for data assimilation is obtained from a short-range forecast and it suffers from all model deficiencies in addition to errors caused by imperfect initial conditions. As a consequence of complex tropical dynamics, a priori information has larger deficiencies in the tropics than in the mid-latitudes. Exact knowledge of the errors in the background state is not available. Instead, the errors are represented by surrogate quantities with statistical properties assumed to be similar to those of the unknown errors. Derived statistical and dynamical relationships are used to spread observed information to nearby grid-points and levels (e.g. RABIER et al., 1998). Moreover, the observed information is also distributed to other variables. In this way observations of the temperature field carry information about the wind field, and vice versa. This is the fundamental reason why relationships between the mass-field and wind-field variables are of such great importance, especially in regions where observations are sparse and in a global observing system dominated by mass-field information.

In the mid-latitudes, the basic balance relationship is geostrophy, which has been extensively used in the analysis procedure (e.g. COURTIER et al., 1998;

GUSTAFSSON et al., 2001). Because of the lack of a similar dominant relationship in the tropics, the analysis here has traditionally been undertaken in a univariate fashion. This means that analyzing available temperature and humidity data does not result in new information about the wind field.

As a consequence, the analysis differences between ECMWF and NCEP in the tropics are still significant (e.g. Fig. 1 in ŽAGAR et al., 2009). Although the differences are partly affected by the way in which use is made of observations, the main underlying factors are the a priori information and assimilation methodology, in particular the statistical and dynamical properties of the background information used in the assimilation. The uncertainty of atmospheric analyses is, furthermore, carried over to ocean models, in which the poorly known surface flux, usually provided by the atmospheric model, is one of the main limiting factors for prediction capabilities (e.g. HAO and GHIL, 1994).

This paper reviews basics of the four-dimensional variational data assimilation (4D-Var) in the tropics based on a simplified tropical assimilation model. An idealized tropical modelling framework is furthermore used to discuss adjustment process in the moist tropical atmosphere in association with the assimilation problem. Namely, the shape of analysis increments is similar to that of the balanced state resulting from the adjustment process, at least in the adiabatic case in mid-latitudes. Tropical dynamics, because of their complexity, do not offer similar simple solutions; rather, the shape of analysis increments is defined by sensitive distribution of the forecast-error variance among various tropical waves, as discussed in the next section.

When moist processes are involved in NWP framework the picture becomes much more complicated because of complex interactions between the dynamics, physics, and the tangent-linear physical approximation in 4D-Var. Nevertheless, assimilation of radiances in precipitating areas in the tropics has now been successful (GEER et al., 2008). However, in a full-scale NWP system it is very difficult to isolate particular processes contributing to changes in the analysis fields. Therefore it is meaningful to apply a simplified tropical model to illustrate some aspects of tropical data assimilation; it is done here by performing single observation assimilation experiments for a moist tropical atmosphere and the results are discussed with regard to previous similar studies in the dry model environment. The discussion of idealized experiments is supplemented by a single observation experiment from a state-of-the-art data-assimilation system.

2. Four-Dimensional Variational Data Assimilation in the Tropics

In four-dimensional variational data assimilation (4D-Var) the analysis solution is a model solution consistent with all the observations over the time period $[t_1, t_2]$. An optimal analysis solution, $\mathbf{x}_a$, minimizes the cost function, J, to both the observations and the background model state, taking into account statistics of their respective errors:

$$J(\mathbf{x}) = J_b + J_o = (\mathbf{x} - \mathbf{x}_b)^T \mathbf{B}^{-1} (\mathbf{x} - \mathbf{x}_b)$$

$$+ \sum_{n=1}^{K} (\mathbf{y}_n - H[\mathbf{x}]_n)^T \mathbf{R}^{-1} (\mathbf{y}_n - H[\mathbf{x}]_n) \qquad (1)$$

In Eq. 1, the background term J_b measures the distance to a background model state. The distance to the observations is measured by J_o. Equation 1 is usually solved for an increment $\delta\mathbf{x}$, which hereafter is added to the background field, $\mathbf{x}_b$, to form the analysis, $\mathbf{x}_a = \mathbf{x}_b + \delta\mathbf{x}$ (e.g. COURTIER et al., 1998). The vector $\mathbf{y}$ contains observations distributed over K different times. With observations available at K discrete times between t_1 and t_2, a model is applied to propagate their information content forward and backward in time. During the model time integration, the information transfer between the wind and the mass fields takes place through the model equations; i.e. observed mass-field information is passed on to the wind field and vice versa.

The matrix $\mathbf{R}$ is the covariance matrix of the observation errors. Conventional observations can be assumed to be statistically independent in which case $\mathbf{R}$ becomes a diagonal matrix including the error variance values for the wind components, temperature, moisture, and surface pressure fields. The model equivalents of the observed quantities at the observation points are generated by the observation operator H. It should be noted that H includes the model time integration, i.e. it is the "forward" operator. The covariance matrix of the errors in the background state is denoted by $\mathbf{B}$.

The most pronounced difficulty when minimizing the cost function (Eq. 1) lies in the specification of the background-error covariances. The truth is not available and dealing with the full $\mathbf{B}$ matrix is intractable in NWP applications. Simplifications have to be introduced and their effects are most important in less well observed regions such as the tropics.

In order to simplify Eq. 1, a set of suitable transformations is applied to the analysis increments, which enables re-definition of the variational optimization problem defined by Eq. 1 in terms of a new variable for which $\mathbf{B}$ becomes a block-diagonal or a diagonal matrix. In an especially simple tropical modelling framework such as the shallow-water equations on the equatorial β-plane, the $\mathbf{B}$ matrix can be simplified to become an identity matrix (e.g. DALEY, 1993). Several idealized studies dealing with the dynamical aspects of tropical data assimilation applied a sequence of transformations based on the idea that forecast-error covariances can be approximated by using the normal modes of the model (PHILLIPS, 1986; DALEY, 1993; ŽAGAR et al., 2004b). In the tropical case, these are eigen solutions of the linearized shallow water equations on the equatorial β-plane. ŽAGAR et al. (2004b) implemented this solution methodology in the spectral space which is also used to solve the prognostic model equations (ŽAGAR et al., 2004a). In this case, the meridional wind-field error is formulated as

$$v(x, y) = \sum_{k=-N_k}^{N_k} \sum_{n=0}^{N_n} \sum_{p=1}^{3} \chi_{knp} v_{knp}(k, y) e^{ikx}, \qquad (2)$$

where k is the zonal wave number, n is the meridional mode number, and p describes the equatorial wave type. The three modal indices define a modal index $v(k, n, p)$ for a single equatorial mode. A meridional structure function, $v_v(k, y)$, for v-wind error is given by the parabolic cylinder function of the order n. The corresponding truncation limits are defined in the following way: N_k is defined by the minimum wave length resolved, three times the horizontal grid spacing, and the elliptic truncation criterion, N_n is defined by the equivalent modal truncation criterion, in order to ensure consistency with the forecast model (GUSTAFSSON *et al.*, 2001). The modified minimization problem for a new control variable χ is:

$$J(\chi) = J_b + J_o = \frac{1}{2}\chi^T\chi$$
$$+ \frac{1}{2}\sum_{i=1}^{K}(\mathbf{y}_i - H(\mathbf{x}_b + L^{-1}\tilde{S}^{-1}\chi_i))^T\mathbf{R}^{-1}$$
$$\times (\mathbf{y}_i - H(\mathbf{x}_b + L^{-1}\tilde{S}^{-1}\chi_i)). \quad (3)$$

The sequence of transformations which leads from Eq. 1 to Eq. 3 is formulated as:

$$\chi = SDP_yF_xF^{-1}\delta\mathbf{x}, \quad (4)$$

where F^{-1} is the inverse Fourier transform used to obtain assimilation increments in grid-point space and F_x is the direct Fourier transform in the zonal direction. The projection on the meridionally dependent part of the eigenmodes in grid point space is denoted by P_y, the normalization by the spectral-variance density by D, and the operator S represents the various means of truncation, i.e. Fourier truncation, elliptic truncation, and frequency cut-off.

The spectrum of variance of forecast errors included in the assimilation is specified through the operator D in Eq. 3. In ŽAGAR *et al.* (2004b) and ŽAGAR (2004) it was assumed that the variance spectrum of forecast errors for various tropical motions can be approximated by their observed variance (e.g. WHEELER and KILADIS, 1999). In ŽAGAR *et al.* (2008), the variance spectrum was based on the tropical forecast errors of the ECMWF model as derived in ŽAGAR *et al.* (2005). About 50% of the variance was represented in terms of the equatorial Rossby waves and the rest was divided among various inertio-gravity modes. In particular, Kelvin and MRG modes each represented between 5 and 10% of the variance and their relevance increased in the stratosphere. In the case of shallow-water equations applied in the aforementioned studies the state vector was $\mathbf{x} = [u\,v\,h]^T$ or $\mathbf{x} = [u\,v\,\theta]^T$, where u, v, h, and θ stand for the zonal wind, the meridional wind, geopotential height, and potential temperature, respectively.

The main result of these studies addressing the importance of tropical waves for the structure of background-error covariances is that the presence of the mass-wind relationships for the MRG wave, Kelvin wave, and other equatorial IG waves in the **B** matrix changes the correlation between the mass field and the wind field, which would exist if the Rossby-wave balance applies. In particular, the Kelvin wave efficiently reduces the Rossby-wave balance between the zonal wind and the geopotential at the equator whereas the MRG wave reduces the balance between the geopotential and the meridional wind component. Other equatorial IG waves additionally reduce the meridional scale of analysis increments making them equatorially trapped (ŽAGAR *et al.*, 2004b, 2005).

Here, we extend the previous studies by including the moisture field as described in the following section.

3. Adjustment and Assimilation Experiments in the Moist Tropical Atmosphere

3.1. Idealized Aqua-Tropics Model

The applied model is the moist model of GILL (1982) which describes the fundamental properties of the large-scale tropical atmosphere by considering the first baroclinic mode associated with the deep convection. The model dynamical equations for (u, v, θ) were presented in ŽAGAR *et al.* (2008) where 4D-Var experiments with the model were performed to study the potential impact of observations from several Doppler wind lidars in the tropics. Here, the system is complemented with the moisture variable, q, which represents the depth of precipitable water per unit area, i.e. the total moisture in a column:

$$q(x, y, t) = \int_0^{\pi H_o} s\,dz/\rho_w. \quad (5)$$

Specific humidity is here denoted s, and ρ_w is the density of water. The lower layer has a depth H_o and the height of the tropopause for the first baroclinic mode is at πH_o. If the water vapour is assumed to be concentrated in the lower layer, the moisture flux is given by:

$$\int\limits_{0}^{\pi H_o} \mathbf{V} s \, dz = q\mathbf{V}, \qquad (6)$$

where $\mathbf{V} = (u, v)$ is the horizontal velocity in the lower layer. Then the moisture conservation equation takes the form:

$$\frac{\partial q}{\partial t} + P = -\frac{\partial}{\partial x}(uq) - \frac{\partial}{\partial y}(vq) + E, \qquad (7)$$

where P is the precipitation rate and E is the evaporation rate.

Evaporation rate (rate of gain of moisture) is modelled as:

$$E = (\hat{q} - q)/\tau,$$

where τ is a characteristic timescale over which q is driven toward its (column-integrated) equilibrium (saturation) value $\hat{q}$. Precipitation occurs when the atmosphere is saturated ($q = \hat{q}$) and there is a convergence of moisture. In such a region, the atmosphere remains saturated and all excess moisture is precipitated. This can be expressed as:

$$\frac{\partial q}{\partial t} = 0, \quad P = \text{RHS},$$

when $q = \hat{q}$ and RHS > 0, otherwise:

$$\frac{\partial q}{\partial t} = \text{RHS}, \quad P = 0,$$

where RHS is the right-hand side of Eq. 7.

The rate at which latent heat is released, Q_{LH}, which appears in the prognostic equation for θ (Eq. 2 in ŽAGAR et al. 2008) is related to precipitation by the following relationship (GILL, 1982):

$$Q_{LH} = \frac{L\rho_w}{C_p \rho_o H_1} P, \qquad (8)$$

where L is the latent heat of condensation and C_p is the specific heat of the air at constant pressure. H_1 is the height over which the heating is distributed and it is taken to be equal to H_o. The heating rate is expressed in Kelvins per second. The gravity wave speed is defined by H_o and the static stability N; for typical values of tropical parameters (Table 1) c is approximately 60 ms^{-1}.

When the diabatic heating is prescribed and the moisture budget included, the model becomes a so-

called "aqua-tropics model". Its variations have been extensively used for understanding various aspects of the tropical circulation (e.g. DAVEY and GILL, 1987; DAVEY, 1989). Setup of the model used for the adjustment experiments presented below follows DAVEY (1989) with the values of the constant parameters chosen typical for the tropics (Table 1). In all figures, winds are representative of the lower troposphere. Winds in the upper layer have the same magnitude but the opposite sign. The model dynamics are illustrated first with an example of the adjustment to the external heating Q centred in the middle of the tropical domain. The amplitude of the heating at the centre is 20 K day^{-1}.

If the simulation is centred on the mid-latitude β-plane, the adjustment to the heating would occur on a time scale of hours by generation of inertio-gravity waves which propagate quickly away from the heated region. A balanced vortex which develops in response to the heating is nearly symmetric with a stronger inflow on its southern side, because of the β-plane approximation. The southern side of the vortex is the location of most intense precipitation (figure not shown).

In the tropics, the solution is not symmetrical (Fig. 1). The region east of the heated area is more affected by the adjustment because of the eastward-propagating Kelvin wave, a first wave emerging from the heated region after inertio-gravity waves. It is followed by the equatorial Rossby wave progressing westward. The amplitude of the mass-field response is significantly smaller than in the mid-latitude case, whereas the lower-layer convergence toward the heated region causes significant vertical motion. In the saturated atmosphere, the lower-level convergence leads to precipitation, first within the heated

Table 1

Values of the constant parameters used for the parameters presented in the paper

Tropopause height	πH_o	17 km
Lapse rate	$d\theta_o/dz$	5 K km^{-1}
Mid-level potential temperature	θ_o	333 K
Horizontal velocity scale	$c = NH_o$	65 ms^{-1}
Time scale for evaporation	τ	12 h
Column water depth of the saturated atmosphere	$\hat{q}$	0.06 m

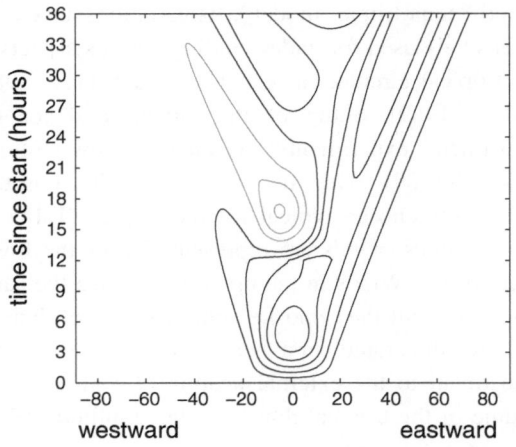

Figure 1

Contours in (x, t) space of the potential temperature θ along the central latitude of the β-plane centred at the equator. Contouring interval is ± 0.4 K. Values used in the experiment are listed in Table 1 and discussed in the text. *Black lines* denote positive and *gray lines* negative values with respect to the initial state

region and with a greater intensity, and afterwards precipitation areas move further away from the equator while the rate of precipitation decreases. There is no precipitation associated with the initial wave front because the initial atmosphere was not saturated (figure not shown).

In saturated regions the slowing down of the equatorial waves occurs because of interaction between the dynamics and convection, in this model represented by the flow convergence in the saturated environment. Smaller phase speeds imply a smaller equivalent depth. The exact nature of the interaction in reality is not yet fully understood, but available equivalent-depth estimates according to linear shallow-water theory are in the range 12–50 m (WHEELER and KILADIS, 1999). Consequently, this range of values was successfully used for modelling the short-range tropical forecast errors of the ECMWF model (ŽAGAR et al., 2005, 2007).

3.2. Adjustment Process in the Moist Tropical Atmosphere

The moist adjustment is presented for the zonal wind perturbation which has the form of a Gaussian function. The same perturbation was studied in

ŽAGAR et al. (2004a) for the adiabatic tropical atmosphere. Here, the initial moisture field takes on various states, as discussed below, whereas the lateral boundary fields are for the saturated and motionless atmosphere. Results for the unsaturated case are shown in Fig. 2. In the unsaturated case the model behaviour is nearly identical with the dry model solution presented in ŽAGAR et al. (2004a). Zonally propagating eastward and westward IG waves after 1 h and after 6 h appear with longitudinally symmetrical amplitudes of opposite sign relative to the initial position. After 12 h in the forecast the balanced temperature perturbation spanning the equator at the initial location has the form of an equatorial Rossby wave. The maximum vertical velocity is at the beginning over the area east of the initial perturbation (Fig. 2a) but after 6 h strongest upward motion is found in the westward-propagating lobe (Fig. 2b). Because the local moisture increase as a result of convergence does not reach the saturated value anywhere, there is no precipitation in the simulation.

When the initial wind perturbation occurs in the saturated atmosphere, adjustment brings about intense precipitation which falls over the region with positive vertical motion (Fig. 3). The released latent heat warms the atmosphere to the east of the initial perturbation; as a consequence, the negative temperature perturbation is almost negligible compared with the unsaturated case (Fig. 3a). A wave with the reduced amplitude moves to the east much more slowly than in the unsaturated case and it also appears more equatorially trapped compared with the unsaturated case. After 12 h the structure of the positive temperature perturbation at the location of initial perturbations is similar to that in the unsaturated case except for a somewhat smaller meridional scale.

The effect of moisture is to reduce stability and thus to slow down waves propagating through the wet regions. The implied smaller equivalent depths help the "equatorial trapping" in the model. (Note that in the previous studies the trapping was imposed explicitly through the equivalent depth constant.) The meridional "trapping scale", also known as the equatorial Rossby radius of deformation, R_e, is defined as

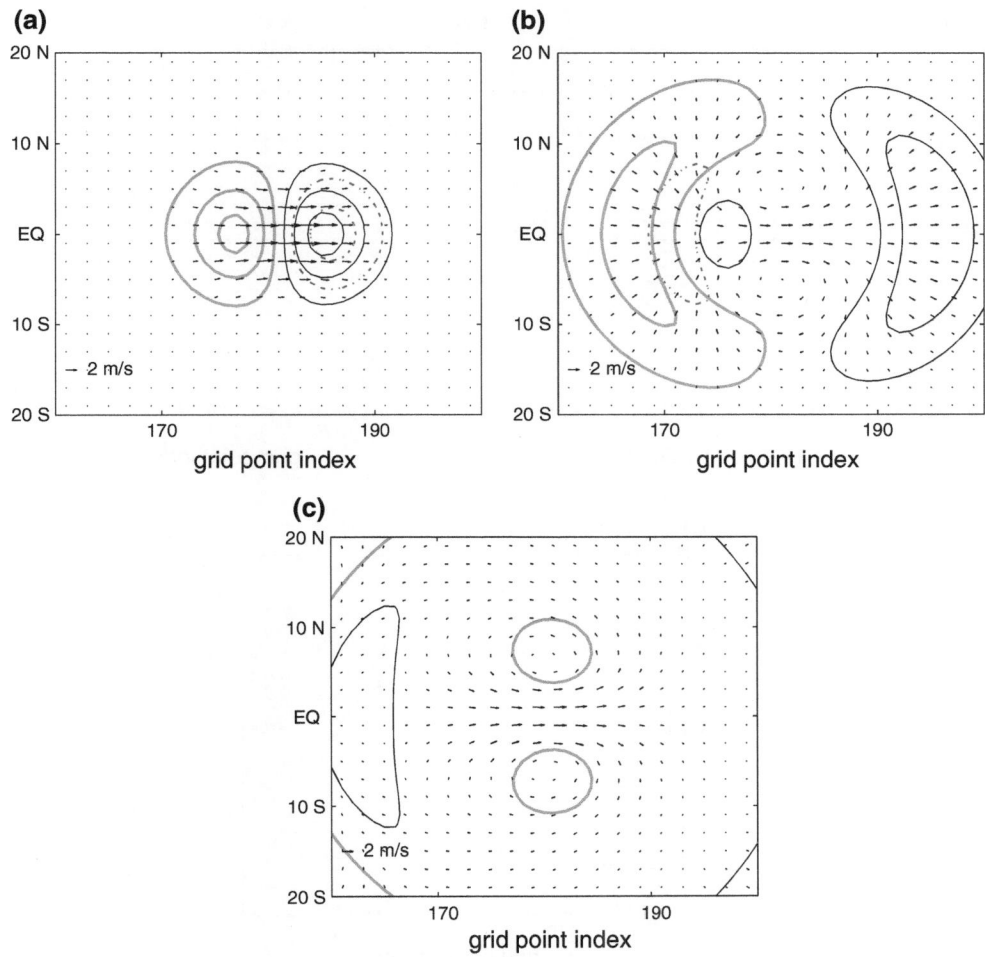

Figure 2
Adjustment in the moist unsaturated atmosphere. **a** after 1 h, **b** after 6 h, **c** after 12 h. Shown are temperature perturbation each 0.2 K, from 0.1 K (*thick lines* for positive and *thin lines* for negative values) and positive vertical velocity each 1 cm s^{-1}, from 1 cm s^{-1} (*dashed–dotted lines*)

$$R_e = \left(\frac{NH_o}{2\beta}\right)^{1/2}.$$

The effect of moisture on reducing stability is further illustrated in Fig. 4 which can be compared with Fig. 3a. The positive temperature perturbation to the west of the heated region has a significantly reduced amplitude compared with the earlier simulation, whereas the eastern perturbation has become positive. Vertical motion is twice as strong and the maximum precipitation is approximately one-third stronger than earlier (46 mm day^{-1} compared with 35 mm day^{-1}). Implied phase speed is also smaller

by approximately 30% (47 m s^{-1} compared with 66 m s^{-1} earlier), as well as the Rossby deformation radius is also smaller (1,000 km compared with 1,200 km). In conclusion, reduction of the static stability has major effects on equatorial trapping of the perturbations as well as on their propagation characteristics.

3.3. Single Observation Assimilation Experiments in the Moist Tropical Atmosphere

The background-error variance spectra for the potential temperature and wind fields are based on the

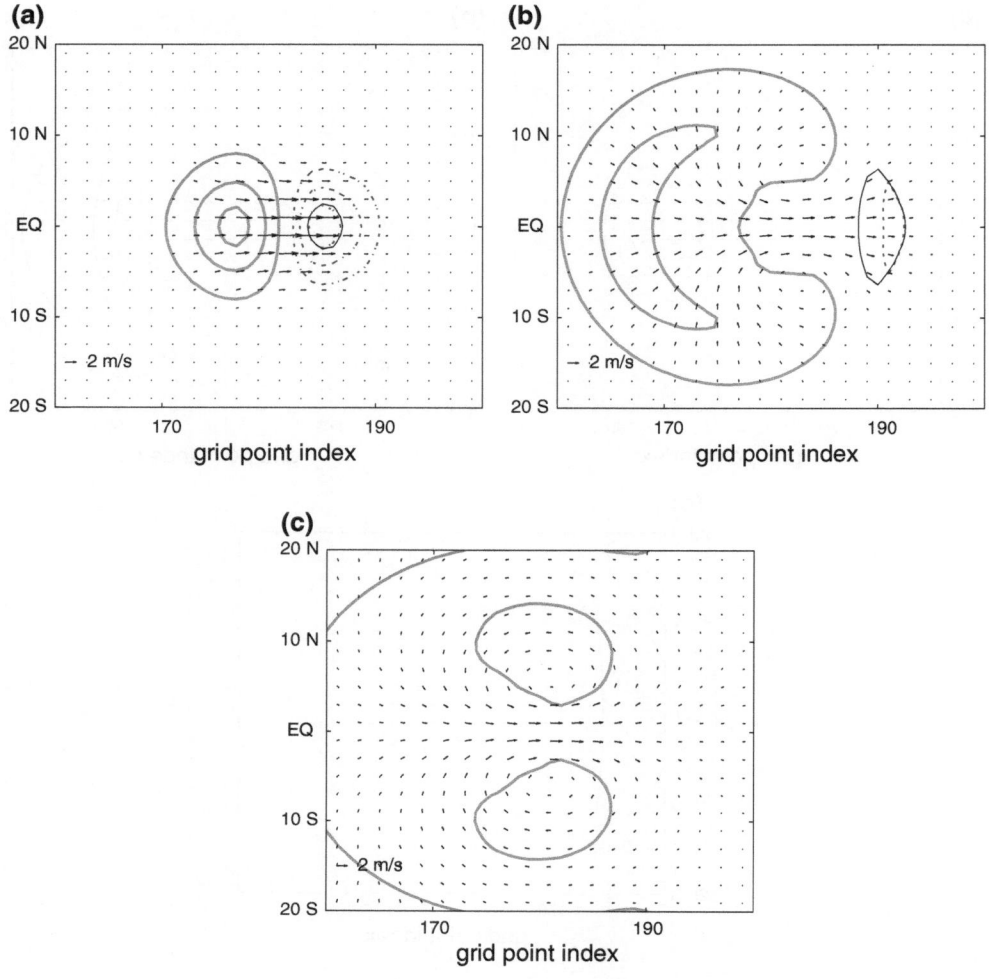

Figure 3
As in Fig. 2 but for the saturated atmosphere

spectra of tropical forecast-error variance of the ECMWF model as presented in ŽAGAR et al. (2005). The difference here is that the potential temperature variable is used instead of geopotential which requires a minor modification of the relationships between the mass-field and wind-field errors. The fundamental dispersion relation remains the same, with $c = (gH)^{1/2}$ replaced by $c = NH_o$. The modified expression for the phase speed applies to all equations. A more important part of this system is the representation of the background errors in the moisture field. A new control variable, ψ, is introduced to describe forecast errors in the moisture field:

$$q(x,y) = \sum_{k=-N_k}^{N_k} \sum_{n=0}^{N_n} \sum_{p=1}^{3} \psi_{knm} q_{knp}(k,y) e^{ikx}, \qquad (9)$$

where q_{knp} is a characteristic structure function for the moisture forecast error. The structure functions q_{knp} are derived assuming that moisture perturbations (and their forecast errors) are correlated with the vertical motion resulting from latent heat release:

$$q_{knp}(x,y) \sim w(x,y) = -H_o \left(\frac{\partial u}{\partial x} + \frac{\partial v}{\partial y} \right) \qquad (10)$$

The distribution of variance among various equatorial waves is based on the variance spectra of forecast

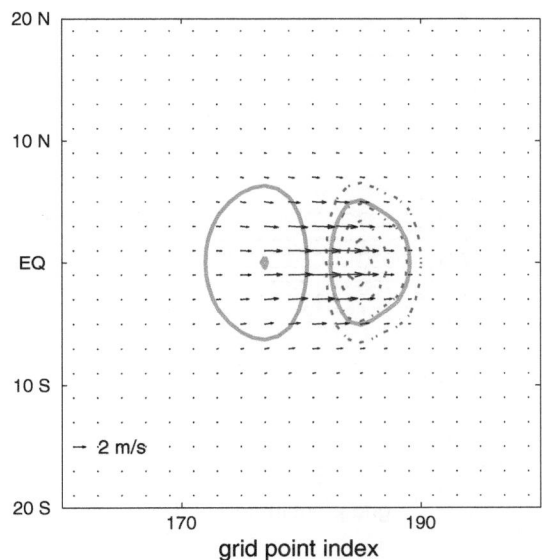

Figure 4

As in Fig. 3a, but with a 50% smaller vertical potential temperature gradient ($d\theta_o/dz = 2.5$ K km^{-1})

errors at the 500 hPa level in the ECMWF model as estimated in ŽAGAR et al. (2007). At the selected level, approximately 43% of the variance is represented in terms of the equatorial Rossby modes, westward and eastward-propagating IG modes together make 39% of the error variance, 8% of the variance belongs to the Kelvin waves, and approximately 10% of the variance pertains to the MRG modes. The shape of the variance spectrum as a function of the zonal wavenumber was presented in Fig. 2 in ŽAGAR et al. (2008) where the same spectrum was applied.

Despite the dynamical argument (Eq. 10) used to derive the moisture background-error variances, in 3D-Var the moisture is analyzed univariately. Thus a new assimilation control vector is twice as long ($[\chi \ \psi]^T$). Equations 3 and 4 apply with a new part of the J_b term in Eq. 1. Furthermore, χ and $\delta\mathbf{x} = [\delta u \ \delta v \ \delta\theta \ \delta q]^T$ in Eq. 3 are replaced by $[\chi \ \psi]^T$ and $\delta\mathbf{x} = [\delta u \ \delta v \ \delta\theta \ \delta q]^T$, respectively. The univariate assimilation means that increments because of the wind or temperature observations produce no increments in the moisture field in 3D-Var and vice versa. In 4D-Var the moisture increments from the wind and/or temperature observations result from the internal model adjustment and other observation types. This is illustrated by the following two examples.

Figures 5 and 6 show analysis increments due to single zonal wind and temperature observations, respectively, at three time instants: at the beginning, in the middle, and at the end of a 12-h window of 4D-Var. These two figures illustrate the effect of moisture processes in the saturated background atmosphere. In both cases there are small negative increments in the moisture field resulting from the local wind convergence. Other than that, increments have shapes as in the dry case (e.g. ŽAGAR et al. 2005); that is, mass–wind coupling at the equator reflects the Kelvin–wave relationship between the zonal wind and temperature fields. However, the coupling is weak, especially for the temperature observation in which case the amplitude of the wind increment is nearly zero. For data-assimilation purposes this result means that extracting information about unobserved variables from the background-error covariances and model dynamics in 4D-Var is inefficient for the applied static (average) distribution of background errors. Furthermore, the increments appear more univariate in the moist case than in the dry cases discussed in previous studies (ŽAGAR et al. 2005; ŽAGAR 2004).

Figures 5 and 6 can be compared with Fig. 3 which illustrated the adjustment process of the zonal wind perturbation. A positive temperature increment at the end of the window with a shape of the equatorial Rossby wave illustrates the fact that the mass field adjusts to the wind field in the tropics. The maximum impact of the temperature observation at the end of the assimilation window (hour 12) is found to the east of the observation location (at time 0). A positive temperature increment is accompanied by a negative moisture increment because of flow convergence toward the observation location in the lower layer.

Finally, we show an analysis increment due to a single temperature observation in the 4D-Var system of ECMWF (Fig. 7). The ECMWF system applies state-of-the-art variational assimilation methodology; however, it is tuned to deal with the global forecast errors that are dominated by mid-latitude errors and the quasi-geostrophic coupling between the wind-field and mass-field errors. In other words, the J_b term in the ECMWF system is very different from that used to produce Figs. 5 and 6. The temperature observation is located at 10 UTC, an hour behind the 4D-Var window

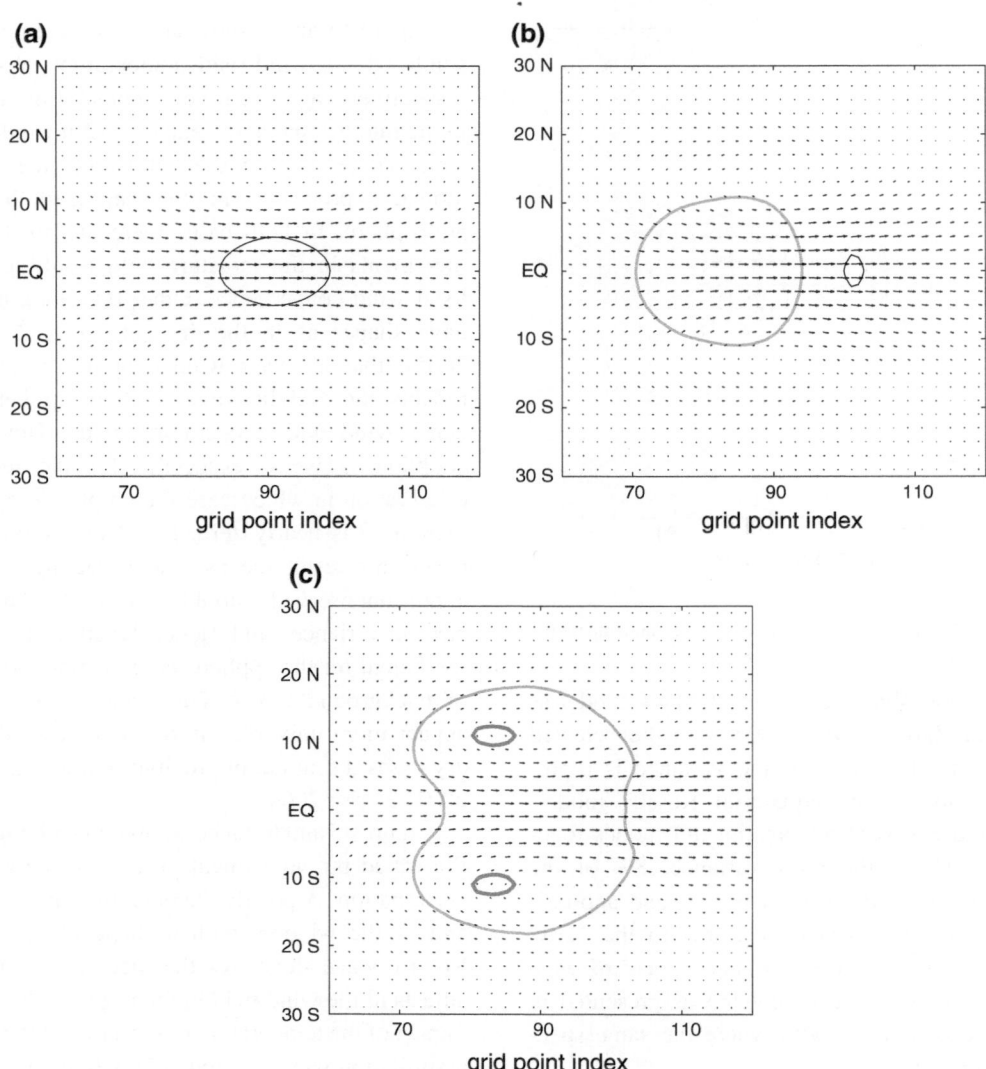

Figure 5

Effect of a single zonal wind observation at the equator in the idealized moist tropical 4D-Var assimilation system with the 12-h assimilation window. **a** analysis increments at time 0; **b** analysis background at time 6 h; **c** analysis background at time 12 h. Observation time is time zero, amplitude is 5 ms^{-1}, and observation error is 1.6 ms^{-1} (taken equal to the error in the background field at the same grid point). The background atmosphere is saturated and motionless. Isoline spacing for temperature is every 0.2 K (*thick lines* for positive and *thin lines* for negative values), starting from ± 0.2 K. *Light gray lines* correspond to the negative moisture increments, drawn every 0.25 mm

start, at the equator and 30°W and at 848.5 hPa (close to model level 77). The increment is drawn at time 9 UTC, the beginning of the 12-h 4D-Var window as a difference between the nonlinear trajectory and 3-h forecast started from the background state at 06 UTC. Increments in the wind field are multiplied by a factor 50 in order to visualize the divergent nature of the weak circulation due to temperature observation. The maximum zonal and meridional wind increments have

absolute amplitudes up to 4 cm/s (Fig. 7b). This result illustrates once again the inefficiency of using data assimilation to extract information about the tropical wind field from the mass-field observations. There is a small effect in the moisture field at the location of maximum convergence some 200 km south of the observation location. In the vertical direction, the observation effect is localized to neighbouring levels and there is a small effect of the opposite sign close to

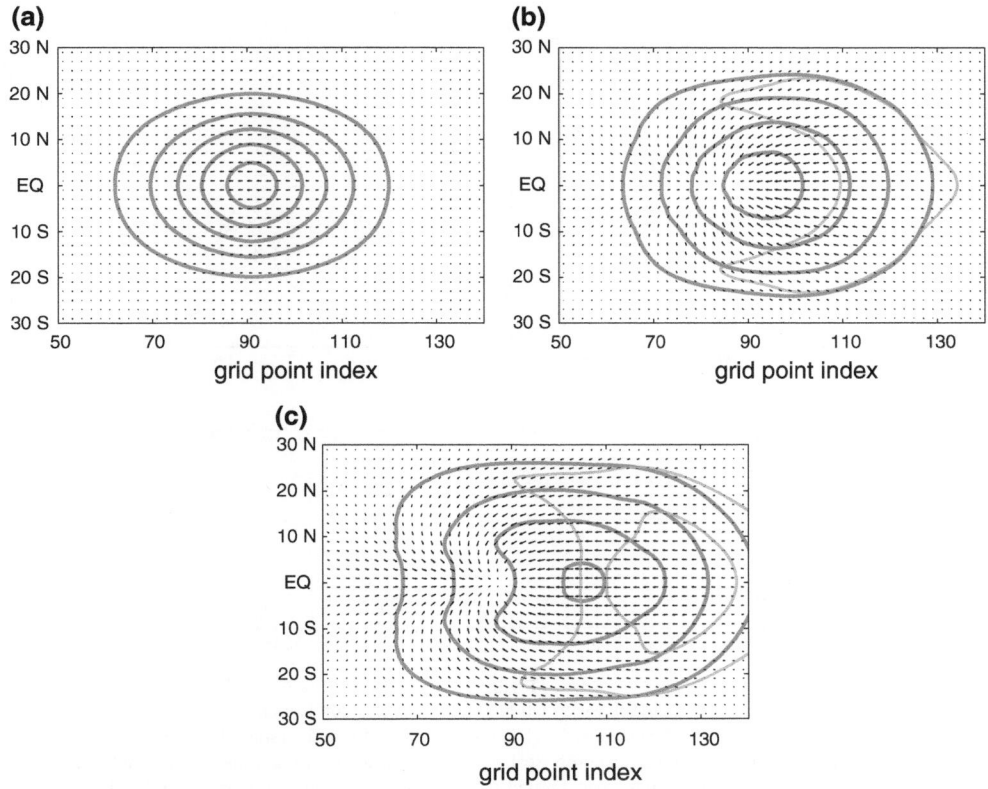

Figure 6
As in Fig. 5 but for a single temperature observation 2 K warmer than the background and with observation error 0.5 K

the surface (Fig. 7c). Although the main result from the temperature observation in the NWP system agrees with the results of the idealized system, it is important to understand differences in the structure of the resulting wind and moisture increments, despite their small magnitudes. Although the differences may partly be in different timing of the observation in the 4D-Var window, the main reason is most probably the imposed forecast-error variance spectra. Further experiments with different variance spectra are needed to better understand the coupling between the wind and mass variables in tropical data assimilation.

4. Conclusions

Opposite to the mid-latitudes, there is no a well-defined approach to tropical data assimilation. This is a consequence of complex tropical dynamics in which various IG waves, each characterized by own coupling

between the mass field and the wind field, play an insufficiently well understood role in coupling between convection and horizontal motion. This is an unknown problem in the mid-latitudes where the quasi-geostrophic theory provides a basic framework for understanding dynamics on large scales.

Recent work in modelling methodology for tropical data assimilation provides better understanding of the relative effects of different equatorially trapped wave solutions in variational analysis in the tropics. Time-averaged information about the variance of tropical waves, built into the background-error covariance matrix for data assimilation, provides analysis increments which appear nearly univariate even though they result from the advanced multivariate assimilation methodology. This applies to both idealized dry and moist tropical systems and a state-of-the-art NWP 4D-Var assimilation system.

This paper presented experiments with a simple moist tropical model as a useful tool to highlight

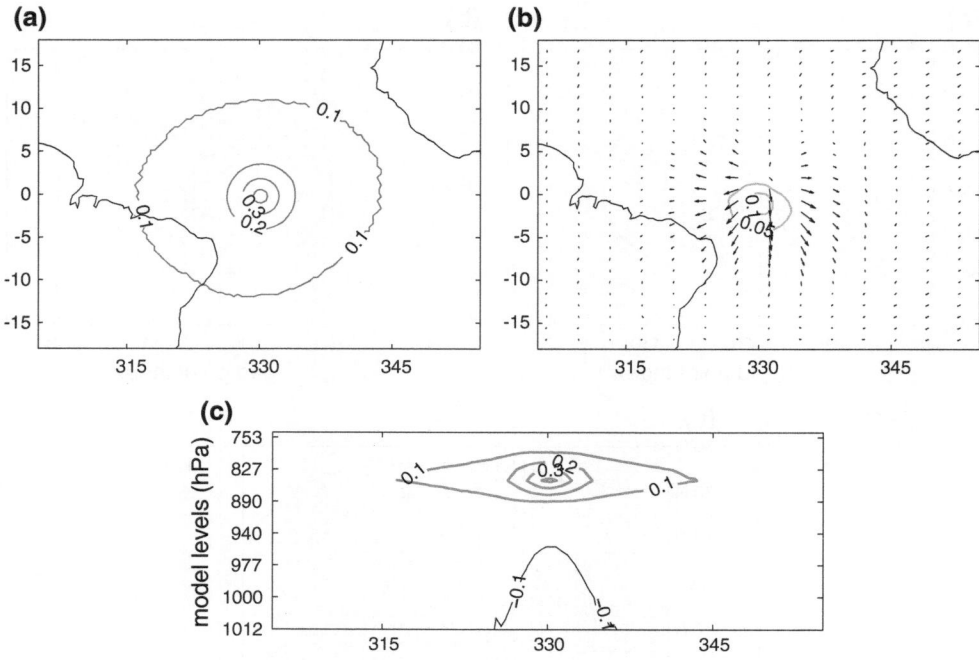

Figure 7

Effect of a single temperature observation in the ECMWF assimilation system. Observation is located at 10 UTC, an hour behind the 4D-Var window start, at the equator, and 30 W and 848.5 hPa (model level 77). Increments of **a** temperature, **b** specific humidity (g/kg) and wind at model level 77 and **c** temperature increments in the vertical cross-section. Increments are drawn at time 9 UTC, the beginning of the 12-h 4D-Var window. Temperature observation amplitude minus background is 1.2 K and the observation error is 1.178 K. Wind components are multiplied by 50. Maximum zonal wind increment has amplitude 0.03 m/s and −0.04 m/s and the meridional wind increment has positive amplitude 0.01 m/s and negative value −0.04 m/s. Temperature increment is drawn every 0.1 K and maximum increment is 0.43 K. Maximum positive specific humidity increment is 0.15 g/kg

aspects of tropical dynamics and 4D-Var that are difficult to understand in NWP systems. The adjustment experiments illustrated that the reduced static stability greatly increases the equatorial trapping of moving perturbations. This is closely coupled with the structure of analysis increments associated with wave perturbations propagating through saturated areas with different vertical stability.

Of major importance for tropical analysis increments are the spectra of the tropical forecast-error variances. In 4D-Var, the variances are static. In the ensemble Kalman filter (e.g. HOUTEKAMER and MITCHELL, 2005), the forecast-error spectra are updated at every analysis step and the derived information is multivariate. Given the importance of moist dynamics in the tropics and its spatio-temporal variability on one side, and increasing amounts of observations and a growing need to improve the extended range-forecasting and reanalyses on the other, the idea of

formulating useful tropical multivariate relations is one worthy of further trials.

Acknowledgments

The author would like to thank David Tan of ECMWF for allowing her to use his assimilation experiment with a single temperature observation in the ECMWF 4D-Var system. The Centre of Excellence for Space Sciences and Technologies SPACE-SI is an operation part financed by the European Union, European Regional Development Fund and Republic of Slovenia, Ministry of Higher Education, Science and Technology.

REFERENCES

COURTIER, P., E. ANDERSSON, W. HECKLEY, J. PAILLEUX, D. VASILJEVIĆ, M. HAMRUD, A. HOLLINGSWORTH, F. RABIER, and M. FISHER,

1998: *The ECMWF implementation of three-dimensional variational assimilation (3D-Var). I: Formulation.* Q. J. R. Meteorol. Soc., **124**, 1783–1807.

DALEY, R., 1993: *Atmospheric data analysis on the equatorial beta plane.* Atmos.-Ocean, **31**, 421–450.

DAVEY, M. K., 1989: *A simple tropical moist model applied to the '40-day' wave.* Q. J. R. Meteorol. Soc., **115**, 1071–1107.

DAVEY, M. K. and A. E. GILL, 1987: *Experiments on tropical circulation with a simple moist model.* Q. J. R. Meteorol. Soc., **113**, 1237–1269.

GAUTHIER, P. and J.-N. THÉPAUT, 2001: *Impact of the digital filter as a weak constraint in the preoperational 4DVAR assimilation system of Meteo-France.* Mon. Wea. Rev., **129**, 2089–2102.

GEER, A., P. BAUER, and P. LOPEZ, 2008: *Lessons learnt from the 1D+4D-Var assimilation of rain and cloud affected SSM/I observations at ECMWF.* Q. J. R. Meteorol. Soc., **134**, 1513–1525.

GILL, A. E., 1982: *Studies of moisture effect in simple atmospheric models: the stable case.* Geophys. Astrophys. Fluid Dyn., **19**, 119–152.

GUSTAFSSON, N., L. BERRE, S. HÖRNQUIST, X.-Y. HUAN, M. LINDSKOG, B. NAVASCUÉS, K. S. MOGENSEN, and S. THORSTEINSSON, 2001: *Three-dimensional variational data assimilation for a limited area model. Part I: General formulation and the background error constraint.* Tellus, **53A**, 425–446.

HAO, Z. and M. GHIL, 1994: *Data assimilation in a simple tropical ocean model with wind stress errors.* J. Phys. Ocean., **24**, 2111–2128.

HOUTEKAMER, P. L. and H. L. MITCHELL, 2005: *Ensemble Kalman filtering.* Q. J. R. Meteorol. Soc., **131**, 3269–3289.

LE DIMET, F.-X. and O. TALAGRAND, 1986: *Variational algorithms for analysis and assimilation of meteorological observations: theoretical aspects.* Tellus, **38A**, 97–110.

MACHENHAUER, B., 1977: *On the dynamics of gravity oscillations in a shallow water model, with applications to normal mode initialization.* Beitr. Phys. Atmos., **50**, 253–271.

MATSUNO, T., 1966: *Quasi-geostrophic motions in the equatorial area.* J. Meteor. Soc. Japan, **44**, 25–42.

PHILLIPS, N., 1986: *The spatial statistics of random geostrophic modes and first-guess errors.* Tellus, **38A**, 314–322.

RABIER, F., A. P. MCNALLY, E. ANDERSSON, P. COURTIER, P. UNDÉN, J. EYRE, A. HOLLINGSWORTH, and F. BOUTTIER, 1998: *The ECMWF implementation of three-dimensional variational assimilation*

(3D-Var). II: Structure functions. Q. J. R. Meteorol. Soc., **124**, 1809–1829.

SMAGORINSKY, J., 1969: *Problems and promises of deterministic extended range forecasting.* Bull. Amer. Meteor. Soc., **50**, 286–311.

STOFFELEN, A., J. PAILLEUX, E. KÄLLÉN, J. M. VAUGHAN, L. ISAKSEN, P. FLAMANT, W. WERGEN, E. ANDERSSON, H. SCHYBERG, A. CULOMA, R. MEYNART, M. ENDEMANN, and P. INGMANN, 2005: *The atmospheric dynamic mission for global wind measurements.* Bull. Amer. Meteor. Soc., **86**, 73–87.

ŽAGAR, N., 2004: *Assimilation of equatorial waves by line of sight wind observations.* J. Atmos. Sci., **61**, 1877–1893.

ŽAGAR, N., E. ANDERSSON, and M. FISHER, 2005: *Balanced tropical data assimilation based on a study of equatorial waves in ECMWF short-range forecast errors.* Q. J. R. Meteorol. Soc., **131**, 987–1011.

ŽAGAR, N., E. ANDERSSON, M. FISHER, and A. UNTCH, 2007: *Influence of the quasi-biennial oscillation on the ECMWF model short-range forecast errors in the tropical stratosphere.* Q. J. R. Meteorol. Soc., **133**, 1843–1853.

ŽAGAR, N., N. GUSTAFSSON, and E. KÄLLÉN, 2004a: *Dynamical response of equatorial waves in four-dimensional variational data assimilation.* Tellus, **56A**, 29–46.

ŽAGAR, N., N. GUSTAFSSON, and E. KÄLLÉN, 2004b: *Variational data assimilation in the tropics: the impact of a background error constraint.* Q. J. R. Meteorol. Soc., **130**, 103–125.

ŽAGAR, N., A. STOFFELEN, G.-J. MARSEILLE, C. ACCADIA, and P. SCHLÜSSEL, 2008: *Impact assessment of simulated Doppler wind lidars with a multivariate variational assimilation in the tropics.* Mon. Wea. Rev., **136**, 2443–2460.

ŽAGAR, N., J. TRIBBIA, J. L. ANDERSON, and K. RAEDER, 2009: *Uncertainties of estimates of inertio-gravity energy in the atmosphere. Part I: intercomparison of four analysis datasets.* Mon. Wea. Rev., **137**, 3837–3857.

WERGEN, W., 1988: *The diabatic ECMWF normal mode initialization scheme.* Beitr. Phys. Atmosph., **61**, 274–302.

WHEELER, M. and G. N. KILADIS, 1999: *Convectively coupled equatorial waves: analysis of clouds and temperature in the wavenumber-frequency domain.* J. Atmos. Sci., **56**, 374–399.

WMO, 2000: Statement of guidance regarding how well satellite capabilities meet WMO user requirements in several application areas. Tech. rep., WMO Satellite Rep. SAT-22, WMO/TD 992, 29 pp., Geneva, Switzerland.

(Received January 10, 2011, accepted March 31, 2011, Published online July 27, 2011)

Pure Appl. Geophys. 169 (2012), 381–399
© 2011 Springer Basel AG
DOI 10.1007/s00024-011-0376-1

A Study on Simulation of Heavy Rainfall Events Over Indian Region with ARW-3DVAR Modeling System

U. C. Mohanty,[1] A. Routray,[2] Krishna K. Osuri,[1] and S. Kiran Prasad[1]

Abstract—An attempt is made to evaluate the impact of the three dimensional variational (3DVAR) data assimilation within the Weather Research Forecasting (WRF) modeling system to simulate two heavy rainfall events which occured on 26–27 July 2005 and 27–30 July 2006. During the 26–27 July 2005 event, the unprecedented localized intense rainfall 90–100 cm was recorded over the northeast parts of Mumbai city; however, southern parts received only 10 cm. Model simulation with the data assimilation experiment is reasonably well predicted for the rainfall intensity (800 mm) in 24 h and with accurate location over Mumbai agreeing with observation. Divergence, vorticity, vertical velocity and moisture parameters are evaluated during the various stages of the event. It is noticed that maximum convergence and vorticity during the mature stage; at the same time the vertical velocity also follows a similar trend during the period in the assimilation experiment. Vorticity budget terms over the location of heavy rainfall revealed that the contribution of the positive tilting term produced positive vorticity which triggered the convection and negative contribution to vorticity from the tilting term to precede the dissipation of the system. Model simulations from the second rain event, the off-shore trough at sea level along the west coast of India, is well represented after assimilation of observations during day-1 and day-2 as compared to the control simulations; the orientation of the off-shore trough is well matched with that of the observed. The intensity and spatial distribution of the rainfall has considerably improved in the assimilation simulation. The statistical skill scores also revealed that the precipitation forecast during the period has appreciably improved due to assimilation of observations. The results of this study indicate a positive impact of the 3DVAR assimilation on the simulation of heavy rainfall events.

Key words: Variational data assimilation, heavy rainfall, vorticity budget terms, off-shore trough.

1. Introduction

The Indian sub-continent receives 80% of its annual rainfall during the southwest monsoon (SWM) period. During the SWM season, the Indian sub-continent often receives widespread heavy rainfall under the influence of off-shore troughs, off-shore vortices, depressions over the Bay of Bengal (BoB)/Arabian Seas, and mid-tropospheric cyclones (MTC). Rainfall amounts of 100–300 mm in a day at and around the weather systems along the west coast of India and other parts of country are common during SWM season. These rainfall events are caused by organized mesoscale convective systems (MCSs) embedded in large scale synoptic systems (Benson and Rao, 1987; Sikka and Gadgil, 1980). Extreme rainfall events result in land slides, flash floods and damage to crops that have major impacts on the society, economy and environment. Although prediction of such extreme weather events is still fraught with uncertainties, a proper assessment of likely future trends would help in setting up infrastructure for disaster preparedness.

The non-hydrostatic mesoscale models are capable for simulation/prediction of high impact weather systems which lead to heavy rainfall episodes over India (Routray et al. 2005, 2010a; Deb et al. 2008; Kumar et al., 2008). However, the forecast skill of these models is very limited, particularly for important variables like rainfall (Roy Bhowmik and Prasad 2001; Rama Rao et al. 2007; Sikka and Rao 2008; Das et al. 2008). Hence, there is a necessity for efforts to improve performance of the mesoscale models in short-range predictions on a real-time basis for the Indian monsoon region particularly for prediction of the MCSs, which lead to heavy rainfall events. Therefore, assimilation approaches that ingest local observations are important to develop improved

---
[1] Centre for Atmospheric Sciences, Indian Institute of Technology Delhi, Hauz Khas, New Delhi 110016, India. E-mail: ucmohanty@gmail.com
[2] National Centre for Medium Range Weather Forecasting, A-50, Sector-62, Noida, India.

analyses which served as initial condition to the mesoscale model (DALEY 1991). Particularly, over the last decade, high-resolution mesoscale models with three/four dimensional techniques (3DVAR/4DVAR) are being increasingly applied for studying meteorological phenomena (KALNAY 2003). Data impact studies over the Indian monsoon region for simulation of extreme weather events have been very sparse and have yielded mixed results. VAIDYA et al. (2004) studied the impact of radiosonde and rawinsonde data on the simulation of monsoon depressions and tropical cyclones and found improvement in the predicted mean sea level pressure and precipitation. ROUTRAY et al. (2010a) study demonstrated, the assimilation of Indian conventional and non-conventional observations using Weather Research and Forecasting (WRF)-3DVAR (hereafter WRF-Var) analysis technique within WRF modeling system has definite impact for simulation heavy rainfall events occurred during monsoon season.

The prime objective of the study is to evaluate the WRF model performance using the WRF-Var data assimilation based improved initial condition by ingesting observations in simulating convective processes leading to intense precipitation events over India during the SWM season. The following section presents an overview of the WRF and WRF-Var modeling systems. The synoptic situations of the events are presented in Sect. 3. Section 4 gives details of the numerical experiments performed in the present study. Section 5 provides the results and discussion while Sect. 6 briefly mentions the broad overall conclusions of this study.

2. Modeling System

2.1. WRF and WRF-Var Systems

The Advanced Research WRF (ARW) modeling system is used in the present study. The model was initially developed at the National Centre for Atmospheric Research (NCAR) in collaboration with other research institutes in the USA. One can find detailed description of model physics, dynamical core and model equations in SKAMAROCK et al. (2005). The 3DVAR data assimilation is built within the ARW

modeling system in such a way that the WRF model can be run with and without data assimilation.

The formulation of 3DVAR developed on the basis of Bayesian probabilities and Gaussian error distributions (LORENC 1986 and LORENC et al. 2000). The basic target of WRF-Var data assimilation system is to produce an optimal estimate of the true atmospheric state at analysis time through iterative solution of a prescribed cost-function $J(x)$.

$$J(x) = J^b + J^o = \frac{1}{2}(x - x^b)^T B^{-1}(x - x^b)$$
$$+ \frac{1}{2}(y - y^o)^T (E + F)^{-1}(y - y^o) \quad (1)$$

where J^b and J^o are the cost functions of background and observation, respectively; x is the state vector; x^b is the background or first guess; B is the background error statistics covariances; y is the observation space $(y = Hx)$; H is the forward (non-linear) operator; y^o is the observations; E is the observational or instrumental error covariances matrix; F is the representivity error covariances matrix. Further, details about the components and real time applications of the 3DVAR system have been reported in BARKER et al. (2004) and JIANFENG et al. (2005). The B statistics are calculated through NMC method (PARRISH and DERBER 1992) over the Indian monsoon region (ROUTRAY et al. 2010b). In the 3DVAR assimilation system, the error statistics are generally calculated off-line and significant tuning is required to optimize performance for a particular application and also specific region.

3. Synoptic Features Associated with Convective Monsoonal Rainfall Events

In this study, two heavy rainfall events, 26–27 July 2005 (Case-1) and 27–30 July 2006 (Case-2), that occurred along the west coast of India due to the presence of meso-convective activities are considered. The brief synoptic situations during the periods are given as below.

3.1. 26–27 July 2005 (Case-1)

The particular heavy rainfall event considered in this study is the Mumbai heavy rain event during the period 26–27 July 2005. On 26 July 2005, an official India Meteorological Department (IMD) rain-gauge at Santacruz International airport (19.11°N, 72.85°E) on the north side of the metropolis of Mumbai (18.93°N, 72.85°E) recorded 944 mm rainfall in 24 h (accumulated from 0300 UTC 26 to 27 July 2005). According to synoptic features, such extreme rain events occurred during the active phase of monsoon all over India, revived after the 4 days (19–22 July) prolonged break/weak phase of the monsoon. Presence of a well marked low pressure area over the northwest Bay of Bengal off West Bengal and Orissa coast, marked off-shore trough at the surface extended from Maharashtra and Kerala coast and also a well marked east–west oriented shear line in the lower troposphere caused the enhanced activity. A western disturbance existing as an upper air circulation up to mid-tropospheric level was over north Pakistan and adjoining Jammu and Kashmir during the period and it moved east-northeast direction. The monsoon trough was shifted towards the south from its original position, which favors good rainfall activity during the period. A cyclonic circulation in the middle troposphere (850–500 hPa levels) was also present north of Mumbai/central India. The record rainfall at Mumbai might be attributed to the interaction of enhanced monsoon westerly current (low level jet speed 60 kts at Mumbai), orientation of orography and the off-shore vortices giving rise to enhanced phenomenal convection of mesoscale dimension.

From the cyclone detection radar at Colaba reported high clouds 5–6 km are observed at 03 UTC and it further extended up to 15 km at 08 UTC on 26 July. The cloud heights decreased gradually as time progressed. The maximum wind speed 78 km/h over Santacruz was observed at 12 UTC 26 July. The sustained wind speed over the region was noticed around 50 km/h and wind direction was mainly from northwest direction.

3.2. 27–30 July 2006 (Case-2)

On 27 July 2006, under the influence of a cyclonic circulation over west-central Bay, a well marked low pressure area formed over the north Bay. The low pressure area over the north Bay persisted throughout the period and moved in a west-northwesterly

direction. During the period, the off-shore trough at sea level extended from south Gujarat to Kerala coasts. A cyclonic circulation observed over west Madhya Pradesh, adjoining east Rajasthan and Gujarat region, extended up to 3.1 kms from the surface during the whole period. Heavy to very heavy rainfall occurred over Gujarat state, Konkan and Goa coast. Another cyclonic circulation extending up to 4.5 kms was observed over the northern parts of Jammu and Kashmir, and it moved in northeast ward direction.

4. Numerical Experiments

The focus of the study is to evaluate the impact of assimilation of conventional and non-conventional data into WRF-Var system on the simulation of heavy rainfall events along the west coast of India which occurred during the summer monsoon period. For this purpose, two numerical experiments (CNTL and 3DV) are carried out using the WRF and WRF-Var modeling systems with nested domain at 30 and 10 km horizontal resolution and 51 sigma levels vertically. The inner domain covered the west coast of India and neighbourhood as shown in Fig. 1. The details of the model configuration along with physical parameterization schemes chosen for this particular study are

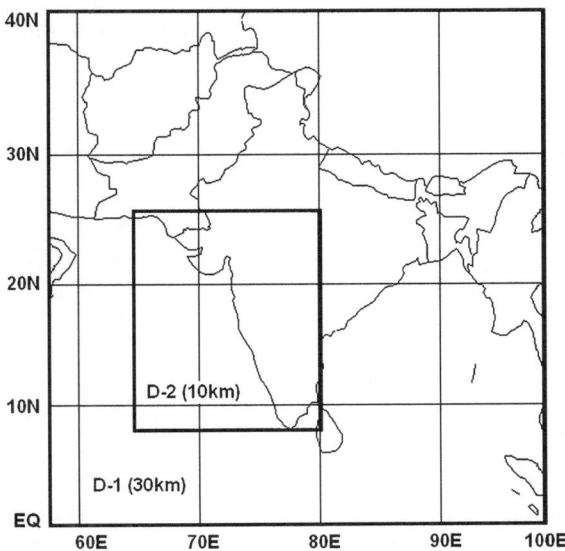

Figure 1
Model domain chosen for the study: horizontal resolution for outer domain (D-1) 30 km and inner domain (D-2) 10 km

given in Table 1. The CNTL experiment is carried out without any data assimilation, the NCEP-FNL ($1^0 \times 1^0$) analyses are provided as initial and boundary conditions to the model. In the 3DV experiment, the data assimilation is performed with a 6-h update cyclic using the conventional and non-conventional observations. In the 6-h update assimilation cycle, the previous cycle's 6-h forecast used as background for the next cycle, a total of four cycles are performed to initialize the model before starting of the actual model integration. The model is integrated for 36 h (Case-1) and 54 h (Case-2) in both experiments from the initial time 0000 UTC of 26 July 2005 and 27 July 2006, respectively. The details of the observation data sets used for the 3DV assimilation experiment are given in Table 2. In the assimilation experiment, the collected observations are absorbed in the WRF-Var system at the analysis time with ±3 h time window. A comparison is made between O-B (observation-first guess) and O-A (observation-analysis) to demonstrate the performance of WRF-Var analysis system. It is clearly noticed that the root mean square error (RMSE) of O-B is higher than the O-A before and after 3DVAR analysis at initial time, respectively, for all parameters from different sources in both the cases. This shows that 3DVAR assimilates the observations properly and

produces an analysis that is consistent with the observations.

4.1. Data Pre-processing

The conventional (synop, radiosonde/rawinsonde, pilot, buoy and ship) and non-conventional (satellite–cloud motion wind and oceanic surface wind) observations obtained from the Global Telecommunication System (GTS)/Internet are processed through observation preprocessor module (3DVAR_OBSPROC) available within WRF-Var system. The observation data sets are obtained from varieties of platforms in different formats. The modules developed to read these decoded observations and make into a suitable format (LITTLE_R; Guo, 2008), which are given as input to the 3DVAR observation preprocessor. In the 3DVAR_OBSPROC module various quality checks are performed within it for each observation, the details can be found in the study by Barker et al., (2004).

5. Results and Discussions

The two heavy rainfall events discussed in Sect. 3 are simulated using the WRF and WRF-Var modeling

Table 1

Overview of the WRF model used in the present study

Number of domain	Double nested domain
Model domain	Domain-1 (57°–100°E; 0°–40°N) Domain-2 (64°–80°E; 8°–25°N)
Central point of the domain	Central Lat and Lon.: 21.5°N and 78.0°E
Integration time step	90 s (D-1) and 30 s (D-2)
Number of grid points	D-1 ($X = 153$ and $Y = 165$ points) D-2 ($X = 175$ and $Y = 208$ points)
Vertical co-ordinate	Terrain-following hydrostatic-pressure co-ordinate (51 levels)
Model top	10 mb
Microphysics	Lin et al. (1983) scheme
Radiation scheme (long-wave)	RRTM scheme
Radiation scheme (short-wave)	Dudhia's short wave radiation
Surface layer physics	Monin–Obukhov scheme
Surface layer parameterization	Thermal diffusion scheme
PBL parameterization	YSU scheme
Cumulus parameterization schemes	Betts-Miller-Janjic scheme
Dynamic option	Eulerian mass
Time integration	3rd order Runge–Kutta
Spatial differencing scheme	6th order centered differencing
Map projection	Mercator
Horizontal grid distribution	Arakawa-C grid
Main prognostic variables	$u, v, w, p', \theta', \Phi'$
Initial and boundary conditions	3-dimensional real-data (FNL:$1^0 \times 1^0$)

Table 2

Number of data used in 3DVAR assimilation system

Serial no.	Name of data set	Description
1	TEMP	Upper air profiles of temperature, humidity and wind from radiosonde
2	PILOT	Wind profiles from optical theodolite
3	SYNOP	Surface observations from land stations
4	SHIP	Voluntary observation from sea
5	BUOY	Drifting and moored buoy observations
6	AMDAR/AIREP/METAR	Upper level wind and temperature, dew point temperature reported by aircrafts
7	SATOB	Satellite observed cloud motion vectors from INSAT, METEOSAT-6, GMS and GOES
8	SATEM	Satellite observed wind and total precipitable water from NOAA series of satellites
9	GEOAMV	Geo-stationary atmospheric motion vectors
10	QSCAT/SSMI	Surface wind fields over oceanic region

system with double interactive nested domain at a 30 and 10 km horizontal resolution to evaluate the impact of modified initial condition. The initial condition of the model is improved through assimilation of conventional and non-conventional observations using WRF-Var analysis system. The impact of the WRF model simulation with the 3DVAR based initial conditions is examined in the section, where the predicted meteorological fields during the acute/intense rainfall events are compared with the available observations. The rainfall obtained from model simulations are compared with the IMD station observations and also with Tropical Rainfall Measurement Mission (TRMM-3B42; HUFFMAN *et al.* 2003). The model simulated results obtained from inner domain is considered for the discussion.

5.1. Case-1

A record-breaking heavy rainfall event occurred on 26–27 July 2005 over Mumbai located on the west coast of India with precipitation of 94.4 cm during the 24 h preceding 0300 UTC of 27 July 2005. The detailed descriptions about the heavy rainfall event are described above in the Sect. 3. The occurrence of such heavy rainfall events along the west coast of India mainly concentrated over Mumbai region; the magnitude of the precipitation was not well predicted by any operational agencies due to limitations of operational numerical models which use coarse resolution and synoptic methods (BOHRA *et al.* 2006; JENAMANI *et al.*, 2006). The spatial variation of precipitation over Mumbai is large due to variations in the surrounding topography. Daily

precipitation records of IMD show that at 0300 UTC of 27 July, Santacruz and Lake Vihar received precipitation amounts of 94.4 and 104.5 cm as compared to 7.3 cm at Colaba and 7.4 cm at Malabar Hill on 26 July (Fig. 2a). Both the measurements as well as the observed accounts indicate that northern Mumbai received torrential rain while southern parts of the city remained relatively dry. The TRMM satellite also captured the rainfall over Mumbai on 26 July 2005 (Fig. 2b) at 1539 local time (1009 UTC). From Fig. 2b, a dark area representing the heavy precipitation zone is concentrated over Mumbai. Figure 2c shows the rainfall accumulation from 25 to 27 July 2005. Both the images show the localized and intense nature of this rainfall event. NOAA's National Environmental Satellite Data and Information System (NESDIS) satellite imagery (Fig. 2d) indicated a cloud-cluster over Mumbai at 0709 UTC on 26 July 2005. The existence of a cloud-cluster with possible vortices over Mumbai on 26 July was also inferred by SHYAMALA and BHADRAM (2006) based on analysis of soundings and satellite datasets. A number of modeling studies have been recently conducted for the 26 July 2005 Mumbai heavy-rain case (DODLA and RATNA, 2009; SIKKA and RAO, 2008; CHANG *et al.*, 2008; RAMA RAO *et al.*, 2007; VAIDYA and KULKARNI, 2007). The meso-convective systems associated with the rainfall event are successfully captured, but the model did not simulate the heavy rain amounts over Mumbai. The distribution and amount of model precipitation is a difficult quantity to verify, particularly for this rain event with its extreme spatial heterogeneity and the limited number of available observations. The model simulated

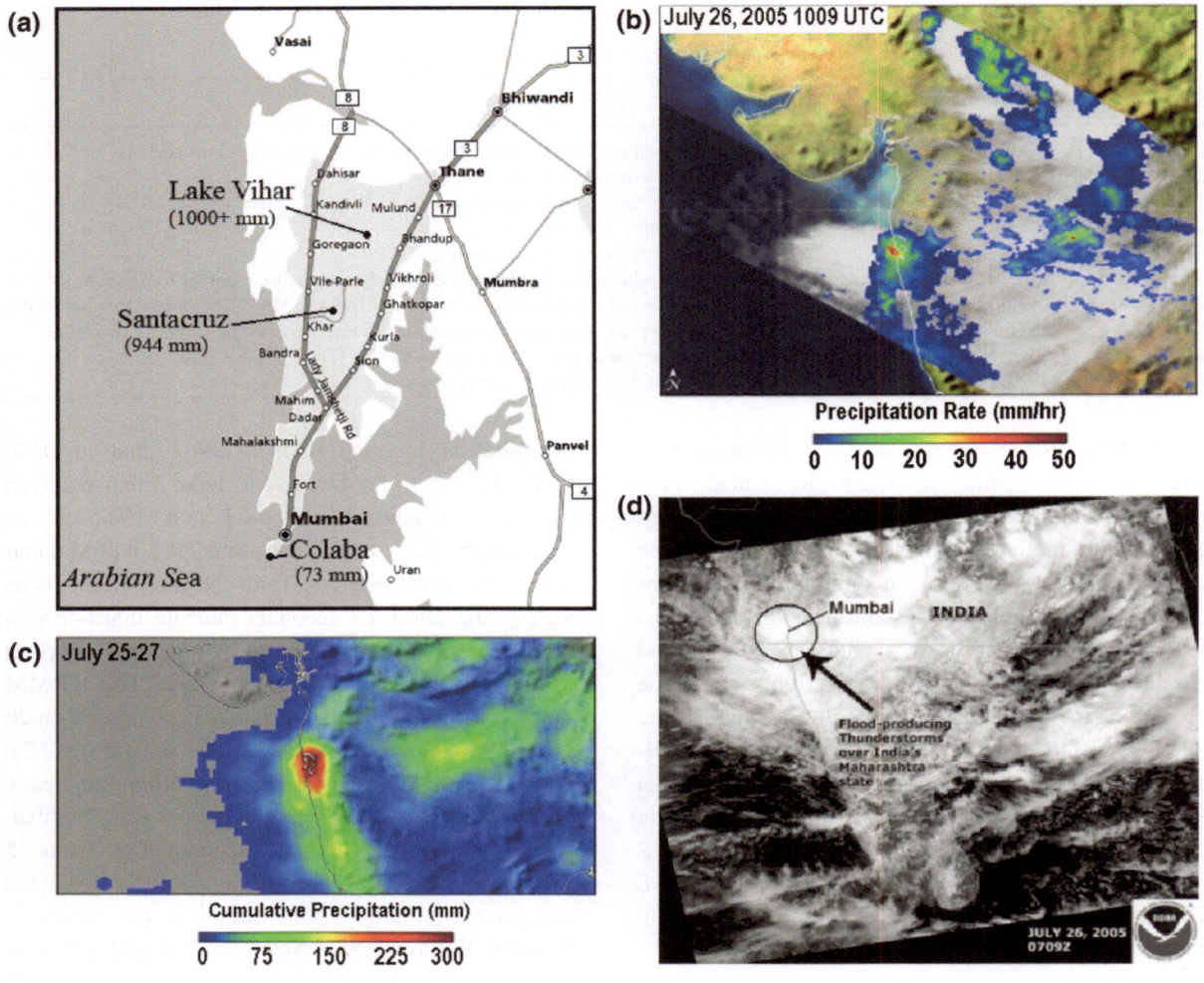

Figure 2

a Accumulated 24 h rainfall from rain-gauges; **b** TRMM rainfall rate for 26 July 2005; **c** TRMM estimated rainfall accumulated total for 25–27 July 2005 and **d** NESDIS satellite imagery at 0709 UTC on 26 July 2005

localized rain event reasonably well as compared to the other simulation (without assimilation) due to improved initial condition. As discussed above Santacruz recorded 944 mm of rainfall and Colaba recorded only 74 mm of rainfall. This shows that the unprecedented rainfall was very much localized.

5.1.1 Initial Wind Fields

The model derived wind fields at 850 and 200 hPa pressure levels from both the simulation CNTL and 3DV are illustrated in Fig. 3a–d. At the 850 hPa level, both the simulations show circulation associated with surface low over Orissa coast. However, the strength of wind is more (5–10 m s^{-1}) in the 3DV simulation (Fig. 3b) surrounding the low pressure area as compared to the CNTL simulation (Fig. 3a). Because of the cyclonic circulation over the northeast part, southwesterly wind flow prevailed over the Arabian Sea to the south of 20°N with strength of 10–20 m s^{-1} (CNTL) and 15–30 m s^{-1} (3DV). The strength of the wind from 3DV simulation over the region is in good agreement with the observations (15–28 m s^{-1}) reported in the Indian Daily Weather Report (IDWR) of IMD. A cyclonic circulation to the north of Mumbai over

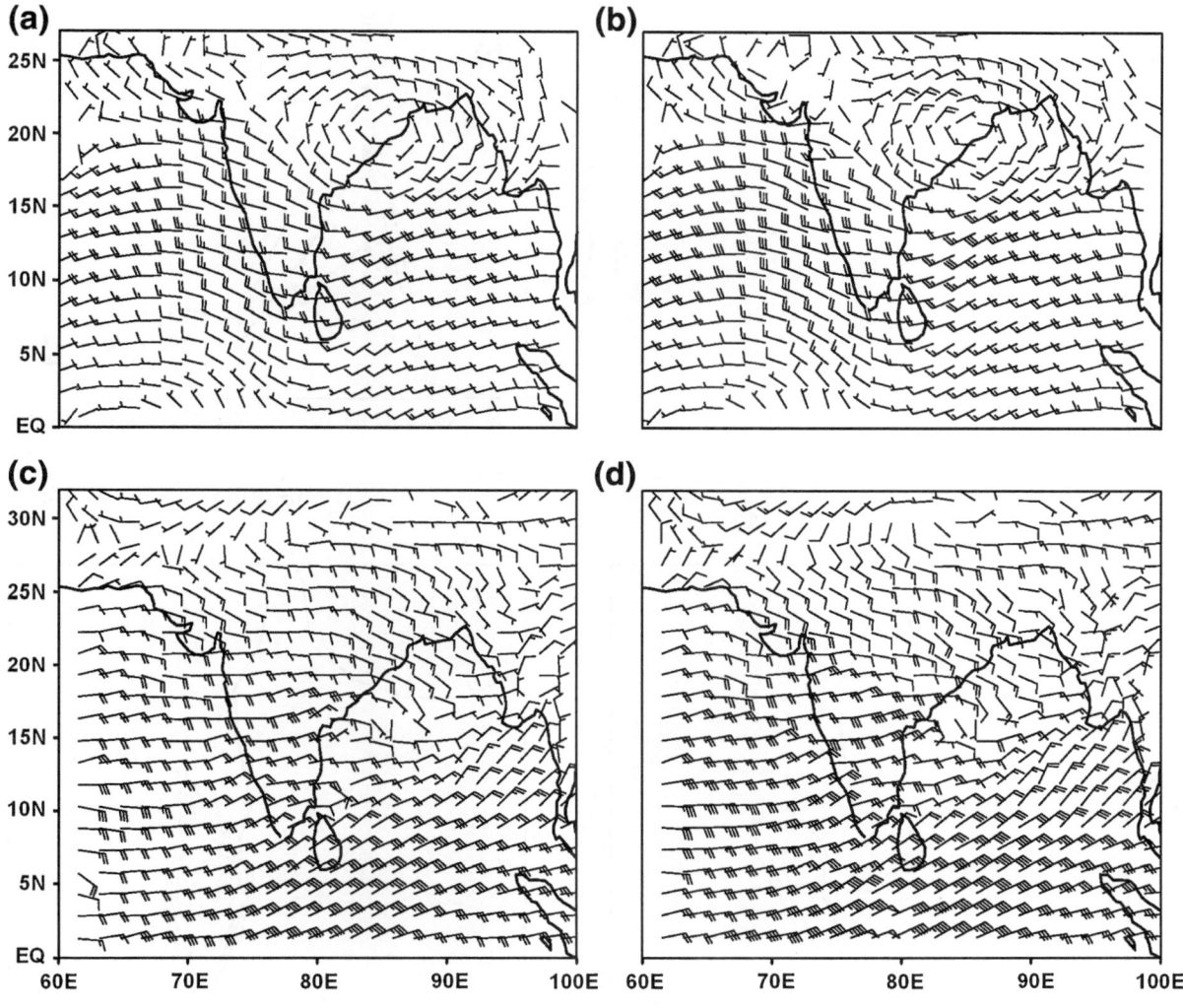

Figure 3
Initial wind fields at 850 hPa (domain-1) for **a** CNTL and **b** 3DV valid at 0000 UTC 26 July 2005. Similarly, (**c–d**) are same as (**a–b**) but at 200 hPa

Gujarat region is reported by IMD which is clearly noticed in the 3DV experiment, as the feature absent in the CNTL experiment. In the Fig. 3d, an anti-cyclonic circulation is clearly observed in the upper atmosphere which indicates the lower level convergence and upper level divergence to trigger the convective activities over the domain. The strong easterlies over the peninsular India are noticed at 200 hPa in the 3DV. This semi-permanent feature is observed during the summer monsoon period (RAO, 1976). These synoptic representations are not found in the CNTL (Fig. 3c).

5.1.2 Precipitation

The observed, model simulated 3-h and 24-h accumulated spatial distribution of rainfall are presented in Fig. 4a–c. Figure 4b shows the 3-h model simulated rainfall along with IMD observation over Santacruz during the period 26–27 July 2005. The observation shows rapid increase of rainfall (around 40 cm) during 3 h of 09–12 UTC 26 July 2005. The occurrence of the sudden increased of intensity of the precipitation rate may be due to the cloud burst during the time period. Subsequently, because of the

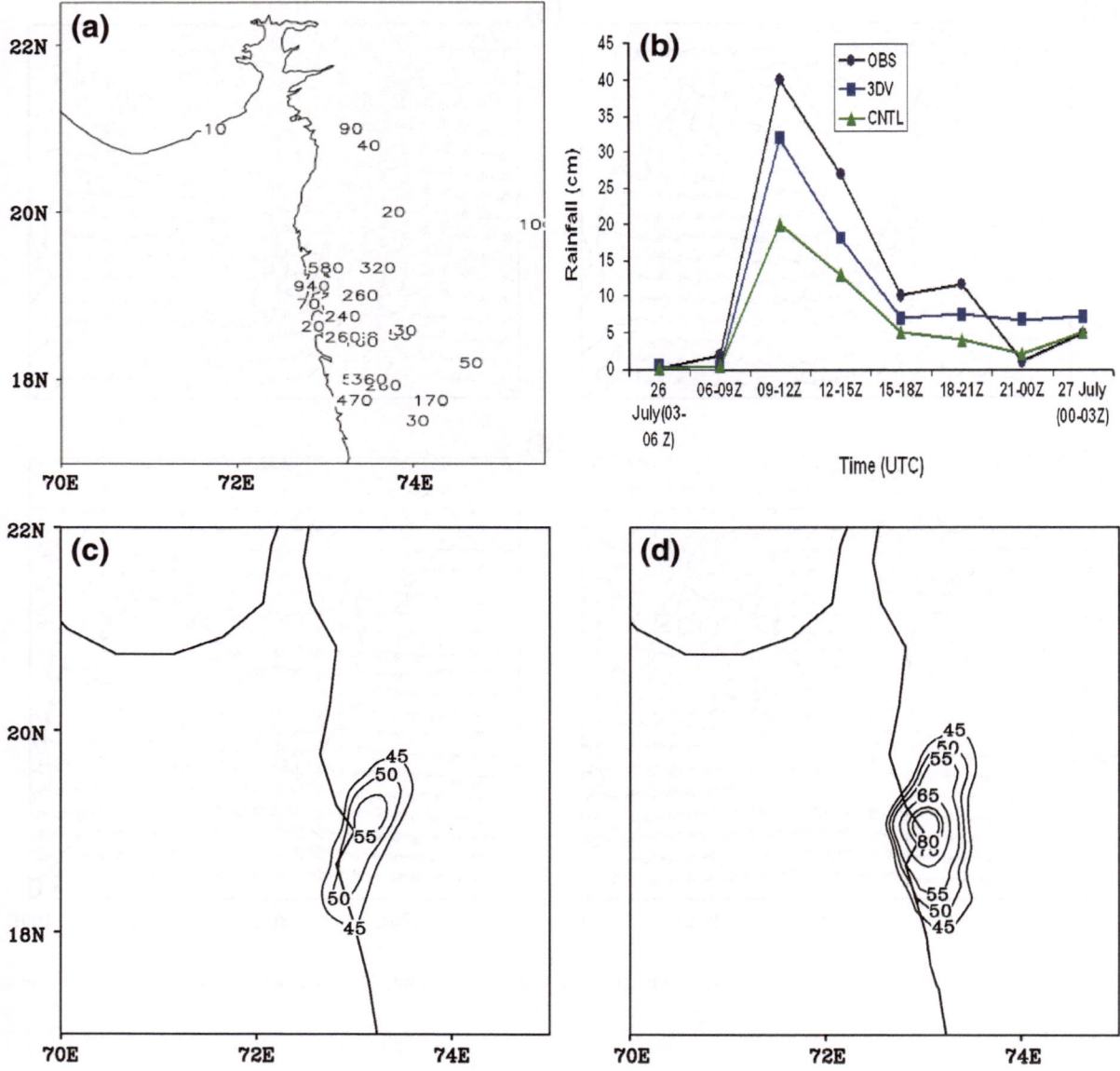

Figure 4
a From rain-gauges at 03 UTC of 27 July 2005 over Mumbai city and surrounding region (Source IMD), **b** observed and model simulated 3hourly rainfall (cm) over Santacruz during 26–27 July 2005. 24 h accumulated rainfall in cm (cint 45 50 55 65 75 80) for **c** CNTL and **d** 3DV valid at 0300 UTC of 27 July 2005

continuous restructuring of the meso-convective activities over the region which influence by the synoptic conditions, the rainfall intensity continued up to 21 UTC of 26 July 2005. The intensity of the rainfall during the period (12–21 UTC) is well simulated in the 3DV experiment as compared to the CNTL simulation. The two maximum rainfall spells are well simulated in the 3DV experiment

around 32 and 18 cm as compared to the CNTL simulation (20 and 13 cm) during 09–12 and 12–15 UTC period, respectively. The rainfall amount from 3DV simulation is reasonably well matched with the observation (40 and 27 cm) during the period. Figure 4c, d illustrates the model simulated 24-h accumulated rainfall from CNTL and 3DV simulations. It is very clearly noticed that the orientation and spatial distribution of rainfall is well simulated in the 3DV simulation (Fig. 4d) over the region as compared with the CNTL simulation (Fig. 4c). The 3DV experiment is simulated the rainfall up to 80 cm over the Mumbai region. However, the CNTL experiment shows the maximum rain (55 cm) away from the observed location (Fig. 2b, c, 4a) with a location error of 40 km northeast of Santacruz, Mumbai. It is worth mentioning here that the location error has been corrected by around 80 km after increasing the model resolution (30–10 km) in the CNTL simulation (KUMAR et al. 2008), whereas the intensity of the rainfall is also increased around 5 cm in both the simulations as compared to the rainfall amount obtained from domain-1 simulations (figure not provided). The spatial distribution and intensity of the rainfall well represented in the assimilation experiment as compared to the CNTL simulation mainly over the Mumbai region.

5.1.3 Divergence, Vorticity and Vertical Velocity

The vertical cross section (time–pressure) of divergence, vorticity and vertical velocity from CNTL and 3DV experiments during the period (0000 UTC 26 to 0600 UTC 27 July 2005) is illustrated in Fig. 5a–f. To keep in mind the intensity of the rainfall, the heavy rainfall period is divided into three stages such as developing stage (06–12 UTC July 26) where accumulated rainfall is moderate; mature stage, where the rainfall intensity is high (12–18 UTC); and, lastly, dissipating stage while the rainfall is started decreasing (18–00 UTC July 27). During the developing stage the convergence starts at a lower level in both the simulations (Fig. 5a, b). The convergence is found in the mature stage (12–18 UTC) up to 850 hPa level with maximum value of 20×10^{-5} s^{-1} at 12 UTC in the 3DV simulation (Fig. 5b) which is not observed in the CNTL simulation (Fig. 5a). At the same time, the maximum divergence found in the 3DV simulation at upper atmosphere is around 200 hPa. The vertical cross section revealed the structure of low level convergence and upper level divergence throughout this period. It is also seen that the divergence (8×10^{-5} s^{-1}) developed around 700 hPa at 2100 UTC which meant that the system started dissipating.

The depth as well as strength of the convergence is gradually decreased as time progressed mainly from the mature stage to dissipating stages. The vertical cross sections of vorticity from CNTL and 3DV simulations are depicted in Fig. 6c, d, respectively. The 3DV simulation (Fig. 6d) exhibits a structure of strong cyclonic vorticity (30–36×10^{-5} s^{-1}) at lower and upper levels which is not found in the CNTL simulation in Fig. 5c (maximum vorticity is 21×10^{-5} s^{-1} only at upper level). In the 3DV simulation, the positive vorticity increased gradually from developing stage to mature stage and then started decreasing after 1800 UTC. The maximum cyclonic vorticity is found during the mature stage and extended to upper troposphere. This shows the gradual growth of the convective system which is well represented in the 3DV simulation. Strong negative vorticity (12–24×10^{-5} s^{-1}) is noted at upper atmosphere around 200 hPa pressure level and above throughout this period. The maximum upward velocity obtained from 3DV simulation (Fig. 5f) is 2.2 m s^{-1} and found at mid-level (around 400 hPa) and upper atmosphere (around 200 hPa) during the mature stage (1200–1800 UTC). However, the maximum vertical velocity from CNTL simulation (Fig. 5e) is 0.8 m s^{-1} at upper atmosphere on 0000 UTC 27 July 2005.

5.1.4 Horizontal, Vertical Moisture Fluxes and Mixing Ratio of Hydrometers

Furthermore, the performance of the 3DV simulation is investigated by analyzing the mesoscale characteristics associated with the convective system which influence such extreme rain events. Figure 6a–f represents the vertical cross section of horizontal and vertical moisture fluxes as well as hydrometers from CNTL and 3DV simulations at Santacruz ($19.11°$N, $72.85°$E) during the period 26–27 July 2005. The vertical cross section of horizontal moisture flux obtained from the 3DV simulation (Fig. 6b) shows gradual increase of positive fluxes (extended up to 600 hPa) from developing stage and maximum at mature stage 51×10^{-4} g kg^{-1} s^{-1}, the flux again commenced decreasing afterward (dissipating stage). In the CNTL simulation (Fig. 6a), the horizontal moisture fluxes also followed the same trend as

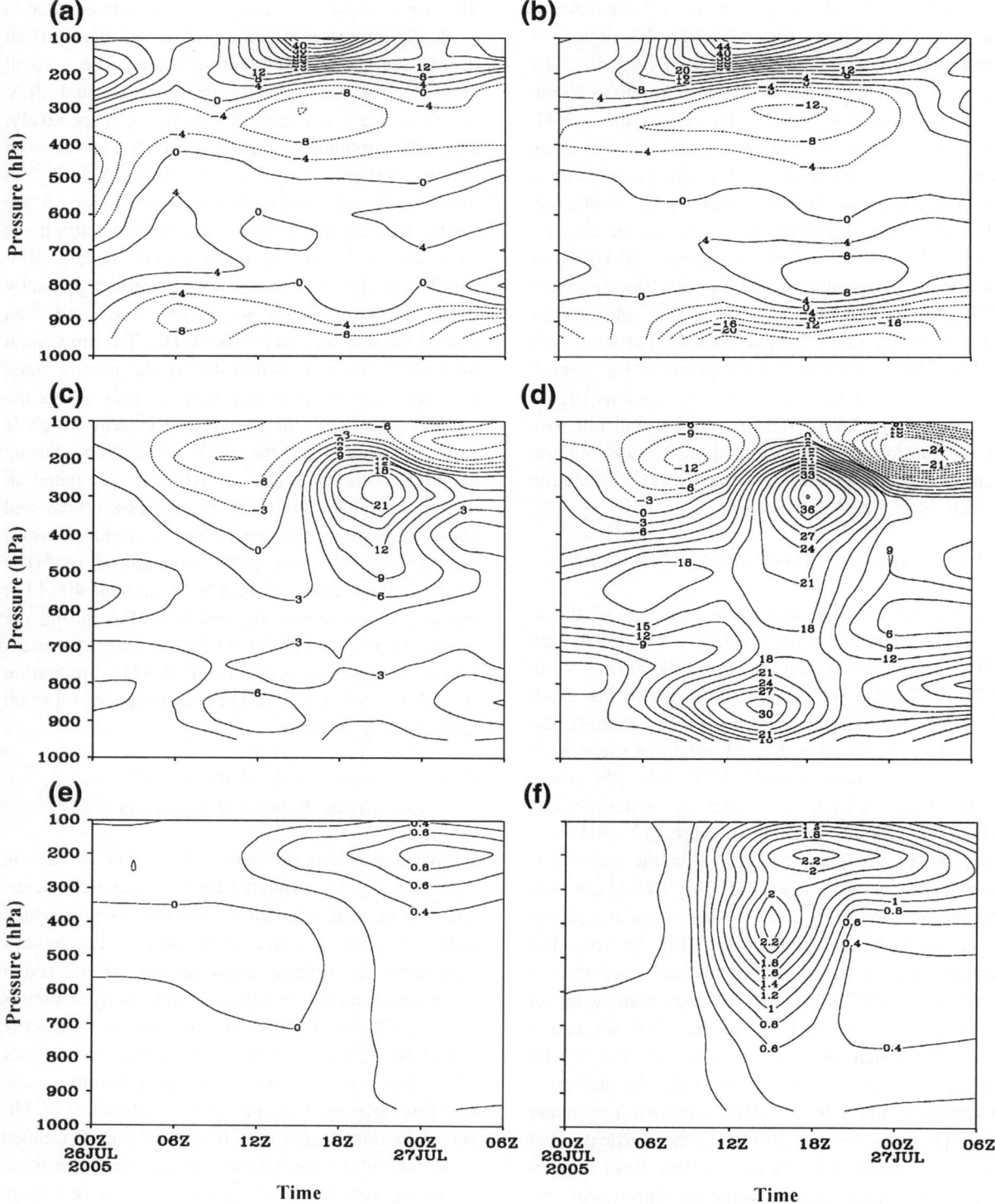

Figure 5
Time–pressure cross section of divergence (10^{-5} s^{-1}) at Santacruz (19.11°N; 72.85°E) for **a** CNTL and **b** 3DV. Similarly, (**c**–**d**) and (**e**–**f**) are same as (**a**–**b**) but for vorticity (10^{-5} s^{-1}) and vertical velocity (m s^{-1}), respectively

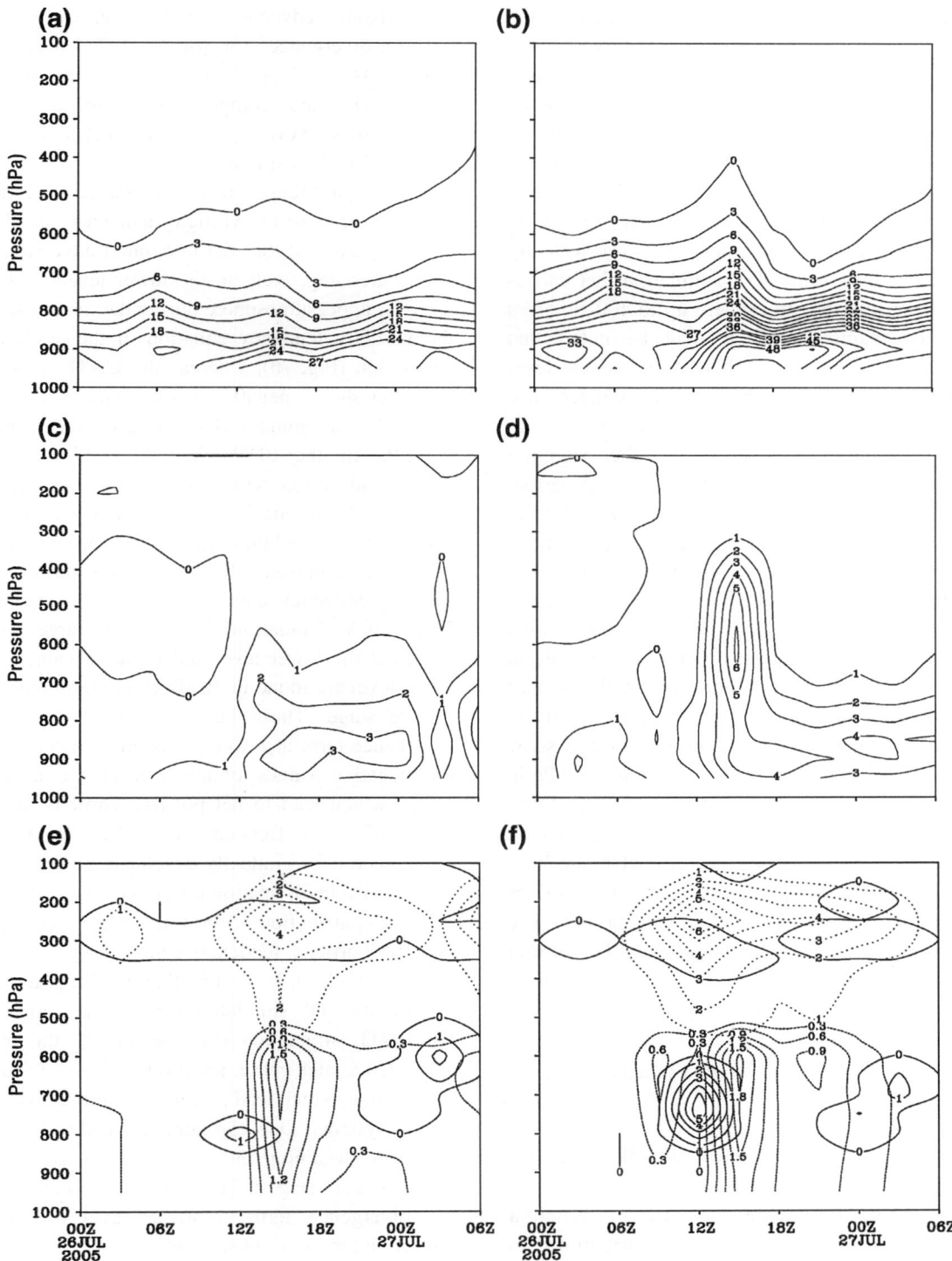

Figure 6

Time–pressure cross section of horizontal moisture flux (10^{-4} g kg^{-1} s^{-1}) at Santacruz (19.11°N; 72.85°E) for **a** CNTL and **b** 3DV. Similarly, (**c–d**) and (**e–f**) are same as (**a–b**) but for vertical moisture flux and mixing ratios for cloud water (*solid line*); rain water (*short dash*) and ice + snow + graupel (*dotted*), respectively (10^{-4} g kg^{-1} s^{-1})

observed in the 3DV simulation. However, the maximum horizontal moisture flux value 27×10^{-4} g kg^{-1} s^{-1} and extended up to 700 hPa at mature stage is simulated in the CNTL experiment. During the mature stage, the vertical moisture fluxes are extended from surface (900 hPa) to 300 hPa in the 3DV simulation (Fig. 6d). The maximum moisture flux is obtained at 600 hPa. This feature is consistent with the maximum upward velocity (Fig. 5f) obtained from 3DV simulation that carries the moisture into the atmosphere up to 400–300 hPa during the mature stage. It may be noted that 400–300 hPa is the average height of freezing level during the period. The depth of moisture convergence and vertical moisture flux extended up to 600–400 hPa during the mature stage, the maximum rainfall (around 35 cm) occurred during the period. The vertical cross section of mixing ratio of hydrometers obtained from CNTL and 3DV simulations are represented in Fig. 6e, f, respectively. In the 3DV simulation (Fig. 6f), the cloud droplets are generated due to the high incursion of moisture and vertical motion during mature stage, which is not noticed in the CNTL simulation (Fig. 6e). During the mature stage, the rain water mixing ratio is extended up to 600 hPa along with the other hydrometers (ice, snow and graupel) at upper atmosphere around 200 hPa in both the simulations. This suggested that the simulated cloud height reached up to upper atmosphere. However, it is noticed that the hydrometers are little stronger in the assimilation experiment (Fig. 6f) as compared to the CNTL simulation (Fig. 6e). The clouds started dissipating as the moisture fluxes, rain water mixing ratio and other hydrometers started decreasing during dissipating stage.

5.1.5 Diagnoses of the Different Vorticity Budget Terms

The model predicted the heavy rainfall event over Mumbai and is associated with sudden increase of the vorticity during the mature stage. For the different vorticity budget terms' relative contributions to trigger such a heavy rain event during the period, we diagnose the role of individual terms in this section. The major contributing terms involved in the vorticity tendency equation considered in this study

are horizontal advection $(-V \cdot \nabla \xi)$, vertical advection $(-w \frac{\partial \xi}{\partial z})$, divergence term $[-(f + \xi) \nabla \cdot V]$ and tilting term $(-\frac{\partial w}{\partial x} \frac{\partial v}{\partial x} + \frac{\partial w}{\partial y} \frac{\partial u}{\partial z})$. Where V (u, v) is zonal and meridional wind components, ξ is the vertical component of vorticity, w is vertical velocity and f is the Coriolis parameter.

Figure 7a–f represents the vertical cross section of different terms of the vorticity tendency equation. It is clearly noticed that the horizontal advection term contributes negatively in the lower levels and positively at upper atmosphere during the mature stage to the vorticity tendency equation in the assimilation simulation (Fig. 7b), whereas the CNTL simulation (Fig. 7a) shows negative values only at the lower level. The maximum horizontal advection is found at 500 hPa in the 3DV simulation. Similarly, the vertical advection terms shows negative values at lower level and positive values at upper atmosphere in the 3DV simulation (figure not provided). Figure 7c, d represent the divergence term of the vorticity tendency equation from both simulations. In the 3DV simulation (Fig. 7d), the large positive values at the lower level and negative values at the upper level are found in the divergent term during the mature stage. These large positive values of the divergence term at lower levels are compensated by the negative values obtained from the advection terms, which lead to net positive vorticity at lower levels (Fig. 5d). Beyond 1800 UTC, the negative divergence term gradually developed at lower levels which aided in dissipation the convective system. The positive contribution of the tilting term increases the positive vorticity tendency which triggers the convective activities. It is found that the tilting terms are more in the 3DV simulation (Fig. 7f) as compared to the CNTL simulation (Fig. 7e) during the mature stage. On a later phase, the tilting terms are found to be negative, which lead to initiation of the dissipation of the systems. One can summarize that the major factor of sustaining the positive vorticity tendency during mature stage is because of the contribution of the divergence term at lower level as well as advection terms at upper levels.

Figure 8 presents the vertical cross section of the zonal (Fig. 8a–b) and meridional (Fig. 8c–d) gradient of w obtained from two experiments. The shear terms play an important role in the initiation and the

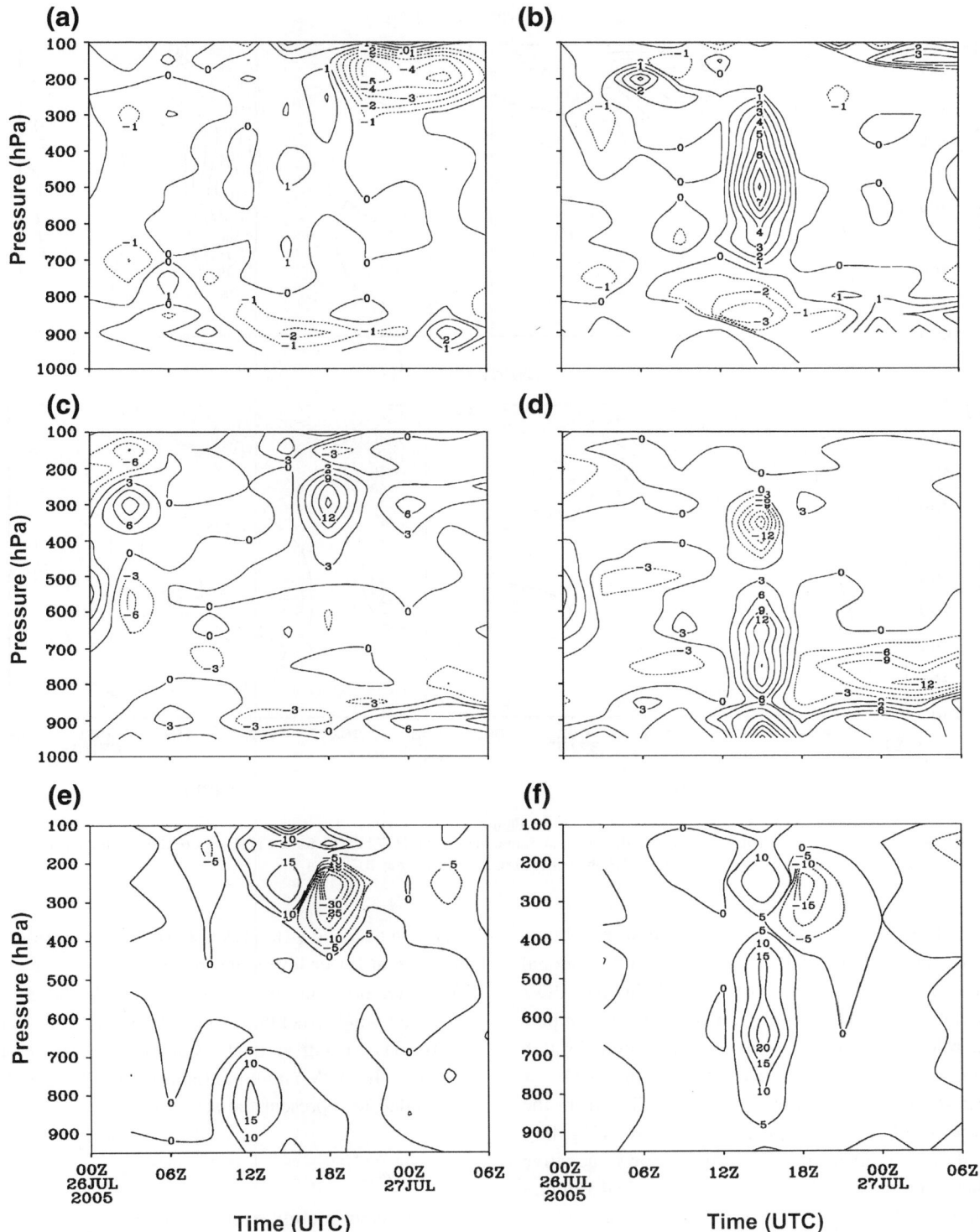

Figure 7
Time–pressure cross section of horizontal of advection term (10^{-8} s^{-2}) at Santacruz (19.11°N; 72.85°E) for **a** CNTL and **b** 3DV. Similarly, (**c–d**) and (**e–f**) are same as (**a–b**) but for divergent term (10^{-8} s^{-2}) and tilting term (10^{-8} s^{-2})

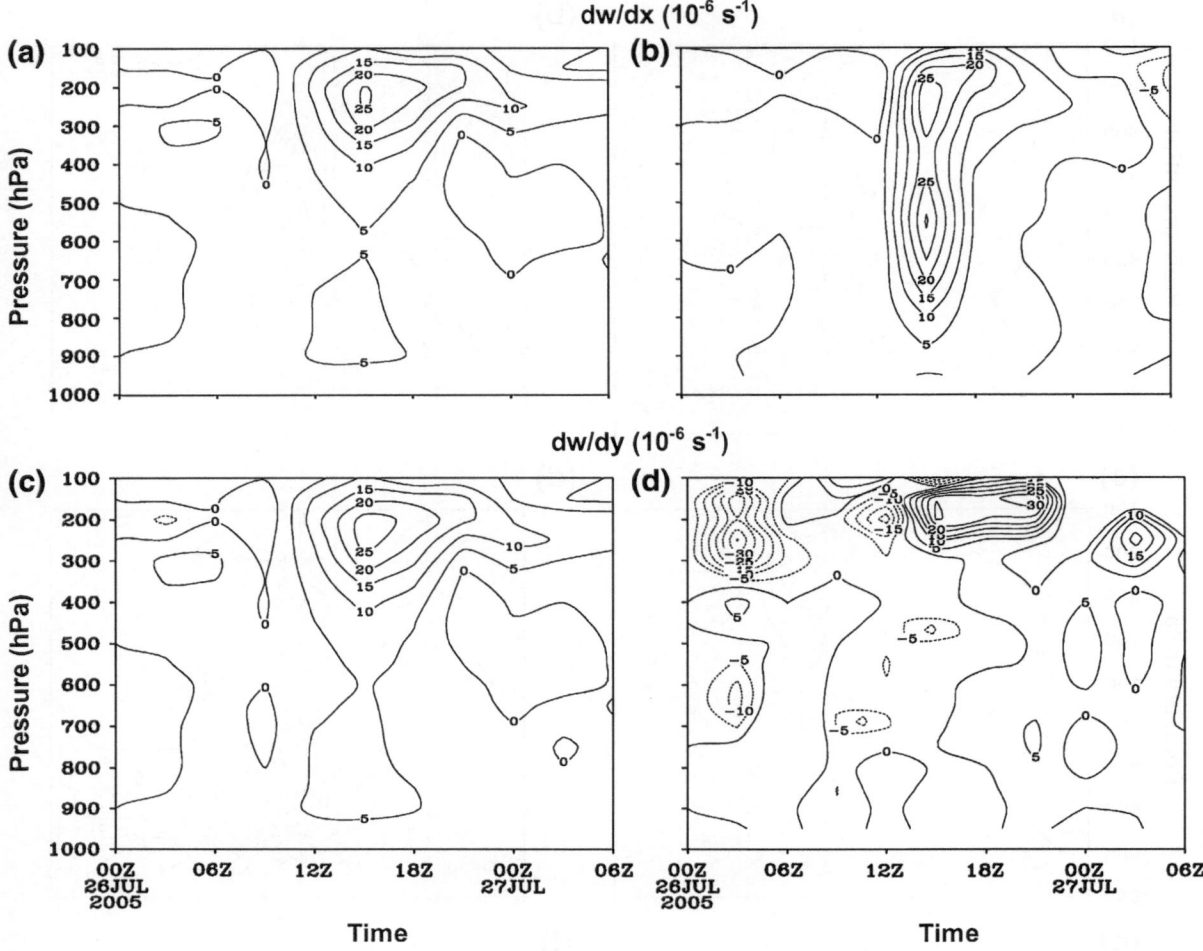

Figure 8

Time-pressure cross section of shear term (dw/dx; 10^{-6} s^{-1}) at Santacruz (19.11°N; 72.85°E) for **a** CNTL and **b** 3DV. Similarly, (**c–d**) are same as (**a–b**) but for shear term (dw/dy; 10^{-6} s^{-1})

dissipation of the system (VAIDYA and KULKARNI, 2007). In the Fig. 8a–d, the zonal and meridional gradient of w is well represented in the 3DV simulation. The horizontal gradient of w play important role in the vorticity tendency equation through tilting term. The individual tilting term, dw/dx, is positive during the mature stage extending from the lower level to the upper atmosphere (Fig. 8b). Similarly, another term, dw/dy, shows negative values which means the existence of vertical velocity in the southeast side of the location. In a later phase of the mature stage, the positive values of dw/dy resulted in decreased vorticity leading to dissipation of the system. It is also noticed that the vertical

variation of u and v (figures not provided) contributed negative at lower levels and positive at upper levels. These features are well correlated with the mean monsoon wind structure over the Indian region (RAO, 1976). The strong vertical gradients of u and v are found in the 3DV simulation and the CNTL simulation failed to represent the structure.

5.2. Case-2

The heavy rainfall event occurred on 27–30 July 2006 with precipitation around 20–30 cm within 24 h over a few stations along the west coast of India. The detailed descriptions about the heavy rainfall event are described above in Sect. 3. The model is integrated up to 54 h from 0000 UTC 27 July 2006 after improved the initial condition through WRF-Var analysis system using conventional and non-conventional observations. In this section, the simulated meteorological fields, mean sea level pressure and rainfall, have been examined in detail. As a measure of forecast accuracy, equitable threat score (ETS), root mean square errors (RMSE) and correlation coefficients (CC) of rainfall are calculated. The details of the formulation of these skill scores can be found in the study by JANKOV *et al.*, (2005).

5.2.1 Mean Sea Level Pressure (MSLP)

Figure 9a–f represent the model simulated (day-1 and day-2) and subjectively analyzed MSLP by IMD valid at 0300 UTC of 28 and 29 July 2006. During the period, the off-shore trough at sea level was observed along the west coast of India extending from south Gujarat to Kerala coasts. The off-shore trough is simulated in all experiments along the west coast of India. However, the off-shore trough is well represented in the 3DV simulations (Fig. 9c, f) during day-1 and day-2 as compared to the CNTL simulations (Fig. 9b, e). The features are reasonably matched with the IMD observations (Fig. 9a, d).

5.2.2 Precipitation

The 24h accumulated precipitation measurements for day-1 and day-2 as obtained from the CNTL and 3DV simulations and the observed rainfall from the TRMM satellite are shown in Fig. 10a–f. During day-1 and day-2, the rainfall is well simulated. However, a well established closed circulation system is found only in the 3DV experiment (Fig. 10c, f). For day-1, the spatial distribution of the precipitation pattern due to the 3DV experiment (Fig. 10c) is similar to the TRMM derived rainfall pattern (Fig. 10a). The maximum amount of rainfall (20–30 cm) is observed along Konkan and Goa coast (Fig. 10a) and the feature is well simulated in the 3DV experiment (Fig. 10c) as compared to CNTL simulation (Fig. 10b). Similarly, in day-2, the rainfall amount is improved significantly in the assimilation experiment (Fig. 10f), whereas a similar rainfall distribution pattern is not found in the CNTL

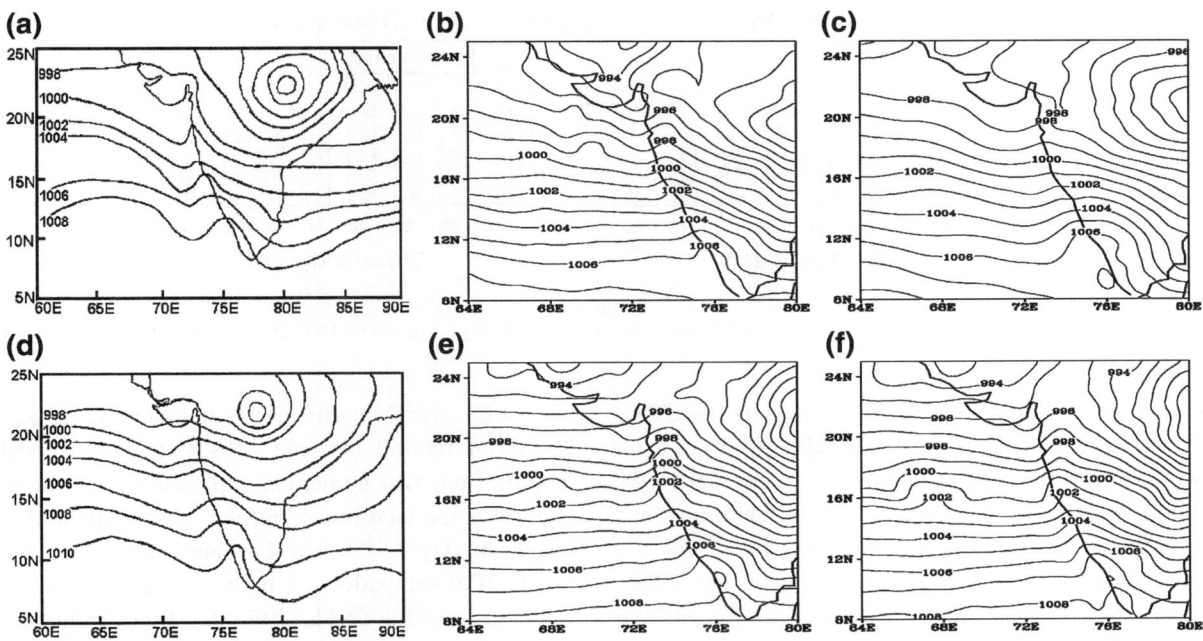

Figure 9
MSLP for **a** observed; **b** CNTL and **c** 3DV valid at 03 UTC 28 July 2006. Similarly, (**a–b**) are same as (**d–f**), respectively, but valid at 03 UTC 29 July 2006

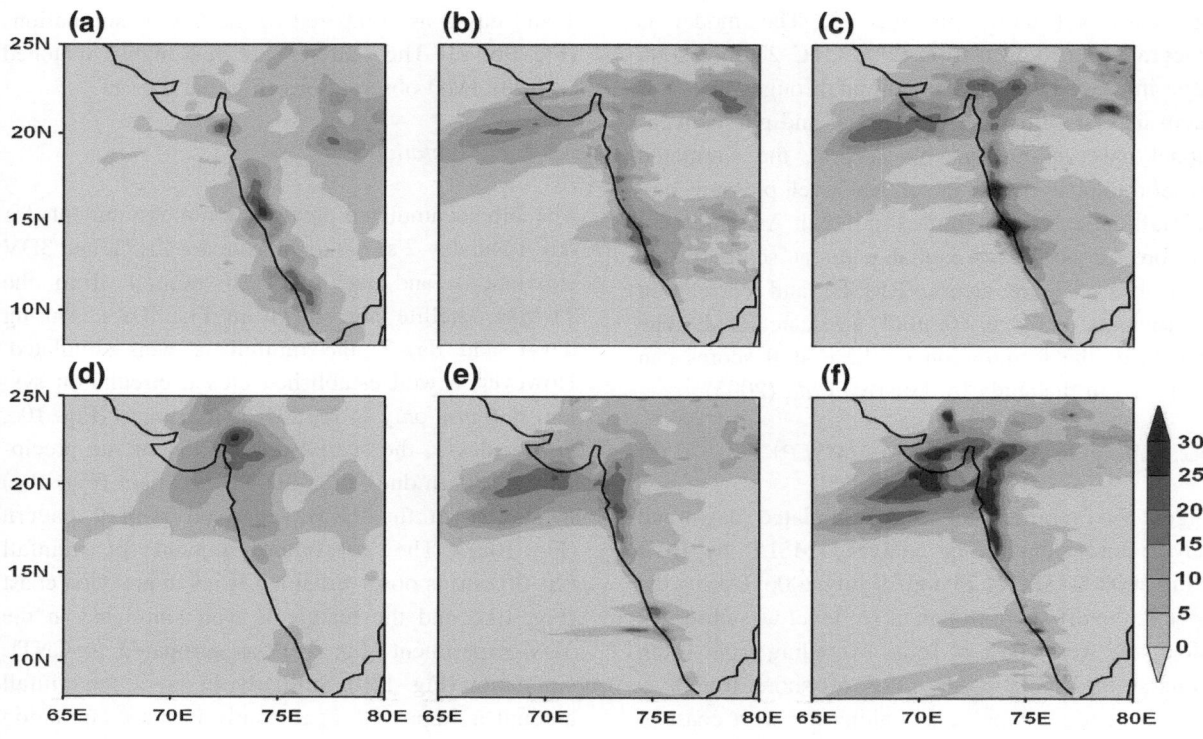

Figure 10
Twenty-four hour accumulated rainfall (cm) for **a** TRMM; **b** CNTL and **c** 3DV valid at 03 UTC 28 July 2006. Similarly, (**a–b**) are same as (**d–f**), respectively, but valid at 03 UTC 29 July 2006

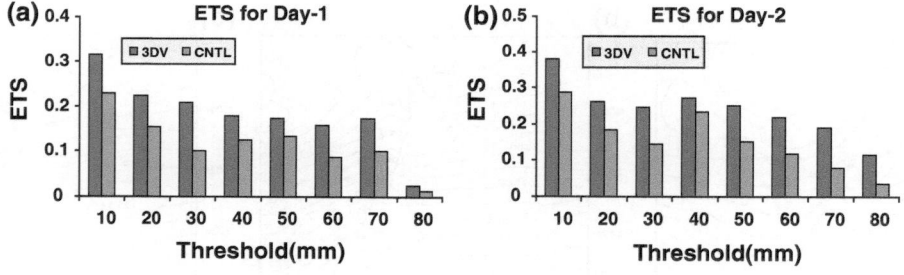

Figure 11
ETS of rainfall with different threshold (mm) from CNTL and 3DV **a** day-1 and **b** day-2 valid at 03 UTC 28 and 29 July 2006, respectively

simulation (Fig. 10e). Figure 11a, b depicts the ETS of rainfall with different threshold values (mm) of the two experiments valid at 0300 UTC 28 and 29 July 2006, respectively. Significant improvements in the precipitation forecast appeared in day-1 and day-2 forecast period due to the 3DV experiment. The ETS scores have considerably improved in the 3DV experiment throughout the period with different rainfall threshold values as compared with the CNTL experiment. The spatial (Lat = 10°–25°N;

Lon = 69°–78°E) correlation coefficient (CC) and RMSE of rainfall between TRMM and the model outputs from two numerical experiments are calculated over the land (mask out the ocean part) during day-1 and day-2. The RMSE values (mm) are found less in 3DV simulation (12 in day-1 and 13 in day-2) as compared with CNTL (20 in day-1 and 18 in day-2) simulations. Similarly, the CC is improved in the 3DV (0.74 in day-1 and 0.67 in day-2) as compared to CNTL (0.37 in day-1 and 0.24 in day-2) experiments.

It is clearly noticed that the rainfall pattern is better simulated after assimilation of the conventional and non-conventional observations as compared to without assimilation.

6. Conclusion

The mesoscale model WRF and 3DVAR data assimilation systems with double nested domain (outer domain 30 km and inner domain 10 km) are used to simulate the two heavy rainfall events that occurred during the summer monsoon season along the west coast of India. The inner domain configured along the west coast of India. The model simulated results obtained from the finer domain is considered for discussion. For this purpose, two detailed numerical experiments (CNTL: without and 3DV: with data assimilation) are carried out to evaluate the impact of high-resolution analysis. The model initial conditions are improved through 3DVAR data assimilation using conventional and non-conventional observation collected from the GTS/Internet. In the light of the "Results and Discussions" section above for simulation of intense rain events, broad conclusions can be put forward as follows.

The unprecedented localized heavy rainfall as observed 944 mm within 24 h over Santacruz, Mumbai region during the period 26–27 July 2005 (Case-1) is reasonably well predicted in the WRF model based on 3DVAR improved initial condition as compared to without assimilation experiments. The initial time wind fields at 850 hPa shows stronger wind around 5–10 m s^{-1} associated with surface low over Orissa coast as compared to the wind obtained from the CNTL experiment over the region. The magnitude of southwesterly wind, 15–30 m s^{-1}, over the Arabian Sea is simulated in the 3DV which is well agreed with the IMD observation (15–28 m s^{-1}). A cyclonic circulation simulated in the 3DV as reported by IMD over the northern part of Mumbai. Simulation of these features is significant, as a lot of moisture incursion at this level was crucial for development of deep convective system. The characteristics of the meso-convective system are well represented in the initial condition, which produced the model simulated intense rainfall over the isolated area of Mumbai region. The 3DV experiment produced reasonably well prediction of the precipitation in terms of intensity, location and spatial distribution over the region. The model simulation produced maximum accumulated precipitation of 80 cm and location of the maximum rainfall was well correlated with the observation. The CNTL experiment produced maximum accumulated rainfall up to 55 cm; however, it is away from the observed location with a location error of 40 km northeast of Santacruz, Mumbai. It is seen that the location error is corrected around 80 km after increasing the model resolution (30–10 km) in the CNTL simulation while not much improvement was found in the intensity of the rainfall, with only 5 cm increase in both the simulations. The intensity and temporal variability of rainfall during 09–12 UTC is well simulated in the 3DV simulation. The model simulated dynamic and thermodynamic features at heavy rainfall location are analyzed to know the structures of the meso-convective system (MCS) which lead to such intense rain events. It is noticed that maximum convergence and vorticity during the mature stage, and at the same time, the vertical velocity, also follow the same trend during the period in the assimilation experiment. The feature is contradictory in the CNTL simulation. In the 3DV simulation, the contribution of moisture fluxes and hydrometeors is maximum during the mature stage, which leads to the development of the cloud up to upper atmosphere around 200 hPa. One can summarize that the major responsibility to sustain the positive vorticity tendency during the mature stage is because of the contribution of the divergence term at the lower level as well as advection terms at upper levels. The contribution of positive tilting term produced positive vorticity which trigger the convection and negative contribution to vorticity from tilting term to precede the dissipation of the system. The dynamic and thermodynamic structures of MCS are well resolved in the 3DV experiment after assimilation of conventional and non-conventional observations through WRF-Var.

The heavy rainfall event during Case-2 along the west coast of India well simulated in the 3DV simulation as compared to the CNTL simulation. After successful insertion of observations, the analyzed data with increased horizontal resolution produced properly the MCS structures, which was consistent with the large scale monsoonal flow. The off-shore trough at sea level along the west coast of India extending from south Gujarat to Kerala coasts as reported by IMD is simulated in all experiments. However, the off-shore trough is well represented in the 3DV simulations during day-1 and day-2 as compared to the CNTL simulations, the orientation of the off-shore trough matches with that of the observed. The intensity and spatial distribution of the rainfall is considerably improved in the 3DV simulation. The statistical skill scores also revealed that the precipitation forecast during the period has appreciably improved in the 3DV simulation due to assimilation of observations.

This study clearly demonstrated the advantages of impact of data assimilation techniques like the 3DVAR analysis system in prediction of heavy rainfall events during the monsoon season such as the unprecedented localized Mumbai rainfall. Furthermore, improvement can be obtained by increasing the model resolution, proper representation of land surface in the model, additional high quality dense surface (ships, buoys, etc. available to IMD in delayed mode)/upper-air observations (MEATR, aircraft reports over Santacruz) and remote sensing observations like radiance, radial velocity and reflectivity from Doppler weather radar with latest version of WRF and data assimilation systems. The BE statistics used in this study is calculated through NMC method. As the BE statistics play an important role in the assimilation cycle, it will be further calculated through available better BE formulations.

Acknowledgments

Authors are thankful to NCEP for providing the FNL analysis and GTS observations from IMD and National Centre for Medium Range Weather Forecasting (NCMRWF) used in the WRF and WRF-Var systems, respectively, to successfully carry out the study. Authors acknowledge the use of WRF and WRF-Var developed by National Centre for Atmospheric Research (NCAR), USA. We express our sincere thanks to anonymous reviewers for their valuable comments and suggestions for improvement of the manuscript.

REFERENCES

BARKER, D. M., W. HUANG, Y.-R. GUO, A. Bourgeois, and X. N. Xiao, 2004: *A Three-Dimensional Variational Data Assimilation System for MM5: Implementation and Initial Results*, Mon. Wea. Rev., *132*:897–914.

BENSON, C. L. and G. V. RAO, 1987: *Convective bands as structural components of an Arabian Sea convective cloud cluster*, Mon. Wea. Rev., *115*:3013–3023.

BOHRA, A.K., BASU, S., RAJAGOPAL, E.N., IYENGAR, G.R., GUPTA, M.D., ASHRIT, R., and ATHIYAMAN, B., 2006: *Heavy precipitation episode over Mumbai on 26 July 2005: assessment of NWP guidance*. Curr. Sci. *90*:1188–1194.

CHANG H., A. KUMAR, D. NIYOGI, U. C. MOHANTY, F. CHEN, and J. DUDHIA, 2008: *The Role of Land Surface Processes on the Mesoscale Simulation of the July 26, 2005 Heavy Rain Event over Mumbai, India*, Global Planet Change, doi:10.1016/j.gloplacha.2008.12.005.

DALEY, R., 1991: *Atmospheric Data Analysis*, Cambridge Univ. Press, New York, pp. 457.

DAS, SOMESHWAR, RAGHAVENDRA ASHRIT, GOPAL RAMAN IYENGAR, SAJI MOHANDAS, M DAS GUPTA, JOHN P GEORGE, E N RAJAGOPAL and SURYA KANTI DUTTA, 2008: *Skills of different mesoscale models over Indian region during monsoon season: Forecast errors*, J. Earth Syst. Sci. *117*, 603–620.

DEB, S. K., T. P. SRIVASTAVA and C. M. KISHTAWAL, 2008: *The WRF model performance for the simulation of heavy precipitating events over Ahmedabad during August 2006*, J. Earth Syst. Sci, *117*, 589–602.

DODLA, V.B.R., and RATNA, S.B., 2009: *Mesoscale characteristics and prediction of an unusual extreme heavy precipitation event over India using a high resolution mesoscale model*, Atmos. Res., doi:10.1016/j.atmosres.2009.10.004.

GUO Y-R, 2008: *Observation pre-processor for WRF-Var*, Summer WRF Tutorial Workshop, July 2008, NCAR. http://www.mmm.ucar.edu/wrf/users/tutorial/tutorial_presentation.htm.

HUFFMAN GJ, ADLER RF, STOCKER EF, BOLVIN DT, and NELKIN EJ, 2003: *Analysis of TRMM 3-hourly multi-satellite precipitation estimates computed in both real and post-real time*. Combined Preprints CD-ROM, 83rd AMS annual meeting. Paper P4.11 in 12th conference on satellite meteorology and oceanography, 9–13 Feb 2003, Long Beach, CA, 6 pp.

JANKOV ISIDORA, WILLIAM A. GALLUS JR., MOTI SEGAL, BRENT SHAW, and STEVEN E. KOCH, 2005: *The Impact of Different WRF Model Physical Parameterizations and Their Interactions on Warm Season MCS Rainfall*, Weath. Fore., *20*, 1048–1060.

JENAMANI, R.K., BHAN, S.C., and KALSI, S.R., 2006: *Observational/forecasting aspects of the meteorological event that caused a record highest rainfall in Mumbai*, Current Science, *90*, 1344–1362.

JIANFENG, G. U., QINGNONG XIAO, YING-HWA KUO, DALE M. BARKER, XUE JISHAN and M. A. XIAOXING, 2005: *Assimilation and Simulation of Typhoon Rusa (2002) using the WRF System*, Adv. in Atm. Sc., *22*, 3, 415–427.

KALNAY, EUGENIA, 2003: *Atmospheric Modeling, Data Assimilation and Predictability*. Published Cambridge University Press, pp. 364.

KUMAR, A., J. DUDHIA, R. ROTUNNO, DEV NIYOGI and U. C. MOHANTY, 2008: *Analysis of the 26 July 2005 heavy rain event over Mumbai, India using the Weather Research and Forecasting (WRF)model*, Q. J. R. Met. Soc., *134*, 1897–1910.

LIN, Y.-L., R. D. FARLEY, and H. D. ORVILLE, 1983: *Bulk parameterization of the snow field in a cloud model*. J. Climate Appl. Meteor., *22*, 1065–1092.

LORENC, A. C., S. P. BALLARD, R. S. BELL, N. B. INGLEBY, P. L.F. ANDREWS, D. M. BARKER, J. R. BRAY, A. M. CLAYTON, T. DALBY, D. LI, T. J. PAYNE and F. W. SAUNDERS, 2000: *The Met. Office global three-dimensional variational data assimilation scheme*, Quart. J. Roy. Meteor. Soc., *126*, 2991–3012.

LORENC, A.C., 1986: *Analysis methods for numerical weather prediction*, Quart. J. Roy. Meteor. Soc., *112*, pp. 1177–1194.

PARRISH, D.F., and J.C. DERBER, 1992: *The National Meteorological Center's spectral statistical interpolation analysis system*, Mon. Wea. Rev., *120*, 1747-1763.

RAMA RAO,Y.V, HATWAR H.R, SALAH A.K, and SUDHAKAR Y. 2007: *An experiment using the high resolution eta and WRF models to forecast heavy precipitation over India*, Pure Appl. Geophys. *164*, 1593–1615.

RAO, Y. P., 1976: *Southwest monsoon, Meteorological Monograph, Synoptic Meteorology, No. 1/1976*, India Meteorological Department, India, pp. 367.

ROUTRAY, A., U. C. MOHANTY, A. K. DAS, and N. V. SAM, 2005: *Study of heavy rainfall event over the west-coast of India using analysis nudging in MM5 during ARMEX-I*, Mausam, *56*, 1, 107–120.

ROUTRAY, A., U. C. MOHANTY, DEV NIYOGI, S. R. H. RIZVI, and KRISHNA K. OSURI, 2010a: *Simulation of Heavy Rainfall Events over Indian Monsoon Region using WRF-3DVAR Data Assimilation System*, Met. and Atm. Phy.,*106*, 107–125.

ROUTRAY, A., U. C. MOHANTY, S. R. H. RIZVI, DEV NIYOGI, KRISHNA K. OSURI and D. PRADHAN 2010b: *Impact of Doppler Weather Radar Data on Numerical Forecast of Indian Monsoon Depressions*, Qut. Jr. Roy. Met. Sco.,doi:10.1002/qj.678.

ROY BHOWMIK, S. K. and PRASAD, K., 2001: *Some characteristics of limited area model precipitation forecast of Indian monsoon and evaluation of associated flow features*, Met. Atm. Phys. *76*, 223-236.

SHYAMALA B and BHADRAM CVV, 2006: *Impact of mesoscale–synoptic scale interactions on the Mumbai historical rain event during 26–27 July 2005*, Current Science, *91*, 1649–1654.

SIKKA, D. R. and SULOCHANA GADGIL, 1980: *On the maximum cloud zone and ITCZ over the Indian longitudes during the southwest monsoon*, Mon. Wea. Rev., *108*, 1122–1135.

SIKKA, D.R., and P. SANJEEVA RAO, 2008: *The use and performance of mesoscale models over the Indian region for two high-impact events*, Nat. Hazards, *44*, 35–372.

SKAMAROCK, W. C., KLEMP, J. B., JIMMY DUDHIA, DAVID O. GILL, DALE M. BARKER, WEI WANG, and JORDAN G. POWERS, 2005: *A description of the Advanced Research WRF Version 2*, NCAR TECHNICAL NOTE, available at www.wrf-model.org.

VAIDYA S.S, and KULKARNI J.R., 2007: *Simulation of heavy precipitation over Santacruz, Mumbai on 26 July 2005, using mesoscale model*, Meteorol. Atmos. Phys. *98*, 55–66.

VAIDYA, S. S., P. MUKHOPADHYAY, D. K. TRIVEDI, J. SANJAY, and S. S. SINGH, 2004: *Prediction of tropical systems over Indian region using mesoscale model*, Mete. Atmos. Phys. *86*, 63–72.

(Received December 24, 2010, accepted March 26, 2011, Published online July 31, 2011)

Pure Appl. Geophys. 169 (2012), 401–414
© 2011 Springer Basel AG
DOI 10.1007/s00024-011-0377-0

Some Physical and Computational Issues in Land Surface Data Assimilation of Satellite Skin Temperatures

SCOTT M. MACKARO,[1] RICHARD T. MCNIDER,[2] and ARASTOO POUR BIAZAR[2]

Abstract—Skin temperatures that reflect the radiating temperature of a surface observed by infrared radiometers are one of the most widely available products from polar orbiting and geostationary satellites and the most commonly used satellite data in land surface assimilation. Past work has indicated that a simple land surface scheme with a few key parameters constrained by observations such as skin temperatures may be preferable to complex land use schemes with many unknown parameters. However, a true radiating skin temperature is sometimes not a prognostic variable in weather forecast models. Additionally, recent research has shown that skin temperatures cannot be directly used in surface similarity forms for inferring fluxes. This paper examines issues encountered in using satellite derived skin temperatures to improve surface flux specifications in weather forecast and air quality models. Attention is given to iterations necessary when attempting to nudge the surface energy budget equation to a desired state. Finally, the issue of mathematical operator splitting is examined in which the surface energy budget calculations are split with the atmospheric vertical diffusion calculations. However, the high level of connectivity between the surface and first atmospheric level means that the operator splitting leads to high frequency oscillations. These oscillations may hinder the assimilation of skin temperature derived moisture fluxes.

Key words: Data assimilation, land surface, satellite, skin temperature, boundary layer.

1. Introduction

As mesoscale models moved to higher resolutions, it became clear that improving surface representation was needed. Over the last few decades, investigators have embraced more physically complete land surface models as viable methods to better represent how a model surface reacts to changes in incoming or outgoing energy (DEARDORFF 1972, DORMAN and SELLERS,

1989, MCCUMBER and PIELKE, 1991). Employing more physically complete land surface models by including vegetative layers and/or multiple soil layers, while more sophisticated, require detailed land use characteristics. This leads to additional model parameters such as stomatal resistance or vegetative layer conductance for which measurements are not routinely conducted. It is for this reason that the recovery of surface parameters through the use of satellite data has been examined (WETZEL *et al.* 1984, WETZEL and WOODWARD 1987; CARLSON *et al.* 1981, CARLSON 1986; MCNIDER *et al.* 1994; ANDERSON *et al.* 1997). One such technique, MCNIDER *et al.* 1994 (hereafter referred to as MCN94), has been directly employed within several mesoscale model frameworks to recover surface moisture availability with improvements in mesoscale forecasts. Its framework spurred other investigations to recover other parameters, e.g. stomatal resistance (JONES *et al.* 1998a, b), grid scale heat capacity (MCNIDER *et al.* 2005), and soil moisture (PASQUI *et al.* 2004). The overarching strategy in MCN94 and MCNIDER *et al.* (2005) is that for weather forecasting and retrospective air quality modeling a simple land use scheme that can be constrained by observations is preferred to a complex scheme with many ill-defined and uncertain parameters.

Skin temperatures that reflect the radiating temperature of a surface observed by infrared radiometers are one of the most widely available products from polar orbiting and geostationary satellites and the most commonly used satellite data in land surface assimilation. Within a pixel of a satellite footprint, the skin temperature is the radiating temperature of everything within that particular field of view (trees, buildings, roads, etc.). While this intrinsic spatial averaging smears out detail in the surface, it is exactly the type of averaging that models need and as such

---
[1] Precision Wind, Inc, Boulder, CO, USA.
[2] Earth System Science Center, University of Alabama in Huntsville, Huntsville, AL, USA. E-mail: mcnider@nsstc.uah.edu

may be better than land surface classification schemes which require weighting or averaging to be consistent with the model grid. As surface skin temperatures have become more widely used in land surface assimilation, boundary layer investigators began to explore the issue of inconsistencies in the use of skin temperatures in similarity flux forms and issues with using radiometric satellite data in land surface models (ZILITINKEVICH 1970; LHOMME et al. 1988; KUSTAS et al. 1989; SUGITA and BRUTSAERT 1990).

BELJAARS and HOLTSLAG (1991) and SUN and MAHRT (1995) recognized that the original similarity forms for surface layer fluxes could not use these skin temperatures directly. Rather than a skin value, the similarity forms required an aerodynamic temperature. The aerodynamic temperature is not directly measured but rather inferred through an extrapolation of the temperature profile to the roughness height. CHOUDHURY et al. (1986) and BELJAARS and HOLTSLAG (1991) showed that the temperature difference between the surface skin temperature and the temperature at the roughness height can range from 2 K to 6 K in stable conditions to -2 to -6 K in unstable conditions. Thus, extrapolations of the Monin-Obukhov temperature profiles to the surface boundary rather than the roughness height can lead to errors in the estimation of the lower boundary temperature used in the surface flux calculations (BELJAARS and HOLTSLAG 1991). This indicates that using skin temperature data directly must be done so in a manner which is physically consistent with the bulk aerodynamic formulation of similarity functions.

The first purpose of this paper is to show the importance of attention to details such as consistent use of skin and aerodynamic temperature and iteration when incorporating satellite assimilation techniques into mesoscale models whether for forecasting or applications such as air quality simulation. Simply using existing model frameworks that do not always have the consistency in skin temperatures and aerodynamic temperatures can lead to problems as satellite data is introduced. This is illustrated through issues encountered in applying the MCN94 satellite assimilation technique within the Community Mesoscale Model MM5. While this is only one model, since other models share similar approaches in their surface formulations, it may be instructive to the general modeling

community. The second purpose is to clarify applications of the MCN94 satellite assimilation technique and follow-on use in light of these subsequent papers on skin temperatures and aerodynamic temperatures. Also included here is the importance in understanding the adjustments that take place when making changes in parameters in the surface energy balance and appropriate computational approaches. The surface energy budget is a highly non-linear system that, though often viewed as a balanced state, is changing quickly with time during the day. Changes to any term means that other terms will change so that a balanced state can be attained. Thus, one cannot simply change one term and assume that other terms stay the same in an attempt to move the system to a desired state.

2. Techniques for Assimilating Skin Temperatures

One of the first methods proposed for using satellite skin temperatures to recover surface variables was that of Carlson (1980) who proposed the use of twice daily data from polar orbiting satellites to recover surface moisture availability and thermal inertia. WETZEL et al. (1984) also hypothesized that skin temperature tendencies from satellites might be used to recover surface moisture availability. As new sensors have been put into space, Carlson and others made continuing progress in coupling surface skin temperatures and greenness measures to refine the recovery of surface moisture availability. However, these studies were diagnostic and did not take the step of using the results in a mesoscale model. One of the first attempts to use satellite data directly in the surface energy budget of a mesoscale model was MCN94 followed by JONES et al. (1998a, b).

In the MCN94 approach, Geostationary Operational Environmental Satellite (GOES)-derived skin temperature tendencies were assimilated into the surface energy budget equation of a mesoscale model so that the simulated rate of temperature change closely agreed with the satellite observations. At approximately the same time NORMAN et al. (1995) were also beginning to assimilate similar skin temperature tendencies in an offline boundary layer model. The adjustments made using the MCN94 technique are performed within the framework of a simple slab model. A composite slab model assumes all of the vegetative and soil properties that dictate land surface forcings can be described with a limited set of parameters. Note that in the context of the scheme in which satellite data is used, to determine moisture availability and surface resistance, the slab is a composite of vegetation and soil. Thus, even though soils are used to specify initial conditions, it is more than a bare soil slab. The slab evaporation includes both soil evaporation and plant transpiration.

As discussed in MCN94 and Norman *et al.* (1995), assimilating skin temperature tendencies rather than absolute temperatures is advantageous in that errors in the assumed emissivity of the surface as well as instrument errors from sensor degradation and drift are largely negated. In the MCN94 assimilation method, a prognostic form of the surface energy budget is defined for both the satellite and model;

$$c_b\left(\frac{dT_G}{dt}\right)_m = (R_N + H + G)_m + E_m \qquad (1)$$

$$c_b\left(\frac{dT_G}{dt}\right)_s = (R_N + H + G)_s + E_S \qquad (2)$$

where R_N is the net radiation, including net shortwave, net longwave, and surface radiance, H, E, and G are, in order, the sensible, latent, and soil heat flux and the subscript $()_m$ and $()_s$ are used to infer the model and satellite derived variables respectively. Here, a critical assumption based on a Wetzel *et al.* (1984) sensitivity analysis is invoked in that during the mid morning hours, when surface heating is most rapid, the largest uncertainty is due to the moisture availability (embedded within the latent heat energy term) and all the other terms can be considered to have the same magnitude. Taking the difference allows us to obtain an equation relating the satellite and model latent heating terms to the difference in the surface temperature heat rates;

$$E_S = C_b\left[\left(\frac{dT_G}{dt}\right)_s - \left(\frac{dT_G}{dt}\right)_m\right] + E_m \qquad (3)$$

The formulation of this technique stems from the use of the prognostic surface energy budget equation for a ground temperature. An inconsistency arises here in that the ground temperature is that for a surface with a finite depth and heat capacity. Fundamentally this is

not the same as the radiating skin observed by the satellite. The temperature tendency retrieved from the satellite is that of the skin temperature, and not the ground temperature.

The ground temperature evolution as dictated by the prognostic equation of (1) and (2) allows storage dependency on soil or canopy characteristics through the bulk heat capacity, c_b but does not produce a skin temperature or an aerodynamic temperature. However, this is the form employed in the slab formulation of MM5 in that the ground temperature is used as the aerodynamic temperature in flux calculations and in determination of the surface longwave outward radiation (Fig. 1). The use of such formulations in MM5 and other models has produced realistic fluxes in real world applications. However, when true skin temperatures are needed for satellite assimilation techniques or in comparison with tower radiometer skin temperatures, the need for consistency is evident.

The original implementation as reported in MCN94 serendipitously avoided part of this inconsistency in that in the mesoscale model, in which they applied the moisture recovery technique, a true skin temperature was calculated from a surface energy balance model. The model energy balance skin temperature was calculated by finding the root of a balanced energy budget for an infinitely thin surface. Thus, the results of MCN94 showed excellent results in recovering soil moisture in comparisons against flux observations under a field campaign in Kansas (FIFE). However, later applications of the MCN94 technique, when used within MM5, indicated

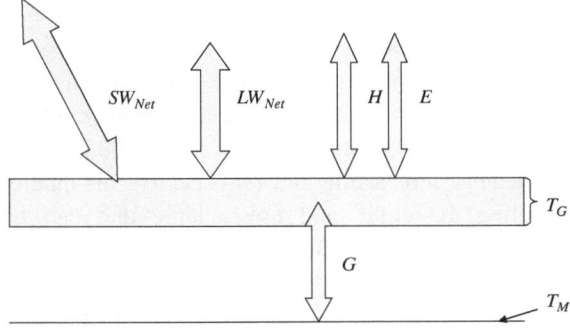

Figure 1
Schematic of the use of ground temperature within the surface to boundary layer interface of the MM5 using a composite slab land surface model

problems with mixing the satellite skin temperatures and the slab temperatures. During the morning hours, skin temperature tendencies from the satellite were often much larger than the ground temperature tendencies resulting in erroneously large adjustments. Results from other case studies indicated that using model slab temperatures mixed with satellite skin temperatures, while often improving model performance, usually produced an over drying of the surface (e.g., MACKARO 2003).

In order to correct the problems with mixing skin, aerodynamic and ground temperatures a new slab formulation for use within the MM5 or other mesoscale model was formulated. This is described in the following sections. The need for this formulation is in light of skin temperature research related to fluxes (BELJAARS and HOLTSLAG (1991) and SUN and MAHRT (1995)) that has made the community aware of issues with its use since the development of the MCN94 technique. We note that since MCN94 there have been several land surface schemes which have consistently computed the skin, ground and aerodynamic temperature, most notably the widely used NOAH land use scheme (EK *et al.* 2003, CHEN *et al.* 1996) which has now been connected to the WRF model. In keeping with the theme of this special issue, the following emphasizes the need and application of these techniques within a satellite data assimilation system.

3. Three-Temperature system

A 1-D boundary layer model (hereafter referred to as 1DBLM) was developed for this study. It uses a surface energy budget assuming that the all vegetative and soil properties can be described using a small set of surface land use parameters. The overarching strategy in MCN94 and MCNIDER *et al.* (2005) is that for weather forecasting and retrospective air quality modeling a simple land use scheme that can be constrained by observations is preferred to a complex scheme with many ill-defined and uncertain parameters. The single slab takes on the characteristics of a composite land surface based on the specification of parameters albedo, slab heat capacity, moisture availability, roughness length, thermal conductivity, emissivity, slab depth. The surface to atmospheric

boundary layer interface is assumed consistent with the traditional bulk aerodynamic formulation of similarity fluxes. See Fig. 2.

Some may argue that the use of a single slab is too simple. However, the radiating skin temperature observed by satellite (generally of order 4 km pixel size from geostationary orbit) gives a composite measure of the radiating surface. This composite radiating temperature may be made up of vegetation, asphalt, roofs, grass, etc., but the satellite sees, in effect, only a composite surface. While there may be some value in understanding the decomposition of the surface, in short-term forecasts or in air quality modeling the most important need is the correct heat and moisture composite flux. In climate modeling this is not the case since here the separate components of the system (plant evapotranspiration, soil evaporation, heat flux into the ground, etc.) are needed to run the model in an unattended mode for hundreds of years.

While it is also true that splitting the system into multiple components such as vegetation or asphalt may help specify parameters such as aerodynamic roughness. It must be remembered that the actual handling of these processes are highly idealized in models. Also, parameters such as the resistance or heat capacity of a vegetative canopy are highly uncertain. Investigators have found differences of several order of magnitudes in these parameters (PLEIM and XIU 1995). In fact they are model heuristics, not fundamental parameters, and their magnitude

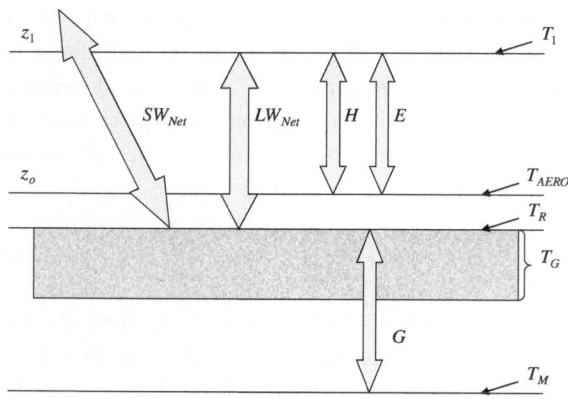

Figure 2
Schematic of a land surface to boundary layer interface that follows the traditional bulk aerodynamic formulation of surface layer fluxes

depends on the model formulation (McNider *et al.* 2005).

The surface energy budget is based on the diagnostic energy balance formulation,

$$
\begin{aligned}
R_S + R_{LW}^{\downarrow} &- \sigma T_R^4 - \rho c_p u_* \theta_* - \rho L_V M u_* q_* \\
&- \rho_s c_s k_s (T_R - T_G) \Delta h_{eff}^{-1} \\
&= 0
\end{aligned}
\tag{5}
$$

where the first two terms represent the net shortwave and incoming longwave radiation, the outgoing longwave radiation is given as a function of the skin temperature to the fourth power, T_R^4, M is the moisture availability which describes the total amount of water available for evaporation from 0 to 100%, ρ is the air density, and $u_* \theta_*$ and $u_* q_*$ are the similarity relationship formulations representing the fluxes of heat and moisture. The slab density, ρ_s, slab specific heat capacity, c_s, slab thermal diffusivity, k_s, and effective slab depth, Δh^{-1}, combine to describe the slabs heat storage capability, while the sign and magnitude of the difference between model skin temperature, T_R, and the ground temperature, T_G, dictate the direction and magnitude of the heat flux from the radiating surface to the slab. For the rate of heat transfer from the slab to the substrate, Blackadar (1979) uses $\kappa_m C_b (T_G - T_M)$, where $\kappa_m = 1.18\omega$. This formulation, for a pure sinusoidal heat input at the surface, provides a realistic representation of the surface temperature's amplitude and phase. When applying this formulation to the heat flow from the surface to the slab in the diagnostic surface energy budget where $\frac{\rho_s c_s \kappa_s}{\Delta h_{eff}} = \frac{\lambda}{\Delta h_{eff}} = \kappa_m C_b$, we must find the effective soil depth, Δh_{eff}, which will give a response consistent with Blackadar's analytical solution. This is accomplished by solving for Δh_{eff}:

$$
\Delta h_{eff} = \frac{\lambda}{\kappa_m C_b}
$$

In the present model for the land surface parameters used the effective slab depth is approximately 0.5 meters.

We convert to a prognostic surface energy balance for the temperature of the slab in the form of Eq. (1). The formulation of the resistance term comes from the simple slab model formulation of Blackadar (1979), where C_b is the thermal heat capacity of the slab per unit area and is related to the thermal inertia of the slab.

To obtain an aerodynamic temperature from information available in 1DBLM, we employ a method suggested by Zilitinkevich (1970) and Deardorff (1972) to relate the temperature at z_o to the model skin temperature;

$$
T_{Aero} = T_{Zo} = T_R + 0.0962(\theta_*/k)(u_* z_o/v)^{0.45}
\tag{6}
$$

where v is the kinematic viscosity of air.

The process of solving for the model skin, ground, and aerodynamic temperature involves the use of the iterative process shown in Fig. 3. T_R is first retrieved by use of a root finding technique. The aerodynamic temperature is then recovered by use of Eq. (6) and is then used to update fluxes each iteration. The final step of the process involves calculating the ground temperature as it is time dependent.

With a physically consistent method of determining aerodynamic, model skin, and ground temperature in place, we now want to invoke the use of the MCN94 assimilation technique. In order to do

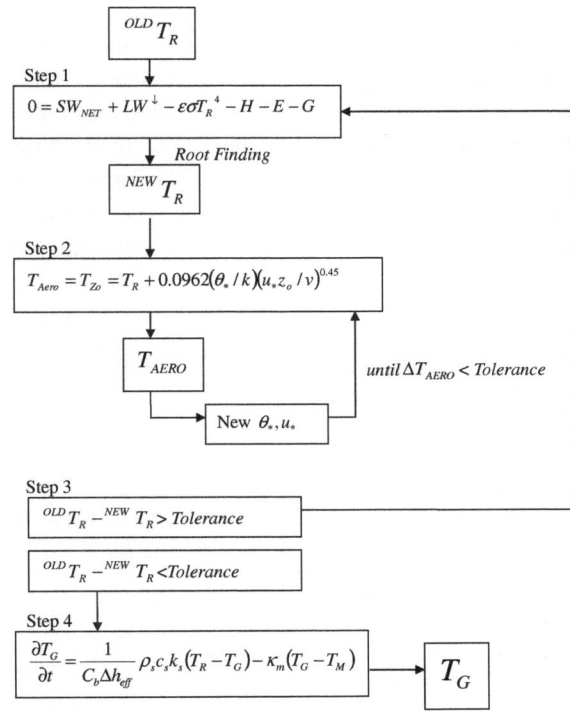

Figure 3
Schematic flow chart of the iteration used to solve the three temperature system used in 1DBLM

so, we must allow for the fact that the only information from the satellite is the satellite skin temperature not the ground temperature. Using the time rate of change of the satellite skin temperature, we can arrive at the tendency used by the MCN94 technique, $\frac{dT_R}{dt}$. The technique was developed from a prognostic form of the surface energy budget and simply replacing $\frac{dT_G}{dt}$ with $\frac{dT_R}{dt}$ is not valid. A relationship between $\left(\frac{dT_G}{dt}\right)_S$ and $\left(\frac{dT_R}{dt}\right)_S$ thus is necessary. Since we are using a closed system of equations for our three temperatures, a relationship then exists between T_G and T_R. We relate the tendency of the ground temperature and skin temperature by α given as,

$$\left(\frac{dT_G}{dt}\right) = \alpha\left(\frac{dT_R}{dt}\right)_S \tag{7}$$

Since both $\frac{dT_G}{dt}$ and $\frac{dT_R}{dt}$ are available from 1DBLM, we assume that the ratio of the two is the same for both the satellite and model, and solve for the relationship factor, α as

$$\alpha = \frac{dT_G}{dt}\Big/\frac{dT_R}{dt} \tag{8}$$

The behavior of α was found to be smooth and non-linear during the mid morning period for most types of land surfaces. While it is recognized that there may be a more elegant method of relating the radiative and ground temperature, the relationship used above is easily determined and is internally consistent at least within the framework of 1DBLM.

The next step is to replace $\left(\frac{dT_G}{dt}\right)_S$ with $\alpha\left(\frac{dT_R}{dt}\right)_S$ in Eq. (3) to arrive at a form which is physically consistent;

$$E_S = c_b\left[\left(\alpha\frac{dT_R}{dt}\right)_S - \frac{dT_G}{dt}_m\right] + E_m \tag{9}$$

From this we can continue as described in Lapenta (1999) with the model latent heat flux defined as,

$$E_m = \frac{M\rho\kappa u_*(q_{VS}(T_{Z_o}) - q_{AIR})}{\ln\left[\frac{z}{z_o}\right] - \psi_H} \tag{10}$$

By replacing E_m with the latent heat flux recovered for the satellite, E_S, we can invert the equation and arrive at an adjusted moisture availability, M_S;

$$M_S = E_S\frac{\ln\left[\frac{z}{z_o}\right] - \psi_H}{\rho\kappa u_*(q_{SAT}(T_G) - q_{AIR})}. \tag{11}$$

The adjusted moisture availability then replaces that specified as the initial condition. The adjustment typically takes place on the hour of each subsequent hour within the assimilation period during the mid morning hours. It is the previous hour's temperature tendency used in the adjustment so that no a priori information is necessary.

4. 1DBLM Test Cases

A series of simulations were performed to evaluate 1DBLM's physically consistent framework and its ability to simulate the surface to boundary layer interactions. Of specific interest here is the ability to reproduce values of observed skin temperature, air temperature, and surface fluxes of heat and moisture. The impact of the MCN94 technique on these variables is also of interest. In order to validate the simulations, surface based observations were obtained from Oklahoma Mesonet (Brock 1995) sites that are part of the Oklahoma Atmospheric Surface Layer Instrumentation System (OASIS; Brotzge 2000) and from the U.S. Department of Energy's Atmospheric Radiation Measurement (ARM) Cloud and Radiation Testbed (CART) (STOKES and SCHWARTZ 1994) Southern Great Plains Central facility (SGP-CF) site. From each site skin temperature and net radiation from an infrared thermometer (IRT; Fiebrich 2003), net radiation, air temperature, wind speed, sensible and latent heat flux from tower mounted instruments, and ground heat flux from ground instruments were collected.

Several OASIS sites were used with results from the Norman, OK (NRMN), and Foraker, OK (FORA) sites valid 4 July 2003 and from the ARM-CART SGP-CF site valid 28 July 2005 described here. Clear sky conditions persisted during the morning hours across the state of OK on both days while clouds developed over many of the sites during the afternoon of 4 July 2003, and remained clear on 28 July 2005. Initial profiles of potential temperature, u- and v-component wind, and specific humidity for 06 UTC were obtained from an MM5 4 km horizontal resolution forecast. Initial conditions for the MM5 were specified at 00 UTC for each day from the NCEP Eta Data Assimilation System 40-km analyses, and the Blackadar boundary layer scheme and a five-layer soil model were implemented during the simulation. The closest grid point to the mesonet station was used. The land use category specified within the MM5 was used to determine the initial land surface specifications for 1DBLM, namely moisture availability, thermal inertia, and roughness length. Table 1 provides values of pertinent parameters for each site. The observed net shortwave radiation, as calculated from the explicit tower measurements of the net radiation components, was used as input to 1DBLM to ensure consistency in the amount of solar radiation input to the system. This also allows us to represent the decrease in incoming radiation resulting from the presence of clouds.

A control (CTRL) simulation was performed at each location for which the model is allowed to run using only the observed net shortwave radiation and the given initial parameters. An assimilation simulation (ASSIM) was then performed for which the MCN94 moisture availability adjustment takes place over a three hour period from 1300UTC to 1500UTC. The surface skin temperatures from the IRT were used for the assimilation.

Figure 4 shows the CTRL and ASSIM-case model predictions for surface skin temperature, 2-m temperature, sensible and latent heat flux, and wind speed as compared with observations from the NRMN site. The net radiation from both the IRT and tower are also plotted along with the model predicted value to show the relationship between the tower measurements and the model. The CTRL, using the physically consistent three-temperature method produced a time series of model skin and 2-m temperature which matches the observations reasonably well.

The two instruments at the site (IRT and the sensors on the tower) show some discrepancy with respect to measured net radiation. This is attributed to a combination of differences in the field of view and emissivity of the measurements (Jeff Basara, 2007 personal communication) and is largely attributed to differences in incident solar insolation. Since we have been using the tower measurements of insolation in this study, this discrepancy also manifests itself in the skin temperature plot. The overestimation of winds near the surface as predicted by the model is responsible for warmer 2-m temperature during the day as the surface shear stress leads to a more well mixed environment. The surface sensible and latent heat fluxes were well represented by the model, deviating from the observations on the order of 30 to 50 W/m^2. It is evident that lowering wind speed would have led to a larger sensible heat flux, and consequently, a lower latent heat flux. The relative success of the model for this case means that the climatological values for the surface parameters

Table 1

Pertinent initial model parameters for the OK and SGP model simulations. A list of symbols can be found on page xiii

Site	Lat	Lon	χ	c_s	ρ_s	λ	MA	z_o	ε	T_M	u_g	v_g
NRMN	35.24	97.46	0.04	1256	1850	1.23	0.30	0.15	1.00	298	−7.24	4.40
FORA	36.84	96.43	0.04	1256	1850	1.23	0.30	0.15	1.00	295.7	−2.69	2.97
SGPCF	36.62	97.50	0.04	1256	1850	1.23	0.30	0.15	1.00	295.0	−2.10	−2.31

χ Thermal inertia (Jm^{-2}K^{-1}s^{-5})

c_s Specific heat capacity of soil (Jkg^{-1}k^{-1})

ρ_s Soil density (kgm^{-3})

λ Thermal conductivity (Jm^{-1}K^{-1}s^{-1})

MA Moisture availability (fraction)

z_O Surface roughness (m)

ε Surface emissivity (fraction)

T_M Sublayer temperature (K)

u_g West-east geostrophic wind component (m/s)

v_g South-north geostrophic wind component (m/s)

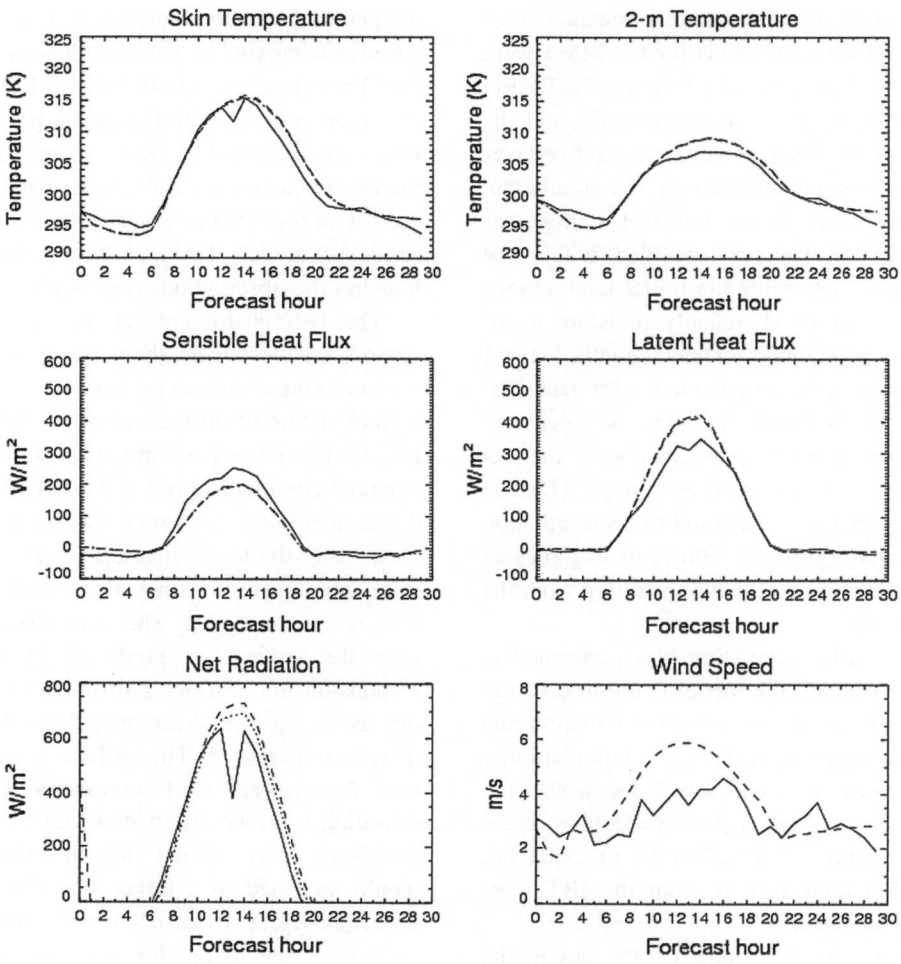

Figure 4

Observed (*solid*), CTRL simulated (*dashed*), and ASSIM simulated (*dash-dot*) variables valid 06 UTC 4 July 2003 to 12 UTC 5 July 2003 for NRMN. Net Radiation as measure by the IRT (*solid*) and Tower instruments (*dotted line*) are given

including roughness length, moisture availability, and thermal inertia are representative for this particular case. Since the observed rise of skin temperature during the assimilation period was well represented by the CTRL run, there was little adjustment made within ASSIM, as indicated by little visible difference between the CTRL and ASSIM time series lines. This case shows that 1DBLM is able to provide a reasonable representation of the surface to boundary layer interactions and boundary layer evolution. It also indicated the limitation of the assimilation technique as it is unable to correct errors in partitioning of fluxes and the Bowen ratio when these errors are due to other factors (e.g., wind speed) than surface moisture.

Figure 5 shows the same parameters as Fig. 4 except for the FORA site. FORA exhibited a surface that was fairly moist, resulting from convective precipitation on the previous days. As a result, the CTRL, using the MM5 specified land use parameters, simulated a skin temperature which was much warmer than observed during the daytime, and a 2-m temperature curve which was slightly warmer than observed. The CTRL, using the MM5 specified land use parameters, simulated skin temperatures which were much warmer than observed, and a 2-m temperature curve which was slightly warmer than observed. The CTRL simulated a maximum skin temperature 6 K warmer than the observation. In response, the CTRL simulated sensible heat flux was over predicted, and the latent

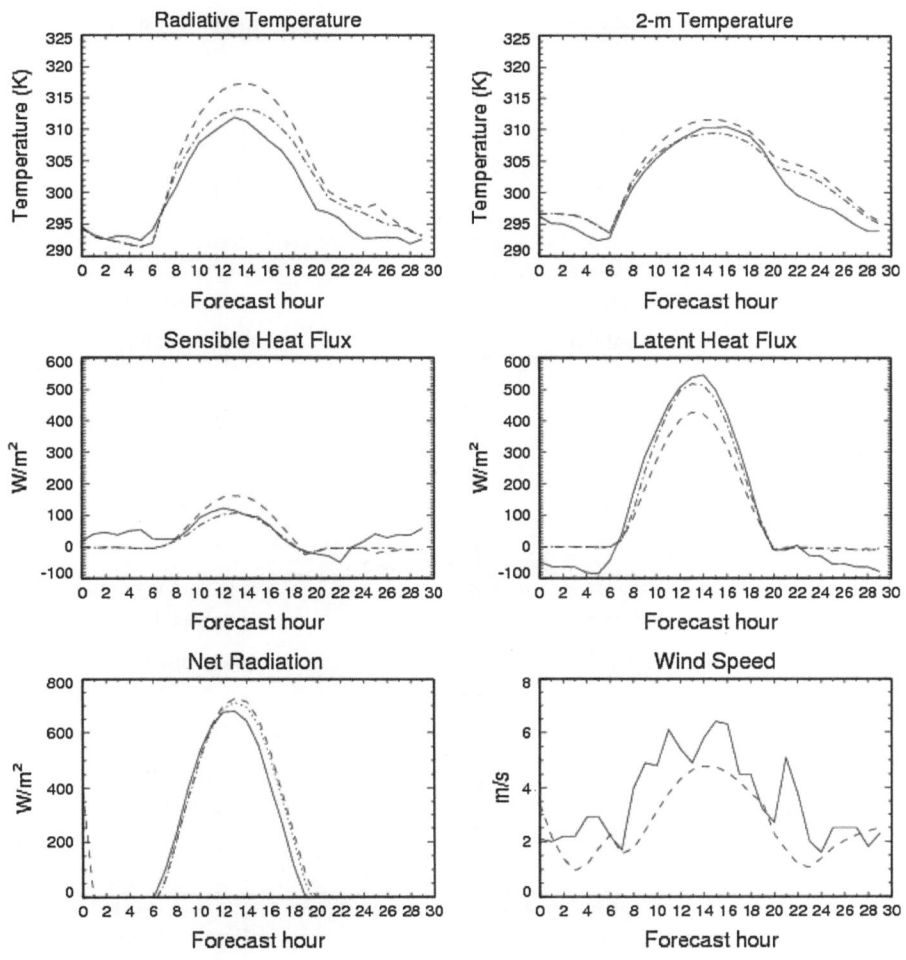

Figure 5
Observed (*solid*), CTRL simulated (*Dashed*), and ASSIM simulated (*dash-dot*) variables valid 06 UTC 4 July 2003 to 12 UTC 5 July 2003 for FORA. Net Radiation as measure by the IRT (*solid*) and Tower instruments (*dotted line*) are given

heat flux under predicted. When the MCN94 technique is applied, the moisture availability is increased to 0.48 and a 5 K improvement in the maximum model skin temperature is observed. In response, the sensible heat flux is reduced and the latent heat flux is increased, nearly matching the observations. The success of the assimilation technique in this case is also due to the validity of the assumptions made in the technique. The model predicted wind speed show a much better agreement with the observations and, therefore, most of the error in the Bowen ratio is due to the specification of the surface moisture in the model.

Figure 6 shows the same information as Figs. 4–5 but for the SGP-CF site. The CTRL simulated a skin temperature which reached a daytime maximum 6 K cooler than observed. This is in part attributed to an over-representation of the initial moisture availability as indicated by the extremely dry surface shown through the measured fractional water index surrounding the SGP CF site for 29 July 2005. Consequently the sensible heat flux was under predicted, by about 50%, and the latent heat flux over predicted, by about twice the observation. When the MCN94 technique is applied, the moisture availability is reduced, resulting in a 2.5 K improvement to the maximum daytime model skin temperature. In addition, both the sensible and latent heat flux is nudged towards the observations. However, the results still leave differences in the key flux values.

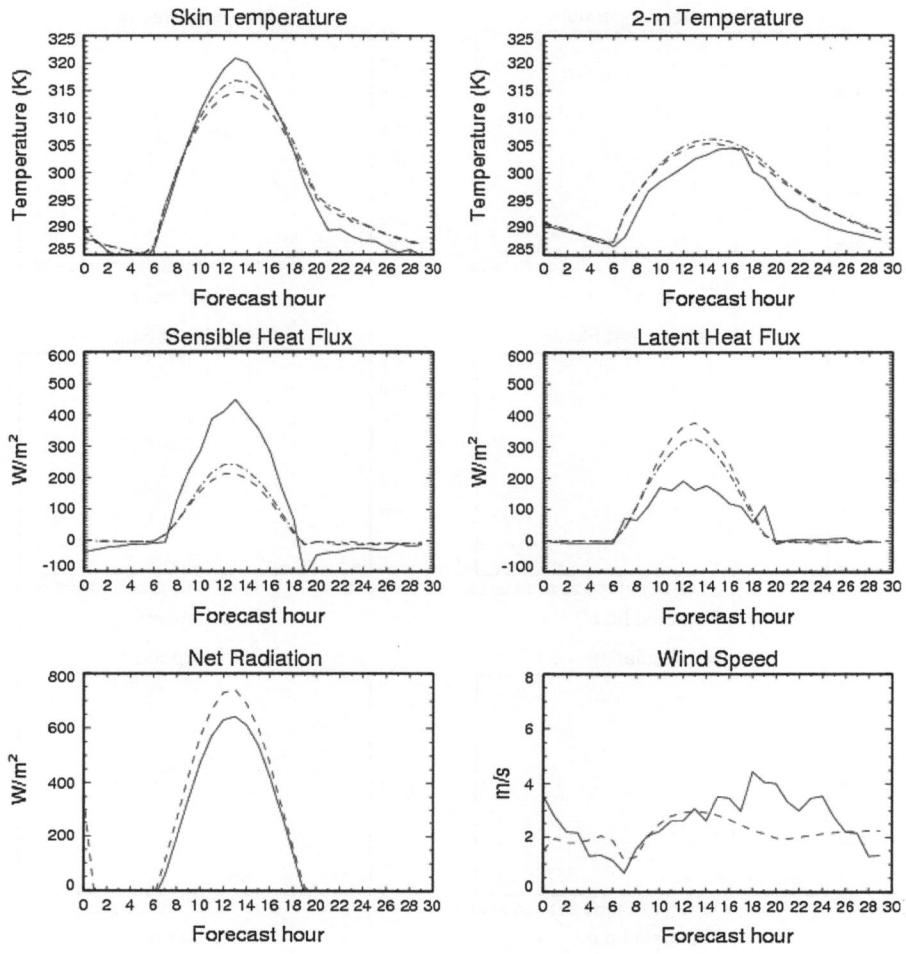

Figure 6
Observed (*solid*), CTRL simulated (*dashed*), and ASSIM simulated (*dash-dot*) variables valid 06 UTC 29 July 2005 to 12 UTC 5 July 2003 for SGP-CF. Net Radiation as measure by the IRT (*solid*) and Tower instruments (*dotted Line*) are given

The assimilation technique as implemented does not retrieve the exact observed skin temperature or even the rate of temperature change, but does nudge the results in the correct direction. In addition, both the predicted sensible and latent heat fluxes partitioning were nudged in the correct direction. It has never been a requirement that the MCN94 technique provide exact results, as one of the major assumptions made in the analytical technique is that all other model parameters remain constant when the required moisture is analytically retrieved. As discussed originally in MCN94 and further below, this assumption does not hold in that new values of moisture lead to new balances in the surface energy budget. MCN94 applied iteration to balance the system.

After the relative success of the 4 July 2003 cases, the inability of 1DBLM with the moisture availability adjustment active to replicate the large heating rates through the entire heating period of 29 July 2005 lead to an investigation that examined how the surface energy balance was reacting to the changes in moisture and whether the iteration strategy employed was working. This is described in the following section.

5. Energy Balance Considerations

The cases above show that the use of observational data can lead to better simulations of an Earth-atmosphere system. The observations are not used directly, but rather indirectly to provide some difference measure. Simply replacing parameters within a model system with observations, while conceptually seems the easiest thing to do, can be problematic since actual values of those parameters are typically a product of a different equilibrium state. A direct replacement can shock a system such that it may lead to numerical instability and ultimately a model crash. For this reason investigators began by attempting to nudge a model state towards observations.

Early problems with attempts to nudge surface temperatures directly into the surface energy budget were recognized by STAUFFER et al. (1991). They encountered difficulties in that the terms in the energy budget no longer supported the temperatures being inserted. Thus, the system became unbalanced leading to numerical problems and erroneous results. MCN94, building on the experience of STAUFFER et al. (1991), recognized that direct insertion of skin temperature would not work. Rather MCN94, following WETZEL and WOODWARD (1987) and CARLSON et al. (1981), took the approach that fundamental parameters within the surface energy budget such as surface moisture availability or heat capacity must be changed to support the desired temperature result. In this regard they analytically inverted a surface energy model to solve for the moisture availability that would produce the desired temperature change in the morning hours. McNIDER et al. (2005) employed a similar technique to solve for the slab heat capacity that would give the desired temperature response in the early evening. Other investigators such as JONES et al. (1998a, b) have used this analytical inversion to solve for other parameters in more complex surface models to solve for stomatal resistance. However, in all of these analytical inversions the assumption is that other terms remain the same as when the desired parameter is recovered. In reality, once one parameter is changed other terms in the budget equation also change. Thus, attention is required in solution techniques and iteration strategies so that the desired recovery is made.

An examination of the observed, control model, and assimilation simulated heating rate for the SGP-CF case uncovered interesting findings. Figure 7 gives a graphic representation of these heating rates along with the subsequent impact on the skin temperature. The high frequency changes of the surface energy balance

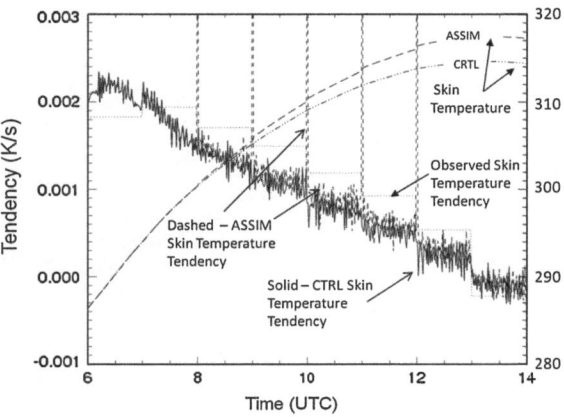

Figure 7

Time series of skin temperature tendency from CTRL (*solid*), ASSIM (*small dash*), and observations (*dotted*) along with time series of skin temperature from CTRL (*dash dot*) and ASSIM (*large dash*). The left axis shows the temperature tendency and the right axis temperature. The assimilation is implemented every 60 minutes

can be seen in the time series of temperature tendencies. As expected, the model simulated heating rate was less than the observations. When the moisture availability adjustment takes place, the result is a jump in the surface temperature tendencies and an increase in skin temperature. However, the model very quickly returns to a heating rate close to its previous one. Even with the jump in tendency, it can be seen that the average hourly tendency would be below the observation. In order to test the impact of increasing the assimilation frequency, a test case was produced where the moisture availability was adjusted every 2 minutes. Figure 8 provides the same information as Fig. 7, but for this higher frequency assimilation run. The result is an hourly average tendency that is more closely matched to the observation. A portion of the hourly tendency is much less than previously found resulting from the fact that the observed tendency was calculated from hourly observations. While the time series of tendencies appears to be quite chaotic, the result on the surface skin temperature and fluxes was much improved over the hourly adjustment. This can be seen in Fig. 9 which indicates skin temperatures that nearly match the observations. In addition to temperature, the other surface fluxes were further nudged towards observation thus the assimilation of new information is improving the model performance. We believe they may be further improved if more attention is given in

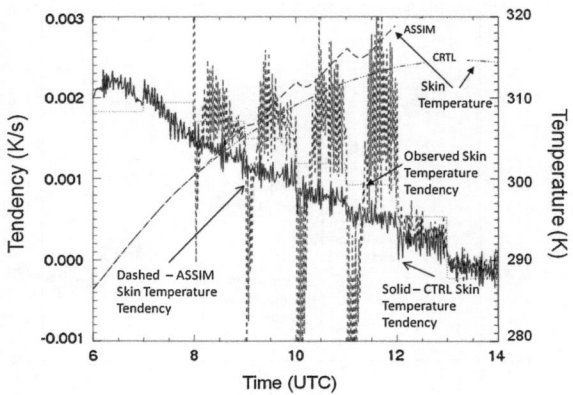

Figure 8
Time series of skin temperature tendency from CTRL (*solid*),
ASSIM (*small dash*), and observations (*dotted*) along with time
series of Skin temperature from CTRL (*dash dot*) and ASSIM
(*large dash*). The left axis shows the temperature tendency and the
right axis temperature. The assimilation is implemented every
2 minutes

ground temperature was calculated with the new
retrieved moisture flux. These new fluxes were then
inserted back into the ground temperature prognostic
equation. In fact, this continued until the sensible heat
flux converged (see the flow chart in MCN94). At this
point the model exited the surface energy budget
calculations. In the MM5 implementation iteration
was not carried out. But, rather than directly inserting
the retrieved moisture flux it was nudged into the
system. In the present implementation we decided not
to nudge and not to iterate. Rather, we assumed with
the relatively small time steps that the system would
adapt over time naturally to the new balances. Figures 7 and 8 show that while the system is apparently
stable, there are substantial oscillations. The lack of
balance in the system may be inhibiting the full
assimilation of the new retrieved moisture fluxes.
These results indicate that either iteration or nudging
may be needed to better balance the system.

6. *Future Work: Operator Splitting*

While iteration within the surface energy budget
calculations may reduce some of the oscillations, it

correcting the boundary layer processes responsible
for the near surface temperature representation and
winds as discussed in the following section.

In the original implementation of the skin temperature assimilation in MCN94, sensible ground and
outgoing radiation fluxes were changed as the new

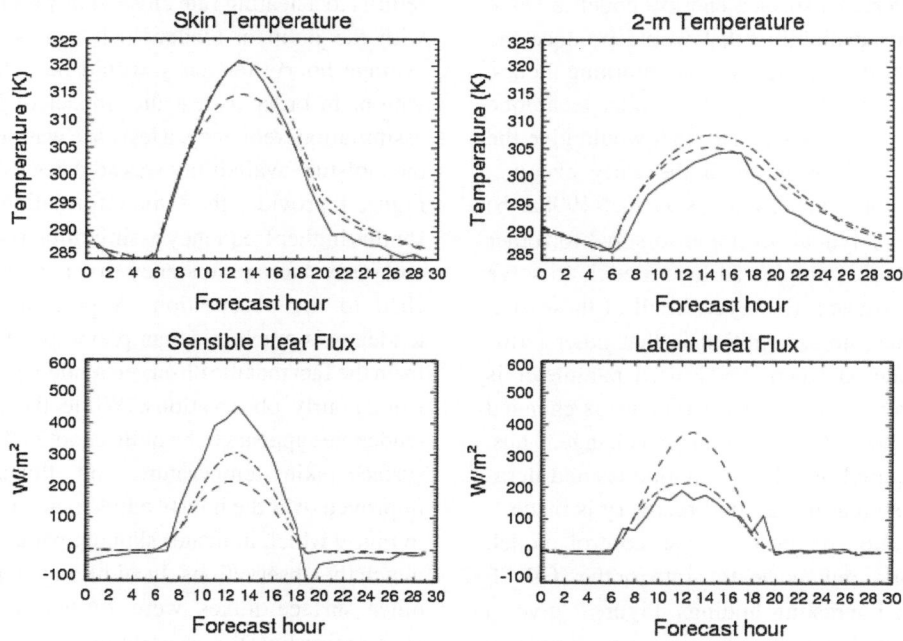

Figure 9
Observed (*solid*), CTRL simulated (*dashed*), and ASSIM simulated (*dash-dot*) variables valid 06 UTC 29 July 2005 to 12 UTC 5 July 2003 for
SGP-CF applying the MCN94 assimilation technique every 2 minutes

appears that the major source of the oscillations shown in Figs. 7 and 8 are a result of operator splitting between the surface energy budget calculations and the atmospheric diffusion step. Compared to mathematicians, atmospheric scientists have been relatively cavalier in carrying out operator splitting in building atmospheric models. Consider the way this operator is split within the 1DBLM and in most all mesoscale boundary layer models. The surface energy budget is used to provide a bottom flux condition in the vertical diffusion step. This is usually written for the first atmospheric level in finite difference form for an explicit difference scheme as

$$T_j^{i+1} = T_j^i + K_z[(T_{j-1}^i - T_j^i)/\Delta z - sfcflux^i)]/\Delta z \tag{15}$$

Here it assumed that the *sfcflux* was calculated using the previous time step. However, since the surface flux calculation involves T_j^i rather than T_j^{i+1} there is an inconsistency. In examining the terms in the surface energy budget there is an oscillation introduced in that in the next time step *sfcflux* changes because T_j^{i+1} has changed. This oscillation then interferes with the calculation of the appropriate moisture flux in the assimilation step.

In order to avoid these numerical oscillations the problem could be split differently. Rather than splitting the problem between the surface and atmosphere, perhaps at least the first layer temperature should be calculated at the same time as the other terms in the surface energy budget. This would allow an iteration to converge to a solution for both the new skin temperature and a new air temperature in the same way iterations (see Fig. 3) are used to adjust the fluxes due to a new skin temperature. This will be the focus of future research.

7. Summary

This paper has outlined the importance of consistent use of skin and aerodynamic temperature representation and iteration processes when incorporating surface information from satellite data into mesoscale models. The three-temperature system used here can be directly implemented within currently forecasting systems in which the boundary layer parameterization can interact with a slab land surface model. While this is a limited set of model configurations, the implications of inconsistent temperature use can be applicable to all situations where surface information from satellites is used.

The MCN94 satellite assimilation technique is applied in this study. In light of research that took place since its development, some of the problems faced when incorporating this technique within the MM5 were alleviated. This was accomplished by using the aforementioned three-temperature system so that satellite observations could be directly compared with a physically consistent model simulated counterpart. The results of the tests look promising, and provide results similar to those originally produced in MCN94 but in a physically consistent system.

The impact of making adjustments to parameters within a surface energy balance on that balance was discussed. Results from test cases indicate how sensitive the system is to these changes and why iteration is necessary. Second, the generally accepted framework of splitting the surface energy budget calculations from the vertical diffusion operator may not be appropriate. Additional work on splitting the operators differently is needed. We believe that the issues discussed here related to balancing the energy budget in conjunction with the fluxes to the atmosphere may be critical in any land surface system in which observations are inserted into the system.

Acknowledgments

This work was supported in part by the Minerals Management Service Prepared under MMS Contract 1435-01-04-CA-37002/M05AC12302.

References

ANDERSON, M. C., J.M. NORMAN, G.R. DIAK, W.P. KUSTAS, and J.R. MECIKALSKI , 1997. *A Two-Source Time-Integrated Model for Estimating Surface Fluxes Using Thermal Infrared Remote Sensing*. Remote Sens. Env. *60*:195-216.

BELJAARS, A. C. M. and A. A. M. HOLTSLAG. 1991. *Flux Parameterization over Land Surfaces for Atmospheric Models*. Journal of Applied Meteorology *30*:327-341.

BLACKADAR, A. K. 1979. *High Resolution Models of the Planetary Boundary Layer*. Adv. Environ. Sci. Eng. *1*:50-85.

BROCK, F. V., K. C. CRAWFORD, R. L. ELLIOTT, G. W. CUPERUS, S. J. STADLER, H. L. JOHNSON, and M. D. EILTS, 1995: *The Oklahoma*

Mesonet: A Technical Overview. Journal of Atmospheric and Oceanic Technology, *12*, 5-19.

BROTZGE, J. A. and C. E. DUCHON, 2000: *A Field Comparison among a Domeless Net Radiometer, Two Four-Component Net Radiometers, and a Domed Net Radiometer.* Journal of Atmospheric and Oceanic Technology, *17*, 1569-1582.

CARLSON, T. N. 1986. *Regional Scale Estimates of Surface Moisture Availability and Thermal Inertia Using Remote Thermal Measurements.* Remote Sens. Rev. *1*:197-246.

CARLSON, T. N., J. K. DODD, S. G. BENJAMIN, and J. N.COOPER. 1981. *Satellite Estimation of the Surface Energy Balance, Moisture Availability and Thermal Inertia.* Journal of Applied Meteorology 20:67-87.

CHEN, F., K. E. MITCHELL, J. SCHAAKE, Y. XUE, H.-L. PAN, V. KOREN, Q. Y. DUAN, M. EK, and A. BETTS, 1996. *Modeling of land-surface evaporation by four schemes and comparison with FIFE observations,* J. Geophys. Res., *101*,7251–7268, 1996.

CHOUDHURY, B. J., R. J. REGINATO, and S. B. IDSO. 1986. *An Analysis of Infrared Temperature Observation of Wheat and Calculation of Latent Heat Flux.* Agric. For. Meteor. *37*:75-88.

DORMAN, J. L., and P. J. SELLERS. 1989. *A global climatology of albedo, roughness length and stomatal resistance for atmospheric general circulation models as represented by the Simple Biosphere model (SiB),* J. Appl. Meteorol., *28*, 833–855.

DEARDORFF, J. W. 1972. *Theoretical expression for the countergradient vertical heat flux.* J. Geophys. Res. 77:5900-5904.

EK, M.B., K. E. MITCHELL, Y. LIN, E. ROGERS, P. GRUNMANN, V. KOREN, G. GAYNO, and J. D. TARPLEY, 2003: *Implementation of Noah land surface model advances in the National Centers for Environmental Prediction operational mesoscale Eta model.* J. Geophy. Res., VOL. *108*, NO. D22, 8851, doi:10.1029/2002 JD003296.

FIEBRICH, C. A., J. E. MARTINEZ, J. A. BROTZGE, and J. B. BASARA. 2003: *The Oklahoma Mesonet's Skin Temperature Network.* Journal of Atmospheric and Oceanic Technology, *20*, 1496-1504.

JONES, A. S., GUCH, I. C., and VONDER HAAR, T. H. 1998a. *Data Assimilation of Satellite-Derived Heating Rates as Proxy Surface Wetness Data into a Regional Atmospheric Mesoscale Model. Part I: Methodology.* Monthly Weather Review 126:634-645.

JONES, A. S., I. C. GUCH, and T. H. VONDER HAAR, 1998b. *Data Assimilation of Satellite-Derived Heating Rates as Proxy Surface Wetness Data into a Regional Atmospheric Mesoscale Model. Part II: A Case Study.* Monthly Weather Review 126:646-667.

KUSTAS, W. P., B.J. CHOUDHURY, M.S. MORAN, R.J. REGINATO, R. D. JACKSON, L.W. GARY, and H.L. WEAVER. 1989. *Determination of Sensible Heat Flux Over Sparse Canopy Using Thermal Infrared Data.* Agric. For. Meteor. 44:197-216.

LAPENTA, W. M., R. J. SUGGS, G. J. JEDLOVEC, and R. T.MCNIDER. 1999: Impact of Assimilating GOES-derived Land Surface Variables Into the PSU/NCAR MM5. *Preprints, Workshop on Land-Surface Modeling and Applications to Mesoscale Models.*

LHOMME, J. P., N. KATERJI, A. PERRIER, and J. M. BERTOLINI. 1988. *Radiative Surface Temperature and Convective Flux Calculation Over Crop Canopies.* Boundary-Layer Meteorology 43:383-392.

MCCUMBER, M. C., and R. A. PIELKE. 1991: *Simulation of the effects of surface fluxes of heat and moisture in a mesoscale numerical model: 1.* Soil Layer, J. Geophys. Res., *86*, 9929–9938.

MACKARO, S. 2003. Applications of Land Surface Data Assimilation to Simulations of Sea Breeze Circulations. M.S. Thesis, Atmospheric Science Department, University of Alabama in Huntsville. 86 pp.

MCNIDER, R.T., A.J. SONG, D.M. CASEY, P.J. WETZEL, W.L. CROSSON, andR.M. RABIN. 1994. *Toward a dynamic-t thermodynamic assimilation of satellite surface temperature in numerical atmospheric models.* Mon. Wea. Rev. *122*:2784-2803.

MCNIDER, R. T., W.M. LAPENTA, A. BIAZAR, G. JEDLOVEC, R. SUGGS, and J. PLEIM. 2005: *Retrieval of grid scale heat capacity using geostationary satellite products: Part I: Case-study application,* J. Appl. Meteor., *88*, 1346-1360.

NORMAN, J. M., W. P. KUSTAS, K. S. HUMES. 1995. *A two-source approach for estimating soil and vegetation energy fluxes in observations of directional radiometric surface temperature.* Agricultural and Forest Meteorology 77:263–293.

PASQUI, M., C. J. TREMBACK, F. MENEGUZZO, G. GIULIANI, and B. GOZZINI. 2004: A soil moisture initialization method, based on antecedent precipitation approach, for regional atmospheric modelingsystem: a sensitivity study on precipitation and temperature,in: Proceedings of the 18th Conference on Hydrology, AMS, Seattle.

PLEIM, J. and A. XIU. 1995: *Development and testing of a surface flux and planetary boundary layer model for application in mesoscale models.* J. Appl. Meteor., *34*, 16-32.

STAUFFER, D. R., N. L. SEAMAN, and F. S. BINKOWSKI. 1991: *Use of Four-Dimensional Data Assimilation in a Limited-Area Mesoscale Model Part II: Effects of Data Assimilation within the Planetary Boundary Layer.* Monthly Weather Review, *119*, 734-754.

STOKES, M., G., and S. E. SCHWARTZ. 1994: *The Atmospheric Radiation Measurement (ARM) Program: programmatic background and design of the cloud and radiation test bed.* Bulletin of the American Meteorological Society, *75*, 1201-1221.

SUGITA, M. and W. BRUTSAERT, 1990: *How Similar are Temperature and Humidity Profiles in the Unstable Boundary Layer.* Journal of Applied Meteorology, *29*, 489-497.

SUN, J. and L. MAHRT. 1995. *Determination of Surface Fluxes from the Surface Radiative Temperature.* Journal of the Atmospheric Sciences 52:1096-1106.

WETZEL, P. J. and R. H. WOODWARD. 1987. *Soil Moisture Estimation Using GOES-VISSR Infrared Data: A Case Study with a Simple Statistical Method.* Journal of Applied Meteorology 26:107-117.

WETZEL, P. J., D. ATLAS, and R. H. WOODWARD. 1984. *Determining Soil Moisture from Geosynchronous Satellite Infrared Data: A Feasibility Study.* Journal of Applied Meteorology 23:375-391.

ZILITINKEVICH, S. S. 1970. Dynamics of the Atmospheric Boundary Layer. Leningrad Gidrometeor. 291 pp.

(Received December 25, 2010, accepted March 2, 2011, Published online July 20, 2011)

Pure Appl. Geophys. 169 (2012), 415–424
© 2011 Springer Basel AG
DOI 10.1007/s00024-011-0378-z

Improving Scatterometry Retrievals of Wind in Hurricanes Using Non-Simultaneous Passive Microwave Estimates of Precipitation and a Split-Step Advection/Convection Model

ALEX FORE,[1] ZIAD S. HADDAD,[1] T. N. KRISHNAMURTI,[2] and ERNESTO RODGRIDEZ[1]

Abstract—One of the current problems in the accurate estimation of over-ocean wind from scatterometry observations is the proper accounting for precipitation. Specific cases such as hurricanes are particularly difficult, because precipitation in the eye wall and rain bands can be quite heavy, and therefore, affect the scatterometer signatures so drastically that a category-4 hurricane can appear, to the scatterometer, to have category-1 winds. We have developed an approach to infer and account for the signature of the precipitation from non-simultaneous passive-microwave measurements of rain, with the help of geostationary IR measurements. In this note, we describe the basic approach, and the results of applying it to the data taken by the Tropical Rainfall Measurement Mission Microwave Imager measurements several hours before and after the QuikSCAT observation of Hurricane Rita in September 2005. We also describe how we are enhancing the approach with more realism in the assimilation of the IR information.

1. Introduction

One of the current problems of ocean-vector-wind estimation from scatterometry observations is the proper accounting for precipitation. Over severe weather systems with areas of high precipitation, such as tropical cyclones, the measurements of the ocean radar backscattering cross-section are affected by the signature of the rain, with backscattering from the leading edge of the precipitation (relative to the radar line of sight) and two-way attenuation of the radar wave on its way to and back from the rainy surface. Instantaneous estimates of the precipitation

This work was performed at the Jet Propulsion Laboratory, California Institute of Technology, under contract with the National Aeronautics and Space Administration, and at Florida State University under National Science Foundation grant number 0533108.

[1] Jet Propulsion Laboratory, California Institute of Technology, Pasadena, CA, USA . E-mail: zsh@jpl.nasa.gov
[2] Department of Meteorology, Florida State University, Tallahassee, FL, USA.

that are necessary to properly account for the rain effects are typically only available significantly earlier or later than the instantaneous scatterometer measurements. One way to use these measurements is to assimilate them into a high-resolution cloud model, whose analyzed fields could then be used as the background for the scatterometry observations. In this paper, we describe our first attempt to implement a simplified approach that replaces the meso-scale model with advection fields derived from all the geo-stationary IR observations that are available between the times of the precipitation estimates and the time of the scatterometer measurements. We then describe how this approach can be enhanced with a one-dimensional vertical model step, with a result which is not as detailed or realistic as a variational assimilation into a cloud-resolving model but which has the advantage of preserving the total water content.

We developed the approach for the Seawinds radar on NASA's QuikSCAT satellite. Seawinds is a Ku-band scatterometer which is in a near-polar orbit, and had a swath width of 1,800 km which covered 90% of the Earth's surface every day from July 1999 until November 2009 (SPENCER *et al.*, 1997). Because of the centimeter-scale wavelength of Seawinds, it is particularly sensitive to capillary waves on the surface of the ocean. These capillary waves can be caused by surface stress at the interface between the water and air due to wind. The signature of these waves in the radar backscattering cross-section is the basis for the retrieval of wind speed and direction from Ku-band measurements taken at different angles, as was the case for Seawinds. Rain, however, can confound the measurement through attenuation of the surface scattering signal and direct scattering from the rain drops.

The next three sections investigate our approach to estimate and remove the effects of rain using non-simultaneous passive radiometer observations. Section 5 describes how this approach can be extended in a split-step procedure, to account for convection as well as advection during the time intervals that separate the scatterometer observation from the rain estimates.

2. Model Function

Ku-band surface back-scattering cross-sections are affected in three different ways due to rain (Draper and Long, 2004; Hilburn et al., 2006; Stiles and Yueh, 2006; Yueh et al., 2003): We expect to see attenuation of the backscatter due to the ocean surface as it propagates through the rain, additional backscatter due to the rain drops, and additional surface scattering due to roughness caused by rain drops (which we will hereafter neglect). Thus we formulate the following radiative transfer model

$$\sigma_0^{obs}(\vec{v}, R) = \sigma_0^{wind}(\vec{v}) e^{-2\int_0^{\rho_s} k(R)d\rho}$$
$$+ e^{-2\int_0^{\rho_s} k(R)d\rho} \int_{\rho_s}^{\rho_1} \sigma_0^{rain}d\rho \qquad (1)$$

where σ_0^{wind} is the backscatter[1] due to the ocean surface, k is the attenuation due to rain, R is the rain rate (understood to be a function of $z = \rho \cos\theta_{inc}$), $\rho_1 = c/t_1$ is the range when the range-gate is opened at t_1, and σ_0^{rain} is the radar cross-section per unit volume due to rain drops. We simplify Eq. (1)

$$\sigma_0^{obs}(\vec{v}, R) = A(R)\sigma_0^{wind}(\vec{v}) + \sigma_0^{rain}(R), \qquad (2)$$

where $A(R)$ is the path attenuation due to rain (Fig. 1).

3. Estimation of A and σ_0^{Rain}

Because of the similar frequency of Seawinds, 13.4 GHz, and the Tropical Rainfall Measurement

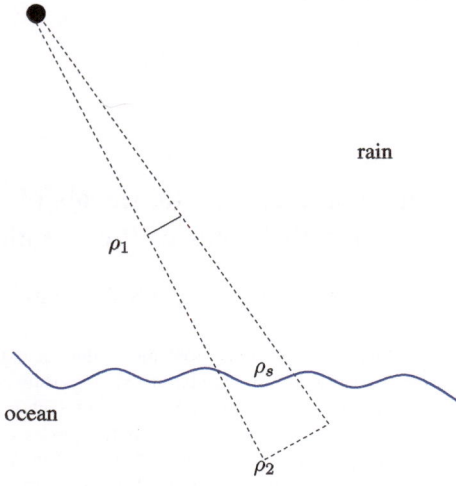

Figure 1
Diagram of Seawinds measurement process. The range gate is opened at $t_1 = \rho_1/c$, and closed at $t_2 = \rho_2/c$. Within the trapezoid between ρ_1 and ρ_2 one expects backscatter due to rain drops, as well as backscatter from the surface of the ocean attenuated by the rain

Mission (hereafter referred to as TRMM) Precipitation Radar (hereafter called PR), 13.8 GHz, we use TRMM data to develop empirical corrections for path attenuation and rain backscatter. The PR radar gives estimates of the two-way path attenuation in the 2A21 data products, and radar reflectivity factors in the 1C21 products. The 2A21 products contain an estimate of the two-way path attenuation and we only need to correct for the different viewing geometry of Seawinds compared to TRMM PR. Since the attenuation is proportional to $e^{\int dl k(z)}$ and $dl = dz/\cos\theta_{inc}$, to correct for the different incidence angle of PR and Seawinds we compute $\ln A_{QS} = (\cos\theta_{PR}/\cos\theta_{QS})\ln A_{PR}$. From the reflectivities we compute an estimate of σ_0^{rain} as

$$\sigma_{0,PR}^{rain} = 10^{-10}\frac{\pi^5}{\lambda^4}|K_w|^2 \int_{\rho_1}^{\rho_s} Z^{1C21}(\rho)d\rho \qquad (3)$$

which has units of inverse distance. Here, $K_w \approx 0.93$ is related to the dielectric properties of water, $\lambda = 0.0217$ m is the wavelength of the radar, and Z^{1C21} is the 1C21 reflectivity factor in mm^6/m^3. This estimate pertains to the particular incidence angle θ_{PR}, and in order to transform it to another incidence angle θ_{QS} we must compute $\sigma_{0,QS}^{rain} = (\cos\theta_{PR}/\cos\theta_{QS})\sigma_{0,PR}^{rain}$.

---

[1] Backscatter relative to a scatterer that reflects the received power uniformly over 4π steradians, thus in this document σ_0 will always be dimensionless.

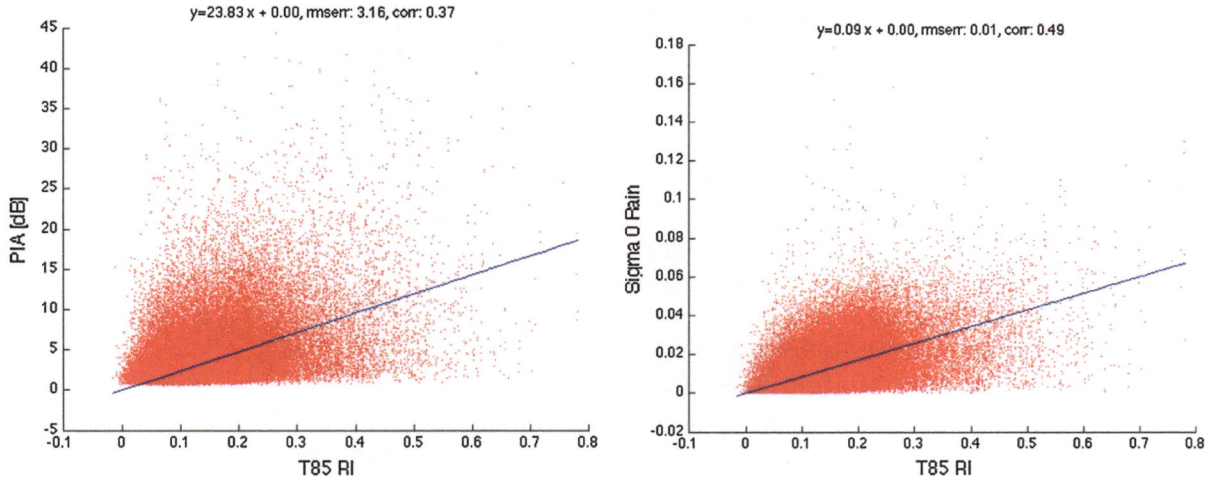

Figure 2
Co-located TRMM TMI observations of T_{85}^{RI} and PR derived estimates of PIA and σ_0^{rain}

The TRMM satellite also hosts the TRMM Microwave Imager (TMI), and we use a database of co-located TRMM/TMI 1B11 and TRMM/PR (2A21, and 1C21) observations of North Atlantic hurricanes to train the TMI brightness temperatures to retrieve σ_0^{rain} and A. Figure 2 illustrates co-located TMI observations and PR derived estimates of PIA and σ_0^{rain}. Here, we use the 85 Ghz rain indicator, defined as

$$T_{85}^{\mathrm{RI}} := 1 - T_{\mathrm{PCT}}^{85}/T_{\mathrm{PCT,clear}}^{85}, \qquad (4)$$

where $T_{\mathrm{PCT}}^{85} = 1.818\,T_v^{85} - 0.818\,T_h^{85}$, and $T_{\mathrm{PCT,\ clear}}^{85}$ is the rain-free T_{PCT}^{85} in the vicinity of the storm. From the data plotted in Fig. 2 we derived the following least-squares fits

$$10\log_{10} A = 23.8 T_{85}^{\mathrm{RI}}, \text{ with r.m.s. error } = 3.16,$$
$$\sigma_0^{\mathrm{rain}} = 0.0859 T_{85}^{\mathrm{RI}}, \text{ with r.m.s. error } = 0.0098. \qquad (5)$$

4. Modified Retrieval

4.1. Affine Transformation

TRMM has a low-inclination non-sun-synchronous orbit. This implies that TRMM's instantaneous rain estimates of a particular weather system almost never coincide with observations by the sun-synchronous polar orbiters such as QuikSCAT. Therefore, the main challenge in making any TRMM estimates relevant to the scatterometer observations is to develop an approach to estimate the evolution of the rain fields (or of their distribution) from the time they were made until the time of the scatterometer observation. The approach that we illustrate here proceeds by advecting the A and σ_0^{rain} fields in two dimensions only, following the estimates of cloud motion the we derive from consecutive geostationary IR images.

Indeed, we transform the TMI observations to the Seawinds time by piecewise affine transformations generated from a sequence of GOES-12 geostationary IR images. The latter have a high temporal (30 min) and spatial resolution. The top left and right panels of Fig. 3 show two GOES-12 images of Hurricane Rita on 21 September 2005, one at 11:15 UTC and the other at 11:45 UTC. The general shape of the storm during the 30 min interval is largely unchanged allowing us to synthesize a continuous transformation between the two images, in a procedure similar to "CMORPH" (JOYCE et al., 2004). From each successive pair of GOES-12 images, one can generate an offset vector for every pixel in the earlier image by maximizing the correlation of a neighborhood around that pixel to a region in the later image. From this set of offset vectors we perform a weighted least-squares fit of the following affine transformation

$$\begin{pmatrix} lat' \\ lon' \end{pmatrix} = \begin{pmatrix} a & b \\ c & d \end{pmatrix} \begin{pmatrix} lat \\ lon \end{pmatrix} + \begin{pmatrix} e \\ f \end{pmatrix}, \qquad (6)$$

to the set of offset vectors. This affine transformation then allows us to continuously "morph" one TMI observation to a later (or earlier) time using a composition of these affine transformations. The lower left plot of Fig. 3 shows the offset vectors that were computed by maximizing the correlation, while the lower right panel shows the result of applying the affine transformation derived from these offset vectors to the GOES-12 image at 1115. Figure 4 shows the result of applying the composition of transformations to a TMI observation at 0911 UTC and another TMI observation at 1541 UTC.

Given all the offset vector fields, one can use the most basic form of Kalman filtering to derive, from an initial observation (or estimate) $A^{(i)}(x, y)$ at the initial time (0911 UTC in our example) and a final one $A^{(f)}(x, y)$ at the final time (1541 UTC in our example), an optimal estimate of A_t at the required

time t. Indeed, if one assumes that the only deterministic dynamics governing A are the horizontal advection—which one can derive (however, approximately) from the correlation-maximizing offset vector fields described above—and that the random behavior of A_t is Brownian, the optimal estimate $\hat{A}_t$ at any time t between the initial and final times is the weighted average

$$\hat{A}_t = \frac{(t_{\text{final}} - t)\sigma_f^2}{(t - t_{\text{initial}})\sigma_i^2 + (t_{\text{final}} - t)\sigma_f^2} A^{(i)}(x_i, y_i)$$
$$+ \frac{(t - t_{\text{initial}})\sigma_i^2}{(t - t_{\text{initial}})\sigma_i^2 + (t_{\text{final}} - t)\sigma_f^2} A^{(f)}(x_f, y_f) \quad (7)$$

where (x_i, y_i) are the coordinates of the pixel one reaches by starting at (x, y) and following the offsets back to the initial time t_{initial}, and where (x_f, y_f) are

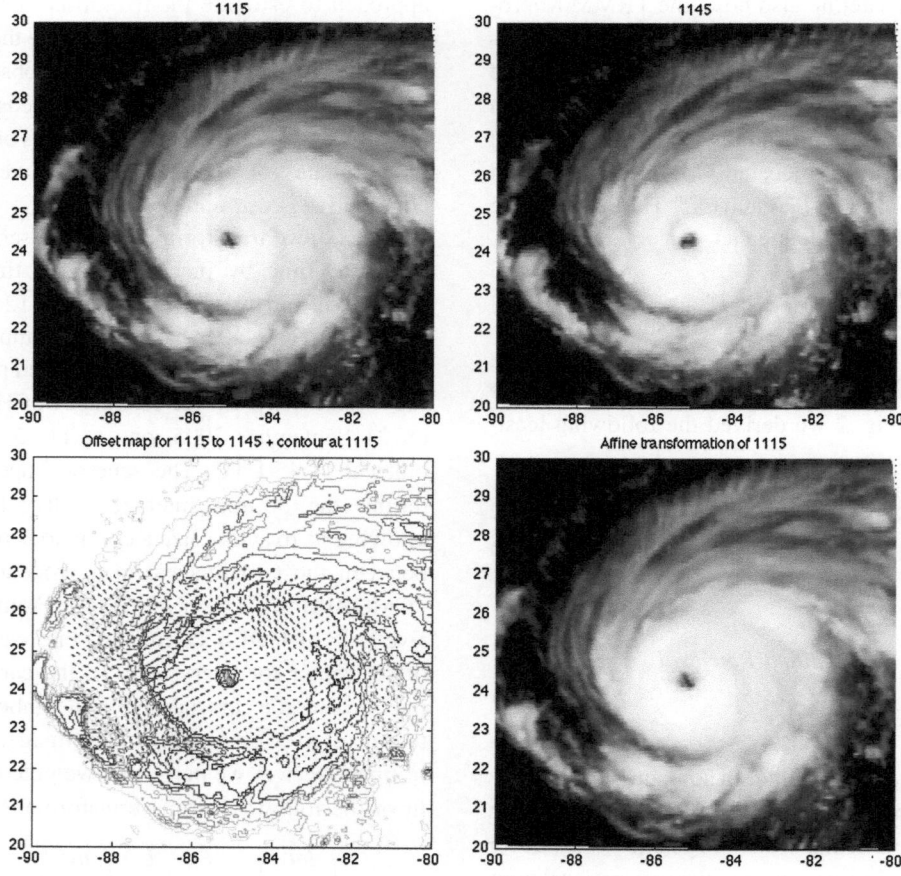

Figure 3

GOES-12 geostationary IR image of Hurricane Rita on 21 September 2005. The *top left* shows the storm at 11:15 UTC, the *top right* at 11:45 UTC. The *bottom left panel* is a contour plot of the storm at 11:15 UTC along with the offset vectors computed with the correlation maximizing routine, while the *lower right panel* shows the affine transformation of the image from 11:15 UTC

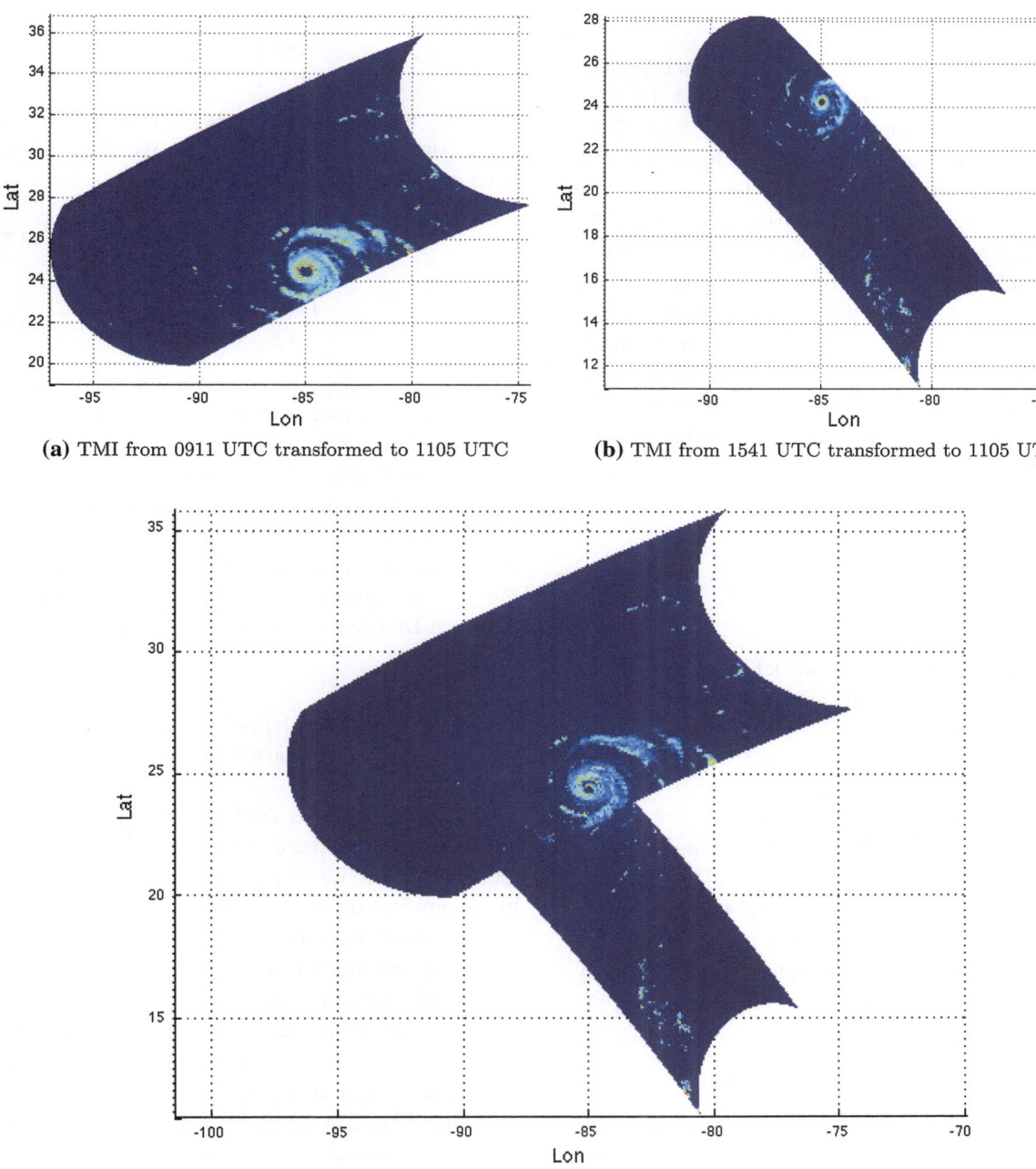

(a) TMI from 0911 UTC transformed to 1105 UTC

(b) TMI from 1541 UTC transformed to 1105 UTC

(c) Both TMI Observations transformed to 1105 UTC

Figure 4
TRMM TMI observations at 0911 and 1541 UTC translated to the Seawinds time 1105 UTC on 21 September 2005 as in Fig. 3

similarly the coordinates of the pixel one reaches by starting at (x, y) and following the offsets on to the final time t_{final}, and where σ_i and σ_f are the r.m.s. uncertainties in the initial and final microwave estimates $A^{(i)}$ and $A^{(f)}$, respectively. Note that (7) is the enhancement that the CMORPH developers applied to their approach to obtain the "Kalman CMORPH".

111

4.2. Retrieval

The Seawinds 25 km level 2A data product contains the σ_0 measurements as well as the wind vector cells. Our approach proceed by finding, for every Seawinds observation σ_0^{obs}, all translated TMI observations that lie within 25 km of σ_0^{obs} and computing the mean and variance of the TMI T_{85}^{RI}. From this mean and variance we use Eq. (5) to derive an estimate of A, and σ_0^{rain} as well as an estimate of the variance A and σ_0^{rain} for this Seawinds observation.

Next we construct the sample probability distribution function for (w, ϕ_w) for every wind vector cell as defined in the 25 km L2A data product. For the kth σ_0^{obs} contained in a wind vector cell we compute

$$p_k(w, \phi_w) \propto \sum_{ij} \exp$$
$$\times \left[-\left(10 \log_{10} \sigma_0^{\mathrm{GMF}}(w, \phi_w) - 10 \log_{10} \sigma_0^{ij}\right)^2 / \left(2\Delta\sigma^2\right)\right], \quad (8)$$

where σ_0^{GMF} is the clear air model function, w is the wind speed, ϕ_w is the wind direction relative to north, $\Delta\sigma = 0.25$, and

$$\sigma_0^{ij} = \frac{\sigma_0^{\mathrm{obs}} - \sigma_0^{\mathrm{rain,j}}}{A_i} \quad (9)$$

is the (ijth) estimate of the rain-free σ_0 due to wind. We consider a few values of A and σ_0^{rain}, estimated using the mean and variance of T_{85}^{RI} to represent the uncertainty of our estimates. We perform this addition for every σ_0 in the wind vector cell, then multiply the results of Eq. (8) together to form the pdf for the particular wind vector cell

$$p_{\mathrm{cell}}(w, \phi_w) = \Pi_{k=1,\ldots,n_{\mathrm{cell}}} p_k(w, \phi_w), \quad (10)$$

where n_{cell} is the number of σ_0 observations in the particular wind vector cell. At this point one could perform a maximum-likelihood retrieval where one picks the 4 points in (w, ϕ_w) space with the highest probability, followed by an ambiguity-removal procedure. We adopted a different approach where for every wind vector cell we assume a distribution of the wind direction based on the location of the wind vector cell relative to the eye (interpolated from the NHC best-track data (KNABB et al., 2006), for example

$$p_{\mathrm{azi}}(\phi_w) = \exp\left[-\frac{(\phi_w - \phi_0)^2}{2\Delta\phi^2}\right], \quad (11)$$

where ϕ_0 is our a priori estimate of the wind direction in the particular cell, and $\Delta\phi = 5°$. We then retrieve wind speed and wind direction as

$$w_{\mathrm{est}} = \frac{\int p_{\mathrm{cell}}'(w, \phi_w) w \, dw}{\int p_{\mathrm{cell}}'(w, \phi_w) dw}, \quad (12)$$

$$\phi_{w,\mathrm{est}} = \frac{\int p_{\mathrm{cell}}'(w, \phi_w) \phi_w \, d\phi_w}{\int p_{\mathrm{cell}}'(w, \phi_w) dw}, \quad (13)$$

where $p_{\mathrm{cell}}' = p_{\mathrm{cell}} p_{\mathrm{azi}}$.

Figure 6 illustrates our results. The wind estimation that does not account for the rain effects (top panel of the figure) misdiagnoses the hurricane as a level-1 system, as did the operational Seawinds level-2 product (see Fig. 5). Our approach diagnoses the system as a level-4 hurricane (bottom panel of Fig. 6), in agreement with NOAA's Hurricane Research Division's H*Wind analysis (Fig. 7).

5. Enhanced Approach: The Split-Step IR-Advection/ 1D-Convection Model

Having verified that the initial implementation of the approach does improve the wind estimates under the precipitation, we propose to enhance the approach by improving on the overly simplistic advection model. Indeed, the main problem with this "horizontal morphing" evolution model is that horizontal advection is only one part of the dynamics of a storm. And while the consecutive IR images track the horizontal evolution of the system, they also provide quantitative information about cloud top heights and their local evolution. How can we incorporate the latter to augment the horizontal advection information?

The evolution equation for the three-dimensional water concentration q is of the form

$$\frac{\partial q}{\partial t} = -u_H \cdot \nabla_H q - \omega \frac{\partial q}{\partial z} - F \quad (14)$$

where u_H is the horizontal component of the velocity field, ∇_H denotes the horizontal gradient operator, ω is the vertical component of the total velocity field, and F is the sources-and-sinks forcing term. A classic

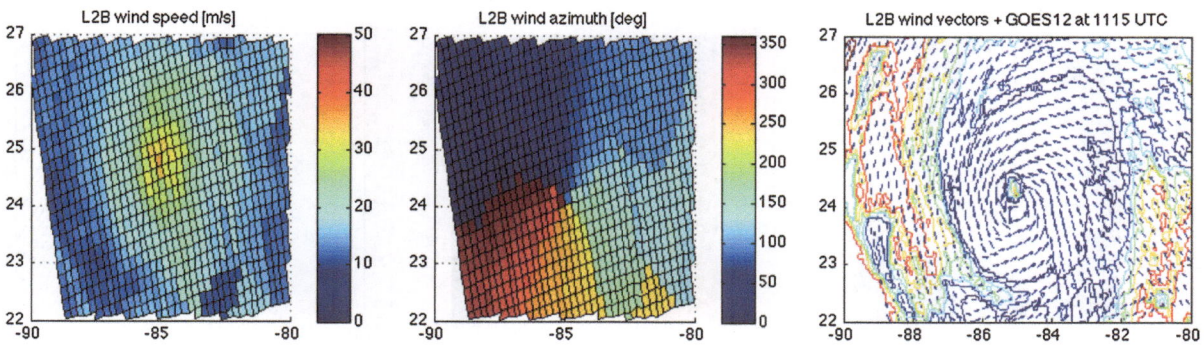

Figure 5
Seawinds level 2B data product for Hurricane Rita as in Fig. 3

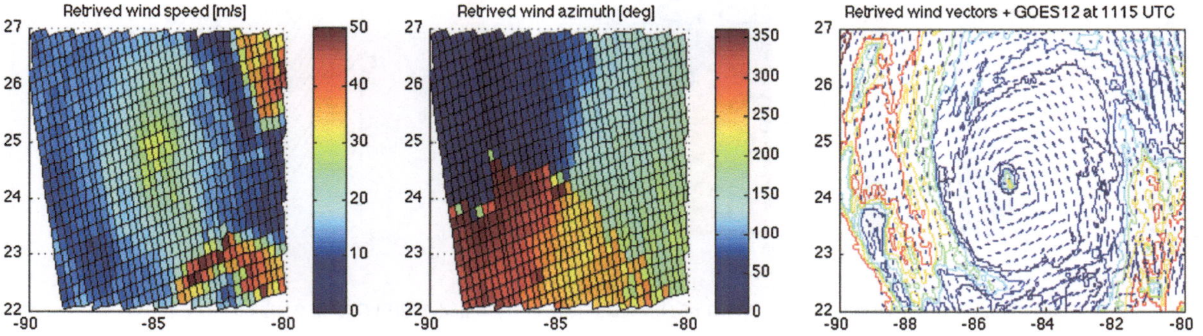

(a) Retrieval without A or σ_0^{rain}

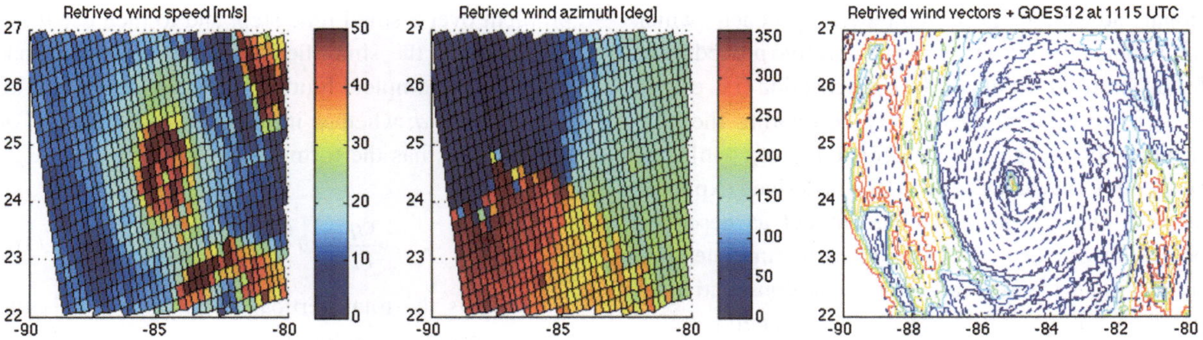

(b) Retrieval with A and σ_0^{rain}

Figure 6
The *top row* shows the wind retrievals with no correction for rain; the *bottom row* shows our retrievals accounting for rain attenuation A and backscattering σ_0^{rain}

result of perturbation theory (TROTTER, 1958) states that if we denote by D_1 and D_2 the operators $D_1 q = -u_H \cdot \nabla_{Hq}$ and $D_2 q = -\omega \frac{\partial q}{\partial z}$, the solution to our evolution equation can be expressed as a limit

$$q = \lim_{n \to \infty} \left(\exp\left(\frac{t}{n} D_2\right) \cdot \exp\left(\frac{t}{n} D_1\right) \right)^n \cdot q_0 \qquad (15)$$

In words, if time t is split into sufficiently small increments t/n, the solution q at time t can be obtained by iteratively tracking the initial q_0 through horizontal advection during a time interval t/n, then through vertical convection for the same duration, and repeating the process until the time of interest t is reached.

113

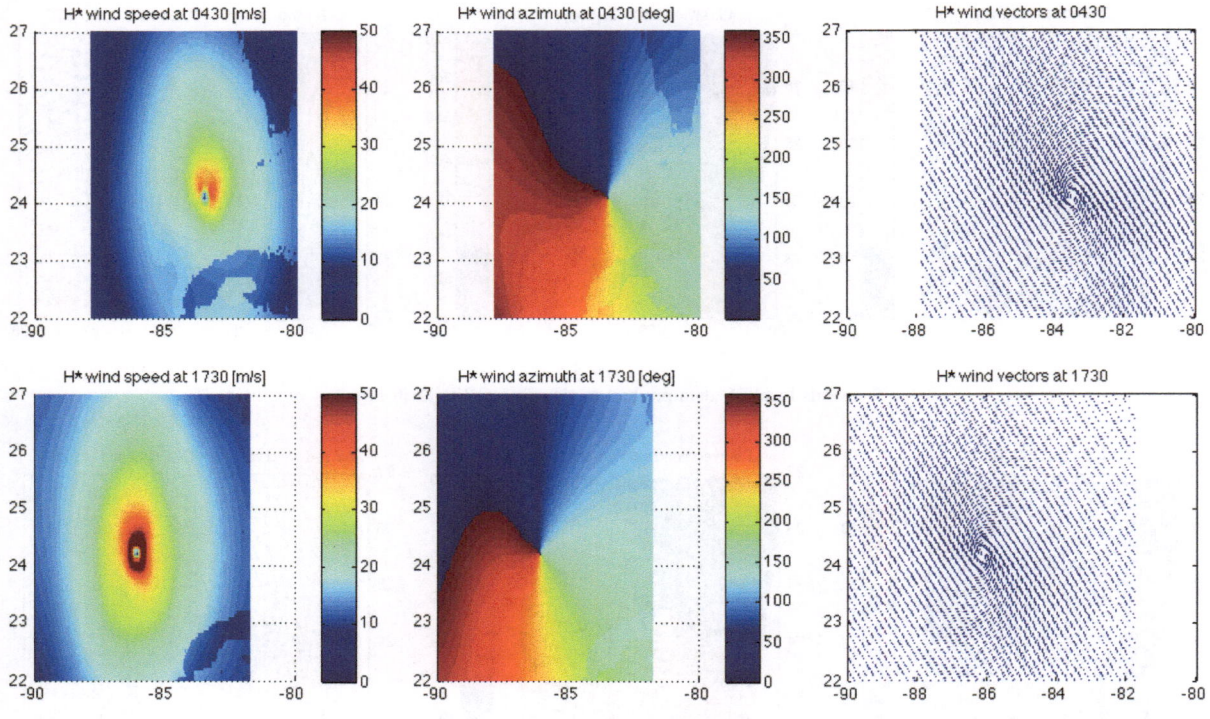

Figure 7
HRD wind analysis (POWELL *et al.*, 1998) of Hurricane Rita at 0430 and 1730 UTC on 21 September 2005

Our proposed enhancement to the two-dimensional advection is a split-step approach which implements the two steps of this iterative procedure as follows: the advection of the condensation (i.e. $\exp[(t/n)D_2]$) is still implemented by correlating the consecutive IR images as described above, while the convection (namely $\exp[(t/n)D_1]$) is derived explicitly using a one-dimensional cloud model as described below. In practice, we are limited to time steps no smaller than the separation between consecutive IR images from the various geostationary satellites.

Call $q(t, z)$ the water concentration (ice, liquid or water vapor), and consider one precipitating column at a time, so that q depends only on the time t and the height z in the atmosphere. Assuming we know the concentration profile $q(0,z)$ at a given point in time $t = 0$, what should q become at the next time step t? To answer the questions perfectly, one would need to solve the four-dimensional problem while optimally assimilating the microwave radiometry and IR data. The reason we reduce the problem to the one-dimensional evolution of the water profiles is to

obtain a reasonably accurate solution to the vertical problem over a small time step, and to then use these solutions (of the simplified problem) as approximations to the complete four-dimensional problem. The equation for q, whether in the case of liquid, ice or water vapor, has the form

$$\frac{\partial q}{\partial t} + \omega \frac{\partial q}{\partial z} = -F \qquad (16)$$

where ω is the total vertical velocity, and F is the forcing term representing sources and sinks, which in our case are the conversions between ice, liquid and water vapor. To solve (16), we make the simplifying assumptions that

- ω can be inferred approximately from the time evolution of the IR measurements for the column in question (cooling being interpreted as deepening of the convection, and warming as detrainment, to first order), and
- F is approximately linear in q, say $F = F_0 + G \cdot q$ with F_0 and G independent of q.

If one then finds the trajectories z'_t which start at a given z at time $t = 0$ and solve

$$\frac{dz}{dt} = -\omega \tag{17}$$

the solution of (16) is then given by

$$q(t, z) = e^{-\int_0^t G(s, z'_{t-s})ds} q(0, z'_t) \\ - \int_0^t e^{-\int_s^t G(\tau, z'_{t-\tau})d\tau} F_0(s, z'_{t-s})ds \tag{18}$$

as one can verify directly by substituting the right-hand-side of (18) for q in (16) and verifying that the latter's left-hand-side is indeed equal to its right-hand-side. In words, (18) says that q at time t is obtained by "stretching" q at the previous time step in a fashion specified by the trajectories z'_t of ω, then "adjusting" the stretched q according to the forcing.

The main advantage of this enhanced approach is seen when one considers the vector $Q(t)$ of water mixing ratios, $Q = (q_i(t), q_l(t), q_v(t))^t$, together with the simplifying assumption that the conversion rates are piecewise constant, i.e. that G is constant over contiguous intervals of z with $F = 0$. At least three intervals need to be considered: an all-ice region aloft, followed by a melting layer, and a rain region near the surface. In each region, because the rate of conversion from species a into species b has to be the same as the rate of conversion from b to a but with opposite sign, the coefficients of G are of the form

$$G = \begin{pmatrix} m + \epsilon' & -f & -d' \\ -m & f + \epsilon & -d \\ -\epsilon' & -\epsilon & d + d' \end{pmatrix} \tag{19}$$

Thus, not only is G singular, we know more specifically that the column vector $u_0 = (1, 1, 1)^t$ is in the nullspace of G^t:

$$(1, 1, 1) \cdot \begin{pmatrix} m + \epsilon' & -f & -d' \\ -m & f + \epsilon & -d \\ -\epsilon' & -\epsilon & d + d' \end{pmatrix} = 0 \tag{20}$$

Therefore, since (18) implies (under our simplifying assumptions) that $Q(t) = (\Pi \exp(-tG)) \cdot Q(0)$, the product being over the different layers that are traversed by z'_t, we can verify that

$$q_i(t) + q_l(t) + q_v(t) = u_0 \cdot Q(t) \\ = u_0 \cdot \exp(-tG) \cdot Q(0) \\ = u_0 \cdot \left(Q(0) + \sum_{n=1}^{\infty} \frac{(-t)^n}{n!} G^n \cdot Q(0) \right) \tag{21}$$

$$q_i(t) + q_l(t) + q_v(t) = u_0 \cdot Q(0) \\ + \sum_{n=1}^{\infty} \frac{(-t)^n}{n!} (u_0 \cdot G) G^{n-1} \cdot Q(0) = u_0 \cdot Q(0) + 0 \tag{22}$$

$$q_i(t) + q_l(t) + q_v(t) = q_i(0) + q_l(0) + q_v(0), \tag{23}$$

i.e. total water is conserved exactly. That is the main advantage of the enhanced approach.

6. Conclusions

Used in conjunction with a simple IR-cloud-motion advection model, current passive microwave retrievals of bulk precipitation can dramatically improve the scatterometry estimates of wind speed and hurricane intensity, even though the former are estimated from satellite data taken several hours before or after the scatterometer measurements. Our initial approach does not account for the convection that takes place in the intervening time intervals, but it can be enhanced to account for the convection using a split-step advection/convection modification. The advantage of the enhanced approach over classical data assimilation is that it conserves total water.

REFERENCES

D.W. DRAPER and D.G. LONG, 2004: *Evaluating the effect of rain on SeaWinds scatterometer measurements*. J. Geophys. Res. - Oceans *109*, doi:10.1029/2002JC001741.

K.A. HILBURN, F.J. WENTZ, D.K. SMITH, and P.D. ASHCROFT, 2006: *Correcting active scatterometer data for the effects of rain using passive radiometer data*. J. Appl. Meteor. Clim. *45*, 382–398.

R.J. JOYCE, J.E. JANOWIAK, P.A. ARKIN, and P. XIE, 2004: *CMORPH: A method that produces global precipitation estimates from passive microwave and infrared data at high spatial and temporal resolution*. J. Hydromet. *5*, 487–503.

R.D. KNABB, D.P. BROWN and J.R. RHOME, 2006: National Hurricane Center Tropical Cyclone Report" Hurricane Rita, 18-26 September 2005. National Hurricane Center, 17 March 2006 (updated 14 August 2006).

115

M.D. POWELL, S.H. HOUSTON, L.R. AMAT, and N. MORISSEAU-LEROY, 1998: *The HRD real-time hurricane wind analysis system.* J. Wind Eng. Indust. Aerodyn. *77-78*, 53–64.

M.W. SPENCER, C. WU and D.G. LONG, 1997: *Tradeoffs in the design of a spaceborne scanning pencil beam scatterometer: application to SeaWinds.* IEEE Trans. Geosci. Rem. Sens. *35*, 115–126.

B.W. STILES and S.H. YUEH, 2006: *Impact of rain on spaceborne Ku-band wind scatterometer data.* IEEE Trans. Geosci. Rem. Sens. *40*, 1973–1983.

H.F. TROTTER, 1958: *Approximation of semi-groups of operators,* Pac.J.Math. *40*, 887–919.

S.H. YUEH, B.W. STILES and W.T. LIU, 2003: *QuikSCAT wind retrievals for tropical cyclones.* IEEE Trans. Geosci. Rem. Sens. *41*, 2616–2628.

(Received November 2, 2010, accepted April 19, 2011, Published online July 31, 2011)

Pure Appl. Geophys. 169 (2012), 425–445
© 2011 Springer Basel AG
DOI 10.1007/s00024-011-0379-y

Impact of ATOVS Radiance on the Analysis and Forecasts of a Mesoscale Model over the Indian Region During the 2008 Summer Monsoon

RANDHIR SINGH,[1] C. M. KISHTAWAL,[1] and P. K. PAL[1]

Abstract—Assimilation experiments are performed with the Weather Research and Forecasting (WRF) models' three-dimensional variational data assimilation (3D-Var) scheme to evaluate the impact of directly assimilating the Advanced Television and Infrared Observation Satellite Operational Vertical Sounder (ATOVS) radiance, including AMSU-A, AMSU-B and HIRS, on the analysis and forecasts of a mesoscale model over the Indian region. The present study is, to our knowledge, the first where the impact of ATOVS radiance has been evaluated on the analysis and forecasts of a mesoscale model over the Indian region. The control (without ATOVS radiance) as well as experimental (which assimilated ATOVS radiance) run were made for 48 h starting at 0000 UTC during the entire July 2008. The impacts of assimilating the radiances from different instruments (e.g., AMSU-A, AMSU-B and HIRS) were measured in comparison to the control run. The assimilation experiments for July 2008 (30 cases) demonstrated a positive impact of the assimilated ATOVS radiance on both the analysis state as well as subsequent short-range forecasts. Relative to the control run, the moisture analysis was improved with the assimilation of AMSU-B and HIRS radiance, while AMSU-A was mainly responsible for improved temperature analysis. The comparison of the model-predicted temperature, moisture and wind with NCEP analysis indicated that a positive forecast impact is achieved from each of the three instruments. HIRS and AMSU-A radiance yielded only a slight positive forecast impact, while AMSU-B radiance had the largest positive forecast impact for moisture, temperature and wind. The comparison of model-predicted rainfall with observed rainfall indicates that ATOVS radiance, particularly AMSU-B and HIRS, impacted the rainfall positively. This study clearly shows that the improved analysis of mid-tropospheric moisture, due to the assimilation of AMSU-B radiances, is a key factor to improve the short-term forecast skill of a mesoscale model.

Key words: ATOVS, AMSU-A, AMSU-B, HIRS, radiance assimilation, WRF.

1. Introduction

Socioeconomic aspects of life in India are highly dependent on both the intensity and distribution of summer monsoon rainfall. Therefore, providing accurate weather forecasts using numerical weather prediction (NWP) models during monsoon season is of prime importance within the scientific community. A continuing difficulty with respect to the improvements in forecast at smaller spatial scales by mesoscale models relates to the fact that observational information is limited, particularly over the oceans and deserts. The Indian subcontinent is affected by the tropical disturbances that form over the data-sparse surrounding oceanic region. The remotely sensed data from satellites can be the best alternative to improve model initial fields, where conventional observations are sparse. The satellite data, particularly temperature and moisture profiles, may be very useful in improving the precipitation forecast associated with Indian summer monsoon. A number of satellite sounders, including the advanced TIROS operational vertical sounder (ATOVS), the high-resolution Atmospheric InfraRed Sounder (AIRS), the moderate resolution imaging spectroradiometer (MODIS) and infrared atmospheric sounding interferometer (IASI), are currently available, and can provide vertical profiles of atmospheric temperature and humidity.

There are two ways of assimilating the satellite radiances into NWP models. In the first case, satellite sounding radiances are converted into atmospheric variables of temperature and humidity using a physical or statistical retrieval method (GOLDBERG, 1999), and then retrieved temperature and moisture data are assimilated (GAL-CHEN *et al.*, 1986; DOYLE and WARNER, 1998; LIPTON and VONDERHAAR, 1990;

---
[1] Atmospheric Sciences Division, Atmospheric and Oceanic Sciences Group, Space Applications Centre (ISRO), Ahmedabad 380015, India. E-mail: randhir_h@yahoo.com

McNALLY AND VESPERINI, 1996; LIPTON et al., 1995; RUGGIERO et al., 1999; ZHU et al., 2002; ZAVODSKY et al., 2004; ZHAO et al., 2005; FAN AND TILLEY, 2005; CHOU et al., 2006; SINGH et al., 2008; REALE et al., 2009; LI and LIU, 2009) in the same way as conventional data. In the second method, EYRE (1989) proposed using radiances directly via three-dimensional variational (3D-Var) assimilation technique (LE DIMET and TALAGRAND, 1986). Here, the model maps conventional meteorological parameters (pressure, temperature and water vapor) into satellite radiance space using radiative transfer models (RTMs). Optimum initial conditions are then determined by minimizing the misfit between observed and the model-generated radiances using a variational approach. Direct radiance assimilation is theoretically superior to retrieved (parameters) assimilation because the observational error statistics are better understood in direct radiance assimilation than in retrieval assimilation (ANDERSSON et al., 1991, 1994; EYRE et al., 1993; ENGLISH et al., 2000; BOUTTIER and KELLY, 2001). This approach has led to a significant improvement in the quality of the NWP analyses and forecasts, particularly in the southern hemisphere and tropics (DERBER and WU, 1998). However, at present only infrared radiances from the clear sky regions are assimilated in most of the NWP systems, while the cloudy infrared radiances are almost completely eliminated from the assimilation process. Some efforts to include cloud-affected infrared radiances (McNALLY, 2009) have been made in recent years, but still a lot has to be done in order to use them in operational assimilation systems. Recent studies show that assimilation of retrieved thermodynamic parameters in partly cloudy conditions led to improved prediction of rainfall and tropical systems (REALE et al., 2009; ZHOU et al., 2010). Hence, a suitable blending of infrared radiance assimilation under clear sky conditions and retrieved parameters assimilation under partly cloudy conditions can be the optimum strategy for data assimilation problem.

Compared to Global NWP, the analysis and forecast sensitivity to radiance assimilation in the mesoscale models has been less rigorously examined (ZAPOTOCNY et al., 2005; MONTMERLE et al., 2007). Recently, increased focus has been put on the operational applicability of the meso-scale analysis and forecast system. The Weather Research and Forecasting (WRF) system, developed at the National Center for Atmospheric Research (NCAR), is intended to improve the forecast accuracy of significant weather features across scales ranging from cloud to synoptic. The WRF model has been widely used for meso-scale weather forecasts by the research community and operational agencies. Recently, direct radiance assimilation capability has been developed and implemented in the WRF-Var assimilation system (LIU and BARKER, 2006), which is expected to improve the forecasts of meso-scale and severe weather events. The majority of research regarding the impact of satellite radiances in mesoscale models is conducted over the USA and Europe. Similar attempts for the Indian region are quite rare (SINGH et al., 2010). Therefore, the aim of this study is to investigate the impact of ATOVS radiance on the analysis and forecasts of a mesoscale model over the Indian region.

2. ATOVS Data

ATOVS instruments are sensors on the National Oceanic and Atmospheric Administration (NOAA) series of polar-orbiting satellites. ATOVS instruments are composed of three independent sensors: a High-resolution Infra-Red sounder (HIRS) and two Advanced Microwave Sounder Units (AMSU-A, AMSU-B). In this paper the radiances measured by HIRS/3 onboard NOAA-15, -16 and -17; HIRS/4 onboard NOAA-18; AMSU-A onboard NOAA-15, -16 and -18; and AMSU-B onboard NOAA-15, -16 and -17 are used.

The AMSU-A is a multi-channel radiometer with 15 channels (Table 1) and is mainly used to provide information on atmospheric temperature profiles. The AMSU-A is a cross-track scanning total power radiometer and has an instantaneous field of view (IFOV) of 3.3° at the half-power points (Swath width 2,343 km), providing a nominal spatial resolution at nadir of 48 km. The antenna provides a cross-track scan, scanning ±48.3° from nadir with a total of 30 earth fields of view per scan line. The instrument completes one scan every 8 s. The AMSU-A measures radiation at microwave frequencies ranging

Table 1

Characteristics of the AMSU-A, AMSU-B and HIRS

Channel	AMSU-A (frequency GHz)	AMSU-B (frequency GHz)	HIRS-3/4 (wavelength μm)
1	23.8	89.0	14.95
2	31.4	150.0	14.71
3	50.3	183.3 ± 1	14.49
4	52.8	183.3 ± 3	14.22
5	53.6	183.3 ± 7	13.97
6	54.4		13.64
7	54.9		13.35
8	55.5		11.11
9	57.2		9.71
10	57.29 ± 0.217		12.47
11	57.29 ± 0.322 ± 0.048		7.33
12	57.29 ± 0.322 ± 0.022		6.52
13	57.29 ± 0.322 ± 0.010		4.57
14	57.29 ± 0.322 ± 0.0045		4.52
15	89.0		4.47
16			4.45
17			4.13
18			4.00
19			3.67
20			0.69

from 23.8 to 89.0 GHz. Atmospheric temperature profiles are obtained using channels near 60 GHz, which is an oxygen band. The AMSU-A sounding channels (3–14) respond to the thermal radiation at various altitudes in the atmosphere, whereas channels 1, 2 and 15 are primarily designed for obtaining the information on surface properties (e.g., emissivity and skin temperature).

The AMSU-B is a cross-track, continuous line scanning and total power radiometer with an instantaneous field of view of 1.1° (at half power points). Spatial resolution at nadir is nominally 16 km. The antenna provides a cross-track scan, scanning ±48.9 from nadir with a total of 90 scan fields of view per scan line. The AMSU-B has five channels (Table 1) where channels 3–5 measure radiances in the 183 GHz absorption band and provide moisture information at different levels of the atmosphere. Channels 1–2 are window channels and are sensitive to the radiation emitted from the surface.

The HIRS (HIRS/3 on NOAA-15, -16 and -17, and HIRS/4 on NOAA-18) is a cross-track line scanning radiometer that measures scene radiance in the infrared and visible spectrum (Table 1). Among

the 20 spectral channels there are 12 longwave (6.7–15 μm), 7 shortwave (3.7–4.6 μm) and 1 visible (0.69 μm) channel. Data from the instrument are used in conjunction with the Advanced Microwave Sounding Unit (AMSU) instruments to calculate the atmosphere's vertical temperature profile from the Earth's surface to about 40 km altitude. The data are also used to determine ocean surface temperatures, total atmospheric ozone, tropospheric moisture information, cloud height and coverage, and surface radiance. The spatial resolution for HIRS/3 and HIRS/4 at nadir is about 20 and 10 km, respectively.

3. The WRF Model and Assimilation Methodology

3.1. WRF model

The forecast model used herein is the Weather Research and Forecasting (WRF; SKAMAROCK et al., 2008) Model version 3.2. It is a limited area, non-hydrostatic primitive equation model with multiple options for various physical parameterization schemes. The WRF physical options used in this study consisted of the WSM 3-class graupel scheme for microphysics, the New Kain-Fritsch (KAIN, 2004) cumulus convection parameterization scheme and the Yonsei University (YSU) planetary boundary layer scheme (HONG and PAN, 1996). The Rapid Radiative Transfer Model (RRTM; MLAWER et al., 1997) and Dudhia scheme (DUDHIA, 1989) were used for long-wave and shortwave radiation, respectively. The model domain consisted (33°E–121°E; 26°S–52°N) of 330 × 330 with 30 km horizontal spacing. The model had 36 vertical levels with the top of the atmosphere located at 10 hPa.

3.2. Assimilation Methodology

The WRF three-dimensional variational (3D-Var) data assimilation system was used in this study. It is capable of assimilating data from many different observational platforms, including radiance measurements obtained from satellites (LIU and BARKER, 2006). The WRF 3D-Var method (BARKER et al., 2004) is based on the minimization of a cost function defined as

$$J(x) = J^b + J^o = \frac{1}{2}\left(x - x^b\right)^T B^{-1}\left(x - x^b\right)$$

$$+ \frac{1}{2}(H(x) - y^o)^T R^{-1}(H(x) - y^o) \qquad (1)$$

Here x is a state vector composed of atmospheric and surface variables, and x^b is a background vector usually composed of values taken from a previous forecast. The observed AMSU-A, AMSU-B and HIRS radiances or brightness temperatures and other observations are contained in y^o; H is the observation operator that transforms the model variables to observation quantities. For radiance assimilation, H contains a fast RTM (Radiative Transfer for TIROS Operational Vertical Sounder, RTTOV) that computes radiance or brightness temperature from the input values of the atmospheric state x, R is the estimated error covariance of the observations, and B is the estimated error covariance of the background field. The background covariance matrix (B) used in this study was estimated using the so-called "NMC" method (PARRISH and DERBER, 1992; WU et al., 2002), which assumes that the background error covariances are well approximated by averaged forecast differences between 24- and 12-h forecasts verifying at the same time. In (1), the analysis $x = x^a$ represents a posterior maximum likelihood (minimum variance) estimate of the true state of the atmosphere given two sources (x^b and y^o) of data. The analyses fitted to this data are weighted by estimates of their errors (B, R). The cost function (1) assumes that observational and background error covariances are described using a Gaussian probability density function with zero mean error.

3.3. Observational Operator

For the assimilation of satellite radiance, an observational operator transforms model state variables (e.g., temperature and moisture) to radiance space. This operator is often viewed as consisting of one model, namely a fast RTM. Two widely used fast RTMs in NWP are RTTOV and the Community Radiative Transfer Model (CRTM). RTTOV (developed by EUMETSAT) and CRTM (developed by the Joint Center for Satellite Data Assimilation at NCEP) have already been implemented in the WRF assimilation system. In this study we used RTTOV for the

computation of model-equivalent ATOVS radiance. RTTOV is a fast radiative transfer model where the layer optical depth is parameterized (SAUNDERS et al., 1999) using linear combinations of profile-dependent predictors. RTTOV was originally developed (SAUNDERS et al., 1999) and maintained by the European Center for Medium Range Weather Forecast (ECMWF), but the improvement and updates are now implemented through the EUMETSAT NWP Satellite Applications Facility (SAF) by the UK Met Office (METO), Météo-France and ECMWF. RTTOV does not perform monochromatic radiative transfer (RT) calculations, but computes channel-specific convolved transmittances for a particular satellite channel. RTTOV can simulate a range of different infrared and microwave sensors.

4. Experimental Design

Five experiments, each starting from 0000 UTC analyses during the entire July 2008, were performed. In the control run (CNT), the major data (Table 2) assimilated are both conventional (synops, metar, buoys, ships and aircraft) and derived from various satellites (atmospheric motions vectors from Meteosat-7, surface winds from the Quick Scatterometer and Sensor Microwave Imager). The EXP-A, EXP-B and EXP-H experiments, which assimilate the AMSU-A, AMSU-B and HIRS radiance (radiance and brightness temperature are interchangeably used), respectively, at 1800 and 0000 UTC, were conducted to assess the impact of individual instruments. Another experiment, EXP-ATOVS, was conducted to determine the combined effects of AMSU-A, AMSU-B and HIRS radiance.

The 6-hourly NCEP analyses with $1° \times 1°$ resolution were used for the WRF model boundary conditions. The 48-h forecast cycles started at 0000 UTC each day, preceded by two 6-h assimilation cycles at 1800 and 0000 UTC, each of them using the previous 6-h WRF forecast as a background. Figure 1 shows the procedure used for CNT and EXP-A experiments. Figure 1 is generated as if it were the 0000 UTC 01 July analysis; thereafter, it was duplicated daily during July 2008. EXP-B, EXP-H and EXP-ATOVS are similar to EXP-A (See

Table 2

Data assimilated in the CNT experiment

Observations	Platform
Ground based	
Upper air	Sonde aircraft
Land surface	SYNOP METAR
Marine surface	Buoy ships
Satellite based	
Atmospheric winds	Meteosat AMVs
Ocean surface winds	QuikSCAT SSM/I

Fig. 1b), except AMSU-B, HIRS and ATOVS radiances were assimilated in place of AMSU-A. In order to include sufficient data, a 3-h time window (1.5-h before and after the model initial time) was used. The experiments (CNT, EXP-A, EXP-B, EXP-H and EXP-ATOVS) were then executed in forecast mode for 48 h daily from 0000 UTC during the entire month of July.

4.1. Quality Control and Channel Selection

Prior to data assimilation, AMSU-A, AMSU-B and HIRS radiances underwent (before bias correction) a quality control process that included a check of the extreme values and a deviation check between the estimated values from the background fields and the observation and the rain (microwave and infrared) and cloud (only infrared) detection. In order to avoid the assimilation of poor quality observations, the following quality checks were applied. Firstly, brightness temperatures within the range of 150–350 K were used, and the others were excluded. Secondly, the

observations influenced by precipitation (microwave and infrared) and cloud (only infrared) were removed. The presence of precipitation is detected using algorithms (GRODY, 2001) that compute the scattering index and cloud liquid water (CLW) amount, respectively. For the AMSU-B, the index is the difference between the observed AMSU-B radiances of channels 1 and 2, while for the AMSU-A index is the difference between the observed ASMU-A radiances of channels 1 and 15. If the index was >3 or CLW was >0.2 mm, the microwave radiances were assumed to be contaminated by precipitation. The infrared threshold technique of McMILLIN and DEAN (1982) was applied on the HIRS radiances to detect the clouds. Finally, a gross error quality control was performed, the observations being rejected when the first guess departure (observed background brightness temperature) was more than three times the observation error. To keep the radiances relatively uncorrelated, observations were thinned to one every 120 km.

The channel selection is decided according to the peak energy contribution level of the sounder channel. Channels 1–4 and 15 of AMSU-A and channels 1–2 of AMSU-B, which have a contribution from the surface, are used only for quality control and not for assimilation. The weighting functions derived for a standard tropical atmosphere of AMSU-A, AMSU-B and HIRS channels used for the assimilation are shown in Fig. 2. Channels 10–14 of AMSU-A are also not assimilated because they peak above the top boundary of the model. Channels 1–2, 6–10 and 13–19 of HIRS are also not assimilated because either

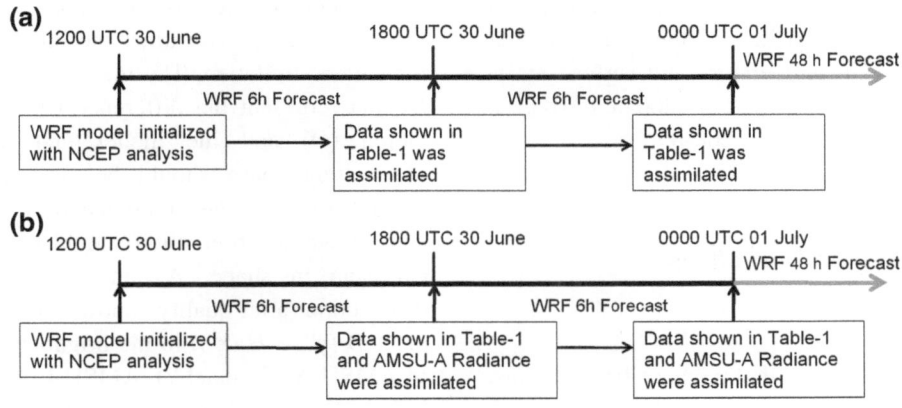

Figure 1

The experimental setup **a** for control run and **b** for AMSU-A radiance run

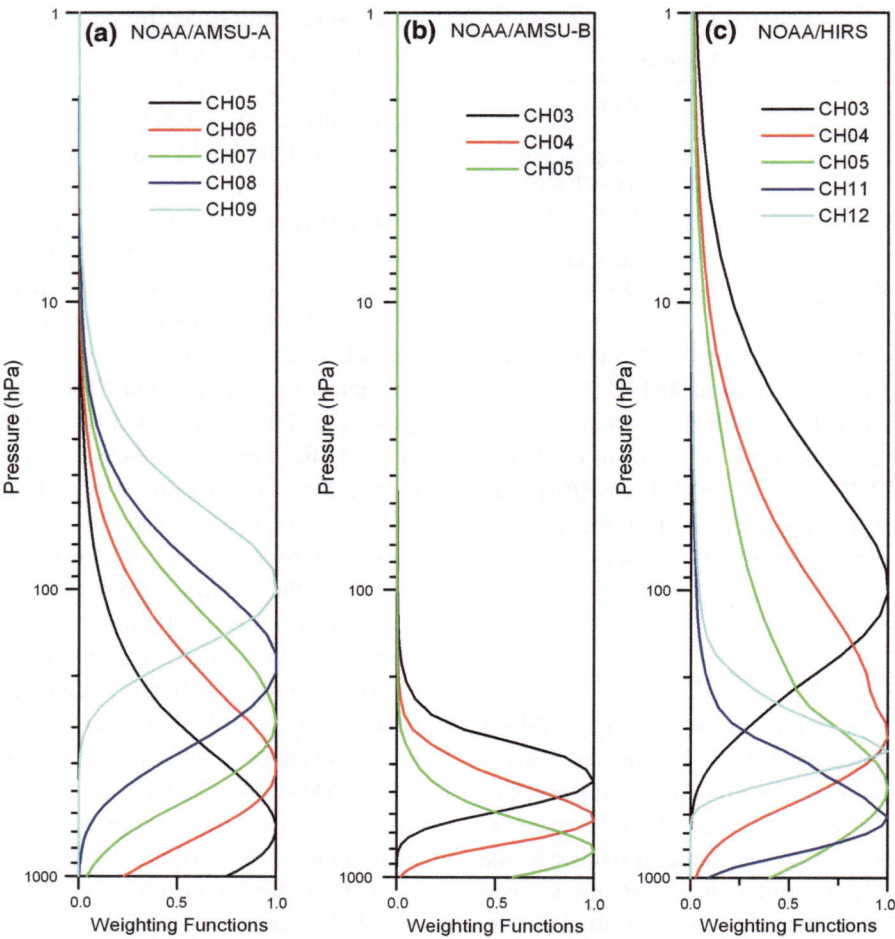

Figure 2

The normalized weighting functions **a** AMSU-A, **b** AMSU-B and **c** HIRS channels. The weighting functions are calculated from the tropical atmosphere using RTTOV

they peak above the model levels or they are influenced by the surface. The reason for eliminating the channels that are sensitive to the surface is that the background skin temperature and surface emissivity have large errors (Hua *et al.*, 2004a, b; Barker *et al.*, 2005; Liu and Barker, 2006).

5. Assimilation Results

5.1. Bias Correction

It is a well-known fact that radiance observations, as well as radiative transfer models, contain a substantial amount of errors. It is essential to remove

the radiance biases prior to its assimilation (data assimilation schemes assume unbiased observations) in order to properly extract the information content for data assimilation. The first guess departures (O–B) that represent the difference between the observation (O) and the model first guess (B) in observation space should be zero centered (e.g., 'model' observations computed from the first guess and the actual observation should be unbiased) and Gaussian in shape. As an example, the spatial distribution (after quality control and bias correction) of O–B ($y^o - H(x^b)$) in the brightness temperatures of AMSU-A (Channel 7), AMSU-B (Channel 3) and HIRS (Channel 12), within ± 1.5 h of the model initial conditions (0000 UTC 15 July), are shown in

Fig. 3a, b, c, respectively. After the quality control and bias correction, the values of O–B are within ±0.4 K, ±4 K and ±3 K for AMSU-A, AMSU-B and HIRS, respectively. The observations are corrected for biases by a variational bias correction scheme (VarBC; DERBER and WU, 1998). Figure 4 shows the mean (BIAS) and root mean square differences (RMSD) of O–B at 0000 UTC 15 July averaged over the model domain, without (w/o) and with (w) bias correction, for different channels. It is evident from Fig. 4a that the O–B mean departures

with the bias correction (Fig. 4a; dark gray colour bar) are smaller than the O–B mean departures without (Fig. 4a; gray colour bar) the bias correction. This shows that the bias correction scheme (VarBC) is successful in reducing the systematic differences between the observed and model first guess radiance. The RMSD (Fig. 4b; Table 3) of O–B is also reduced with the bias correction. On average, the BIAS and RMSD in the moisture-sensitive channels (AMSU-B and HIRS) are higher than in the temperature-sensitive channels (AMSU-A and channels 3, 4 and 5 of HIRS).

5.2. Overview of the Fit to Radiance

In a successful assimilation, the analysis departures (known as O–A) are smaller than the first guess departures (known as O–B); hence the analysis better matches the observations. As an example, this is illustrated in Fig. 4b and Table 3, which shows the domain-averaged first guess (before and after bias correction) and analysis departures for 0000 UTC 15 July 2008. The first guess fit (O–B; dark gray colour bar) for AMSU-A channels 5–9 has an RMSD of 0.48, 0.31, 0.25, 0.36 and 0.36 K, while the analysis fit (O–A; black colour bar) has an RMSD of 0.27, 0.19, 0.15, 0.24 and 0.18 K, respectively. The first guess fit (O–B; dark gray color bar) for AMSU-B channels has an RMSD of 3.6, 3.6 and 2.9 K, while the analysis fit (O–A; black color bar) has an RMSD of 2.4, 2.2 and 2.1 K, respectively. For HIRS, the first guess fit (O–B; dark gray color bar) has an RMSD of 0.71, 0.55, 0.47, 1.5 and 2.2 K, while the analysis fit (O–A; black colour bar) has an RMSD of 0.61, 0.45, 0.35, 0.70 and 1.3 K, respectively. Overall, Fig. 4b and Table 3 confirm that the analysis (in observation space) is closer to the observations than that of the background.

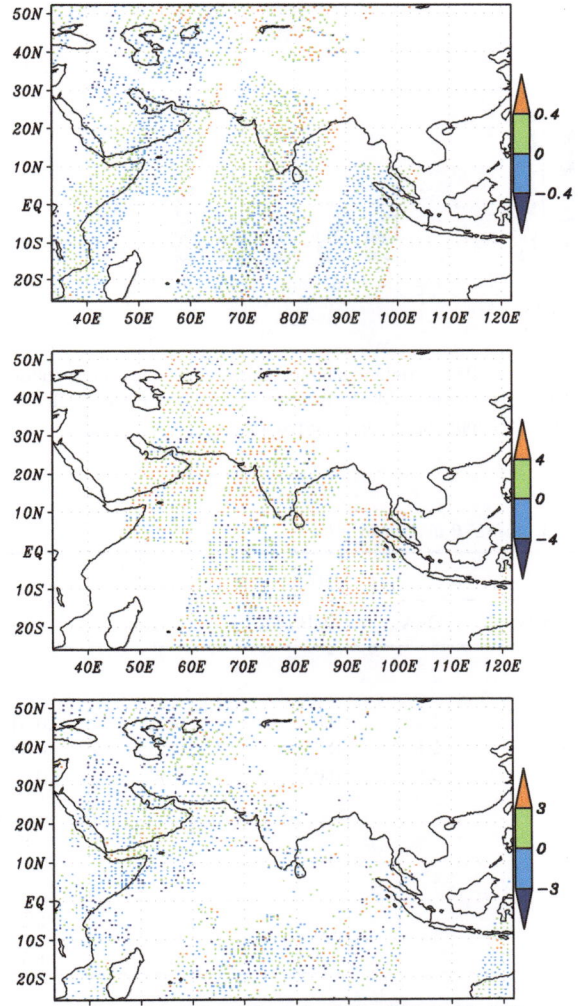

Figure 3
Spatial distribution of bias-corrected observed minus first guess brightness temperatures (O–B; K), AMSU-A channel 7 (*upper panel*), AMSU-B channel 3 (*middle panel*) and HIRS channel 12 (*lower panel*) at 0000 UTC 15 July 2008

5.3. Differences in the Initial Conditions with and without the Use of Radiance

To illustrate the structure of increments imposed on the model fields (e.g., temperature and moisture) by the assimilation of AMSU-A, AMSU-B and HIRS radiance, we computed the RMSD in temperature and humidity between the control (CNT; without

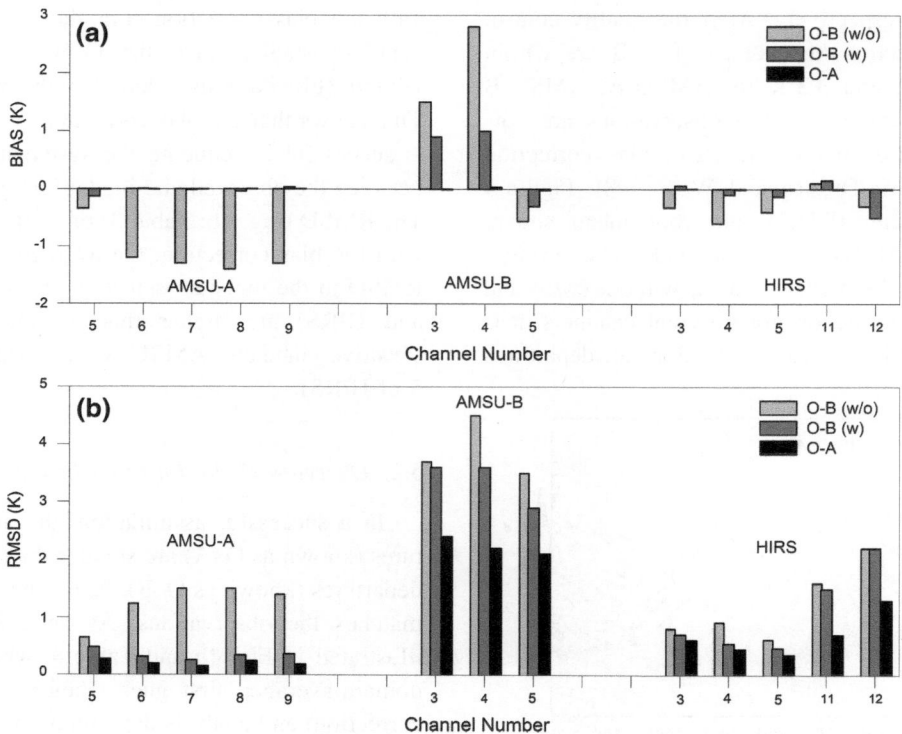

Figure 4
Mean (**a**) and root mean square (**b**) difference of ATOVS (AMSU-A, AMSU-B and HIRS) brightness temperatures minus the corresponding WRF first guess (O–B) and analyzed (O–A) brightness temperature at 0000 UTC 15 July 2008. Observed minus first guess brightness temperature (O–B) is shown before (w/o) and after (w) applying the bias correction

Table 3

O–B and O–A statistics for AMSU-A, AMSU-B and HIRS

	AMSU-A			AMSU-B			HIRS		
	O–B (w/o)	O–B (w)	O–A	O–B (w/o)	O–B (w)	O–A	O–B (w/o)	O–B (w)	O–A
1									
2									
3				3.7	3.6	2.4	0.81	0.71	0.61
4				4.5	3.6	2.2	0.92	0.55	0.45
5	0.65	0.48	0.27	3.5	2.9	2.1	0.62	0.47	0.35
6	1.24	0.31	0.19						
7	1.50	0.25	0.15						
8	1.50	0.34	0.24						
9	1.40	0.36	0.18						
10									
11							1.6	1.5	0.7
12							2.2	2.2	1.3

radiance) and different radiance assimilation (EXP-A, EXP-B, EXP-H, EXP-ATOVS) experiments. These results were obtained by comparing 30 samples of control analysis with the corresponding experimental analysis. Figure 5 shows the vertical profiles

of RMSD of specific humidity (q; g kg^{-1}) and temperature (T; K) between CNT and different radiance assimilation experiments. The largest change, due to the assimilation of radiance, in the specific humidity (Fig. 5a) is found in the layer

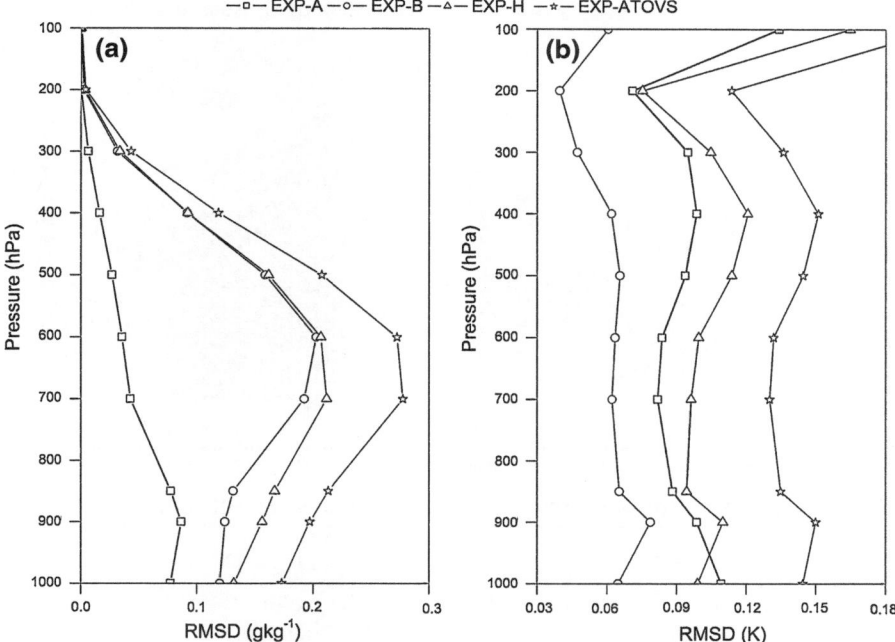

Figure 5
The vertical profiles of RMSD between CNT and different radiance assimilation experiments analyzed, **a** specific humidity (g kg^{-1}) and **b** temperature (K). A set of 30 analyses at 0000 UTC during July 2008 is compared to obtain the RMSD between CNT and different radiance assimilation experiments

between 800 and 500 hPa, and is mainly due to the AMSU-B and HIRS radiance. From Fig. 5b, it is clear that assimilation of AMSU-A, AMSU-B, HIRS and ATOVS impacted the vertical structure of temperature in a similar (maximum RMSD at 100, 400 hPa and near the surface and smaller RMSD in the layer between the 850 to 500 and at 200 hPa) way. However, the magnitude of the RMSD in the temperature is quite different in each experiment. The largest change for a single instrument in the temperature is seen with the assimilation of HIRS followed by AMSU-A, while negligible change in the temperature is seen with the assimilation of AMSU-B radiance. Overall, the changes in the temperature (maximum RMSD = 0.18 K), due to radiance assimilation, are much smaller than the humidity (maximum RMSD = 0.28 g kg^{-1}). This is because the O–B radiances that were used in 3D-Var of the temperature-sensitive channels of AMSU-A and HIRS were much smaller than the O–B radiance of moisture-sensitive channels of AMSU-B and HIRS.

The geographical distributions of RMSD of CNT minus EXP-B-analyzed temperature (at 100 hPa) and specific humidity (at 600 hPa) are shown in Fig. 6. The RMSD in temperature due to the assimilation of AMSU-A and HIRS radiance is of the order of 0.2–0.3 K (with some isolated pockets of 0.4 K) and is mainly observed over the equatorial Indian Ocean, the Bay of Bengal (only in AMSU-A), Arabian Sea and Southern Indian Ocean (only in HIRS). The RMSD in the specific humidity due to assimilation of AMSU-B and HIRS radiance is in the range of 0.4–0.6 g kg^{-1}, and large differences occur over the Arabian Sea, Saudi Arabia, western equatorial Indian Ocean and western part of India. Assimilation of AMSU-B and HIRS (channels 11 and 12 are sensitive to moisture) radiance shows a similar impact on the humidity, particularly over the northern Arabian Sea and over the northwestern part of India and adjoining Pakistan. However, over the western equatorial Indian Ocean, the differences between CNT and HIRS are much smaller compared to CNT and AMSU-B. This may be due to the assimilation of cloudy radiances in case of AMSU-B, which were kicked off from the assimilation system in case of the HIRS. It is clearly evident from the O–B distribution

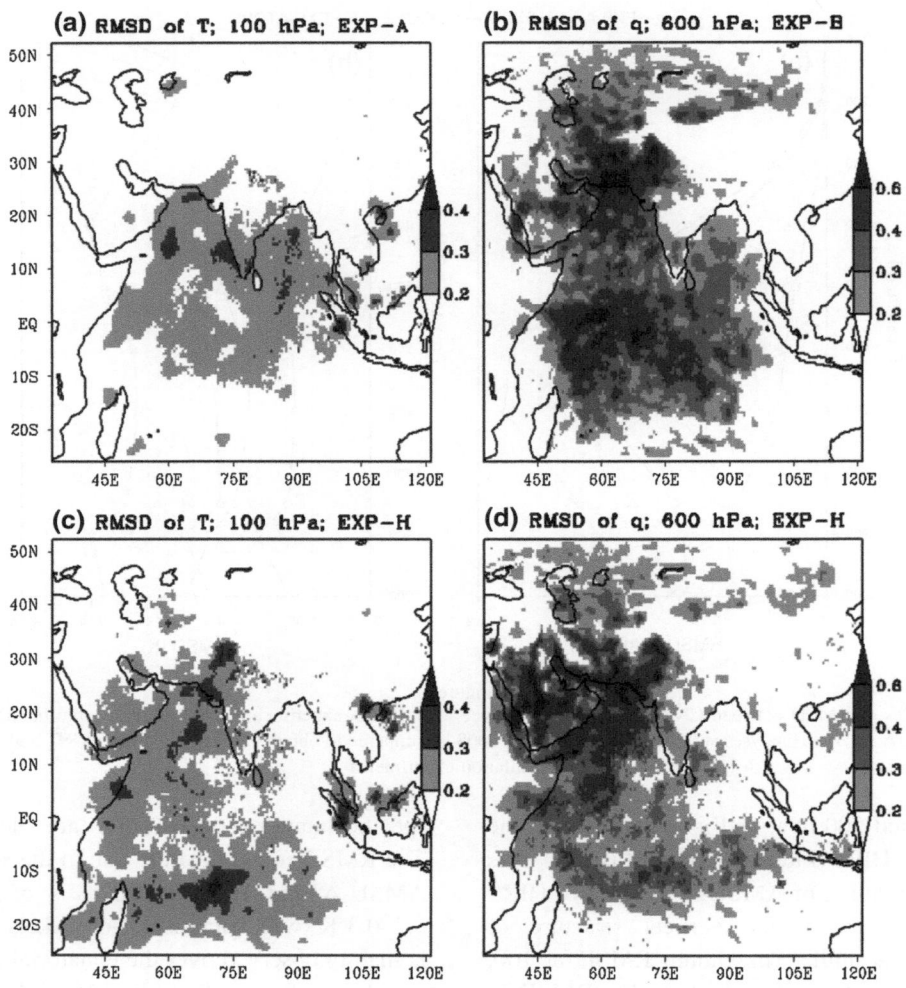

Figure 6
The spatial distribution of RMSD between CNT and **a** EXP-A-analyzed 100 hPa temperature (K), **b** EXP-B-analyzed 600 hPa specific humidity (g kg^{-1}), **c** EXP-H-analyzed 100 hPa temperature (K) and **d** EXP-H-analyzed 600 hPa specific humidity (g kg^{-1}). A set of 30 analyses at 0000 UTC during July 2008 is compared to obtain the RMSD between CNT and radiance assimilation experiments

on a particular day (Fig. 3) that HIRS radiances are kicked off particularly from the equatorial region, while microwave radiances are assimilated. One of the major advantages of AMSU-B over HIRS is that it provides the crucial observation over the convectively active regions and hence can be expected to be a little more beneficial for assimilation.

5.4. Validation of the Analysis

Comparison of the analysis performance between CNT and radiance assimilation experiments (EXP-A, EXP-B, EXP-H and EXP-ATOVS) was carried out by computing RMSD of different analyses against the NCEP analysis at 0000 UTC daily for the entire month of July 2008. The statistics were computed as:

$$\text{RMSD}^{\text{CNT}} = \overline{\sqrt{\frac{1}{n}\left(\sum_{n}\left(\text{Var}^{\text{NCEP}} - \text{Var}^{\text{CNT}}\right)^2\right)}}^{\text{Analysis}}$$

(2)

$$\text{RMSD}^{\text{EXP}} = \overline{\sqrt{\frac{1}{n}\left(\sum_{n}\left(\text{Var}^{\text{NCEP}} - \text{Var}^{\text{EXP}}\right)^2\right)}}^{\text{Analysis}}$$

(3)

$$\alpha = 100 \left\{ \frac{\overline{\sqrt{\frac{1}{n} \left(\sum_{n} \left(\text{Var}^{\text{NCEP}} - \text{Var}^{\text{CNT}} \right)^2 \right)}}^{\text{Analysis}} - \overline{\sqrt{\frac{1}{n} \left(\sum_{n} \left(\text{Var}^{\text{NCEP}} - \text{Var}^{\text{EXP}} \right)^2 \right)}}^{\text{Analysis}}}{\overline{\sqrt{\frac{1}{n} \left(\sum_{n} \left(\text{Var}^{\text{NCEP}} - \text{Var}^{\text{CNT}} \right)^2 \right)}}^{\text{Analysis}}} \right\} \qquad (4)$$

where RMSD^{CNT} is the root mean square difference between NCEP and CNT analysis of temperature and moisture; RMSD^{EXP} is the root mean square difference between NCEP and EXP (EXP denotes EXP-A, EXP-B, EXP-H, EXP-ATOVS) analysis of temperature and moisture. We also computed an analysis impact parameter (α; defined in Eq. 4), which indicates the improvement of the analysis after assimilation of radiance. Division by the RMSD^{CNT} and multiplication by 100 normalize the results and provide the improvement or degradation in percentage terms. In Eqs. 2 and 3, Var can be temperature, relative humidity or wind.

The vertical variation of domain-averaged RMSD (defined in Eqs. 2 and 3) between the NCEP analysis and the analysis (for temperature and moisture) obtained from CNT and the improvement parameter obtained for different radiance assimilation experiments is shown in Fig. 7. Compared to NCEP analysis, CNT-analyzed relative humidity showed (Fig. 7a) an RMSD of the order of 5–6% at the surface and at 100 hPa and 8–9% between 900 and 200 hPa. Compared to NCEP analysis, CNT-analyzed temperature showed (Fig. 7b) an RMSD of the order of 0.7–0.8, 0.4–0.5 and 1 K in the lower, middle and upper troposphere, respectively.

The moisture analysis is improved as much as 4% with the assimilation of AMSU-B radiance (Fig. 7c). The largest improvement in the moisture analysis is seen in the layer between 800 and 400 hPa. Assimilation of HIRS radiance also impacted the moisture analysis positively over most of the levels. The maximum positive impact is of the order of 1–2%, and is seen at 300 hPa and in the layer between 600 and 500 hPa. Assimilation of AMSU-A radiance did not show much impact on the moisture analysis.

When the AMSU-A, AMSU-B and HIRS radiances were assimilated together over most of the levels the moisture analysis was similar to that in AMSU-B. With the assimilation of AMSU-A radiances the temperature analysis was improved at most of the levels (Fig. 7d). The maximum improvement was of about 3% and was seen at 300 hPa.

Assimilation of AMSU-B radiance did not show much impact on the temperature analysis, while assimilation of HIRS degraded the temperature analysis (as large as 3%), particularly in the mid to upper troposphere. This negative impact from HIRS radiance may be due to the assimilation of cloud-contaminated HIRS observations. Some cloud-contaminated observations might have been flagged as clear radiance. The combined assimilation of AMSU-A, AMSU-B and HIRS impacted the temperature analysis positively in the layer between 850 and 600 hPa and above the 350 hPa, while negative impact was seen around 400 hPa. Overall, the moisture analysis (Fig. 7c) is improved with the use of AMSU-B and HIRS radiances, whereas temperature analysis (Fig. 7d) is improved with the use of AMSU-A radiance.

5.5. Impact of the ATOVS Radiance on the Forecast

5.5.1 Comparison with NCEP Analysis

The RMSD computed with respect to NCEP analysis is used as a measure to quantitatively compare the 6- to 48-h forecasts from the control and the experimental runs. Figure 8 shows the RMSD for the control run, averaged for the entire July at different pressure levels for 6- to 48-h forecasts of relative humidity (Fig. 8a), temperature (Fig. 8b) and wind

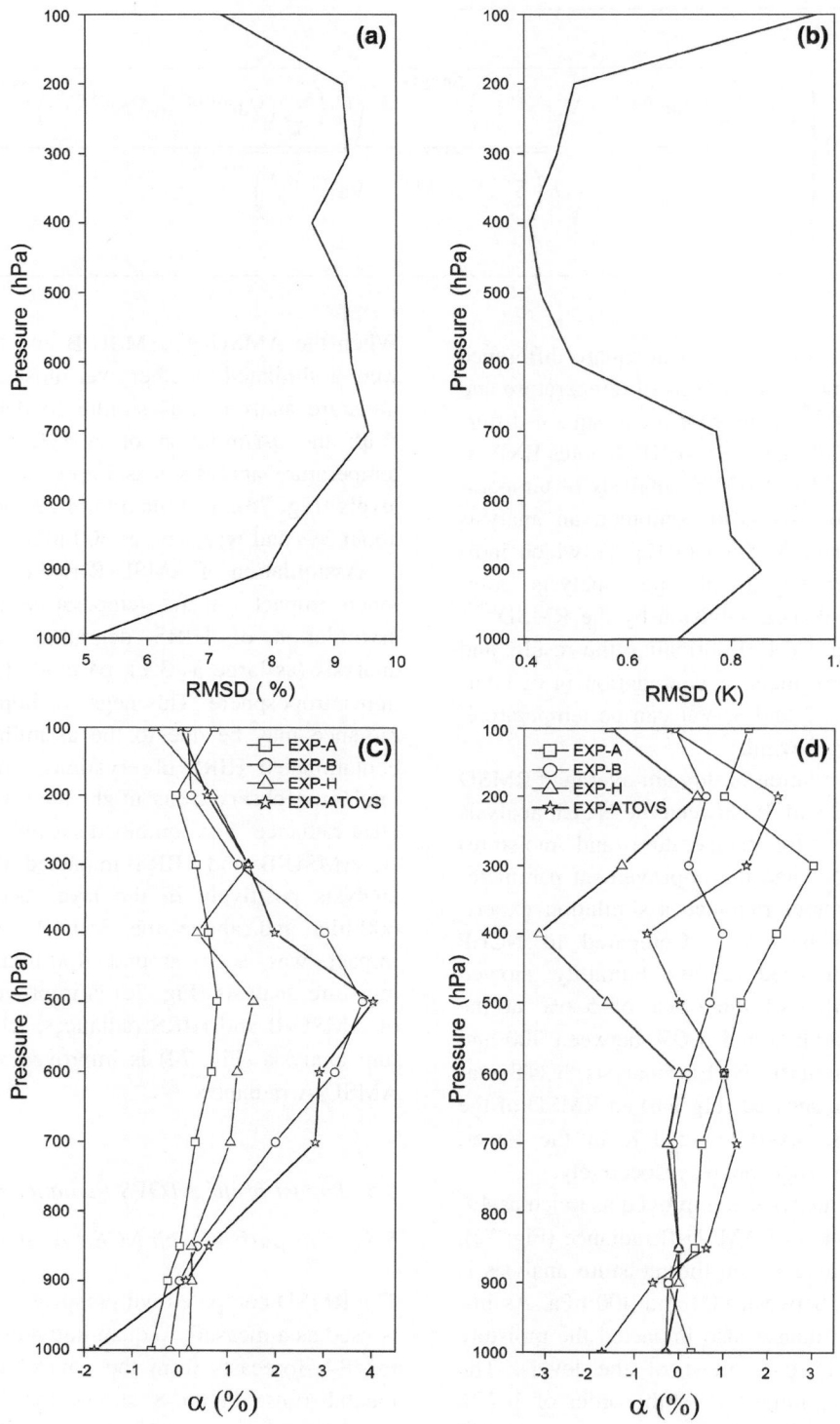

Figure 7
The vertical profile of **a** RMSD of NCEP minus CNT-analyzed relative humidity (%), **b** RMSD of NCEP minus CNT-analyzed temperature (K), **c** percentage improvement in the analysis of relative humidity with the use ATOVS radiance, **d** percentage improvement in the analysis of temperature with the use of ATOVS radiance. A set of 30 analyses during July 2008 from CNT and EXP is compared to the NCEP analysis

speed (Fig. 8c). The improvement parameter (α), which shows the improvement/degradation with the use of ATOVS radiances for relative humidity, temperature and winds, is shown in Figs. 9, 10 and 11.

An RMSD of the order of 6 (lower and upper troposphere) to 10% (middle troposphere) is observed

in the 6-h forecast of relative humidity by WRF CNT. As expected, the RMSD is increased to 17% (middle troposphere) at the 48-h forecast (Fig. 8a). An RMSD of the order of 0.8 K (lower troposphere), 0.5 K (middle troposphere) and 1 K (upper troposphere) is observed at the 6-h forecast of temperature by WRF

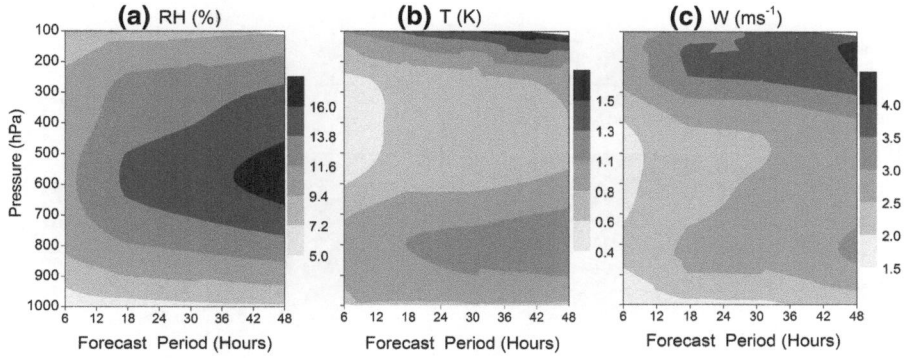

Figure 8
RMSD of 6- to 48-h CNT predicted **a** relative humidity (%), **b** temperature (K) and **c** wind speed (m s^{-1}). A set of 30 forecasts during July 2008 from CNT is compared to the NCEP analysis

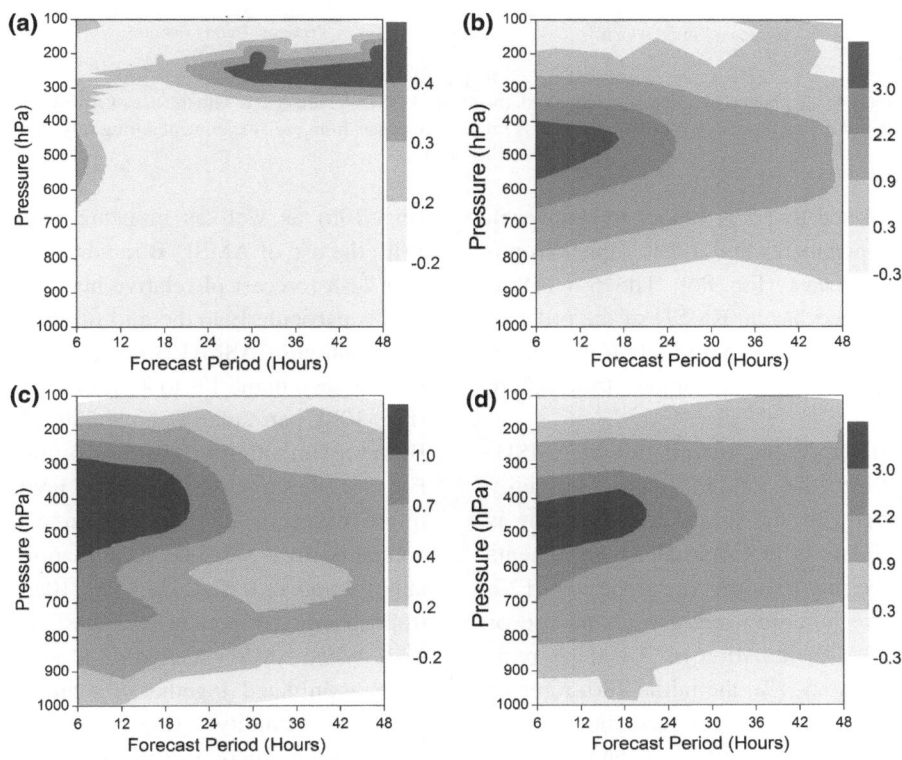

Figure 9
Percentage improvement in the prediction of relative humidity **a** with the use of AMSU-A radiance, **b** with the use of AMSU-B radiance, **c** with the use of HIRS radiance and **d** with the use of ATOVS radiance. A set of 30 forecasts from each experiment during July 2008 is compared to the NCEP analysis

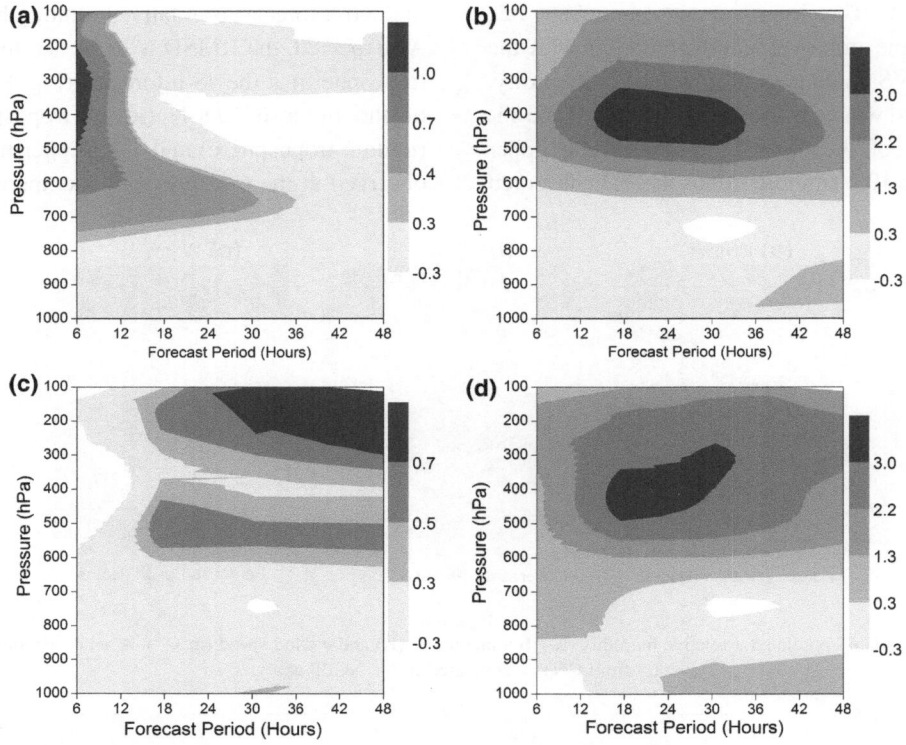

Figure 10
Percentage improvement in the prediction of temperature **a** with the use of AMSU-A radiance, **b** with the use of AMSU-B radiance, **c** with the use of HIRS radiance and **d** with the use of ATOVS radiance. A set of 30 forecasts from each experiment during July 2008 is compared to the NCEP analysis

CNT, which increased to 1.2 K (upper troposphere), 0.7 K (middle troposphere) and 1.5 K (upper troposphere) at 48-h forecast (Fig. 8b). The 6-h (48-h) forecast of wind speed has an RMSD of the order of 2.2 (2.8) m s^{-1} in the lower and middle troposphere, and 2.7 (3.7) m s^{-1} in the upper troposphere (Fig. 8c).

Compared with CNT, the assimilation of AMSU-A radiance improved the 300-hPa relative humidity forecast (Fig. 9a). The maximum improvement is seen in the later hours of the forecast. Improvements are also seen in the mid to upper tropospheric 6–12-h forecast of relative humidity because of the use of ASMU-A radiance. Assimilation of ASMU-A radiance also improved (0.5–1%) the initial 18-h forecast of mid-upper tropospheric temperature, and a small degradation (0.5%) is seen in the 100 and 400 hPa temperature forecast, particularly after 18 h (Fig. 10a). Compared to CNT, the impact of AMSU-B radiance is positive on the temperature

(Fig. 10b) as well as moisture (Fig. 9b) forecasts. With the use of AMSU-B radiance, the RMSD in the first 24-h forecast of relative humidity is reduced by 3–4%, particularly in the mid-troposphere. Due to the assimilation of ASMU-B radiances, a large improvement is seen in the 12- to 42-h forecast of temperature (Fig. 10b), particularly above 500 hPa. Like AMSU-B, the assimilation of HIRS also improved (1–1.5%; Fig. 9c) the relative humidity prediction, but the improvement is slightly less in case of HIRS as compared to AMSU-B. Assimilation of HIRS radiances improved (0.3–1%; Fig. 10c) the mid-upper tropospheric temperature forecast, particularly after 12 h. When AMSU-A, AMSU-B and HIRS radiances were assimilated together, the improvement in the relative humidity (Fig. 9d) and temperature (Fig. 10d) prediction was very similar to that in AMSU-B (Fig. 9b; Fig. 10b). The assimilation of ATOVS, particularly AMSU-B, also led to a positive impact on wind forecast. The assimilation of AMSU-B

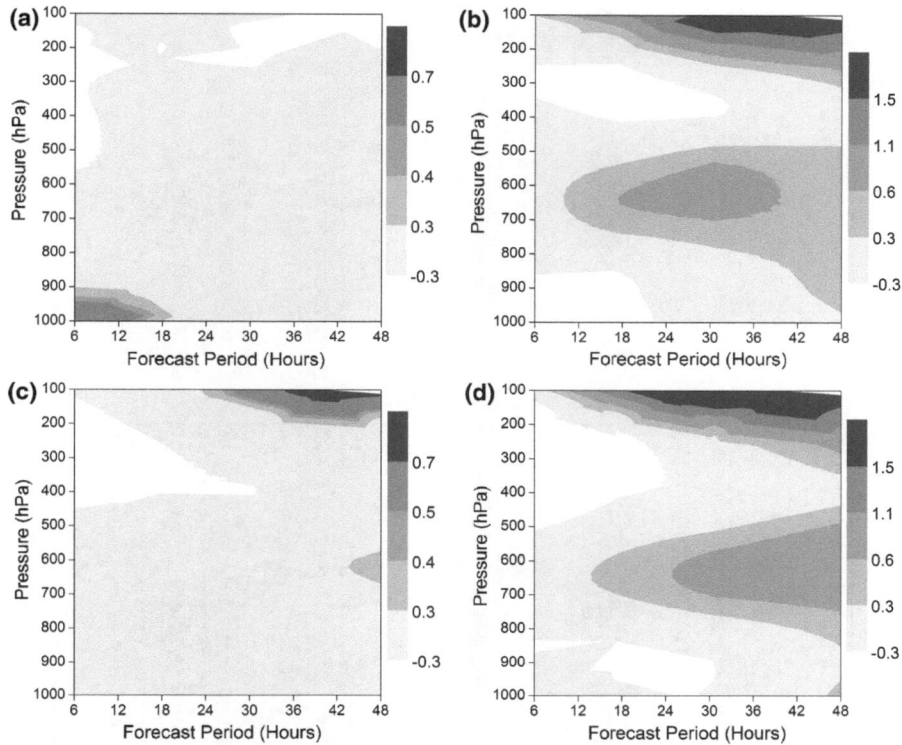

Figure 11
Percentage improvement in the prediction of wind speed **a** with the use of AMSU-A radiance, **b** with the use of AMSU-B radiance, **c** with the use of HIRS radiance and **d** with the use of ATOVS radiance. A set of 30 forecasts from each experiment during July 2008 is compared to the NCEP analysis

radiance improved (Fig. 11b) wind forecast at 100 hPa and in a layer between 750 and 550 hPa, while HIRS improved (Fig. 11c) the 100-hPa winds forecast. The first 18-h forecast of lower level winds is improved (Fig. 11a) with the assimilation of AMSU-A radiance. This improvement in the wind field can be viewed as a consequence of mass-wind adjustment forced by model dynamics.

The largest improvement in the mid and upper tropospheric humidity, temperature and wind forecast is observed because of the assimilation of AMSU-B radiance. Furthermore, in order to identify the regions where improvement/degradation is caused by the assimilation of AMSU-B, we analyzed the spatial distribution of the forecast improvement parameter on the selected levels. The spatial distribution of the forecast improvement parameter (defined in Eq. 4) procured with the assimilation of AMSU-B radiance for 6-, 12-, 24- and 48-h forecasts of 400 hPa relative

humidity is displayed in Fig. 12. The assimilation of AMSU-B radiance showed significant improvement in humidity prediction at 400 hPa. The improvements of the order of 15–25% in 06 to 24-h forecast are seen over the Indian Ocean, Middle East and higher northern latitudes. Although pockets of negative values, particularly at the 48-h forecast, are seen, the domain averaged value is still positive. A similar impact of the assimilated AMSU-B radiance on the forecast is shown in Fig. 13, but this time for the temperature forecast at 400 hPa. The improvements of the order of 10–20% in 6- to 24-h forecasts are seen over the eastern equatorial Indian Ocean, western equatorial Indian Ocean (only in the 12- and 48-h forecast) and Indian region (particularly in the 24-h forecast). The pockets of negative values are seen over the northern Arabian Sea, southwestern Indian Ocean and western equatorial Indian Ocean (in the 24-h forecast). Spatial distribution of

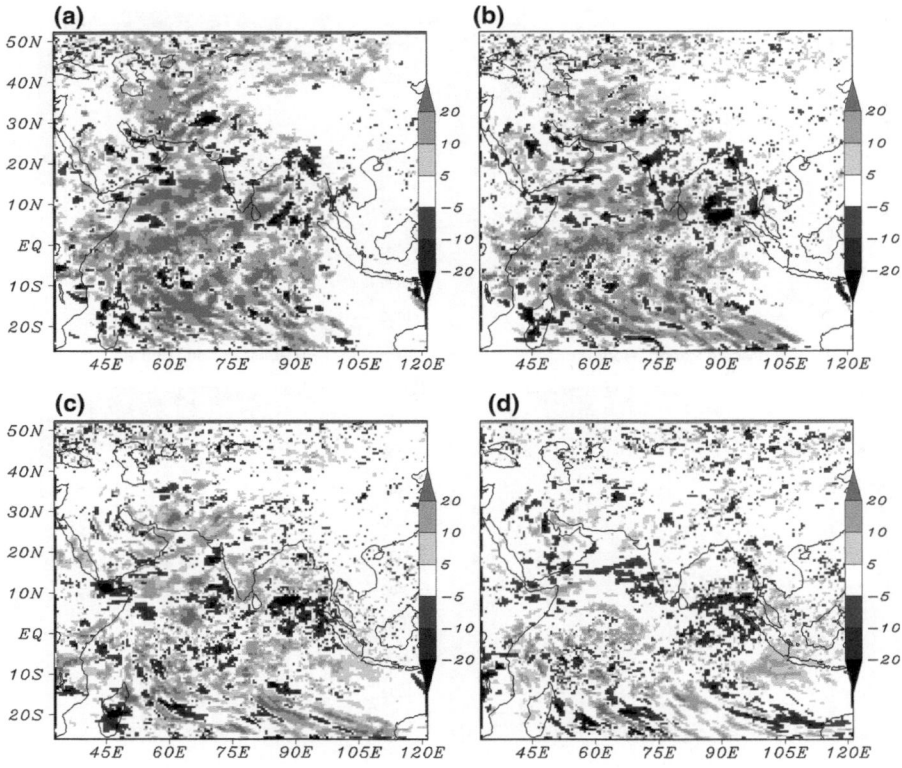

Figure 12
Horizontal maps of forecast improvement parameters (%) for 400-hPa relative humidity **a** at 6-h forecast, **b** 12-h forecast, **c** 24-h forecast and **d** 48-h forecast. A set of 30 forecasts from control and radiance assimilation experiments (EXP-B) is compared to the NCEP analysis

improvement parameters in the 100-hPa winds showed (not shown) that AMSU-B radiance impacted the wind forecast positively (particularly after the 12-h forecast) over the Arabian Sea, south of Sri-Lanka and the eastern equatorial Indian Ocean.

5.5.2 Rainfall Validation

Another measure of performance for evaluating the impact of satellite data on numerical weather forecasts is by verifying the qualitative precipitation forecast (QPF) resulting from control and radiance assimilation experiments. Here we have examined the spatial distribution of monthly mean rainfall during July 2008. The observed monthly mean rainfall data of India Meteorological Department (IMD) (RAJE-EVAN *et al.*, 2006) showed (Fig. 14a) the rainfall maxima over the west coast of India and eastern part of India. Much less precipitation is observed over semi-arid regions of northwest India and the rain-

shadow areas of the eastern coast of southern peninsular India. The observed rainfall maxima and rain-shadow regions were qualitatively reproduced by rainfall prediction (based on the first 24-h accumulated rainfall) from the control run (Fig. 14b). The difference plot (Fig. 14c) of accumulated rainfall prediction by WRF (from control run) with IMD-observed rainfall showed that WRF overpredicted rainfall over the western Ghat, eastern part of India and northern part of central India. The rainfall over the southern part (except western Ghat) of India and over the northern part of India is strongly underpredicted. For the quantitative assessment of improvement/degradation due to the assimilation of satellite radiance as compared to CNT, we have computed the improvement parameter η = [absolute (IMD-CNT)—absolute (IMD-EXP-B)] in monthly accumulated rainfall prediction. It is clear from the spatial distribution of η (Fig. 14d) that the assimilation of AMSU-B radiance improved (positive area is

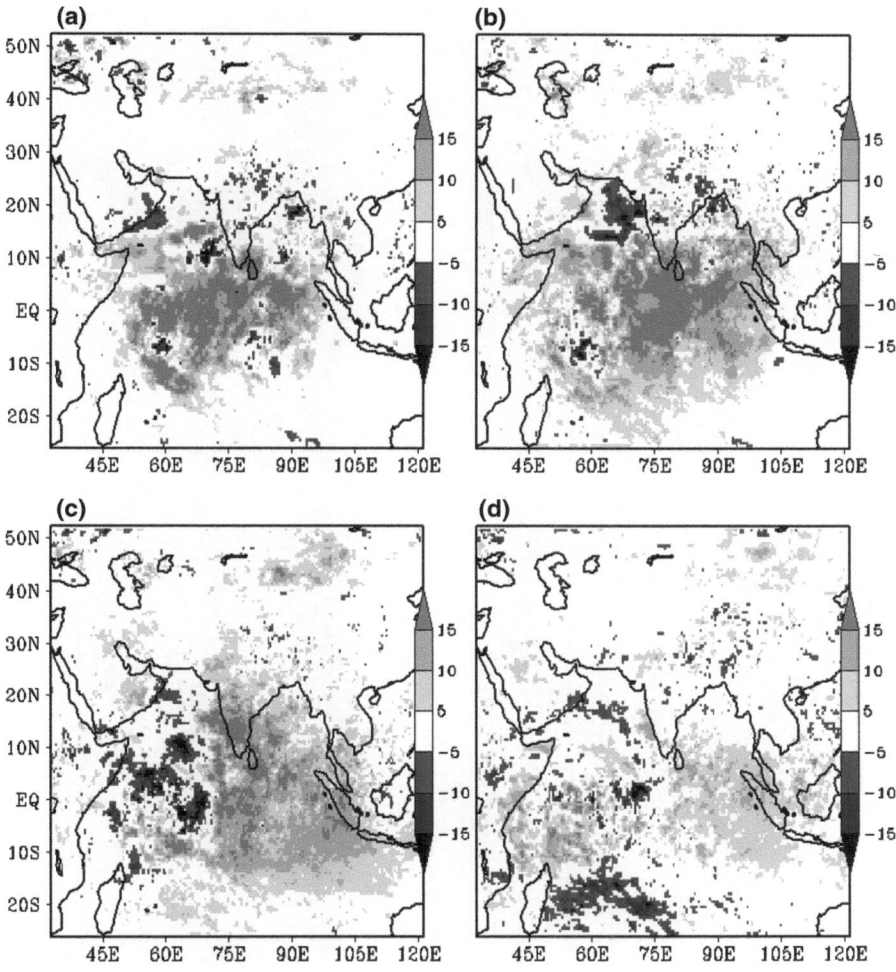

Figure 13

Horizontal maps of forecast improvement parameters (%) for 400 hPa temperature **a** at 6-h forecast, **b** 12-h forecast, **c** 24-h forecast and **d** 48-h forecast. A set of 30 forecasts from control and radiance assimilation experiments (EXP-B) is compared to the NCEP analysis

much larger than the negative area) the monthly accumulated rainfall prediction over India. The rainfall prediction improved (improvement parameter is not shown) with the use of HIRS and AMSU-A radiance, but the improvement was less as compared to the AMSU-B.

To examine the performance of CNT and EXP in reproducing the frequency of occurrence of rainfall events at or above a precipitation threshold, we also computed statistical skill scores (bias scores; BSs and equitable threat scores; ETSs; SINGH *et al.*, 2008) for the first 24-h accumulated rainfall predictions. These statistics are obtained by comparing 30 samples of daily accumulated rainfall predictions from CNT and radiance assimilation experiments with corresponding IMD observed rainfall at various rainfall thresholds. Figure 15 shows the ETS (Fig. 15a) and bias (Fig. 15b) scores for different thresholds for the control and radiance assimilation runs. The model with the radiance, particularly AMSU-B and HIRS, data assimilation shows better skill (higher ETS) in predicted rainfall over most of the thresholds than without radiance data assimilation. The positive impact is seen with the use of AMSU-B and HIRS radiance, while impact is neutral for the AMSU-A. This may be due to the fact that AMSU-B and HIRS contain moisture-sensitive channels, while AMSU-A channels are mostly sensitive to the temperature.

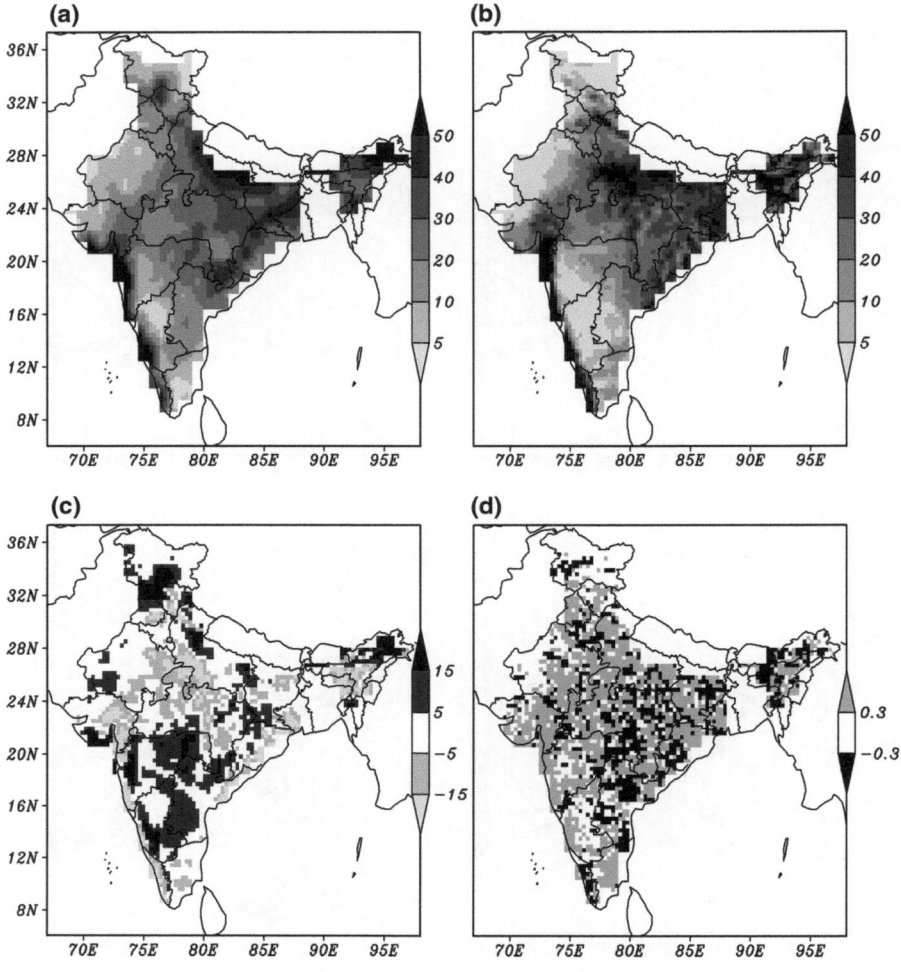

Figure 14
Accumulated rainfall (cm) for July 2008; **a** IMD observed, **b** WRF forecast from CNT, **c** IMD observed minus WRF forecast from CNT and **d** improvement parameter

6. Conclusions

This article presents a study assessing the impact of assimilation of ATOVS radiance, including AMSU-A, AMSU-B and HIRS, on the analysis and forecasts of a mesoscale model over the Indian region during the 2008 summer monsoon. The numerical experiments are conducted using Weather Research and Forecasting (WRF-ARW) and its three-dimensional variational data assimilation scheme. The control (without radiance) and experimental (with radiance) forecasts are made for 48 h each day starting at 0000 UTC from 1 to 31 July 2008. The

control run served as a baseline for verifying the assimilation experiments.

The verifications against NCEP analyses, obtained from control and radiance assimilation experiments, indicated that ATOVS radiance, particularly AMSU-B, has great potential for improving the mesoscale analysis and forecast over the Indian region. Compared to the control run, improved mid tropospheric temperature analysis was obtained with AMSU-A radiance, while AMSU-B and HIRS produced improved moisture analysis. The assimilation of AMSU-A radiance showed a positive impact in the first 24-h forecast of temperature and in 300-hPa

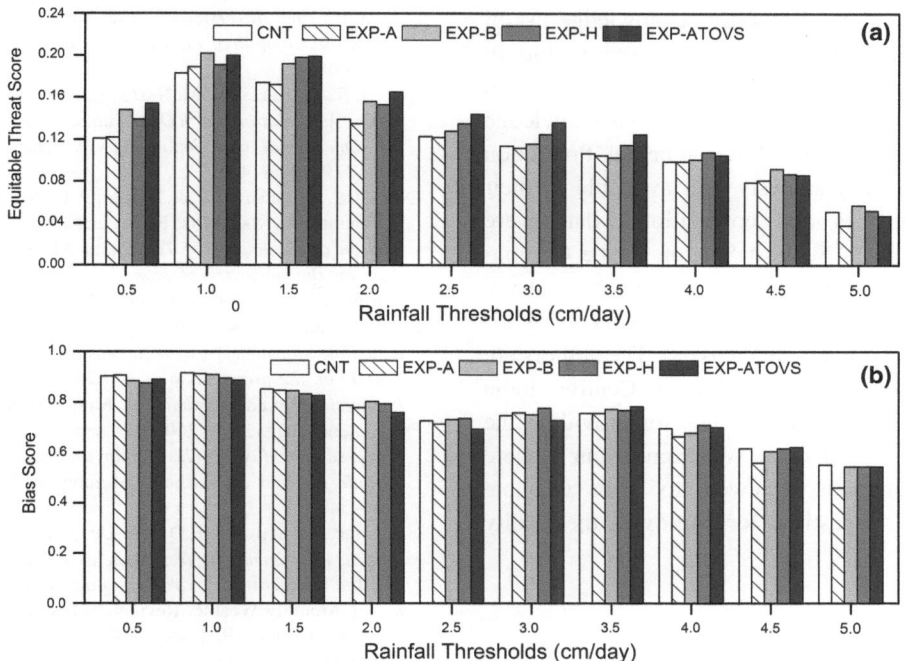

Figure 15
a Equitable threat scores and **b** bias scores in rainfall prediction from CNT and EXP. These scores are calculated by verifying 30 samples each of 24-h accumulated rainfall during July 2008

relative humidity beyond the 18-h forecast. The assimilation of AMSU-B radiance showed a large positive impact on the mid-tropospheric humidity and mid-upper tropospheric temperature forecast. The ATOVS (particularly AMSU-B) radiance also impacted the wind forecast, particularly upper and middle levels, in the later hours of the forecast. The assimilation of AMSU-B and HIRS significantly improved the rainfall forecast skill. This study clearly indicates that over this region the forecast quality is more strongly influenced by the moisture-sensitive channels than by temperature-sensitive channels. Overall, when the performance of individual instruments was compared, AMSU-B radiance consistently outperformed the control as well as other radiance assimilation experiments. In spite of the negligible impact on the temperature and wind analysis, the temperature and wind forecast was improved drastically with the assimilation of AMSU-B radiance.

Recently, FABRY and SUN (2010) concluded that uncertainties in the mid-tropospheric moisture caused the greatest uncertainties in the forecast, especially for short lead time. They suggested that by improving the mid-tropospheric moisture analysis, the short-

term forecast skill of a mesoscale model can be improved. In the present study, the assimilation of AMSU-B radiance improved the mid-tropospheric moisture analysis, which led to the improved prediction of moisture, temperature and winds. The result of our study thus confirms the findings of FABRY and SUN (2010).

In the future, the information of more humidity-sounding channels from the microwave sounder of Megha Tropiques and infrared sounder of INSAT-3D, to be launched by the Indian Space Research Organization (ISRO), will provide more vertical information useful for the definition of the humidity fields. Prior to the launch of these satellites, we plan to assess the impact of the radiance from multi-spectral sounders such as AIRS on the prediction of tropical weather systems.

Acknowledgments

WRF is made publicly available and supported by the Mesoscale and Microscale Meteorology Division at the National Center for Atmospheric Research

(NCAR). The authors gratefully acknowledge useful discussions regarding the radiance assimilation in WRF model with Dr. Zhiquan Liu and Dr. S Rizvi, NCAR, USA. The authors would like to acknowledge the National Centers for Environmental Prediction (NCEP) for making analysis data available at their site. The radiances and conventional data were obtained from Data Support Section of the Computational and Information Systems Laboratory at the National Center for Atmospheric Research in Boulder, CO (http://dss.ucar.edu/datasets). The authors are thankful to the National Climate Centre, India Meteorological Department, Pune, for providing the daily gridded rainfall data. We express our sincere thanks to anonymous reviewers for their valuable comments and suggestions for improving the quality of the article.

References

ANDERSSON, E., A. HOLLINGSWORTH, G. KELLY, P. LONNBERG, J. PAILLEUX and Z. ZHANG, 1991: *Global observing system experiments on operational statistical retrievals of satellite sounding data*. Mon. Wea. Rev., *119*, 1851-1864.

ANDERSSON, E. PAILLEUX J, THEPAUT J N, EYRE J, MCNALLY A P. KELLY G A, COURTIER P, 1994: *Use of cloud cleared radiances in three-four dimensional variation data assimilation*. Q. J. R. Meteor. Soc. *120*, 627-654.

BARKER, D. M., W. HUANG, Y.-R. GUO, A. J. BOURGEOIS, and Q. N. XIAO, 2004: *A three-dimensional variational data assimilation system for MM5: Implementation and initial results*. Mon. Wea. Rev., *132*, 897–914.

BARKER, N.L., T. F. HOGAN, W. F. CAMPBELL, R. L. PAULEY, and S. D. SWADLEY, 2005: The Impact of AMSU-A Radiance Assimilation in the U.S. Navy's Operational Global Atmospheric Prediction System (NOGAPS), A technical Report NRL/MR/7530–05-8836 pp 18.

BOUTTIER F. and G. KELLY, 2001: *Observing-system experiments in the ECMWF 4D-Var data* assimilation *system*, Q. J. Royal Meteorol. Soc., *127*, 1469-1488.

CHOU, S.-H., ZAVODSKY, B., JEDLOVEC, G. and LAPENTA, W., 2006: *Assimilation of Atmospheric Infrared Sounder (AIRS) data in a regional model*. Preprints, 14th Conference on Satellite Meteorology and Oceanography (Atlanta, GA: American Meteorological Society), CD-ROM, P5.12.

DERBER, J. C. and W. S. WU, 1998: *The use of TOVS cloud-cleared radiances in the NCEP SSI analysis system*. Mon. Wea. Rev., *126*, 2287-2299.

DOYLE, J.D. and WARNER, T.T., 1998: Verification of mesoscale objective analyses of VAS and rawinsonde data using the March 1982 AVE/VAS Special Network Data. Monthly Weather Review, *116*, pp. 358–367.

DUDHIA, J., 1989: *Numerical study of convection observed during the winter monsoon experiment using a mesoscale two-dimensional model*. J. Atmos. Sci., *46*, 3077-310.

EYRE, J. R., 1989: *Inversion of cloudy satellite sounding radiances by non-linear optimal estimation*. Quart. J. Roy. Meteor. Soc., *115*, 1001-1037.

EYRE, J. R., G. KELLY, A. MCNALLY, E. ANDERSON, and A. PERSSON, 1993: *Assimilation of TOVS radiance information through one-dimensional variational analysis*. Quart. J. Roy. Meteor. Soc., *119*, 1427-1463.

ENGLISH, S. J., RENSHAW, R. J., DIBBEN, P. C., SMITH, A. J., RAYER, P. J., POULSEN, C., SAUNDERS, F. W., and EYRE, J. R, 2000: *A comparison of the impact of TOVS and ATOVS satellite sounding data on the accuracy of numerical weather forecasts*. Quart. J. Roy. Meteorol. Soc., *126*, p. 2911-2931

FABRY, F., and J. SUN, 2010: *For how long should what data be assimilated for the mesoscale forecasting of convection and why? Part 1: on the propagation of initial conditions errors and their implications for data assimilation*, Mon. Wea. Rev., *138*, 242-255.

FAN, X. and TILLEY, S.J., 2005: *Dynamic assimilation of MODIS-retrieved humidity profiles within a regional model for high-latitude forecast applications*. Monthly Weather Review, *133*, 3450–3480.

GAL-CHEN, T., SCHMIDT, B.D. and UCCELLINI, L.W., 1986: *Simultaneous experiments for testing the assimilation of geostationary satellite temperature retrievals into a numerical prediction model*. Monthly Weather Review, *114*, 1213–1230.

GRODY, N., J. ZHAO, R. FERRARO, F. WENG, and R. BOERS, 2001: *Determination of precipitable water and cloud liquid water over oceans from the NOAA-15 advanced microwave sounding unit*, J. Geophys. Res., *106*, 2943-2953.

GOLDBERG, M. D., 1999: Generation of retrieval products from AMSU-A: Methodology and validation. *Tech. Proc. 10th Int. TOVS Study Conf.*, Boulder, CO, Bureau of Meteorological Research Centre, 215–229.

HONG, S. -Y., AND H. -L. PAN, 1996: *Nonlocal boundary layer vertical diffusion in a medium-range forecast model*, Mon. Wea. Rev., *124*, 2322-2339.

HUA, Z., C., JIFAN, and Q., CHONGJIAN, 2004a: *Assimilation analysis of Rammasun typhoon stricture over Northwest Pacific using satellite data*. Chinese science Bulletin *49* (4), 389-395.

HUA, Z., J.-S. XUE, G.-F. ZHU, S.-Y. ZHUANG, X.-B WU, and F.-Y ZHANG, 2004b: *Application of direct assimilation of ATOVS microwave radiances to typhoon track prediction*. Advan. Atmos. Sci., *21*(**2**), 283-290.

KAIN, J. S., 2004: *The Kain-Fritsch convective Parameterization: An update*. J. Apply. Meteor., *43*, 170-181.

LE DIMET, F.-X., and O. TALAGRAND, 1986: *Variational algorithms for analysis and assimilation of meteorological observations*: Theoretical aspects. Tellus, *38A*, 97–110.

LIU, Z.-Q. and BARKER, D. M., 2006. Radiance Assimilation in WRF-Var: Implementation and Initial Results. 7th WRF user's workshop, Boulder, Colorado. 19-22 June 2006. http://www.mmm.ucar.edu/wrf/users/workshops/WS2006/abstracts/Session04/4_2_Liu.pdf.

LIPTON, A.E., MODICA, G.D., HECKMAN, S.T. and JACKSON, A.J., 1995: *Satellite-model coupled analysis of convective potential in Florida with VAS water vapor and surface temperature data*. Monthly Weather Review, *123*, 3292–3304.

LIPTON, A.E. and VONDERHAAR, T.H., 1990: *Mesoscale analysis by numerical modeling coupled with sounding retrieval from satellites*. Monthly Weather Review, *118*, 1308–1329.

LI, J., and H. LIU, 2009: *Improved hurricane track and intensity forecast using single field-of-view advanced IR sounding measurements*, Geophys. Res. Lett., *36*, L11813, doi:10.1029/2009GL038285.

MCNALLY, A. P., and M. VESPERINI, 1996: *Variational analysis of humidity information from TOVS radiances at ECMWF*. Quart. J. Roy. Meteor. Soc., *122*, 1521-1544.

MCNALLY, A. P, 2009: *The direct assimilation of cloud-affected satellite infrared radiances in the ECMWF 4D-Var*. Quarterly Journal of the Royal Meteorological Society, *135*:1214–1229. doi:10.1002/qj.426.

MCMILLIN, L. M., and C DEAN, 1982: *Evaluation of a new operational technique for producing clear radiances*. Journal of Applied Meteorology, *21*, 1005-1014.

MLAWER, E. J., S. J. TAUBMAN, P. D. BROWN, M. J. IACONO, and S. A. CLOUGH, 1997: *Radiative transfer for inhomogeneous atmosphere: RRTM, a validated correlated-k model for the long-wave*. J. Geophy. Res., *102*(D14), 16663-16682.

MONTMERLE T, F. RABIER and C FISCHER, 2007: *Relative impact of polar-orbiting and geostationary satellite radiances in the Aladin/France numerical weather prediction system*. Quart. J. Roy. Meteor. Soc., *133*, 655-671.

PARRISH, D. F., and J. C. DERBER, 1992: *The National Meteorological Centre's spectral statistical interpolation analysis system*. Mon. Wea. Rev., *120*, 1747-1763.

RAJEEVAN, M., J. BHATE, J. D. KALE, and B. LAL, 2006: *High resolution daily gridded rainfall data for the Indian region: Analysis of break and active monsoon spells*. Current Science, *91*, 296-306.

REALE, O., W. K. LAU, J. SUSSKIND, E. BRIN, E. LIU, L. P. RIISHOJGAARD, M. FUENTES, and R. ROSENBERG, 2009: *AIRS impact on the analysis and forecast track of tropical cyclone Nargis in a global data assimilation and forecasting system*, Geophys. Res. Lett., *36*, L06812, doi:10.1029/2008GL037122.

RUGGIERO, F.H., SASHEGYI K.D., LIPTON, A.E., MADALA, R.V. and RAMAN, S., 1999: *Coupled assimilation of geostationary satellite sounder data into a mesoscale model using the Bratseth analysis approach*. Monthly Weather Review, *7*, 802–820.

SKAMAROCK, W. C., KLEMP, J. B., DUDHIA, J., GILL, D. O., BARKER, D. M., WANG W., and POWERS, J. G, 2008: A description of the Advanced Research WRF Version 3. NCAR Tech Notes-475 + STR (http://www.mmm.ucar.edu/wrf/users/docs/arwv3.pdf).

SAUNDERS, R., M. MATRICARDI and P. BRUNEL, 1999: *An improved fast radiative model for assimilation of satellite radiance observations*. Quart. J. Roy. Meteor. Soc., *125*, 1407-1426.

SINGH, R., P. K. PAL, C. M. KISHTAWAL, and P. C. JOSHI, 2008: *Impact of Atmospheric Infrared Sounder Data on the Numerical Simulation of a Historical Mumbai Rain Event*. Wea. Forecasting, *23, 892-913*.

SINGH, R. P K PAL, and P C JOSHI, 2010: *Assimilation of Kalpana Very High Resolution Radiometer Water Vapor Channel Radiances into a Mesoscale Model*. J. Geophys. Res., *115*, D18124, doi:10.1029/2010JD014027.

WU, W.-S., R. J. PURSER, and D. F. PARRISH, 2002: *Three-dimensional variational analysis with spatially inhomogeneous covariances*. Mon. Wea. Rev., *130*, 2905-2916.

ZHAO, Y., B. WANG, Z. JI, X. LIANG, G. DENG, and X. ZHANG, 2005: *Improved track forecasting of typhoon reaching landfall from four-dimensional variational data assimilation of AMSU-A retrieved data*. J. Geophys. Res., *110*, D14101, doi:10.1029/2004JD005267.

ZHU, T., D.-L. ZHANG, and F. WENG, 2002: *Impact of the Advanced Microwave Sounding Unit measurements on hurricane prediction*, Mon.Weather Rev., *130*, 2416–2432.

ZAPOTOCNY, T. H., W. P. MENZEL, J. A. JUNG, and J. P. Nelson III, 2005: *A four-season impact study of rawinsonde, GOES, and POES data in the Eta Data Assimilation System. Part II: Contribution of the components*. Wea. Forecasting, *20*:178–198.

ZAVODSKY, B.T., LAZARUS, S.M., BLOTTMAN P.F. and SHARP, D.W., 2004: Assimilation of MODIS temperature and water vapor profiles into a mesoscale analysis system. In 20th Conference on Weather Analysis and Forecasting/16th Conference on Numerical Weather Prediction, Seattle, WA, 11–15 January 2004. Available online at: http://srh.noaa.gov/mlb/PDFs/ams_preprints_lazarus.pdf.

ZHOU, Y. P., K.-M. LAU, O. REALE, and R. ROSENBERG, 2010: *AIRS impact on precipitation analysis and forecast of tropical cyclones in a global data assimilation and forecast system*, Geophys. Res. Lett., *37*, L02806, doi:10.1029/2009GL041494.

(Received October 29, 2010, accepted January 14, 2011, Published online July 20, 2011)

Pure Appl. Geophys. 169 (2012), 447–465
© 2011 Springer Basel AG
DOI 10.1007/s00024-011-0380-5

❙Pure and Applied Geophysics

Direct and Inverse Problems in a Variational Concept of Environmental Modeling

VLADIMIR PENENKO,[1] ALEXANDER BAKLANOV,[2] ELENA TSVETOVA,[1] and ALEXANDER MAHURA[2]

Abstract—A concept of environmental forecasting based on a variational approach is discussed. The basic idea is to augment the existing technology of modeling by a combination of direct and inverse methods. By this means, the scope of environmental studies can be substantially enlarged. In the concept, mathematical models of processes and observation data subject to some uncertainties are considered. The modeling system is derived from a specially formulated weak-constraint variational principle. A set of algorithms for implementing the concept is presented. These are: algorithms for the solution of direct, adjoint, and inverse problems; adjoint sensitivity algorithms; data assimilation procedures; etc. Methods of quantitative estimations of uncertainty are of particular interest since uncertainty functions play a fundamental role for data assimilation, assessment of model quality, and inverse problem solving. A scenario approach is an essential part of the concept. Some methods of orthogonal decomposition of multi-dimensional phase spaces are used to reconstruct the hydrodynamic background fields from available data and to include climatic data into long-term prognostic scenarios. Subspaces with informative bases are constructed to use in deterministic or stochastic-deterministic scenarios for forecasting air quality and risk assessment. The results of implementing example scenarios for the Siberian regions are presented.

Key words: Variational principle, weak-constraint formulation, sensitivity, uncertainty, 4D-Var data assimilation, uncertainty assessment, environmental risk, inverse problems.

1. Introduction

Variational principles are useful and universal tools for a combined treatment of mathematical models and observational data for various scientific and practical purposes. When a variational approach is used the influence of the various kinds of uncertainties in models and data is diminished.

The first applications of variational principles with adjoint problems to the construction of mathematical models and methods for numerical modeling of atmospheric thermodynamics were made at Novosibirsk Computing Center, Russia. These were studies related to assessment of model sensitivity to variations in the parameters, input data, and sources (PENENKO 1975), 4D-Var data assimilation (PENENKO and OBRAZTSOV 1976), and the organization of equations for feedback relations between the model parameters and variations in goal functional in which sensitivity functions (SF) were used (PENENKO 1981). Later, the studies on variational data assimilation were continued by LEWIS and DERBER (1985), LE DIMET and TALAGRAND (1986), and TALAGRAND and COURTIER (1987).

The above methods can be referred to as methods of the first generation of variational data assimilation. The goal functional is a measure of deviation of the calculated values of state functions from the observed ones. In this approach, it is important that the mathematical models of processes are used as strong constraints. Implicitly, it means that the models are assumed to be exact. Hence, the differences between the calculated and observed data are produced by uncertainties in the initial data. Therefore, the initial data values for the models are considered as unknown parameters. The basic calculation cycle includes the solving of a direct problem followed by specially constructed adjoint problems generated by a variational principle. The available measurement data are used to correct the initial conditions using SFs. The SFs are calculated from the solutions of the direct and adjoint problems. It is obvious that the adjoint problems are not in full use here, since the refined solution is obtained by fitting to the initial data only.

[1] Institute of Computational Mathematics and Mathematical Geophysics (ICM&MG), Siberian Branch of the Russian Academy of Sciences, 6, Prospekt Lavrentieva, Novosibirsk 630090, Russia.
[2] Research Department, Danish Meteorological Institute (DMI), Lyngbyvej 100, 2100 Copenhagen, Denmark. E-mail: alb@dmi.dk

Taking this into account, PENENKO (1985) proposed a method of data assimilation with weak constraints. In this case, additional "uncertainty" terms are explicitly included in some elements of the models of processes to show the character of "imperfection." The uncertainty is determined from the goal functional minimum conditions. To this purpose, the goal functional is augmented to assess all uncertainties in the models of processes. This augmented functional is built by a weighted least-squares method. To implement the variational principle, an algorithm of direct and inverse modeling is constructed. This algorithm possesses some additional features as compared with the strong-constraint methods. These algorithms can be called the second generation of variational methods. Later VAN LEEUVEN and EVENSEN (1996) and EVENSEN and FARIO (1997) presented algorithms with weak constraints as well. By now the weak constraint approach has been intensively developed by many authors.

Some merits of the second generation methods are: the length of the assimilation window is not essential, and the sensitivity to initial conditions is reduced. This is because the discrepancy between the calculated values and measured data can be distributed by the assimilating system through estimations of the model uncertainty function (UF) over the whole time interval of the assimilation window (PENENKO 2009). From the control theory point of view, the UFs are a control introduced into the system to reach a minimum of the observational functional. In the first generation methods, the control is introduced at the initial time. The second generation differs from the first one in that the control is used not only at the initial time, but distributed over all space-time domain and for those elements of the modeling system that are provided with additional uncertainty terms.

The above methods of data assimilation have now come into the operational practice of numerical weather forecasting as well as into theoretical studies of natural processes. Many publications exist on various theoretical and applied aspects of data assimilation. The 5th WMO International Symposium on Data Assimilation[1] (5–9 October 2009, Melbourne,

Australia) has shown recent progress in all aspects of atmospheric, oceanographic, and hydrological data assimilation.

In this article, a concept of environmental modeling and the results of some scenario calculations for the Siberian regions are presented. The problem of data assimilation is considered here as a part of a methodology for assessing the atmospheric state and solving the direct and inverse problems produced by mathematical models used in environmental studies.

2. The Concept of Environmental Modeling

2.1. General Remarks

The presented environmental modeling concept is based on variational principles and adjoint sensitivity theory (PENENKO 1975, 1981, 2010; PENENKO and BAKLANOV 2001; PENENKO et al. 2002; PENENKO and TSVETOVA 2007; BAKLANOV 2000, 2007). An advantage of this concept is that the variational principles give a possibility to develop an integral technology for the models of various processes (hydrothermodynamic, chemical, biological, economical, etc.) on a common basis using available data about the processes. The variational principles are implemented with optimal numerical schemes and universal algorithms of direct and inverse modeling.

Our approach is essentially based on methods of adjoint equations, which have been continuously developing since pioneering papers in mathematical physics and control theory (LIONS 1968; BRYSON and HO 1969), and atmosphere and ocean dynamics (MARCHUK 1974, 1975, 1995).

In the suggested concept (Fig. 1), some objects in the form of functionals describing some characteristics of the processes, data, and models are considered together with basic model components. Ecological restrictions on environmental quality, results of measurements, control and design criteria, model quality criteria, etc., can be represented by some functionals. The restriction functionals are introduced to optimize environmental protection, management of environmental quality, ecological design, etc. The constraints are built on conditions of environmentally sustainable development and ecological safety. Typical space-time distributed restrictions are mathematically

---

[1] http://www.cawcr.gov.au/staff/pxs/wmoda5/abst_and_pdf2.htm.

expressed requirements of atmospheric quality standards. The characteristics and constraints can be global, local, distributed, or point-wise. With respect to the state variables, the functionals can be linear as well as nonlinear. To estimate the functionals written in a generalised form, the sensitivity, optimisation, and inverse modeling methods based on a variational principle are used. It is assumed that each element of the technology of modeling can have an uncertainty.

In this case, it is natural to build the variational principle in a weak-constraint formulation, where the uncertainties play the role of control parameters. For this purpose, the total measure of the uncertainties should be minimized. The variational principle being proposed ensures that all numerical models of processes are formulated in consistency to one other and a whole.

The basics of the variational principle are given in PENENKO (1981); PENENKO and TSVETOVA (2009b). In essence, the goal characteristics, formulated as functionals defined on the phase spaces of model state variables, can be studied in the spaces of input parameters and forcings. The models considered here play two roles: (1) they describe the relations between the state variables and parameters, and (2) they are the restrictions on the class of functions where the functionals are defined. To formulate the variational principle, additional adjoint functions are introduced. They are calculated by solving the corresponding adjoint problems. Each goal functional generates its own adjoint problem. A sensitivity relation (SR) for the chosen functional is produced using a variational technique. By means of SFs, it

explicitly links the functional variation with variations of all parameters described in the statement of the problem studied.

The SR forms a constructive basis for creating algorithms of direct and inverse relations between the parameters and general type functional. The SFs are calculated from the solutions of direct and adjoint problems. All internal degrees of freedom in the models of processes and observations are retained in the SFs. Thus, the functional variations are dependent on the variations of parameters solely. It turns out that they are independent of the state function variations.

The traditional methods of direct modeling are used for the simulation of spatial and temporal behavior of state functions with given input parameters. To account for the variability of model parameters and forcings, one has to make as many model runs as the number of parameters being changed. Only after that, the goal functional can be estimated. A recent common example is the design of scenario ensembles.

In our approach, the direct problem is only solved once with a given basic set of parameters. From results obtained, the adjoint problem for the goal functional is provided with input data. Using the SR, one can calculate the values of the goal functional variations produced by variations in each parameter *within the desired range with only one run of the direct problem.* Therefore, our formulation of direct modeling is cost-effective especially in multi-parametric studies.

For the inverse modeling, some variational methods can be used to calculate the backward

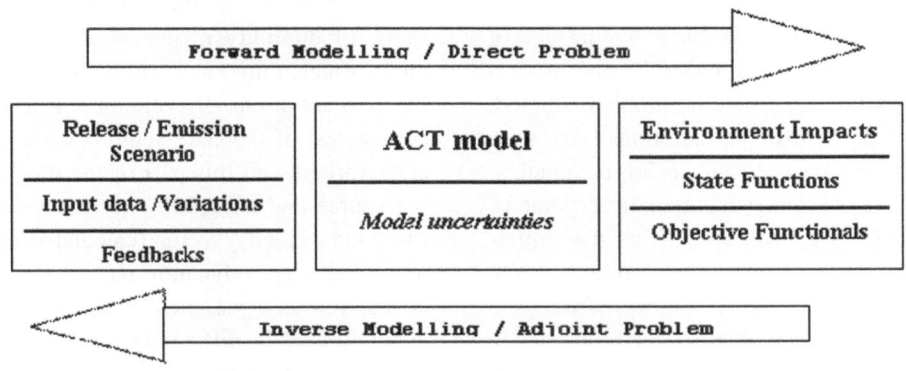

Figure 1
Simplified scheme of variational methods and control theory approach for environmental risk assessments and mitigation strategy optimization

propagation of information from the target functional (for example, observational functional) to the parameters using the SF. There are special constructions, based on the variational principle, to create the feedback algorithms and methods of control theory.

2.2. Models of Processes

The processes considered can be described by hydrodynamic models combined with models of atmospheric transport and transformation of moisture, as well as chemically and optically active gaseous and aerosol pollutants.

In the models, the following types of mathematical objects/functions are introduced: the state vector $\boldsymbol{\varphi} = (\varphi_1, \varphi_2, \ldots, \varphi_{n_s})^T \in Q(D_t)$ with elements $\{\varphi_i(\mathbf{x}, t), i = \overline{1, n_s}, n_s \geq 1\}$, and the vector of model parameters $Y = (Y_1, Y_2, \ldots, Y_N)^T \in R(D_t)$ with elements $\{Y_k(\mathbf{x}, t), k = \overline{1, N}, N \geq 1\}$, where T-upper index stands for transposition; $i = \overline{1, n}$ denotes the ordered set of integers $i = 1, 2, \ldots, n$; n_s is the number of state variables $\varphi_i(\mathbf{x}, t)$, which are 4D fields; $D_t = D \times [0, \bar{t}] \subset R^4$ is the domain of definition of spatial variables $\mathbf{x} = (x_1, x_2, x_3) \in D \subset R^3$; t is the time, and $[0, \bar{t}]$ is the solution time interval; $Q(D_t)$ is the space of sufficiently smooth state functions satisfying some boundary conditions (depend on problem studied) at the boundary Ω_t of the domain (can be a sphere, hemisphere or limited area) D_t; N is the number of parameter categories; $R(D_t)$ is the domain of admissible values of the parameters; x_1, x_2 are the horizontal variables; x_3 is the vertical variable (extended from the Earth's surface to a height/pressure level of 10 hPa).

It should be noted that the above definitions of "state functions" and "parameters" are just a convention depending on a specific problem studied. For the direct problems the solutions are state functions dependent on a given set of parameters, i.e., $\boldsymbol{\varphi} = \boldsymbol{\varphi}(\mathbf{x}, t, Y)$. From a mathematical point of view, such problems are referred to as the initial-boundary-value problems. The inverse problems are formulated as follows: some objects from the category of "parameters" have to be found using additional information. In this sense, for example, the data assimilation can be considered as an inverse problem.

The parameters are divided into two categories as well. One category contains controlling parameters, and the others are in a routine group. Clearly, such division is again just a convention.

In general operator form L the models can be written as follows:

$$L(\boldsymbol{\varphi}, \mathbf{Y}) \equiv \frac{\partial \boldsymbol{\varphi}}{\partial t} + G(\boldsymbol{\varphi}, \mathbf{Y}) - \mathbf{f} - \mathbf{r} = 0, \quad (1)$$
$$\boldsymbol{\varphi}(\mathbf{x}, t) \in Q(D_t),$$

where $G(\boldsymbol{\varphi}, \mathbf{Y})$ is a nonlinear operator, $\mathbf{f}$ and $\mathbf{r}$ are source and uncertainty terms, respectively; their formal structures correspond to that of the state vector. The model parameters may also contain an uncertainty ζ,

$$\mathbf{Y} = \mathbf{Y_b} + \zeta \in \mathbf{R(D_t)}, \quad (\mathbf{x}, t) \in \mathbf{D_t}. \quad (2)$$

The initial conditions for the state variables at $t = 0$ may contain an unknown initial error $\xi(\mathbf{x})$:

$$\boldsymbol{\varphi}(\mathbf{x}, 0) = \boldsymbol{\varphi}_a(x) + \boldsymbol{\xi}(\mathbf{x}), \quad \mathbf{x} \in D. \quad (3)$$

The subscripts a and b denote a priori information for the initial state and basic values for parameters, respectively.

To be more specific, let us consider the part of the model set (1)–(3) that describes processes of heat, moisture, and pollutant transport and transformation in the atmosphere. They are described by equations of the same type:

$$\frac{\partial \pi \varphi_i}{\partial t} + \operatorname{div} \pi(\varphi_i \mathbf{u} - \boldsymbol{\mu}_i \operatorname{grad} \varphi_i) + \pi\big((H(\boldsymbol{\varphi}))_i - f_i - r_i\big)$$
$$= 0, \quad i = \overline{1, n}, \quad (4)$$

where φ_is are the elements [i.e., temperature, mixing ratios of humidity in the atmosphere (water vapor, cloud water, rain water, snow, and ice crystals), and the concentrations of pollutants in gaseous and aerosol states] of the state vector $\boldsymbol{\varphi}$; n is the number of state variables in this part of the model set; f_i and r_i are source and uncertainty terms; $\mathbf{u} = (u_1, u_2, u_3)$ is the wind velocity vector (calculated as a state variable in the hydrodynamic part of the model set and considered as a parameter); $\boldsymbol{\mu}_i = (\mu_1, \mu_2, \mu_3)_i$ is the symmetric eddy diffusivity tensor for the substance φ_i in all coordinate directions $\mathbf{x} = \{x_i, i = \overline{1, 3}\}$.

The non-negative function π satisfies the continuity equation in the hydrodynamic part of the model set:

$$\frac{\partial \pi}{\partial t} + \text{div}(\pi \mathbf{u}) = 0. \qquad (5)$$

The form of π depends on the coordinate system. For the non-hydrostatic models, π is an air density. When a vertical σ-coordinate of pressure is introduced into a hydrostatic model, $\pi = p_s - p_t$, where p_s and $p_t = \text{const}$ are the surface and top level pressures, respectively.

The second term in (4) is a convection-diffusion operator that acts on each constituent φ_i in the whole domain D_t independently of other state variables.

The non-linear operator $H(\varphi)$ describes transformation processes. In a general case it expresses the relations between some state variables in the domain D_t. It is essential that this operator does not contain derivatives with respect to the independent variables. $(H(\varphi))_i$ is the result of action of the general operator $H(\varphi)$ in the i-th equation. For example, for the hydrological cycle equations, $H(\varphi)$ presents the sum of various microphysical conversion rates in each water category. For equations with reactions of chemical species, it describes an outcome of chemical reactions for multi-component species. For aerosols, it gives a description of production and loss mechanisms due to interaction between particles of various types and sizes. For the heat transfer equation it shows the impact of phase transformations of water as a function of air temperature.

Note that the operator $H(\varphi)$ is assumed to be bounded and Gâteaux differentiable with respect to all components of the state vector φ. These properties are required for linearization of the operator.

Let us impose boundary conditions written in the following form:

$$\left\{ \mu_n \frac{\partial \varphi}{\partial \mathbf{n}} + \beta \varphi - \mathbf{b} = 0 \right\}_i, \quad i = \overline{1,n};\ (\mathbf{x}, t) \in \Omega_t, \qquad (6)$$

where β_is are given functions; $\mathbf{b}$ is a vector function describing sources on the boundary Ω_t of the domain D_t; $\mathbf{n}$ is the outer unit normal to Ω_t.

Physically, all state variables in (4)–(6) are non-negative. This is a fundamental property of the space $Q(D_t)$, in addition to those mentioned above.

In (2)–(6) the following seven objects are considered as "parameters" in $\mathbf{Y} = \{\mathbf{Y}_i(\mathbf{x}, t),\ i =$

$\overline{1, N}\} : \mathbf{u}, \boldsymbol{\mu}, \mathbf{f}, \varphi(\mathbf{x}, 0), \boldsymbol{\beta}, \mathbf{b}, \mathbf{Y}_H, N = 7$. The physical meaning of the first six objects is obvious from (2)–(6). The object $\mathbf{Y}_H$ is the set of inner parameters of various kinds of transformation operators $H(\varphi)$ described above.

One aspect of the concept is that the modeling system provides initial fields and spatial–temporal distributions of the state functions that are optimally consistent with available observations.

To incorporate observational data into the modeling system, the following functional dependence between the measured data and the model state variables is formulated:

$$\boldsymbol{\Psi}_{\text{obs}} = \mathbf{W}(\varphi) + \eta, \quad (\mathbf{x}, t) \in D_t^{\text{obs}}, \qquad (7)$$

where $\boldsymbol{\Psi}_{\text{obs}}$ denotes the vector of available observations; $\mathbf{W}(\varphi)$ is the operator of the observation model; η is the vector of observation uncertainty (due to errors of measurements and observation models); $D_t^{\text{obs}} \subset D_t$ is the set of observation locations.

The location of an observation is a point in a 4D domain with coordinates $(\mathbf{x}, t) \in D_t^{\text{obs}}$. The vector $\boldsymbol{\Psi}_{\text{obs}}$ has a block structure. The number of blocks is equal to the number of the various kinds of observations. The elements of each block contain the measured data of a particular kind. In the general case, $\mathbf{W}(\varphi)$ is a nonlinear operator following the $\boldsymbol{\Psi}_{\text{obs}}$ block structure. This operator transforms the model state variables into observed quantities, including interpolation of the calculated data into the observation locations. Similar to the transformation operators, the operators $\mathbf{W}(\varphi)$ are constructed to be bounded and differentiable with respect to the state vector components. It is possible that various kinds of observations can be considered in the set of models (7) and included in the data assimilation system. Possible ways of including ground-based and remote sensing observation models into the modeling system had been discussed by PENENKO (1981, 2000). In Sect. 5.2 one such model will be described.

2.3. Variational Formulation of the Model

A wide variety of processes of different spatial-temporal scales can be described by mathematical models of atmospheric dynamics and chemistry. The variational principles are useful to match the models to available measurement data.

Let us introduce a weighted inner product as follows:

$$(\boldsymbol{\varphi}, \boldsymbol{\psi}) = \int\limits_{D_t} \left(\sum_{i=1}^{n} a_i(\varphi_i(\mathbf{x}, t)\psi_i(\mathbf{x}, t)) \right) dD \, dt,$$

$$\boldsymbol{\varphi}, \boldsymbol{\psi} \in L_2(D_t), \tag{8}$$

where dD is a volume element in D; $L_2(D_t)$ is the Hilbert space of real, sufficiently smooth, and square-integrable functions on the domain D_t. Because the vectors being considered are multivariate, the scaling factors a_i are used in (8).

The next step is to create a variational form of the model set (1–4). Following standard variational technique (see, for example, PEARSON 1974), introduce a vector $\boldsymbol{\varphi}^*$ composed of arbitrary, sufficiently smooth test functions from the space of adjoint functions $Q^*(D_t)$ and form the following integral identity for the model set:

$$I(\boldsymbol{\varphi}, \mathbf{Y}, \boldsymbol{\varphi}^*) = (L(\boldsymbol{\varphi}, \mathbf{Y}), \boldsymbol{\varphi}^*) = 0, \quad \forall \boldsymbol{\varphi}^* \in Q^*(D_t). \tag{9}$$

The structure of $\boldsymbol{\varphi}^*$ is formally identical to that of $\boldsymbol{\varphi}$. When the integral identity (9) is deduced, the boundary and initial conditions are taken into account.

After executing all necessary transformations in (9) for the model set of (4)–(6), we finally obtain:

$$I(\boldsymbol{\varphi}, \mathbf{Y}, \boldsymbol{\varphi}^*)$$

$$= \sum_{i=1}^{n} a_i \left\{ (\Lambda\varphi, \varphi^*)_i + \int\limits_{D_t} (H(\boldsymbol{\varphi})_i - f_i - r_i)\varphi_i^* \pi \, dD \, dt \right\}$$

$$= 0, \tag{10}$$

$$(\Lambda\varphi, \varphi^*)_i = \left\{ \int\limits_{D_t} \left\{ 0.5 \left[\left(\varphi^* \frac{\partial \pi\varphi}{\partial t} - \varphi \frac{\partial \pi\varphi^*}{\partial t} \right) \right. \right. \right.$$

$$+ (\varphi^* \operatorname{div} \pi\varphi\mathbf{u} - \varphi \operatorname{div} \pi\varphi^*\mathbf{u}) \Big]$$

$$+ (\pi\mu \operatorname{grad}\varphi, \operatorname{grad}\varphi^*) \Big\} dD \, dt$$

$$+ \int\limits_{D} 0.5\varphi\varphi^* \pi \, dD \Big|_0^{\bar{t}}$$

$$+ \int\limits_{\Omega_t} (0.5\varphi u_n + (\beta\varphi - q - \varepsilon)\varphi^* \pi \, d\Omega \, dt \Bigg\}_i. \tag{11}$$

Formula (11) is a generalized form of the convection and diffusion operators with initial (3) and boundary (6) conditions; u_n is the normal component of the velocity vector on the boundary Ω_t. The variational form (11) is symmetrized with the help of function π and Eq. (5). The convection part and the terms with time derivatives are transformed to a skew-symmetric form, whereas the turbulent diffusivity operators are written in a symmetric form. These properties are valid for any sufficiently smooth functions $\boldsymbol{\varphi}$ and $\boldsymbol{\varphi}^* \in L_2(D_t)$ and for any (including constant) non-negative function π satisfying the continuity Eq. (5). If $\boldsymbol{\varphi}^* = \boldsymbol{\varphi}$, (11) is transformed to a balance relation of "energy" type. If $\boldsymbol{\varphi}^* = \text{const}$, a balance relation of the first order for the function $\boldsymbol{\varphi}$ itself is obtained. From these properties, we obtain corresponding balance relations for (10) as well. In essence, the properties of symmetry and balance determine the functional structure of the variational model description (9)–(11). Note, that similar operators in the models of atmospheric hydrodynamics possess the same properties (PENENKO 1981; PENENKO and ALOYAN 1985).

The next remarkable property of the integral identity is additivity. This property is essentially used in the construction of numerical schemes using decomposition and splitting methods.

It is required that the basic properties of integral identity (10), (11) remain valid in discrete approximations.

2.4. Variational Principles in Environment Protection Problems

To formulate variational principles, a set of functionals describing some characteristics of the processes and mathematical models under consideration is introduced. For the purposes of monitoring, forecasting, management, and design, the following set of general type functionals can be used:

$$\Phi_k(\boldsymbol{\varphi}) = \int\limits_{D_t} F_k(\boldsymbol{\varphi})\chi_k(\mathbf{x}, t) \, dD \, dt, \tag{12}$$

$$k = 1, \ldots, K, K \geq 1,$$

where $F_k(\varphi)$s are functions that are integrable and differentiable on the set of model state functions $Q(D_t)$; K is the number of the functionals, $\chi_k(\mathbf{x}, t) \geq 0$ are weight functions such that $\chi_k \in Q^*(D_t)$ and $\chi_k \in Q^*(D_t)$. The expression $\chi_k(\mathbf{x}, t)dDdt$ is the definition of the Radon and Dirac measures in D_t (see SCHWARTZ 1967). In a general case, the Radon measure is distributed in space. The Dirac measure with delta functions is a particular case of the Radon measure. The former is introduced for an integral representation of point-wise and other characteristics defined on manifolds of dimension less than that of D_t.

The region $D_{rk} \in \{(\mathbf{x}, t), \chi_k(\mathbf{x}, t) \neq 0\} \subset D_t$ is a receptor or verification region. Using weight functions and measures of such type one has a possibility to uniformly describe a variety of global, distributed, local, and point-wise characteristics of the system as integrals of type (12) in the domain D_t. This remark makes it possible to present the different point-wise and distributed characteristics of observations in (7) as functionals (12) for assimilation problems. This facilitates the construction of numerical algorithms.

3. Algorithms to Implement the Variational Principle

3.1. Augmented Cost Functional

Let us construct an augmented cost functional that combines all elements of the modeling system as follows:

$$
\tilde{\Phi}_k^h(\varphi, \varphi^*, Y, \eta, r, \xi, \zeta) = \left[I^h(\varphi, Y, \varphi^*) \right]_{D_t^h} + \alpha_0 \Phi_k^h(\varphi)
$$
$$
+ 0.5 \Big\{ \alpha_1 (\eta^T M_1 \eta)_{D_t^{obs}} + \alpha_2 (r^T M_2 r)_{D_t^h}
$$
$$
+ \alpha_3 (\xi^T M_3 \xi)_{D^h} + \alpha_4 (\zeta^T M_4 \zeta)_{R^h(D_t^h)} \Big\}^h,
$$
$$
k = \overline{1, K}, \ K \geq 1. \tag{13}
$$

The index h denotes discrete analogues of corresponding objects, $D_t^h \subset D_t$ is the grid domain, K is the number of goal functionals, α_i and M_i are weight coefficients and matrices, respectively.

The first term in (13) is a numerical model in the form of integral identity (9)–(11). The second term is a goal (response/cost) functional. The next four terms

are weighted least-squares functionals for uncertainty measures. The first of them is a functional of observation quality mentioned above.

The weight matrices M_1, M_2, M_3, M_4 are given to be diagonal with diagonal elements of a block structure. In each block the diagonal elements are equal to scaling factors corresponding to evaluated vectors in the functionals of uncertainty measures. The matrix M_1 in the observational functional is an empirically defined observation-dependent weight matrix. It is also of a block-diagonal structure corresponding to that of Ψ_{obs}. Its diagonal elements are dependent on the scaling factors, instrumental errors, as well as a priori estimates of the observational model representativeness. The elements of M_1 are given in the locations where data are measured. With this functional one gets a possibility to organise a procedure of data assimilation for the solution of inverse problems and uncertainty assessment.

The weight coefficients $\alpha_i \geq 0$, $i = \overline{0, 4}$ are used in some combinations to control the augmented functional structure and, as a consequence, the process of solution. For example, if some of α_i, $i = 2, 3, 4$ is equal to 0, the corresponding uncertainty is excluded from consideration.

The above formulation of the variational principle is convenient for the solution of direct and inverse problems with a given goal functional $\Phi_k^h(\varphi)$. In a class of problems related to atmospheric quality assessment and risk control, the functionals may represent different aspects of human life, from health to social organization. In such cases a set of specific functionals should be used. Sometimes it is possible and convenient to synthesize one generalized criterion. This option is foreseen by condition $K \geq 1$ in (13).

To approximate the functionals and operators in (8)–(10), some decomposition and splitting methods are used. With these methods, the additive and other above-mentioned properties of the functionals, operators, and integrands can be correctly exploited. The integrals in functionals (8)–(10), (13) are approximated by multi-dimensional quadratures in the framework of finite volume methods.

Discrete analogues for the advection-diffusion operators are constructed with a local adjoint problem technique. It is applied in each finite volume of the grid domain D_t^h. As a result, some numerical schemes of parallel splitting are obtained. They are of second order in time and exact in space. They also exactly satisfy the boundary conditions in complex domains and possess the properties of monotonicity and transportivity without use of flux-correction procedures (PENENKO and TSVETOVA 2009a).

3.2. Universal Algorithm for Direct and Inverse Modeling

To design the above-mentioned variational modeling technique, the following definition of variation of the augmented cost functional (13) is used:

$$
\delta \tilde{\Phi}_k^h = \left(\frac{\partial \tilde{\Phi}_k^h}{\partial \varphi^*}, \delta \varphi^* \right) + \left(\frac{\partial \tilde{\Phi}_k^h}{\partial \varphi}, \delta \varphi \right) + \left(\frac{\partial \tilde{\Phi}_k^h}{\partial r}, \delta r \right)
$$
$$
+ \left(\frac{\partial \tilde{\Phi}_k^h}{\partial \xi}, \delta \xi \right) + \left(\frac{\partial \tilde{\Phi}_k^h}{\partial Y}, \delta Y \right). \quad (14)
$$

Hereafter the symbol δ denotes variation of an object.

The aim is to design a set of numerical algorithms suitable for practical implementation. For this purpose we use stationary conditions for the augmented functional (13) with respect to variations of all its functional arguments at each point of the grid domain $D_t^h \subset D_t$. As a result, three types of problems are generated: direct, adjoint, and uncertainty assessment:

$$
\partial \tilde{\Phi}_k^h / \partial \varphi^* = 0, \quad \forall \varphi^* \in Q^{*h}(D_t^h), \quad (15)
$$

$$
\left\{ \partial \tilde{\Phi}_k^h / \partial \varphi = 0, \ \varphi_k^*(x)|_{t=\bar{t}} = 0 \right\}, \quad \forall \varphi \in Q^h(D_t^h), \quad (16)
$$

$$
\partial \tilde{\Phi}_k^h / \partial U = 0,
$$
$$
\forall U = \{r, \xi, \zeta, \varepsilon\} \in \{ Q^h(D_t^h), R^h(D_t^h) \}. \quad (17)
$$

With the help of (15)–(17), the first four terms in the right-hand side of (14) go to zero. Then the variation $\delta \tilde{\Phi}_k^h$ becomes $\delta \tilde{\Phi}_k^h = \left(\frac{\partial \tilde{\Phi}_k^h}{\partial Y}, \delta Y \right)$. This is a SR for the augmented cost functional with respect to variations of the model parameters.

The operations in (15)–(17) form the following system of equations, which originates from a universal algorithm of direct and inverse modeling:

$$
\frac{\partial \tilde{\Phi}_k^h}{\partial \varphi^*} = \frac{\partial \varphi}{\partial t} + G^h(\varphi, Y) - f - r = 0, \quad (18)
$$
$$
\forall \varphi^* \in Q^{*h}(D_t^h);
$$

$$
\varphi(x, 0) = \varphi_a(x) + \xi(x), \quad x \in D, \ t = 0; \quad (19)
$$

$$
\frac{\partial \tilde{\Phi}_k^h}{\partial \varphi} = -\frac{\partial \varphi_k^*}{\partial t} + A^T(\varphi, Y) \varphi_k^* + d_k = 0, \quad (20)
$$
$$
\forall \varphi \in Q^h(D_t^h);
$$

$$
d_k = \frac{\partial}{\partial \varphi} \left(\alpha_0 \Phi_k^h(\varphi) + 0,5 \left(\eta^T M_1 \eta^* \right)^h \right)
$$
$$
= \alpha_0 \frac{\partial \Phi_k^h(\varphi)}{\partial \varphi} + \left[\frac{\partial W^h(\varphi)}{\partial \varphi} \right]^T M_1 \eta^h, \quad (21)
$$
$$
\forall \varphi \in Q^h(D_t^h);
$$

$$
\varphi_k^*(x)|_{t=\bar{t}} = 0; \quad (22)
$$

$$
r(x, t) = \alpha_2^{-1} M_2^{-1} \varphi_k^*(x, t); \quad (23)
$$

$$
\xi(x) = \alpha_3^{-1} M_3^{-1} \varphi_k^*(x, 0); \quad (24)
$$

$$
A(\varphi, Y) \delta \varphi \equiv \frac{\partial}{\partial \alpha} \left[G^h(\varphi + \alpha \delta \varphi, Y) \right]_{\alpha=0}. \quad (25)
$$

Equations (15)–(25) are the Euler-Lagrange equations for our problem of minimization of (13). Here, the derivatives of non-linear operators and functionals are meant in the sense of Gâteaux. They are calculated in a vicinity of the current trajectory of the dynamical system studied. These trajectories are defined in D_t^h by a set of the functional arguments of the augmented functional (13). An algorithm for the construction of the tangent linear operator $A(\varphi, \mathbf{Y})$ is given by (25), where α is a real parameter. Linearization is sequentially implemented at each time step.

The initial conditions (22) for the function $\varphi_k^*(\mathbf{x})$ ensure the fulfillment of stationary conditions for (13) to the variations of φ at $t = \bar{t}$ [see (16)]. It should be emphasized that the construction of numerical schemes using the variational principle produces consistency of direct (18)–(19) and adjoint (20)–(22) problems, because they are obtained from the same functional $I^h(\varphi, \mathbf{Y}, \varphi^*)$ by the same approximation rules. The direct problems are solved forward in time, whereas the adjoint ones are solved backward in time. Here, some splitting schemes can be used. All available information about the observed data is included into the adjoint problem (20) by means of

(21). Following (17), the expressions (23) and (24) are used for the calculation of the two UFs from (18) and (19), respectively.

3.3. Sensitivity

As our problem is non-linear, and the initial conditions (19) and (22) are related to different times, the cycle of direct-adjoint problems is solved iteratively. To work out the problem of data assimilation and functional assessment in a sliding mode in time and to use splitting, the system (18)–(24), as a splitting scheme at each time step, can be solved by a direct algorithm without iterations. This version is implemented by a technique of local adjoint problems in time (PENENKO 2009).

Now we turn back to the basic scheme. In this case the first guess is given by

$$r^{(0)} = 0, \quad \boldsymbol{\varphi}^{(0)}(x,0) = \boldsymbol{\varphi}_a^0(x), \quad \boldsymbol{\xi}^{(0)} = 0, \qquad (26)$$
$$Y^{(0)} = Y_a.$$

At each iteration there are three stages:

1. Solution of the main problem (18)–(19) forward in time;
2. Solution of the adjoint problems (20)–(22) for each functional $\Phi_k^h(\varphi)$ backward in time;
3. Calculation of the UFs from system (23)–(24).

Then the SRs for the functional $\tilde{\Phi}_k^h$ (13) with respect to variations of the vector of parameters are produced by the algorithm:

$$\delta\tilde{\Phi}_k^h \equiv \frac{\partial}{\partial\alpha} I^h\left(\boldsymbol{\varphi}, Y+\alpha\delta Y, \boldsymbol{\varphi}_k^*\right)|_{\alpha=0} \equiv (\Gamma_k, \delta Y)$$
$$= \sum_{i=1}^{N} \langle\Gamma_{ki}, \delta Y_i\rangle. \qquad (27)$$

Here $\delta\tilde{\Phi}_k^h$ is variation of the functional $\tilde{\Phi}_k^h$, $\{Y, Y+\alpha\delta Y\} \in R^h(D_t^h)$. Relation (27) defines an inner product in the spaces of SFs, Γ_k, and variations of parameters, $\delta\mathbf{Y}$. The last term in (27) defines an inner product, $(\Gamma_k, \delta\mathbf{Y})$, as a sum of the inner products $\langle\Gamma_{ki}, \delta\mathbf{Y}_i\rangle$ for the various categories of parameters and corresponding sensitivity functions.

To calculate the SFs, the following algorithm is used:

$$\Gamma_k = \partial I^h(\boldsymbol{\varphi}, Y, \boldsymbol{\varphi}_k^*)/\partial Y$$
$$= \frac{\partial}{\partial\delta Y}\left\{\frac{\partial}{\partial\alpha} I^h(\boldsymbol{\varphi}, Y+\alpha\delta Y, \boldsymbol{\varphi}_k^*)|_{\alpha=0}\right\}, \qquad (28)$$
$$\forall \delta Y \in R^h(D_t^h).$$

Taking into account the form of functional (10)–(11), (27) can be written as

$$\delta\tilde{\Phi}_k^h = \left\{\sum_{i=1}^{n} a_i\left\{\int_{D_t} \left\{\delta\mu_i\text{grad }\varphi_i\text{grad }\varphi_{ik}^*\right.\right.\right.$$
$$\left.-\delta f_i\varphi_{ik}^* + \delta u\varphi_{ik}^*\text{grad }\varphi_i + \delta(H(\varphi))_i\varphi_{ik}^*\right\}\pi\,\mathrm{d}D\,\mathrm{d}t$$
$$-\int_D \delta\varphi_{ai}(x,0)\varphi_{ik}^*(x,0)\pi\,\mathrm{d}D$$
$$\left.\left.\left.+\int_{\Omega_t} (\delta u_n + \delta\beta_i\varphi_i - \delta q_i)\varphi_{ik}^*\pi\,\mathrm{d}\Omega\,\mathrm{d}t\right\}\right\}^h\right. . \qquad (29)$$

The term $\delta(H(\boldsymbol{\varphi}))_i$ is the part of the transformation operator variation produced by the inner parameter variations, $\delta\mathbf{Y}_H$. The SFs are the coefficients at the variations of parameters. They are the contributions of the corresponding parameter variations to the total variation $\delta\tilde{\Phi}_k^h$.

It should be mentioned that estimates of source effects are of great interest for environmental studies. In (29), such estimates are the terms with δf_i and δq_i. For data assimilation problems, the variations in the initial data, described by the term with $\delta\varphi_{ai}$, are important. The above-mentioned characteristics define "external" degrees of freedom. The other characteristics describe the "internal" properties of the model.

To obtain final discrete forms of the SRs, the integrals in (29) are replaced suitable multidimensional quadratures, and the integrands are approximated by discrete aggregates according to the finite volume method in D_t^h. All these approximation formulae are automatically produced by the algorithms of (27) and (29) with a given discrete approximations of the functional $\left[I^h(\boldsymbol{\varphi}, Y, \boldsymbol{\varphi}^*)\right]_{D_t^h}$.

A feedback equation for the minimization of the goal functional follows from relations (27)–(29) (PENENKO 1981, 2009),

$$\frac{\partial\mathbf{Y}_i}{\partial t} = -\aleph\Gamma_{ki}, \quad i = \overline{1, N}, \qquad (30)$$

where $\Gamma_k = \partial \tilde{\mathbf{\Phi}}_k^h / \partial \mathbf{Y}$ is the vector of all SFs, Γ_{ki} is the component corresponding to the parameter $\mathbf{Y}_i$, $\aleph$ is a gradient descent parameter. Equation (30) and coefficient $\aleph$ are obtained from minimum conditions for functional (13) in the direction of SF vector Γ_k. It is important to retain the property of nonnegativity of the functional along the trajectories of model parameters.

Thus, the algorithm of formation of SRs, calculation of SFs and UFs, and organization of feedback equations is completed. The solutions of (18)–(24) reduce the number of inner degrees of freedom of the modeling system. Besides, the UFs in Eqs. (23) and (24) play the role of control variables inserted into (18) and (19) to improve the convergence of the goal functional minimization algorithm. Equations (27), (29) are transformations of the goal functional variations being made to operate in the space of parameters, i.e., to study only the external degrees of freedom of the system.

3.4. Inverse Problems and Data Assimilation

The SRs of (27), (29) and feedback Eq. (30) form a more advanced system of modeling as compared with the traditional methods of direct modeling. The time t, as an independent variable, is the same in (1)–(13) and (30). It means that once the cycle of calculations over one time step Δt is fulfilled, the new values of parameters can be obtained from (30) by an explicit Euler method for the next time step. A more accurate solution at the same time step can be produced by the iterative refinement

$$Y^{(m+1)} = Y^{(m)} + \Delta Y^{(m)}, \quad \Delta Y^{(m)} = -\aleph_m \Gamma_\kappa^{(m)} \Delta t. \quad (31)$$

Here, $\delta \mathbf{Y}$ is a correction of the parameter vector, $\Delta \mathbf{Y}^{(m)}$ is a weight parameter, m is the number of iterations. In scheme (31), the sign and value of $\aleph_m$ are chosen from minimum conditions for the goal functional $\tilde{\mathbf{\Phi}}_k(\boldsymbol{\varphi})$ in accordance with a generalized approximation procedure of Newton's type in the space of parameters (PENENKO 1981, 2009).

Another possible way to refine the parameters is the use of recurrent multi-stage methods of Runge-Kutta type for (30). In this case, the SFs are recalculated as many times as the number of stages chosen for these methods.

Thus, the algorithms (18)–(29) for the calculation of SRs, SFs, and UFs can serve as a basis for a successive solution of equations (30) in inverse problems on parameter and source identification in the domain D_t.

Our variational technique together with some decomposition and splitting methods gives an efficient and consistent implementation of direct (18), (19) and adjoint (20)–(22) problems to solve the above inverse problems. The introduction of UFs to the modeling system simplifies the structure of 4D-Var algorithms. Under the formulation like this, observational data are included in the equations of the direct model (4), (18), (19) by means of both the adjoint problem solutions and the UFs (23), (24). As was mentioned above, the Ufs can be considered as control parameters.

PENENKO (1985, 1996) showed that the 4D-Var algorithm can be transformed to a form analogous to the extended Kalman-like filter (KALMAN 1960; KALMAN and BUSY 1961) by eliminating the adjoint function from the system (18)–(22). Unlike the Kalman filter, the 4D-Var algorithm is not based on solving a matrix equation for the second statistical moments of errors. In our case, the UFs of (23) and (24) give a natural way to take into account discrepancies between the measured values and those calculated with the direct observation model (7). This is of particular importance for real-time data assimilation.

A new version of the 4D-Var method for solving inverse problems by (18)–(24) and for successive data assimilation was proposed by PENENKO (2009). It is based on use of a local adjoint problem technique in time together with the algorithm of UFs calculation in (23) and (24) by using sliding assimilation windows.

Generally speaking, the range of an assimilation window can be regulated. In the above version of the algorithm, the assimilation window is equal to one time step of the numerical model of processes. First, the direct and adjoint problems run one step in the time interval $[t_{j-1}, t_j]$, and then the UFs of (23) are obtained. At this time step, the measured data are taken into account in the solution of the adjoint problem. They are included in the algorithm by using the UFs. The algorithm of (30) for the calculation of parameters is implemented without iterations in the same time interval.

With some decomposition and splitting schemes, the computational kernel of the assimilation procedure can be implemented within each time step for all splitting stages independently in a parallel regime. Finally, the solution of the assimilation problem at time t_j is combined from the results of all splitting stages. The algorithm is cost-efficient, since its implementation takes just two times more operations than the solution of the direct problem without data assimilation. A detailed description of the algorithm is given by PENENKO (2009).

Note that nudging-based data assimilation techniques have been being developed since 1974 (ANTHES 1974; DAVIES and TURNER 1977; STAUFFER and SEAMAN 1990; SCHRAFF and HESS 2003) parallel to variational methods. In these methods, the difference η between observed and modeled values of state functions (the observation increment), being calculated with (7) and multiplied by a phenomenological coefficient, is explicitly introduced into the model of type (4). Formally, these methods are very simple and convenient for implementation. But the most important question of how to describe nudging coefficients correctly is still open.

A comparative analysis of the 4D-Var scheme (PENENKO 2009) with the nudging algorithms showed that under the same conditions both methods have almost the same number of calculations. However, the 4D-Var is optimal in a variational sense, and it allows the possibility to take into account various types of data (not only measured values of state variables) in the same assimilation cycle. The use of adjoint problems in a successive assimilation of observations gives an optimal way of including the data into the direct problem.

3.5. Algorithms for Risk Assessment

The SRs and SFs can be used for qualitative analysis and quantitative assessment of ecological risks. The SFs can be interpreted as risk/vulnerability functions for receptor regions D_{rk} prescribed in the goal (response) functional (12). By definitions (27)–(29), the SFs are 4D functions constructed as derivatives of the goal functional with respect to the parameters. Physically, their values at each point $(\mathbf{x}, t) \in D_t^h$ are contributions of parameter variations into the total functional variation.

An example of a goal functional Φ_k^h in environmental applications is the amount of pollution in the atmosphere of a receptor-region D_{rk} in D_t. As the model of pollutant transport (4) is a part of the total model set, the SR for the functional is expressed by (29).

Theoretically, the goal functional variation (29) can be interpreted as a measure of risk caused by a factor represented in the model (1)–(6) through a corresponding parameter. The significance level can be evaluated using corresponding SF. For environmental risk assessment, the SFs calculated with respect to source parameter variations are of particular interest.

For quantitative estimation of ecological risks, some limiting values for the functional variations (29) are introduced. Denote them as Δ_k^s, $k = \overline{1, K}$. In our example these are admissible pollution amounts of "normal life" in a receptor region. It is natural that such values are empirically obtained as environmental quality standards, which can be different for different countries.

A situation, described by state variables in D_t, is considered ecologically safe if the following inequalities are held

$$|\delta\mathbf{\Phi}_k| \leq \Delta_k^s. \tag{32}$$

Otherwise, the situation is an environmental hazard. It follows from (29) that it is easy to check the inequalities of "ecological well-being" (32) by calculating the SFs and using the available quantitative information on variations of the parameters.

Note that problems (18)–(24) are solved with undisturbed input data without any information on variation of the parameters $\delta\mathbf{Y}$. Hence, the solutions φ, φ_k^*, and the SFs are calculated by formulae (29) deterministically. Sometimes the variations of parameters, initial and boundary conditions and sources may be both deterministic and stochastic. Therefore, the risk assessment algorithms are different in these two cases.

If variations of sources and parameters are known, an estimate of the functional variation magnitude can be obtained from (27), (29) as

$$|\delta\mathbf{\Phi}_k| = |(\mathbf{\Gamma}_k, \delta\mathbf{Y})| \tag{33}$$

Using (33), it is possible to check inequality (32).

If variations of parameters and sources have a stochastic character, than estimations become more complicated compared with the deterministic case. Then it is necessary to use multidimensional spaces of deterministic SFs and parameters having stochastic disturbances of $\delta\mathbf{Y}$ elements at the same time. Variations in the parameters can be produced by independent events. In this case functionals (27) and (29) can be considered as linear combinations of independent stochastic values $\delta\mathbf{Y}_j\,(\mathbf{x},\,t)$, $(\mathbf{x},\,t) \in D_t^h$, where j denotes a category in (29). Under these conditions, the Central Limit Theorem (PEARSON 1974) states that $\delta\tilde{\Phi}_k^h$ has an asymptotically normal distribution.

For the case of stochastic variations of the parameters, the following formulae present an algorithm for risk assessment (PENENKO 1981):

$$E(\delta\mathbf{\Phi}) = (\Gamma, E(\delta Y)),$$
$$D(\delta\mathbf{\Phi}) = (D(\delta Y)\Gamma, \Gamma),$$

$$P(|\delta\mathbf{\Phi}| \leq \Delta^s) = \int_{\Delta^s} f(x)dx, \quad x \equiv \delta\Phi(\varphi),$$

$$f(x) = \left(\sqrt{2\pi D(x)}\right)^{-1}\exp(-(x - E(x))^2/2D(x)),$$
$$R^s = P(|\delta\mathbf{\Phi}| \leq \Delta^s), \tag{34}$$

where $E(\cdot)$ is the mean value (expectation), $D(\cdot)$ is the variance, Γ is the vector of SFs, $P(\cdot)$ is the probability of being in an interval, and $f(x)$ is the probability density function; R^s is the probability of being in a safety region. It is assumed that the vector $E(\delta\mathbf{Y})$ and covariation matrix $D(\delta\mathbf{Y})$ are given.

A list of typical conditions of environmental problems that can be solved in the framework of the proposed concept includes the following:

(a) All parameters of sources (emissions, locations, times, etc.) and some a priori data on variations of the parameters are known;
(b) The parameters of sources are known, but the variations are unknown;
(c) Neither the parameters of sources nor their variations are known.

In all the cases it is assumed that necessary meteorological fields (calculated in a modeling system or given by an observation system) are available. Some goal functionals are formulated as well.

In case (a), a variety of admissible problem formulations is defined by capabilities of the set of algorithms (18)–(34). In case (b), the formulations with (33) and (34) are excluded because they need some additional information on the variations. The goal functionals may be linear and nonlinear in both cases.

Even in case (c), some problems of risk assessment for receptor regions to be polluted by all possible sources can be formulated. With necessary data given on the measured concentrations, the problems of finding the locations of sources and assessing their parameters are also possible.

For risk assessment in case (c), a linearized version of the transport model is used. Linear versions of the goal functionals with respect to the state variables are chosen. The weight functions are nonzero in the receptor regions D_{rk} in (12). With $\alpha_i = 0$, $i = \overline{1,4}$ in (13) all uncertainties are disregarded. In this case only the adjoint problems (20)–(22) from the whole set (18)–(24) are solved. Then only the SFs for sources from (29) are obtained. Examples of such calculations are given in Sect. 5.

3.6. Solvability

To complete the theoretical part of the paper, the solvability of our problems will be briefly discussed. The system (2)–(6) is a set of problems described by equations of the convection-diffusion-reaction type. It is related to a class of linear and quasi-linear problems of the parabolic type. Solvability, existence, and some properties of the solutions of this class of problems are studied in LADYZHENSKAYA et al. (1967).

The algorithms considered here are implemented with discrete approximations of all objects in the above technology of modeling. The change from differential statements to their discrete analogs is fulfilled to retain major properties of the original operators by a technique of variational calculus. The discrete problems thus obtained are solvable, and the set of their solutions with various input parameters is a bounded set of functions in $Q^h(D_t^h)$. Equations (15)–(25) express necessary conditions for the stationary state of the augmented goal functional with respect to variations of its functional arguments. These equations are well-known Euler-Lagrange equations. Such equations are constructed, for example, in PEARSON (1974). Equations (18)–(25) have a unique solution under a given set of input data. The set of solutions with different input data generates the bounded spaces $Q^h(D_t^h)$ and $Q^{*h}(D_t^h)$. The space $R^h(D_t^h)$ is bounded by definition.

In the finite-dimensional spaces of discrete approximations, the augmented functional (13) is a real and continuous function defined on the elements of a bounded, closed, and non-empty set of functions from $\{Q^h(D_t^h), Q^{*h}(D_t^h), R^h(D_t^h)\}$. By virtue of the Weierstrass theorem of continuous functions (COURANT and HILBERT 1931), such a function reaches its minimum and maximum values on this set. The augmented functional is non-negative by definition and construction. Its minimum is greater or equal to 0. Thus, the continuity theorem ensures the existence of an optimal solution giving minimal values to the augmented functional (13). In order to find such a solution, a Newton-type gradient algorithm (KANTOROVICH and AKILOV 1982) is used.

It should be noted that, due to non-linearity, the uniqueness of solutions of optimization and inverse problems is not guaranteed in the general case. Multiple local minima are possible. The Newton-type methods are locally convergent. They converge in a vicinity of some state of the sought-for parameters. This means that a good initial guess is a key factor to get the desired result.

A detailed description of the above variational methodology for the construction of numerical models and proofs of existence and boundedness for some statements related to atmospheric dynamics and pollution transport are given in PENENKO (1981).

4. Scenario Analysis and Use of Climatic Data

4.1. Principal Factors on a Global Scale

Scenario analysis is widely used in numerical modeling for assessment of current and future states of environmental and ecological systems. The results of modeling essentially depend on the description of the basic hydrodynamic processes. A methodology for a quantitative description of the

behavior of a dynamic system in a long time interval in a compressed, generalized form was proposed by PENENKO and TSVETOVA (2008). The main aim is to develop a method to incorporate climatic information into environmental studies. In this method, the necessary information used for the construction of deterministic/stochastic scenarios is extracted from a database of measured and/or calculated data on some hydrodynamic state functions. The methodology is a development of ideas of principal component and factor analyses (PREISENDORFER 1988). The method of orthogonal decomposition of the phase spaces formed by multivariate, multidimensional state functions from the database is used. The method is cost-effective and applicable to phase spaces of any size.

The results of an analysis of the global climatic system behavior for a period of 56 years (1950–2005) are shown in Fig. 2. For the method of orthogonal decomposition, the NCEP/NCAR reanalysis database (KALNAY et al. 1996) was used. This database is a well-structured universal-purpose information system containing a basic set of atmospheric characteristics. Two types of bases are calculated: principal components (PC) and orthogonal base vectors (OBV). PCs characterize a year-to-year variability of an initial sample (reanalysis data) with respect to a OBV system. The system consists of 4D (space-time) orthogonal base vectors. In calculations, there are $2m$ time fragments for each month with a time step of 12 h in each vector, where m is the number of days of the current month. The spatial resolution is taken in accordance with the reanalysis data: horizontal resolution of $2.5° \times 2.5°$ in spherical coordinates and standard p-levels in the vertical direction. In Fig. 2 the fragments of leading OBVs corresponding to the maximum eigenvalues of the Gram matrix for the 500-hPa geopotential height are presented for four seasons. These fragments are taken at 00:00 UTC on 15 January, 15 April, 15 July and 15 October (Fig. 2a–c), respectively. Note that areas of the local extrema indicate regions of increased energy activity in the atmosphere. The leading basic vectors contain the so-called "long-term memory" of the climatic system.

Thus, the calculated informative bases have been used for the construction of hydrodynamic scenarios

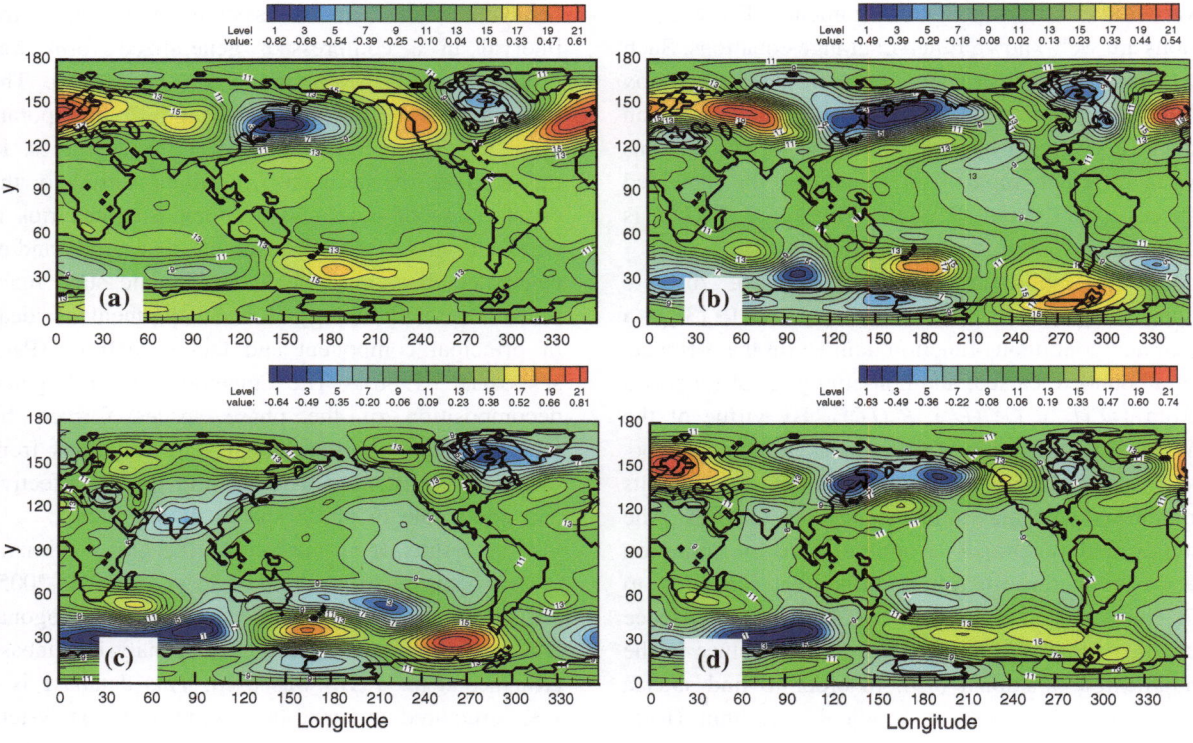

Figure 2
One of 62 fragments of the leading OBV-1 corresponded to 00:00 UTC on **a** 15 January, **b** 15 April, **c** 15 July, and **d** 15 October. The basis is constructed on reanalysis data of the 500-hPa geopotential height for a period of 56 years (1950–2005)

for environmental applications. Moreover, basis subspaces of PCs and OBVs can help to analyze the climatic data in order to detect typical and extreme situations for the formation of targeted scenarios.

The use of orthogonal bases in combination with assimilation problems gives effective algorithms for the reconstruction of the space-time fields of the state variables and, in particular, for the initial data $\varphi_a(\mathbf{x})$ in Eq. (3). Relevant algorithms are discussed in PENENKO (1981), PENENKO and TSVETOVA (2008).

4.2. Regional Principal Factors for Scenario Formation

The Siberian region is of special research interest for studies of fundamental characteristics of the formation of environmental dynamics. An analysis of the above-mentioned basic subspaces (with fragments presented in Fig. 2) showed that Siberia and especially the Lake Baikal region are situated in domains that separate circulation systems of high energy activity. In winter, there are Pacific and Atlantic energy-active zones, whereas in summer there are Arctic and South-Asian zones. With these factors the high dynamic activity in the Siberian region can be explained. During the autumn-winter season there is instability as a rapid change of weather cycles. The formation of the Altai-Sayan cyclogenesis, having the same intensity as the Mediterranean one, is specific for warm seasons. CHUNG *et al.* (1976) and CHEN *et al.* (1991) called it a lee-type cyclogenesis. It is a large-scale phenomenon in the climatic system of the central part of the Eurasian continent, which has a large effect on environmental dynamics.

In the Eurasian region, areas of increased activity are seen (Fig. 3a) in a leading OBV near the Lake Baikal region (in a vicinity of 110°E). In a second leading OBV (Fig. 3b), areas of local extrema

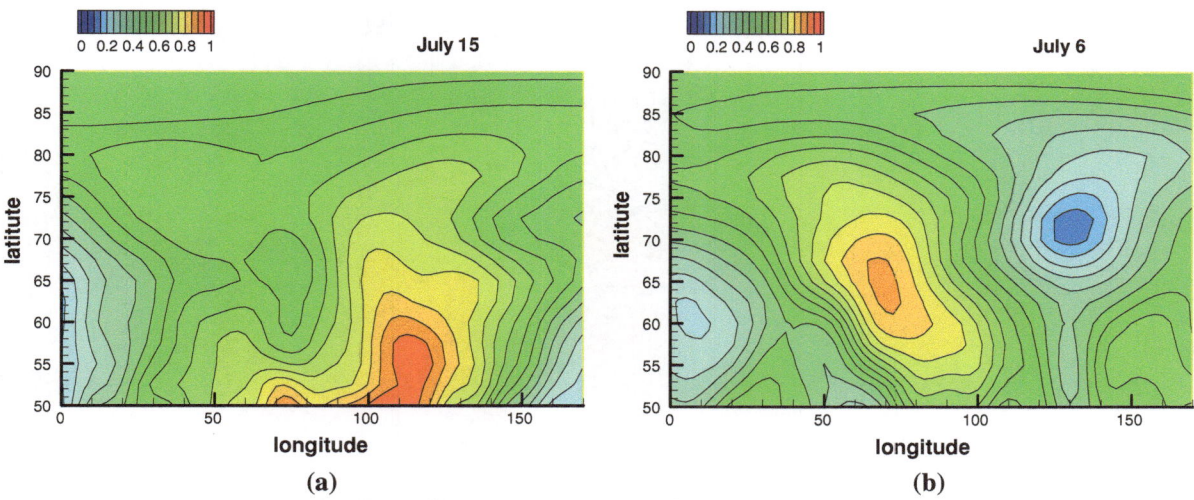

Figure 3

One of 62 fragments of the **a** leading OBV-1 for 15 July and **b** the second OBV-2 for 6 July. The basis is constructed on reanalysis data of the 500-hPa geopotential height for a period of 56 years (1950–2005) for the Eurasian region (50°–90°N, 0–170°E)

indicate the West Siberian wetlands province (about 75°E) and Verkhoyansk region (Yakutia) (around 135°E).

Thus, the orthogonal subspaces show areas of high energy climatic zones both on global and regional scales. This information can be taken into account in studying the environmental processes. Calculated by means of the above inverse modeling technique, the risk/vulnerability areas for anthropogenic objects placed in these active zones are characterized by their large temporal-spatial scales as compared with the outside objects (PENENKO and TSVETOVA 2005).

5. Examples of Environmental Studies for the Siberian Region

The industrial activity in Siberia has a high level of anthropogenic impact on the environment. Besides, recently the emissions of aerosols have increased because of dust storms in Central and Southeast Asia, forest fires, etc. The current climate change in Western Siberia produces large methane emissions from wet soils of peatbogs and wetlands. Increased concentrations of formaldehydes and other chemically active species are also produced through photochemical reactions.

Several examples of environmental scenarios performed with a set of basic mathematical models developed in ICMMG SD RAS and DMI are given in the following sections.

5.1. Probabilistic Risk Assessment

A methodology for multidisciplinary probabilistic environmental risk and vulnerability assessment was proposed in (BAKLANOV et al. 2006, 2007; MAHURA et al. 2006). A Danish Emergency Model for the Atmosphere (DERMA, SØRENSEN et al. 2007) was employed to perform simulations of the transport of atmospheric pollution, and dry and wet deposition by continuous emissions from selected chemical risk sites. Several Siberian chemical and metallurgical enterprises situated near the major industrial cities were selected as representative sources of such emissions. To perform yearly simulations, 3D meteorological fields from the European Center for Medium-Range Weather Forecasts (1985 is a climatologically typical year, and 1983 is a year with a significant deviation of the atmospheric circulation patterns for the North Atlantic Oscillation) were used. For each daily release the pollution transport in the atmosphere and deposition on the underlying surface due to dry and wet removal processes were estimated by forward modeling at an interval of 2 weeks. It was

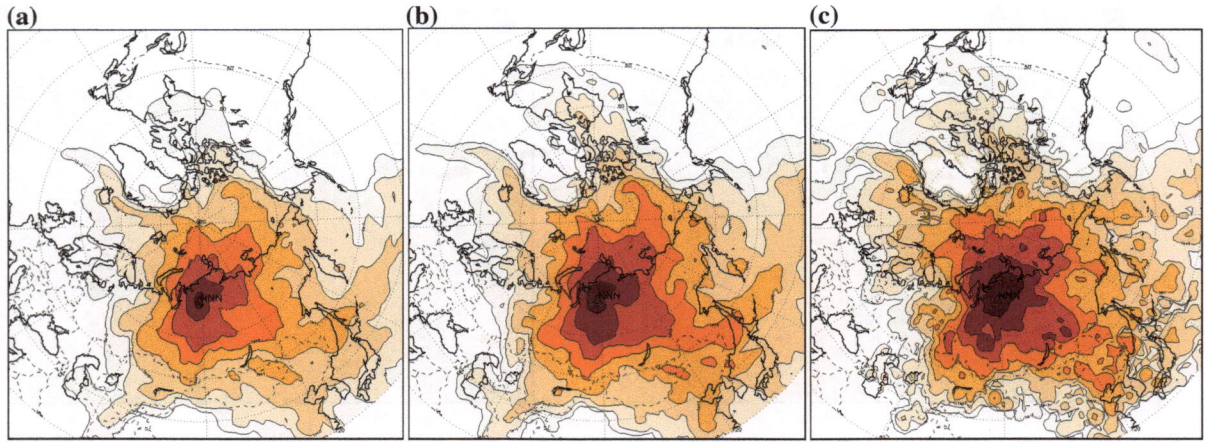

Figure 4
Results of long-term transport modeling—annual total **a** time integrated air pollution concentration [μg/m³], **b** dry deposition [μg/m²], and **c** wet deposition [μg/m²] patterns for sulfates resulted from continuous emissions from the Norilsk nickel plant (NNN)

done for every day for the selected years, and then all fields were summed up. The simulated concentration and deposition fields for each site show the temporal variability of these patterns on both regional and hemispheric scales (an example is given in Fig. 4). These results can be used in GIS integration and evaluation of the doses, impacts, risks, short- and long-term effects from potential sources of continuous emissions on the population and environment.

5.2. Risk Assessment by Inverse Methods

With inverse modeling methods some estimates of the current environmental situations can be obtained even without information on the emissions.

Specifically, some zoning of the domain D_t with respect to degrees of danger for a receptor to get pollution from various kinds of sources can be designed by means of risk functions.

The sensitivity-risk-observability functions Γ_k (27)–(29) calculated with respect to sources of emission for selected industrial regions-receptors in Siberia [(1) Khanty-Mansiisk, (2) Yakutsk, and (3) Ussuriisk] are presented in Fig. 5. All these objects are located in the high energy zones. For each scenario, a functional describing the results of observations by means of ground-based sun/sky photometers located at sites of the Siberian part of the Aeronet observational system (http://aeronet.gsfc.nasa.gov) was used as a response functional $\Phi_k^h(\varphi)$ in (12) in the form

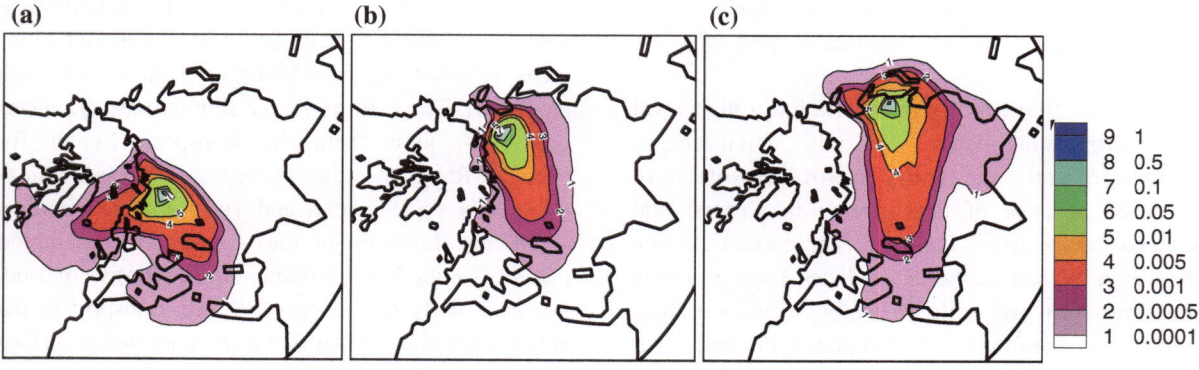

Figure 5
Risk-observability functions (reference values) for Siberian industrial regions: **a** Khanty-Mansiisk, **b** Yakutsk, and **c** Ussuriisk

$$\Phi_k(\varphi) = \int\limits_{D} \int\limits_{t_{in}}^{t_f} F_k(\varphi)\chi_k(x, t)\,\mathrm{d}D\,\mathrm{d}t,$$

where $F_k(\varphi) = \tau(\lambda)$, $\chi_k(\mathbf{x}, t) = \delta(\mathbf{x} - \mathbf{x}_k)$, $\delta(\mathbf{x} - \mathbf{x}_k)$ is the Dirac delta function, $\mathbf{x}_k$s are the locations of receptors, $k = 1, 2, 3$, $[t_{in}, t_f]$ is the time interval of observations;

$$\tau(\lambda) = \sum_{\alpha=1}^{n_\alpha} \frac{3}{4} \int\limits_{p_t}^{p_s} \gamma_{ext}(\lambda, \alpha)q_a(\alpha, x, t)\,\mathrm{d}p, \qquad (35)$$

where $\tau(\lambda)$ is the total aerosol optical thickness in treating all major aerosol substances $q_a(\alpha, \mathbf{x}, t)$ in the model (4) simultaneously as components of the state vector φ, α is an index of particle size, n_α is the number of sizes, γ_{ext} is an extinction efficiency factor at wavelength λ and particle size α for aerosols. Formula (35) is an observational model for the presentation of observed column-integrated data included into (7) if an assimilation problem with an aerosol transport model is stated.

Note, that the term "observability domain" means that there is an area defined by the adjoint problem solution (with or without data assimilation) in which it is possible to find the sources of pollution depending on the data produced by an observational functional $\Phi_k^h(\varphi)$ measured in a receptor $D_{rk} \in \{(\mathbf{x}, t), \chi_k(\mathbf{x}, t) \neq 0\} \subset D_t$ in correspondence with (12). In other words, this is the area from which a pollutant can potentially come to a receptor. It means that the source is "visible" for the receptor. The observable area is described by the support of a SF generated for the functional $\Phi_k^h(\varphi)$ on a predefined significance level. Hence, such data assimilation procedure can be used for source parameter identification.

Figure 6 shows another example of an important forecast element: a calculated field of total (during a "climatic" month of October in terms of orthogonal decomposition) estimation of the degree of risk for Lake Baikal as a receptor D_r. This is the total air column over the lake from the water surface to the 10 hPa level. The linear goal functional $\Phi^h(\varphi) = \int_{t_{in}}^{t_f} \int_D \varphi\chi_\kappa\,\mathrm{d}D\,\mathrm{d}t$, $\chi_k \neq 0$ in the receptor area D_r during a time interval $[t_{in}, t_f]$ describes the amount of pollution passed through the receptor during the month. A time-dependent SF with respect to δf is calculated for October. It gives the relative contribution of the emission of pollutants from

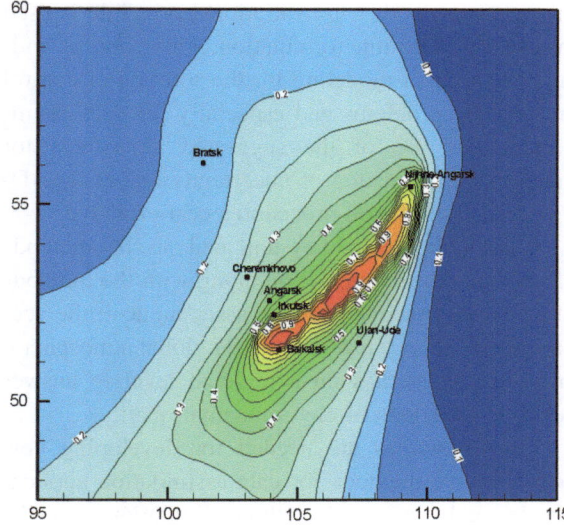

Figure 6
October total estimates of relative contribution of pollutant emission from acting and potentially possible sources in the region (47.5°–60°N, 95°–115°E) to an atmospheric quality functional over the Lake Baikal surface area

acting and potentially possible sources located on the earth's surface in the domain D_t to the functional of atmospheric quality in the receptor area D_r. The scenario is carried out backward in time for October. The fragment shown in Fig. 6 corresponds to a lower model level, which coincides with the upper boundary of the surface layer in the used coordinate system. This fragment is a 2D cross section of the 3D function obtained by integration in time of 4D fields of a prognostic SF calculated with respect to variations of the source function δf from (29) of model (4).

The values on isolines are obtained by normalization of SF to its maximum in the entire 3D domain of integration. The larger the SF values, the higher is the risk for the receptor area to be polluted from sources located in the areas that produce these values. To estimate the risk of pollution of a receptor area due to transboundary transport, it is necessary to take into account the information contained in the SFs on the boundaries of the domain D_t.

6. Conclusion

The presented concept of environmental modeling is based on a variational technique. An advantage of this approach is that the variational principle is used in a weak-constraint formulation. This concept widens the scope of modeling for the solution of control and inverse problems and especially for data assimilation. The use of the suggested algorithms for quantitative estimation of uncertainty makes the 4D-Var data assimilation procedures cost-effective.

From a viewpoint of system analysis, the methods of orthogonal decomposition along with the methods of sensitivity theory and risk assessment in the scenario approach are useful tools for global atmospheric and climatic studies on a regional level to answer concrete questions of environmental quality.

Further studies are needed for developing new cost-effective methods for data assimilation applicable in real time and without limitations on the quantity, as well as finding ways to obtain and present observational data on higher dimensions problems.

Acknowledgments

We thank the anonymous reviewers for their constructive criticism and useful comments. This study was supported by the 6th and 7th Framework Programmes of the European Commission (CA-project Enviro-RISKS: Man-induced Environmental Risks: Monitoring, Management and Remediation of Man-made Changes in Siberia, http://risks.scert.ru, INCO-CT-2005-013427; MEGAPOLI: http://megapoli.info; grant no: 212520; and TRANSPHORM: http://transphorm.eu projects), the Danish Council for Strategic Research (Center for Energy, Environment and Health—CEEH; http://ceeh.dk), and the Russian Foundation for Basic Research (grants RFBR 10-05-01083-a, 11-01-00187-a). We are also grateful to of the Presidium of the Russian Academy of Sciences (program nos. 4 and 20) and the Department of Mathematical Sciences of RAS (program no 3). The DMI supercomputing facilities have been used extensively in this study. We also thank Jens H. Sørensen (DMI) for active participation in the Enviro-RISKS project.

REFERENCES

ANTHES, R.A. (1974), *Data assimilation and initialization of hurricane prediction model.* J. Atmos. Sci. *31*, 702-719.

BAKLANOV, A. (2000), *Modeling of the atmospheric radionuclide transport: local to regional scale.* Numerical Mathematics and Mathematical Modeling, Moscow, INM RAS 2, 244-266.

BAKLANOV, A. (2007), *Environmental risk and assessment modeling–scientific needs and expected advancements.* In: Ebel A. and Davitashvily T. (eds.), Air, Water and Soil Quality Modeling for Risk and Impact Assessment, Springer, 29-44.

BAKLANOV, A, SØRENSEN, J.H., MAHURA A. (2006), *Long-Term Dispersion Modeling. Part I: Methodology for Probabilistic Atmospheric Studies*, J. Comp. Technologies, *11*, 136-156.

BAKLANOV, A., J.H. SØRENSEN and A. MAHURA (2007), *Methodology for Probabilistic Atmospheric Studies using Long-Term Dispersion Modeling.* Environmental Modeling and Assessment *13*, 541-552; doi:10.1007/s10666-007-9124-4.

BRYSON, Jr., A.E. and HO, Y. (1969), *Applied optimal control,* Waltham, Massachusetts, Toronto, London, Blaisdell Publishing Company, A Division of Ginn and Company, 544 pp.

CHEN S.-J., Y.-H. KUO, P.-Z. ZHANG, Q.-F. BAI (1991), *Synoptic climatology of ciclogenesis over East Asia, 1958-1987,* Mon. Wea. Rev. *119*, 1407-1418.

CHUNG Y.-S., HAGE, K.D., REINELT, E.R. (1976), *On lee cyclogenesis and airflow in the Canadian Rocky mountains and the East Asian mountains.* Mon. Wea. Rev. *104*, 879-891.

COURANT, R., HILBERT, D, (1931), *Methoden der mathematischen Physik, V.1, Springer.*

DAVIES, H.C. and TURNER, R.E. (1977), *Updating prediction models by dynamical relaxations: an examination of the technique.* Quart. J. Roy. Meteor. Soc. *103*, 225-245.

LE DIMET, F.X., and TALAGRAND O. (1986), *Variational algorithms for analysis and assimilation of meteorological observations: theoretical aspects,* Tellus, *38A*, 97-110.

EVENSEN, G. and FARIO, N. (1997), *Solving for the generalized inverse of the Lorenz model.* J. Meteor.Soc. Japan *75*, 229-243.

KALMAN, R.E. (1960), *A new approach to linear filtering and prediction problems,* Trans. ASME, J. Basic Eng., *82*, 34-35.

KALMAN, R.E., BUCY, R.S. (1961), *New results in linear filtering and prediction theory,* Trans. ASME, Ser.D, J. Basic Eng., *83*, 95-107.

KALNAY, E., KANAMITSU, M., KISTLER, *et al.* (1996), *The NCEP/NCAR 40-year reanalysis project,* Bull. Amer. Meteor. Soc. *77*, 437-471.

KANTOROVICH, L. V. and G. P. AKILOV (1982), *Functional Analysis,* Pergamon Press; 2nd Revised edition.

LADYZHENSKAYA, O.A., SOLONNIKOV, V.A., and URALTSEVA, N.N (1967), *Linear and quasi-linear equations of parabolic type,* M., Nauka; English transl.: Transl. Math. Monogpaphs, 23, Amer. Math.Soc., Providence, RI, 1968.

LEWIS, J.M., and DERBER, J.C. (1985), *The use of adjoint equations to solve a variational adjustment problems with advective constraints,* Tellus, *37A*, pp. 309-322.

LIONS, J.L. (1968), *Controle optimal de systemes gouvernes par des equations aux derivees partielle.* Dunod Gauthier-Villars, Pasis.

MAHURA A., BAKLANOV A, SØRENSEN J.H. (2006), *Long-term dispersion modeling. part II: Assessment of atmospheric transport and deposition patterns from nuclear risk sites in Euro-Arctic region,* J. Comp. Technologies, *10*, 112-134.

MARCHUK, G.I. (1974), *The numerical solution of problems in the dynamics of the atmosphere and oceans,* Gidrometeoizdat, Leningrad (in Russian).

MARCHUK, G.I. (1975), *Formulation of the theory of perturbation of complicated models. Applied Mathematics and Optimizations.* 2(1), 1-33.

Marchuk, G.I. (1995), *Adjoint Equations and Analysis of Complex Systems*, Kluwer, Dordrecht.

Pearson, K.E. (1974), *Handbook of Applied Mathematics,* Van Nostrand Reinhold Company, N.Y.

Penenko, V.V. (1975), *Computational aspects of modeling the dynamics of atmospheric processes and of estimating the influence of various factors on the dynamics of the atmosphere,* In: Some problems of Computational and Applied mathematics, Nauka, Novosibirsk, 61-77. (in Russian).

Penenko, V.V. (1981), *Methods of numerical modeling of atmospheric processes*, Gidrometeoizdat. Leningrad, 352 pp. (in Russian).

Penenko, V.V. (1985), *System organization of mathematical models for the problems of atmospheric physics, ocean and environmental protection,* Novosibirsk Computing Center, Preprint No 619, 43 pp. (in Russian).

Penenko, V.V. (1996), *Some aspects of mathematical modeling using the models together with observational data,* Bull Nov. Comp. Center, Series Num. Model. in Atmosph., etc. *4,* 31-52.

Penenko, V.V. (2000), *Variational principles in the problems of ecology and climate,* In: Dymnikov V. (Ed), Numerical Mathematics and Mathematical Modeling: Proceedings of International Conference, Publ: Institute of Numerical Mathematics of RAS, V.*1,* 135-148 (in Russian).

Penenko, V.V. (2009), *Variational methods of data assimilation and inverse problems for studying the atmosphere, ocean, and environment*, Numerical Analysis and Applications, 2009, *2,* No 4, 341-351.

Penenko, V.V. (2010), *On a concept of environmental forecasting,* Atmospheric and Oceanic Optics, *23,* No 6, 432-438.

Penenko, V.V., and Aloyan, A.E., (1985) *Models and Methods for Environmental Protection Problems,* Nauka, Novosibirsk, 276 pp. (in Russian).

Penenko, V. and A. Baklanov (2001), *Methods of sensitivity theory and inverse modeling for estimation of source term and nuclear risk/vulnerability area.* Lecture Notes in Computer Science, V. *2074,* No 2, 57-66.

Penenko, V., Baklanov, A. and Tsvetova, E. (2002*), Methods of sensitivity theory and inverse modeling for estimation of source term.* Future Generation Computer Systems, *18,* 661-671.

Penenko, V.V., and Obraztsov, N.N. (1976), *A variational initialization method for the fields of meteorological elements,* Soviet Meteorology and Hydrology, No *1,* 1-11.

Penenko, V.V., Tsvetova, E.A., (2005), *Energy centers of the climate system and their connection with ecological risk domains,* SPIE, V. *6160,* part 2, 61602U.

Penenko, V.V., Tsvetova, E.A., (2007), *Variational technique for environmental risk/vulnerability assessment and control,* In: Ebel A. and Davitashvily T. (eds.), Air, Water and Soil Quality Modeling for Risk and Impact Assessment. Springer, 15-28.

Penenko, V. and Tsvetova, E. (2008), *Orthogonal decomposition methods for inclusion of climatic data into environmental studies,* Ecol. modeling, *217,* 279–291.

Penenko, V. and Tsvetova, E. (2009a), *Discrete-analytical methods for the implementation of variational principles in environmental applications,* Journal of Computational and Applied Mathematics, *226,* 319-330.

Penenko, V. V. and Tsvetova, E. A. (2009 b), *Optimal forecasting of natural processes with uncertainty assessment,* Journal of Applied Mechanics and Technical Physics, *50,* No. 2, 300–308 .

Preisendorfer, R.W. (1988), *Principle component analysis in meteorology and oceanography.* Amstedan-Oxford-New York-Tokio, Elsevier.

Schraff, C. and Hess, R. (2003), *A description of the nonhydrostatic regional model LM. Part III. Data assimilation,* www.cosmo-model.org.

Schwartz, L. (1967). *Analyse mathematique,* I. Hernman.

Stauffer, D.R. and Seaman, N.L. (1990), *Use of four-dimensional data assimilation in a limited-area mesoscale model. Part I. Experiments with synoptic scale data,* Mon.Wea. Rev., *118,* 1250-1277.

Sørensen, J. H., A. Baklanov and S. Hoe (2007), *The Danish Emergency Response Model of the Atmosphere.* J. Envir. Radioactivity, *96,* 122–129.

Talagrand, O. and Courtier, P. (1987), *Variational assimilation of meteorological observations with the adjoint vorticity equation. I: Theory.* Quarterly Journal of the Royal Meteorological Society, *113,* 1311-1328.

van Leeuwen, P.J. and Evensen, G. (1996), *Data assimilation and inverse methods in terms of a probabalistic formulation,* Mon. Wea. Rev, *124,* 2898-2913.

(Received December 30, 2010, accepted March 26, 2011, Published online September 15, 2011)

Pure Appl. Geophys. 169 (2012), 467–482
© 2011 Springer Basel AG
DOI 10.1007/s00024-011-0381-4

Identification of a Point of Release by Use of Optimally Weighted Least Squares

JEAN-PIERRE ISSARTEL,[1] MAITHILI SHARAN,[2] and SARVESH KUMAR SINGH[2]

Abstract—This paper addresses the parametric inverse problem of locating the point of release of atmospheric pollution. A finite set of observed mixing ratios is compared, by use of least squares, with the analogous mixing ratios computed by an adjoint dispersion model for all possible locations of the release. Classically, the least squares are weighted using the covariance matrix of the measurement errors. However, in practice, this matrix cannot be determined for the prevailing part of these errors arising from the limited representativity of the dispersion model. An alternative weighting proposed here is related to a unified approach of the parametric and assimilative inverse problems corresponding, respectively, to identification of the point of emission or estimation of the distributed emissions. The proposed weighting is shown to optimize the resolution and numerical stability of the inversion. The importance of the most common monitoring networks, with point detectors at various locations, is stressed as a misleading singular case. During the procedure it is also shown that a monitoring network, under given meteorological conditions, itself contains natural statistics about the emissions, irrespective of prior assumptions.

Key words: Atmospheric tracer, concentration measurement, inverse problem, unknown errors, apparent geometry.

1. Introduction

This work is a theoretical clarification motivated by results described and partly explained elsewhere for identification, from atmospheric concentration measurements, of the point of release of some trace species. In these papers, various sets of concentrations prepared from experimental releases at continental (ISSARTEL 2005) or local (SHARAN *et al.* 2009) scale were analysed by use of a recent assimilative technique originally developed for estimation of smoothly distributed emissions. Despite an undetermined level of noise in the data, the release was reasonably identified as corresponding to the maximum of the smooth estimation. The theoretical justifications proposed in the second cited paper did not address the effect of noise in the measurements: this is mainly the focus of this work.

Identification of a point of release is a parametric inverse problem (RAO 2007) with a limited number of parameters for location, time, and intensity. The inverse problem is necessarily addressed by use of a dispersion model. When the model is linear, the measurements really observed are compared with those ideally associated with any combination of source parameters and computed from adjoint equations. The comparison with a least-squares criterion is classically weighted by the inverse of the measurement errors covariance matrix (NORLEN 1975). The resulting estimator has been widely advocated in view of its best linear unbiased estimator (BLUE) property for minimising the errors in the estimated source data (MILLIKEN and ALBOHALI 1984). This paper discusses the weighting of the least squares, and proposes and justifies an alternative strategy.

The BLUE property is advantageous in a natural way supporting use of the covariance matrix. However, there is a crucial weakness in this strategy of weighting least squares: it presupposes statistical knowledge of the measurement errors between observed and modelled concentrations. These errors consist of two parts:

1. detector errors arising because of instrumental limitations, which can be tested; and
2. representativity errors arising from the imperfect representativity of the model, which are certainly present and cannot be addressed simply.

In this context, a variety of assumptions have been made by researchers without supporting them by

---
[1] Centre d'Etudes du Bouchet, 5, rue Lavoisier, BP 3, 91710 Vert le Petit cedex, France. E-mail: jean-pierre.issartel@dga.defense.gouv.fr
[2] Centre for Atmospheric Sciences, Indian Institute of Technology Delhi, Hauz Khas, New Delhi 110 016, India.

systematic mathematical reasoning. It is shown here that the most popular assumption of normally distributed representativity errors is not consistent with the linear nature of dispersion.

This paper shows that, behind the successful identification of points of release from renormalisation theory in our previous work, a non-BLUE approach to least squares is operating. This approach is applicable when the statistical behaviour of the measurement errors is not known. To show this, it is first explained that a least-squares technique always amounts to finding the adjoint vector of the model measurements at the minimum angle from the vector of the observations. This angle is evaluated depending on the geometry or norm associated with the least squares. If the least squares are weighted by the identity matrix (ordinary least squares) the angle between any two adjoint vectors is always less than 90° because the components of both are all positive. The same happens with least squares weighted by the inverse covariance matrix under the common assumption that this matrix is diagonal, i.e. the errors in the various measurements are uncorrelated. It is shown here that the alternative weighting matrix is so chosen that the adjoint vectors corresponding to all possible release data become more widely and optimally distributed over the sphere. Thus, the correct one is distinguished more easily.

This study highlights the consistency and strength of renormalisation theory for addressing both the parametric and assimilative inverse problems. During the procedure a point of interest in both cases is clarified: the weighting matrix and renormalised geometry are independent of any prior statistics about the distribution of the emissions.

2. Formulation and Preliminaries

Let μ_1, μ_2, ..., μ_n be a finite set of mixing ratio measurements taken by a monitoring network. It is intended to identify from this set the point of release of some atmospheric species. This point of release is described by a finite number of values, most generally five for intensity, location, and time. This presentation is specialised for constant releases from the surface of the ground. This facilitates illustration

of examples by eliminating the vertical and temporal dimensions. The compactness requirement of renormalisation (in the Sect. 2.2) is fulfilled either by considering a limited portion of the ground or by considering the ground as a planetary sphere. The theory, however, is more general with any number of dimensions. When the time dimension is considered, the point source of a release is intended to be at exactly one location and time.

A constant point release at ground level is characterized by three values: an intensity q_0, in unit amount of tracer per unit time, and a pair $\vec{x}_0 = (x_0, y_0)$ of horizontal coordinates. This is described by a release function $s_0(x,y) = q_0 \delta(x - x_0)\, \delta(y - y_0)$ in which δ is the Dirac delta function. Formally $s_0(x,y)$ represents a rate of release in unit amount of tracer per unit area. The identification of q_0 and $\vec{x}_0$ from the measurement vector $\vec{\mu} = (\mu_1, \mu_2, \ldots, \mu_n)^{\mathrm{T}}$ (the supercript "T" denotes the transposition) is a parametric inverse problem. If the measurements are noiseless and free from exceptional degeneracy, the problem is clearly stated, and has a unique solution that can be found by various numerical techniques. However, the real measurements are noisy. As a result, the problem cannot be solved exactly and this raises issues about the optimum technique for solving the problem as exactly as possible. The purpose of this paper is to introduce new arguments in this direction.

A noticeable feature of this development is the consistency established between the parametric inverse problem of identifying point emissions and the more general ill-posed assimilative inverse problem of estimating emissions distributed smoothly. In particular, it will often be convenient here to consider that the release function $s_0(x,y) = q_0\, \delta(x - x_0)\, \delta(y - y_0)$ is a special case among more general distributions $s(x, y)$.

2.1. Constant Source Observed Under Constant Meteorological Conditions

For introducing the adjoint formulation of the inverse problem, it is convenient to first consider emissions of a trace species distributed through the whole atmospheric volume. Such emissions are represented by a function σ of space coordinates such that $\sigma(x, y, z)$ is the rate of release per unit mass of air and per unit time around the location of horizontal coordinates x and y, at altitude z. If these emissions and the meteorological conditions do not vary, the field of mixing ratio $\chi\,(x,\,y,\,z)$ from which the measurements μ_i are sampled is steady and subject to the equation:

$$\vec{u}\cdot\vec{\nabla}\chi-\frac{1}{\rho}\nabla(\rho\mathbf{K}\vec{\nabla}\chi)=\sigma,\quad \mu_i=\int_{\Omega}\rho\chi\pi_i\mathrm{d}x\,\mathrm{d}y\,\mathrm{d}z$$
$$(1)$$

in which $\vec{u}$, $\mathbf{K}$ and ρ are known fields of wind, diffusion tensor and dry air density respectively, Ω is the part of the atmospheric domain taken into account; and $\pi_i(x,\,y,\,z)$ is a sampling function describing where the air analysed for the ith measurement was sampled. For measurement at given location, π_i is a product of Dirac delta functions (Issartel *et al.* 2007).

On the boundary of the domain $\partial\Omega$ of outward normal vector $\vec{v}$, the following conditions apply to χ:

1. the diffusion flux vanishes through the ground and top of Ω at respective altitudes $z_g(x,\,y)$ and H, i.e. $(\mathbf{K}\vec{\nabla}\chi)\cdot\vec{v}=0$; and
2. at the lateral boundaries, if any, an "outer clean air" condition is implemented by distinguishing whether the wind is entering or leaving Ω.

If $\vec{u}\cdot\vec{v}>0$ then $\chi=0$ (Dirichlet condition), if $\vec{u}\cdot\vec{v}<0$ then $(\mathbf{K}\vec{\nabla}\chi)\cdot\vec{v}=0$ (non Neumann condition). The measurement is equally computed from:

$$-\vec{u}\cdot\vec{\nabla}r_i-\frac{1}{\rho}\nabla(\rho\mathbf{K}\vec{\nabla}r_i)=\pi_i,\quad \mu_i=\int_{\Omega}\rho\sigma r_i\mathrm{d}x\,\mathrm{d}y\,\mathrm{d}z$$
$$(2)$$

which is classically deduced from Green's theorem[1] (e.g. Pudykiewicz 1998). The retrograde dispersion equation, Eq. 2, for r_i is complemented by boundary conditions similar to those for χ except at the lateral boundaries where the roles of $\vec{u}$ and $-\vec{u}$ are interchanged. $r_i(x,\,y,\,z)$ represents the sensitivity of μ_i with respect to an emission at $(x,\,y,\,z)$.

For emissions from the ground, the correspondence $\sigma(x,y,z)=\frac{s(x,y)\delta(z-z_g)}{\rho(x,y,z)}$ between the three and two-dimensional representations transforms Eq. 2 as:

---

[1] $\mu_i=\int_{\Omega}\left(\vec{u}\cdot\vec{\nabla}\chi-\nabla(\mathbf{K}\vec{\nabla}\chi)\right)r_i\rho\,\mathrm{d}V=\int_{\Omega}\left(-\vec{u}\cdot\vec{\nabla}r_i-\nabla(\mathbf{K}\vec{\nabla}r_i)\right)\chi\rho\,\mathrm{d}V+\int_{\partial\Omega}(\chi r_i\vec{u}+\chi\mathbf{K}\vec{\nabla}r_i-r_i\mathbf{K}\vec{\nabla}\chi)\rho\vec{v}\,\mathrm{d}S$, where $\mathrm{d}V=\mathrm{d}x\,\mathrm{d}y\,\mathrm{d}z$, and $\mathrm{d}S$ are the volume and surface elements in Ω and $\partial\Omega$ respectively.

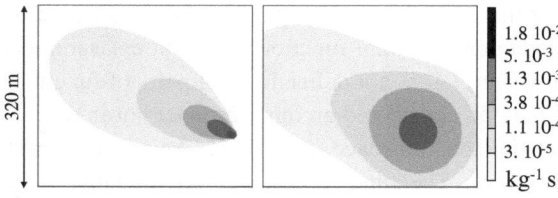

Figure 1
Adjoint functions corresponding to a point detector (*left*) or to a detector averaging the concentration within 50 m (*right*). These correspond to the examples of monitoring networks described by Fig. 4 and discussed in the Sect. 4

$$\mu_i=\int_{\Sigma}sa_i\,\mathrm{d}x\,\mathrm{d}y=(s,a_i)_1\quad\text{with}$$
$$a_i(x,y)=r_i\big(x,y,z_g(x,y)\big)$$
$$(3)$$

in which $(,)_1$ denotes the scalar product of functions $(\xi,\zeta)_1=\int_{\Sigma}\xi(\vec{x})\zeta(\vec{x})\,\mathrm{d}\vec{x}$, with $\vec{x}=(x,y)$, and a_i denotes the adjoint functions corresponding to this product (Fig. 1). A point release of intensity q_0 from location $\vec{x}_0$ ideally generates a measurement vector:

$$\vec{\mu}=q_0\vec{a}(\vec{x}_0)\quad\text{with}\quad \vec{a}(\vec{x}_0)=(a_1(\vec{x}_0),a_2(\vec{x}_0),\ldots,a_n(\vec{x}_0))^{\mathrm{T}}$$
$$(4)$$

This simple result will be helpful in the identification of the point of release from the observed mixing ratios. It is, however, necessary to further consider the general inverse problem of estimating smooth emissions.

2.2. Fundamental and Renormalised Geometries

Renormalisation theory was originally introduced for estimation from measurements $\mu_1,\mu_2,\ldots,\mu_n$ of emissions distributed smoothly. This theory prescribes that the properties of the emissions known a priori are described by choosing:

1. a vector space of acceptable source functions; and
2. a scalar product among these, called the fundamental product.

In our earlier publications, we used the vector space $\mathcal{C}(\Sigma,\mathbb{R})$ of the real functions continuous through the ground with the fundamental product $(\xi,\zeta)_1=\int_{\Sigma}\xi\,\zeta\,\mathrm{d}\vec{x}$ seen in the Sect. 2.1. It is defined by an integral equally weighted through the ground indicating that, a priori, all positions are equally likely to be the seat of the emissions. Owing to the fact that $\mu_i = (s, a_i)_1$, the function $a_i(\vec{x})$ is called the fundamental adjoint function of the ith measurement. However, $(,)_1$ is not defined for all pairs of functions. It may be considered an inner product not in $\mathcal{C}(\Sigma, \mathbb{R})$ but in the subspace $\mathcal{L}^2(\Sigma)$ of square integrable functions. The fundamental adjoint functions a_i do not belong, in general, to $\mathcal{L}^2(\Sigma)$ (ISSARTEL *et al.* 2007). Thus, the fundamental product is not used directly for source estimation. A renormalised scalar product is derived from it by means of a positive weight function $f(\vec{x})$. The multiplication of f by a positive constant is indifferent and thus, conventionally, the total weight is fixed corresponding to the number of measurements:

$$f > 0, \int_\Sigma f \, d\vec{x} = n, (\xi, \zeta)_f = \int_\Sigma \xi(\vec{x}) \, \zeta(\vec{x}) \, f(\vec{x}) \, d\vec{x}$$

(5)

Equation 3 may be rewritten:

$$\mu_i = \int_\Sigma s \, a_{fi} \, f \, d\vec{x} = (s, a_{fi})_f \quad \text{with} \quad a_{fi}(\vec{x}) = \frac{a_i(\vec{x})}{f(\vec{x})}$$

(6)

The weight function should be so chosen that the renormalised adjoint functions a_{fi} belong to the subspace $\mathcal{L}_f^2(\Sigma)$ of the functions that are square integrable with the weights. Any source function s in this space may then be decomposed as $s = s_{\| f} + s_{\perp f}$ into parts respectively parallel and perpendicular, in the meaning of $(,)_f$, to the vector space of functions spanned by the a_{fi}. The component $s_{\perp f}$ of the emission function does not contribute to the measurements whereas $s_{\| f}$ is retrieved from them as:

$$s_{\| f}(\vec{x}) = \vec{\mu}^T \mathbf{H}_f^{-1} \vec{a}_f(\vec{x}) \quad \text{with}$$
$$\mathbf{H}_f = \int_\Sigma \vec{a}_f(\vec{x}) \, \vec{a}_f(\vec{x})^T f(\vec{x}) \, d\vec{x}$$

(7)

in which $\vec{a}_f(\vec{x})$ is the column vector of the $a_{fi}(\vec{x})$ and $\mathbf{H}_f = \left[(a_{fi}, a_{fj})_f \right]$ is the Gram matrix of the modified adjoint functions. This matrix is symmetric and, if the measurements are independent, it is positive. Because $s_{\| f}$ has the least norm $(s_{\| f}, s_{\| f})_f^{1/2}$ among the source functions compatible with the observation vector, it is reasonable to use it as an estimator of the smooth emissions sought.

Among all possible weight functions, there is one, denoted φ, which optimally avoids inversion artifacts (ISSARTEL *et al.* 2007). It is characterised by:

$$\vec{a}_\varphi(\vec{x})^T \mathbf{H}_\varphi^{-1} \vec{a}_\varphi(\vec{x}) \equiv 1 \quad \text{for all} \quad \vec{x} \in \Sigma$$

(8)

The existence and uniqueness of φ fulfilling this requirement is essentially proven by ISSARTEL (2005), despite insufficient attention paid to technicalities. In particular, the proof is based on discretisation of the domain Σ and weight functions f so that its validity is limited by a convergence requirement fulfilled if Σ is topologically compact (Heine's theorem). The compactness requirement was not signalled in this early proof. In practice, it is always fulfilled.

The following fact, also, has not yet been noticed, despite its simplicity and importance. In many situations, for instance when investigating the pollution from a city, the emissions are not expected to be equally likely from all positions on the ground. Then, instead of $(,)_1$, a weighted product $(\xi, \zeta)_w = \int_\Sigma \xi \, \zeta \, w \, d\vec{x}$ is chosen as fundamental with weights $w(\vec{x}) > 0$ proportional to the prior probability density of emissions at location $\vec{x}$. The fundamental adjoint functions become $a_{wi}(\vec{x}) = \frac{a_i(\vec{x})}{w(\vec{x})}$. Equation 5 characterising the weight functions acceptable for renormalisation becomes:

$$f > 0, \int_\Sigma f \, w \, d\vec{x} = n,$$
$$(\xi, \zeta)_{fw} = \int_\Sigma \xi(\vec{x}) \, \zeta(\vec{x}) \, f(\vec{x}) \, w(\vec{x}) \, d\vec{x}$$

(9)

Equation 8 for the best weights φ_w becomes $\vec{a}_{w\varphi_w}(\vec{x})^T \mathbf{H}_{w\varphi_w}^{-1} \vec{a}_{w\varphi_w}(\vec{x}) \equiv 1$. If w is everywhere positive, this implies that $\varphi_w w = \varphi$ and thus:

Proposition 1 *The renormalised scalar product is the same for all fundamental products $(,)_w$.*

On a fixed domain (corresponding to $w > 0$) the renormalised product and geometry are independent of the prior expectations.

2.3. *Classical Least Squares*

We now describe the classical technique utilised to address the identification of the point of release. Let us consider in the n-dimensional real space $\mathbb{R}^n$ a scalar product $(,)_{\mathbf{H}}$ and norm $\|\|_{\mathbf{H}}$ labelled according to the $n \times n$ symmetric positive matrix $\mathbf{H}$ that may always be associated in such a way that, if $\vec{m}, \vec{m}'$ are any two vectors,

$$(\vec{m}, \vec{m}')_{\mathbf{H}} = \vec{m}^{\mathrm{T}} \mathbf{H}^{-1} \vec{m}' \quad \text{and} \quad \| \vec{m} \|_{\mathbf{H}}^2 = (\vec{m}, \vec{m})_{\mathbf{H}} \tag{10}$$

Let $\vec{\mu}$ be the vector observed by the monitoring network. If the measurements are noise-free, in view of Eq. 4, $\| \vec{\mu} - q\vec{a}(\vec{x}) \|_{\mathbf{H}}$ vanishes, when q and $\vec{x}$ are taken as the intensity q_0 and location $\vec{x}_0$ of the release. This allows its identification, and, in this, all norms or symmetric positive matrices are equivalent. However, when $\vec{\mu}$ is noisy, the various norms are no longer equivalent when estimating an intensity q_H and location $\vec{x}_H$ by minimising $\| \vec{\mu} - q_H\vec{a}(\vec{x}_H) \|_{\mathbf{H}}$. This raises the issue of determining a best matrix $\mathbf{H}$.

The classical least-squares approach privileges the norm associated with the covariance matrix of the measurement errors $\vec{\varepsilon} = \vec{\mu}_o - q_0\vec{a}(\vec{x}_0)$ between the measurements effectively observed $\vec{\mu}_o$ and model measurements $\vec{\mu}_m = q_0\vec{a}(\vec{x}_0)$ that could be computed for the point emission if it was known. This error decomposes into an instrumental error $\vec{\varepsilon}_d$ because of the limited quality of the detectors and a representativity error $\vec{\varepsilon}_r$ because of the imperfection of the dispersion model:

$$\vec{\mu}_o = \vec{\mu}_m + \vec{\varepsilon}, \quad \mathbf{Q} = \overline{\vec{\varepsilon}\,\vec{\varepsilon}^{\mathrm{T}}}, \quad \vec{\varepsilon} = \vec{\varepsilon}_d + \vec{\varepsilon}_r \tag{11}$$

In general, $\vec{\varepsilon}$ is supposed to follow a normal law with zero mean and a covariance matrix $\mathbf{Q} = \overline{\vec{\varepsilon}\,\vec{\varepsilon}^{\mathrm{T}}}$ (the bar denotes the statistical average). By minimising $\| \vec{\mu}_o - q\vec{a}(\vec{x}) \|_{\mathbf{Q}}$, an estimator $q_{\mathbf{Q}}, \vec{x}_{\mathbf{Q}}$ of release data is obtained with a remarkable property. In the linear approximation close to q_0 and $\vec{x}_0$ when the measurement error is small, the estimator has the BLUE property: it falls, on average, at the real point of release (unbiased) with, on average, a distance from it that is minimum (best) among all linear estimators for any choice of a norm in the $(q, \vec{x})$-space.

Owing to the BLUE property, the classical approach has been widely advocated. However, as discussed later in the Sect. 5.1, its requirement for a statistical description of the measurement errors is a fragility. New arguments are proposed hereafter leading to a different choice of the matrix $\mathbf{H}$.

3. Identification of a Point of Emission

The elements introduced in the previous section enable proposal and justification of an alternative approach to identification of the point of a release. This approach relies on an angular interpretation, described in the Sect. 3.1, of least-squares techniques showing that the choice of a weighting matrix ultimately amounts to distinguishing various directions over a sphere corresponding to different release locations. In the Sect. 3.2 a criterion is stated which can be used to optimally distinguish. This criterion is shown to be consistent with renormalisation theory utilized for estimating smooth emissions.

3.1. Angular Interpretation of the Least Squares

A simple but original interpretation of the least squares arises from the following analysis of the cost functional associated with a symmetric positive matrix $\mathbf{H}$ (Sect. 2.3):

$$\| \vec{\mu}_o - q\vec{a}(\vec{x}) \|_{\mathbf{H}}^2 = \| \vec{\mu}_o \|_{\mathbf{H}}^2 + q^2 \| \vec{a}(\vec{x}) \|_{\mathbf{H}}^2 - 2q\, \vec{\mu}_o^{\mathrm{T}} \mathbf{H}^{-1} \vec{a}(\vec{x}) \tag{12}$$

For fixed $\vec{x}$, this expression is minimal for an intensity $q_{\vec{x}}$:

$$q_{\vec{x}} = \frac{(\vec{\mu}_o, \vec{a}(\vec{x}))_{\mathbf{H}}}{\| \vec{a}(\vec{x}) \|_{\mathbf{H}}^2},$$
$$\| \vec{\mu}_o - q_{\vec{x}}\vec{a}(\vec{x}) \|_{\mathbf{H}}^2 = \| \vec{\mu}_o \|_{\mathbf{H}}^2 (1 - \mathcal{J}_{\mathbf{H}}(\vec{x})^2) \tag{13}$$

in which $\mathcal{J}_{\mathbf{H}}$ is defined as:

$$\mathcal{J}_{\mathbf{H}}(\vec{x}) = \frac{\vec{\mu}_o^{\mathrm{T}} \mathbf{H}^{-1} \vec{a}(\vec{x})}{\| \vec{\mu}_o \|_{\mathbf{H}} \| \vec{a}(\vec{x}) \|_{\mathbf{H}}} \tag{14}$$

Because $\| \vec{\mu}_o \|_{\mathbf{H}}$ is fixed as observed, $\| \vec{\mu}_o - q\vec{a}(\vec{x}) \|_{\mathbf{H}}$ is minimum when $\mathcal{J}_{\mathbf{H}}(\vec{x})^2$ is maximum. Thus, $\mathcal{J}_{\mathbf{H}}$ may in turn be regarded as a cost function. This first enables estimation from the observations of a source location $\vec{x}_{\mathbf{H}}$. The source intensity $q_{\mathbf{H}}$ is estimated afterwards by substituting $\vec{x}_{\mathbf{H}}$ in Eq. 13. When the measurements are noise-free, the measurement vector $\vec{\mu}_o = q_0\vec{a}(\vec{x}_0)$ in Eq. 14 is proportional to $\vec{a}(\vec{x}_0)$ so that $\mathcal{J}_{\mathbf{H}}(\vec{x}_0)^2 = 1$ is maximum at $\vec{x}_0$.

In Eq. 14, both $\frac{\vec{\mu}_o}{\|\vec{\mu}_o\|_{\mathbf{H}}}$ and $\frac{\vec{a}(\vec{x})}{\|\vec{a}(\vec{x})\|_{\mathbf{H}}}$ are unit vectors for $\|\|_{\mathbf{H}}$. Accordingly, $\mathcal{J}_{\mathbf{H}}(\vec{x})$ is the cosine of the angle between $\vec{\mu}_o$ and $\vec{a}(\vec{x})$ as seen by the geometry of $\mathbb{R}^n$

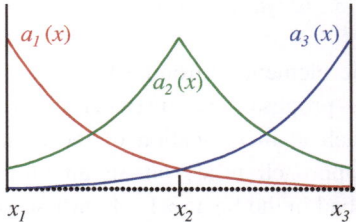
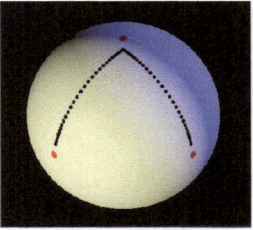
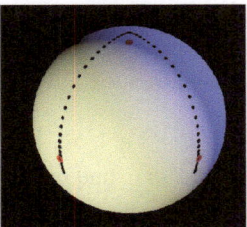

Figure 2

Least dimensional illustration of renormalisation for a linear dispersion along the x axis; the emissions $\sigma(x)$ may occur within the interval $[x_1, x_3]$ with three detectors observing mixing ratios $\chi(x_1)$, $\chi(x_2)$, and $\chi(x_3)$ at each end and in the middle. Dispersion corresponds to constant diffusion and decay, without advection, according to the equation $-\kappa \frac{\partial^2 \chi}{\partial x^2} + \lambda \chi = \sigma$; a point emission of intensity q_0 at x_0 leads to $\chi(x) = q_0 e^{\sqrt{\frac{\lambda}{\kappa}}|x - x_0|}$. The functions $a_1(x)$, $a_2(x)$, $a_3(x)$ (vector $\vec{a}(x)$) adjoint to the measurements assume a similar form (*left panel*). The emission interval is sent over the sphere of $\mathbb{R}^3$ in terms of the adjoint unit vector $\frac{\vec{a}(x)}{\|\vec{a}(x)\|}$ confined within the positive octant indicated by *red dots* at its summits (*middle panel*) or the renormalised vector $\hat{a}_\varphi(\vec{x}) = \sqrt{\mathbf{H}_\varphi}^{-1} \vec{a}_\varphi(\vec{x})$ (*right panel*). The *black dots* are in one-to-one correspondence between the three panels to show the enhanced ability of the renormalised technique to distinguish them

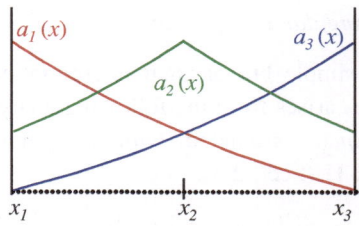
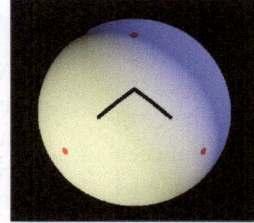
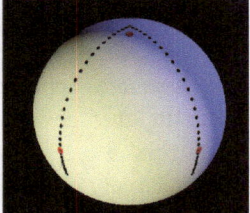

Figure 3

This figure is similar to Fig. 2 except that a stronger diffusion coefficient has been utilized to reduce the contrast between the three measurements (*left panel*). As a result, the classical adjoint unit vectors become indistinct near the axis of the positive octant (*middle panel*), whereas their renormalised counterparts are still clearly distinguished

associated with $(,)_{\mathbf{H}}$. Minimising $\| \vec{\mu}_o - q\vec{a}(\vec{x}) \|_{\mathbf{H}}$ involves first finding the location $\vec{x}_{\mathbf{H}}$ such that $\vec{a}(\vec{x}_{\mathbf{H}})$ is mostly parallel to $\vec{\mu}_o$. To better discriminate the location of the release, the positive definite matrix $\mathbf{H}$ should be chosen in order to maximise the $(,)_{\mathbf{H}}$-angles between the various $\vec{a}(\vec{x})$. This is equivalent to saying that on the unit sphere of the geometry associated with $(,)_{\mathbf{H}}$, the vectors $\frac{\vec{a}(\vec{x})}{\|\vec{a}(\vec{x})\|_{\mathbf{H}}}$ should be distributed as widely as possible (Figs. 2, 3). To facilitate our understanding of the geometry associated with the scalar product $(,)_{\mathbf{H}}$, let us rewrite the cosine of the angle between two vectors $\vec{a}(\vec{x})$, $\vec{a}(\vec{x}')$:

$$\left(\frac{\vec{a}(\vec{x})}{\| \vec{a}(\vec{x}) \|_{\mathbf{H}}}, \frac{\vec{a}(\vec{x}')}{\| \vec{a}(\vec{x}') \|_{\mathbf{H}}} \right)_{\mathbf{H}} = \hat{a}_{\mathbf{H}}(\vec{x})^{\mathrm{T}} \, \hat{a}_{\mathbf{H}}(\vec{x}') \quad \text{with}$$

$$\hat{a}_{\mathbf{H}}(\vec{x}) = \frac{\sqrt{\mathbf{H}^{-1}} \vec{a}(\vec{x})}{\| \vec{a}(\vec{x}) \|_{\mathbf{H}}}$$

(15)

in which the symmetric positive matrix $\sqrt{\mathbf{H}^{-1}}$ is the square root of $\mathbf{H}^{-1}$. The vectors $\hat{a}_{\mathbf{H}}(\vec{x}), \hat{a}_{\mathbf{H}}(\vec{x}')$ are unit

vectors for the ordinary geometry, and the ordinary angle between them is the $(,)_{\mathbf{H}}$-angle between $\vec{a}(\vec{x})$, $\vec{a}(\vec{x}')$. $\mathbf{H}$ should be so chosen that the vectors $\hat{a}_{\mathbf{H}}(\vec{x})$ are widely distributed over the ordinary unit sphere $\mathbb{S}^{n-1}$ of $\mathbb{R}^n$ (FISHER 1953).

Before proceeding, it is possible to understand a drawback of two classical choices of $\mathbf{H}$ as the identity matrix, associated with the ordinary norm, or as the diagonal covariance matrix of noise supposed to be uncorrelated in the various μ_i. Because the $a_i(\vec{x})$ are positive functions, the unit vectors $\hat{a}_{\mathbf{H}}(\vec{x})$ are distributed in the positive octant $\mathbb{S}^{n-1}_+$, 2^n times smaller than $\mathbb{S}^{n-1}$ (MARDIA and JUPP 2000). In fact, the potential advantage is only 2^{n-1} because vectors $\hat{a}_{\mathbf{H}}(\vec{x})$ and $-\hat{a}_{\mathbf{H}}(\vec{x})$ are equivalent at designating the same emissions $q\hat{a}_{\mathbf{H}}(\vec{x}) = -q(-\hat{a}_{\mathbf{H}}(\vec{x}))$, implying that the positive and negative octants are equivalent. Such distribution seems anyway suboptimum if the number n of measurements is large.

3.2. Renormalised Least Squares

Among all symmetric positive matrices, one has special properties. This matrix is $\mathbf{H}_\varphi$, presented in the Sect. 2.2, associated with the optimum renormalising weight function φ subject to Eq. 8. The same matrix can be used for weighting a least-squares functional and identify point emissions. These renormalised least squares have two advantages, explained below in detail. First, the cost function $\mathcal{J}_{\mathbf{H}_\varphi}$ computed from observations $\vec{\mu}_o$ coincides with the estimate of the distributed emissions. The parametric inverse problem for identifying the points of emissions and the assimilative inverse problem for estimating distributed emissions become so consistent that they may be regarded as variants of the same inverse problem. Second, the unit vectors $\hat{a}_{\mathbf{H}_\varphi}(\vec{x}) = \sqrt{\mathbf{H}_\varphi}^{-1}\vec{a}_\varphi(\vec{x})$ (Eq. 15) are widely distributed through the unit sphere so as to maximise the angles between them. Thus, they are optimally discriminated.

3.2.1 Consistency

First, the renormalising condition $\vec{a}_\varphi(\vec{x})^{\mathrm{T}}\mathbf{H}_\varphi^{-1}\vec{a}_\varphi(\vec{x}) = 1$ with vectors $\vec{a}_\varphi(\vec{x}) = \frac{\vec{a}(\vec{x})}{\varphi(\vec{x})}$ (Eq. 8) can be rewritten $\varphi(\vec{x}) = \sqrt{\vec{a}(\vec{x})^{\mathrm{T}}\mathbf{H}_\varphi^{-1}\vec{a}(\vec{x})} = \parallel \vec{a}(\vec{x}) \parallel_{\mathbf{H}_\varphi}$. In view of the fact that $\vec{a}_\varphi(\vec{x}) = \frac{\vec{a}(\vec{x})}{\parallel\vec{a}(\vec{x})\parallel_{\mathbf{H}_\varphi}}$, comparison of Eqs. 14 and 7 leads to:

$$\mathcal{J}_{\mathbf{H}_\varphi}(\vec{x}) = \parallel \vec{\mu}_o \parallel_{\mathbf{H}_\varphi}^{-1} s_{\parallel\varphi}(\vec{x}) \qquad (16)$$

in which $s_{\parallel\varphi}(\vec{x}) = \vec{\mu}_o^{\mathrm{T}}\mathbf{H}_\varphi^{-1}\vec{a}_\varphi(\vec{x})$ (Eq. 7) is the distributed estimate of emissions associated with $\vec{\mu}_o$. Thus, $\mathcal{J}_{\mathbf{H}_\varphi}$ coincides with the same estimate when the norm of the observation vector is reduced to unity.

The cost function $\mathcal{J}_{\mathbf{H}}$ associated with another matrix $\mathbf{H}$ cannot be interpreted so simply. According to Eq. 14, $\mathcal{J}_{\mathbf{H}}$ is a linear combination of $a_{f_{\mathbf{H}}i}(\vec{x}) = \frac{a_i(\vec{x})}{\parallel\vec{a}(\vec{x})\parallel_{\mathbf{H}}}$ that are adjoint to the measurements for the scalar product $(,)_{f_{\mathbf{H}}}$ (Eqs. 5, 6) modified with weights $f_{\mathbf{H}}(\vec{x}) = \sqrt{\vec{a}(\vec{x})^{\mathrm{T}}\mathbf{H}^{-1}\vec{a}(\vec{x})} = \parallel \vec{a}(\vec{x}) \parallel_{\mathbf{H}}$. The only emission function consistent with the observations $\vec{\mu}_o$ that can be obtained as a linear combinations of $a_{f_{\mathbf{H}}i}$ is the estimate $s_{\parallel f_{\mathbf{H}}}$ associated with the scalar product $(,)_{f_{\mathbf{H}}}$ (Eq. 7):

$$s_{\parallel f_{\mathbf{H}}}(\vec{x}) = \vec{\mu}_o^{\mathrm{T}}\mathbf{H}_{f_{\mathbf{H}}}^{-1}\vec{a}_{f_{\mathbf{H}}}(\vec{x}) \quad \text{with}$$
$$\mathbf{H}_{f_{\mathbf{H}}} = \int\limits_\Sigma \vec{a}_{f_{\mathbf{H}}}(\vec{x})\,\vec{a}_{f_{\mathbf{H}}}(\vec{x})^{\mathrm{T}}f_{\mathbf{H}}(\vec{x})\,\mathrm{d}\vec{x} \qquad (17)$$

Notice that all matrices $\alpha\mathbf{H}, \alpha > 0$, are equivalent at defining the same cost function and estimation function and, thus, there is no loss of generality in assuming that $\int_\Sigma f_{\mathbf{H}}\,\mathrm{d}\vec{x} = n$. Then, $\mathcal{J}_{\mathbf{H}}$ and $s_{\parallel f_{\mathbf{H}}}$ are proportional for every observation vector if, and only if, $\mathbf{H}_{f_{\mathbf{H}}} = \mathbf{H}$. This, by definition of $f_{\mathbf{H}}$, leads to $\vec{a}_{f_{\mathbf{H}}}(\vec{x})^{\mathrm{T}}\mathbf{H}_{f_{\mathbf{H}}}^{-1}\vec{a}_{f_{\mathbf{H}}}(\vec{x}) \equiv 1$ characterising $f_{\mathbf{H}} = \varphi$ (Eq. 8). Thus, if, and only if, $\mathbf{H}$ is chosen equal or proportional to $\mathbf{H}_\varphi$, for all observation vectors $\vec{\mu}_o$ generated from a point source:

1. the cost function $\mathcal{J}_{\mathbf{H}}$ coincides with the distributed estimation (up to some constant factor) so that
2. $s_{\parallel f_{\mathbf{H}}}^2$ is maximum at release location.

The second conclusion is of course relaxed if the measurements are noisy.

3.2.2 Wide Distribution

The inconvenience of choosing $\mathbf{H}$ as the identity or a diagonal covariance matrix has been explained at the end of the Sect. 3.1: the unit vector $\hat{a}_{\mathbf{H}}(\vec{x}) = \sqrt{\mathbf{H}^{-1}}\vec{a}_{f_{\mathbf{H}}}(\vec{x})$ is distributed within a narrow octant of the sphere (Figs. 2, 3, central panel) and this reduces the discriminating ability of the directional comparison achieved by the cost function (MARDIA 1975). The discussion of the previous section (Sect. 3.2.1) enables this argument to be placed on a more formal basis, further highlighting the role of the matrix $\mathbf{H}_\varphi$. This is done in two steps, by first showing the role of matrix $\mathbf{H}_{f_{\mathbf{H}}}$ (Eq. 17) for characterising the good or bad distribution of the unit vector and then by proposing a physical interpretation.

For the first step, it is preferable to represent the geometry associated with $\mathbf{H}$ in terms of the unit vectors $\hat{a}_{\mathbf{H}}(\vec{x})$ (Eq. 15) of the ordinary geometry. The counterpart of $\mathbf{H}_{f_{\mathbf{H}}}$ is then:

$$\int\limits_\Sigma \hat{a}_{\mathbf{H}}(\vec{x})\,\hat{a}_{\mathbf{H}}(\vec{x})^{\mathrm{T}}f_{\mathbf{H}}(\vec{x})\,\mathrm{d}\vec{x} = \sqrt{\mathbf{H}^{-1}}\,\mathbf{H}_{f_{\mathbf{H}}}\sqrt{\mathbf{H}^{-1}} \qquad (18)$$

If $\mathbf{H}$ is chosen diagonal, because $f_{\mathbf{H}}$ and the components of all $\hat{a}_{\mathbf{H}}(\vec{x})$ are positive, the off-diagonal terms of $\sqrt{\mathbf{H}^{-1}}\,\mathbf{H}_{f_\mathbf{H}}\sqrt{\mathbf{H}^{-1}}$ may not vanish. On the contrary, if $\mathbf{H}$ is chosen as $\mathbf{H}_\varphi$, Eq. 18 becomes:

$$\int_\Sigma \hat{a}_\varphi(\vec{x})\,\hat{a}_\varphi(\vec{x})^\mathrm{T}\varphi(\vec{x})\,\mathrm{d}\vec{x} = \mathbf{I} \qquad (19)$$

in which $\hat{a}_\varphi(\vec{x}) = \sqrt{\mathbf{H}_\varphi}^{-1}\vec{a}_\varphi(\vec{x})$ is denoted shortly for $\hat{a}_{\mathbf{H}_\varphi}(\vec{x})$ and $\mathbf{I}$ is the identity matrix. This indicates that $\hat{a}_\varphi(\vec{x})$ and the equivalent vectors $-\hat{a}_\varphi(\vec{x})$ (cf. the end of the Sect. 3.1) are distributed in all the octants of the unit sphere (Figs. 2, 3, right panel) and not only the positive or negative octants (MARDIA and JUPP 2000). The relevance of matrix $\sqrt{\mathbf{H}^{-1}}\,\mathbf{H}_{f_\mathbf{H}}\sqrt{\mathbf{H}^{-1}}$ for characterizing the spread of $\hat{a}_\mathbf{H}(\vec{x})$, or $\mathbf{H}_{f_\mathbf{H}}$ for the spread of $\vec{a}_{f_\mathbf{H}}(\vec{x})$, is justified by the following interpretation. The choice of a positive symmetric matrix $\mathbf{H}$ for identifying point emissions induces:

1. the choice of a scalar product among the distributed emission functions; and
2. the choice of a probability distribution of the emissions over the ground Σ.

To see the first point, notice that the form of the cost function (Eq. 16) involves a rescaling of the vector $\vec{a}(\vec{x})$ as $\vec{a}_{f_\mathbf{H}}(\vec{x}) = \frac{\vec{a}(\vec{x})}{\|\vec{a}(\vec{x})\|_\mathbf{H}}$. Whereas the components of $\vec{a}(\vec{x})$ are the functions adjoint to the measurements for the scalar product $(,)_1$ (Eq. 3), the components of $\vec{a}_{f_\mathbf{H}}(\vec{x})$ are adjoint for the weighted product $(,)_{f_\mathbf{H}}$ (Eq. 6). To show the second point, we observe that the matrix elements $\mathbf{H}_{f_\mathbf{H}\,ij} = (a_{f_\mathbf{H}\,i}, a_{f_\mathbf{H}\,j})_{f_\mathbf{H}}$ measure, for the weighted geometry, the overlapping of the functions adjoint to μ_i and μ_j. This overlapping is large if μ_i and μ_j are sensitive to the almost same regions. It is small if the regions to which they are jointly sensitive have a small weighted extent. Thus, up to a factor corresponding to the total weight of $f_\mathbf{H}$, we may see $\frac{1}{n}\mathbf{H}_{f_\mathbf{H}}$ as the covariance matrix of the measurements. In other words, the choice of a matrix $\mathbf{H}$ is consistent with the assumption that the observations will originate from emissions having the probability distribution $\frac{f_\mathbf{H}}{n}$ over the ground Σ.

The integral in the left hand side of Eq. 18 is thus the covariance matrix of $\hat{a}_\mathbf{H}(\vec{x})$ weighted according to this distribution. Equation 19 indicates that when $\mathbf{H}$ is chosen as $\mathbf{H}_\varphi$, this covariance matrix becomes an identity which is the classical criterion of a wide spherical distribution. In particular, the components of $\hat{a}_\varphi(\vec{x})$ are mutually independent.

3.2.3 Apparent Geometry

The privileged nature of $\mathbf{H}_\varphi$ and probability density $\frac{\varphi}{n}$ may seem arbitrary, because the emissions of a species may assume, a priori, different probability distributions over Σ. Carbon monoxide is more likely to be emitted from a city, sulfur from an industrial area, yperite from a battlefield. The point is that a distinction is established between the probability distribution of the emissions independently from any monitoring system, and the probability distribution of the emissions observed by a monitoring system. Even if the emissions are equally likely from all parts of the ground, those occurring close to the detectors or upwind of them will be detected with an enhanced probability compared with remote emissions downwind of the detectors. So, φ is interpreted as the probability distribution of the emissions apparent to the monitoring network. It was shown elsewhere (ISSARTEL 2005), that φ characterised by Eq. 8 leads to the minimum value of $\det\mathbf{H}_\varphi$ among all positive functions of total weight n, in particular among functions $f_\mathbf{H}$. This implies that the entropy $\log\det\mathbf{H}_\varphi$ of the normal distribution with covariance matrix $\mathbf{H}_\varphi$ is minimum. Thus, the weight function φ is less arbitrary than any other for predicting the expectable observation vectors $\vec{\mu}$. This apparent probability distribution depends only on:

1. the arrangement of the monitoring network;
2. the prevailing meteorological circumstances; and
3. the choice of a possible emission domain Σ.

It is noticeable that it be independent of a prior probability distribution of the emissions through Σ (Proposition 1 in the Sect. 2.2).

The authors wish to point out that $\mathbf{H}_\varphi$ is interpreted, up to some factor, as the covariance matrix of the measurement vector $\vec{\mu}$, not as the covariance matrix of the measurement errors $\vec{\varepsilon}$. This is one of the most original features of the proposed approach: the measurements are attributed a statistical status irrespective of any error. In the Bayesian framework, the statistics of the measurements are exclusively related to the errors. This point and the whole theoretical framework are illustrated and clarified by examples.

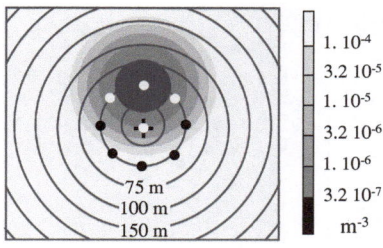

Figure 4
Schematic diagram of the three numerical examples discussed in the Sect. 5.2. Each is based on a set of eight concentration measurements used to monitor continuous emissions at ground level. The *points* on the circle of radius 50 m indicate the location of the point detectors of example 1 and the Gaussian detector of example 2; the sampling distribution is represented for one detector. The *point* at the centre is the location of the single detector used with eight different wind direction in example 3

4. Illustrations of the Alternative Approach

To illustrate the practical meaning of the theory, three examples of a monitoring system have been prepared based on a dispersion model. The reader is reminded that this development is supported by earlier results with real data (ISSARTEL 2005; SHARAN *et al.* 2009). Accordingly, the purpose of the following examples is not to validate the theory but to explain it. They will also enable understanding of a conceptual difficulty associated with the most popular monitoring systems made of point detectors settled at

various locations. This difficulty is related to what can be seen as a degeneracy of the matrix $\mathbf{H}_\varphi$.

Each of the following examples is based on a set of eight measurements related to constant emissions at ground level on a terrain at a local scale (Fig. 4). The proposed approach is described in each example by a figure (Figs. 5, 6, 7) consisting of three panels showing:

1. the overlapping of the adjoint functions involved in the monitoring network;
2. a classical representation of the renormalising weight function φ; and
3. a more geometrical representation of φ called the renormalised framework. In this framework, the various regions are extended or shrunk in proportion to their weights. This is done by means of a transformation in polar coordinates $(l, \theta) \rightarrow (l', \theta)$ such that $l' \, dl' = \varphi(l, \theta) \, l \, dl$. The angular coordinate is unchanged. The pole is chosen at the centre of the monitoring network.

The computational results presented here are based on the dispersion model described below.

4.1. The Dispersion Model

An analytical low wind dispersion model (SHARAN *et al.* 1996) developed for point-source emissions, based on steady dispersion, is used in backward mode

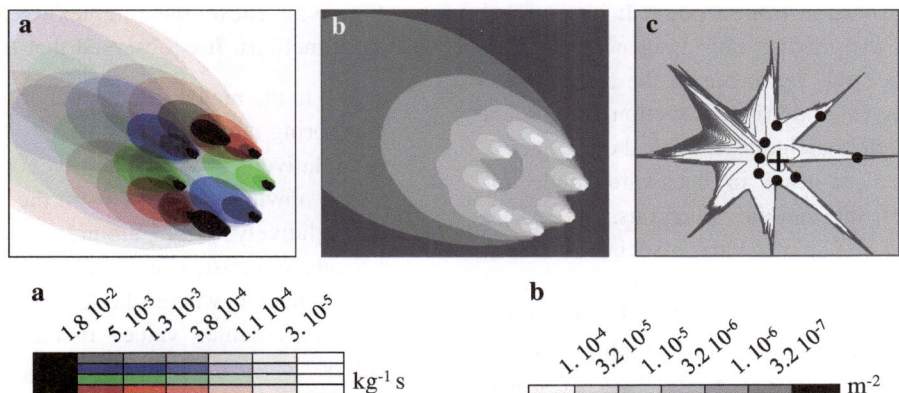

Figure 5
Description of the proposed geometric approach to the monitoring network of eight point detectors arranged on a circle as shown in Fig. 4. **a** Shows the overlapping of the corresponding eight adjoint functions drawn with *four colours*. Two equivalent representations are given of the renormalised geometry associated with this monitoring network and meteorological conditions. **b** Classical representation of the renormalising function φ. **c** More geometrical vision: the various regions are extended or shrunk in proportion to φ by extending or shrinking their radial coordinate with respect to the centre (*plus symbol*) of the circle of detectors (*filled circles*). The angles with respect to the pole are preserved. The *curves within this star-shaped* representation correspond to the circles of Fig. 4

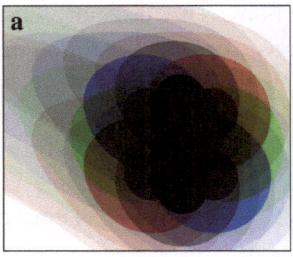 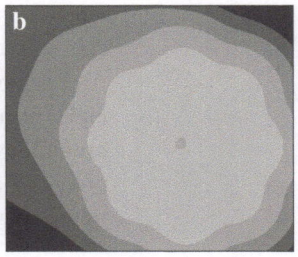 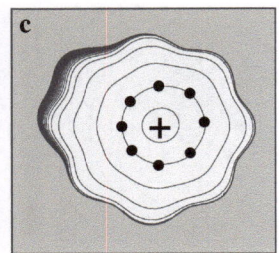

Figure 6
Description of the proposed geometric approach to the monitoring network of eight Gaussian detectors arranged in a *circle* as shown in Fig. 4. The *legend* and *colour scales* are similar to those in Fig. 5

 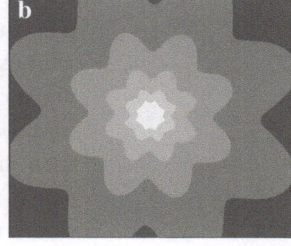 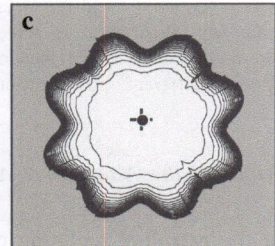

Figure 7
Description of the proposed geometric approach to the monitoring network of eight point detectors measuring at a single location for eight wind directions as shown in Fig. 4. The legend and colour scales are similar to those in Fig. 5

for computation of the functions adjoint to the measurements. This model is an analytical function $\kappa(\vec{x})$ corresponding to the steady field of mixing ratio generated from a unit point source at the origin. It describes more easily than differential equations the successive involvement of larger and larger turbulent scales in the dispersion from a point. It is extended for computing the mixing ratio χ generated by a distributed source $s(\vec{x})$ by means of the convolution: $\chi(\vec{x}) = \int_{\Sigma} \sigma(\vec{y}) \kappa(\vec{y} - \vec{x}) d\vec{y}$ (in this section only, $\vec{x}$ and $\vec{y}$ are location vectors including a vertical component). This model can be used in backward mode for computation of the adjoint functions. Indeed, a measurement $\mu = \int_{\Sigma} \pi(x) \chi(\vec{x}) d\vec{x}$ associated with a sampling function π may be written $\mu = \int_{\Sigma} \pi(x) d\vec{x} \int_{\Sigma} \sigma(\vec{y}) \kappa(\vec{y} - \vec{x}) d\vec{y} = \int_{\Sigma} \sigma(\vec{y}) d\vec{y} \int_{\Sigma} \pi(\vec{x}) \kappa(\vec{y} - \vec{x}) d\vec{x}$. This implies that the function adjoint to μ for the scalar product in Eq. 1 is $r(\vec{y}) = \int_{\Sigma} \pi(\vec{x}) \kappa(\vec{y} - \vec{x}) d\vec{x}$ which corresponds to changing wind direction in the model.

4.2. Point Measurements with Constant Meteorology

The first example of a monitoring system consists of eight detectors evenly distributed, starting from

north, on a circle of radius 50 m and used in unstable conditions with a wind of 1.6 m s^{-1} from 295°N constant through space and time. Owing to these constant conditions, the functions $a_i(\vec{x})$ adjoint to the measurements are superposable on each other with reference to the detector at origin (Fig. 1, left).

Figure 5a, shows the complex overlapping of the adjoint functions. It is observed that their values:

1. vanish in the region immediately downwind of the monitoring network;
2. are relatively low and poorly contrasted in the region upwind and away from the network; and
3. are relatively large around the network and strongly contrasted, especially in the vicinity of each detector where the corresponding adjoint function assumes values much larger than the others.

This is consistent with the distribution of the renormalising weights φ (Fig. 5b) strongly focussed at the detectors, decaying and becoming negligible within 300 m upwind of network. The focus at the detectors is confirmed in the renormalised framework (Fig. 5b) with its characteristic star shape (SHARAN *et al.* 2009). The weight focussed around each detector appears in the form of a branch developed in the directions of the detectors with respect to the pole at the centre.

The focus of φ occurs in the vicinity of the detectors where the overlapping of the adjoint functions is very unbalanced. In view of the statistical interpretations in the Sect. 3.2.2, this suggests that the various measurements μ_i are poorly correlated. This is verified by noticing that the computed $\mathbf{H}_\varphi$, interpreted as the apparent covariance matrix of the expectable measurements, is almost proportional to the identity matrix with diagonal terms between 12,073 and 13,617, and off-diagonal terms between 567 and 2,286 (in units $\mathrm{kg}^{-2}\,\mathrm{s}^2$).

4.3. Gaussian Measurements with Constant Meteorology

The second example of a monitoring network is almost same as the first. However, the point measurements are replaced by Gaussian measurements $\bar{\mu}_i = \int_\Sigma \chi(\vec{x})\pi_i(\vec{x})\mathrm{d}\vec{x}$ centred at eight locations $\vec{x}_i$ on a circle of radius 50 m. The sampling distributions $\pi_i(\vec{x}) = \frac{1}{2\pi l^2} exp\left(-\frac{(\vec{x}-\vec{x}_i)^{\mathrm{T}}(\vec{x}-\vec{x}_i)}{l^2}\right)$ are normal, decaying away from the $\vec{x}_i$ within a standard distance $l = 50\,m$. The Gaussian sampling functions may look artificial in view of the detectors actually being available for local applications. On larger scales, they might correspond to airborne observations having a limited horizontal resolution. However, they are acceptable theoretically. The adjoint functions $a_i(\vec{x})$ are again superposable on each other with reference to $\vec{x}_i$ at the origin (Fig. 1, right).

The wide overlapping of the adjoint functions (Fig. 6a) is very different from the pattern in example 1 (Fig. 5a). In most regions, the various a_i assume values of similar magnitude, relatively high around the network and decaying in the upwind direction. The renormalising weights φ (Fig. 6b) are distributed more homogeneously and on a wider area, without the focus previously related to point detectors (Fig. 5b). The transformation in the renormalised framework (Fig. 5c) of the circles in the Fig. 4 is indicative of homogeneous visibility, without relative distortions, until the circle 100 m from the centre. Beyond this circle, visibility decays rapidly.

The matrix $\mathbf{H}_\varphi$ is very different from $\mathbf{I}$. In the computation, its diagonal terms vary between 4,347 and 5,780, the off-diagonal terms corresponding to neighbouring detectors vary between 3,390 and 4,588, the remaining terms have a minimum value of 1,335.

4.4. Point Measurements at the Same Location with Changing Meteorology

The third example of a monitoring network consists of a single detector utilized eight times, at the same location $\vec{x}_0$, to measure the mixing ratio $\mu_i = \chi(\vec{x}_0, t_i)$ generated by the same constant release at ground level during eight different meteorological episodes around the time t_i. The meteorological conditions in these episodes are constant and the same as in examples 1 and 2 (in the Sects. 4.4, 4.3) except that wind direction assumes eight different values at even intervals of 40°. The episodes last long enough for a steady state of the mixing ratio $\chi(\vec{x}, t) = \chi(\vec{x}, t_i)$ to be established in each. The functions adjoint to the measurements are, then, fully determined by the conditions of each respective episode. With reference to the detector at the origin and wind direction, they are superposable on the adjoint functions of the first example (Fig. 1, left). Near the single detector, the adjoint functions a_i all have the almost same singularity dominated by the diffusion processes, irrespective of wind direction. Thus, they overlap (Fig. 7a) with relatively high values, poorly contrasted around the detector and, in contrast, relatively small values, strongly contrasted away from it. The balance of the opposite tendencies leads to renormalising weights φ clearly focussed around the detector. This is visible in Fig. 7b and, even more clearly, in Fig. 7c, where the counterpart of the circle at 25 m from the detector occupies one third of the renormalised framework thus representing one third of the total visibility. However, φ is more widely distributed and less focussed than in example 1 (Fig. 5b). This may be understood in view of the fact that in example 1, the advantages of strong values and strong contrast of the adjoint functions were cumulated near the detectors.

Theoretically, the diagonal terms of $\mathbf{H}_\varphi$ are all equal, however, depending on their parallel or

169

oblique orientation of corresponding wind relative to the numerical meshes, their value is either $8,246 \pm 5$ or $8,581 \pm 5$. The other terms vary between $4,288$ and $5,715$. As in example 2, $\mathbf{H}_\varphi$ is very different from $\mathbf{I}$, implying that renormalised least squares should be preferred to ordinary least squares.

4.5. Degeneracy of Renormalisation

The first example points out a circumstance that obscures, in most experiments, the nature of the inverse problem. In this example, $\mathbf{H}_\varphi$ is approximately proportional to the identity, implying that least squares, either renormalised or weighted by the identity, should retrieve point emissions similarly well: this is indeed described by SHARAN et al. (2009). Good results associated with ordinary least squares are described elsewhere (MULHOLLAND and SEINFELD 1994; MATTHES et al. 2005; KRYSTA et al. 2006) and interpreted as far as possible, in view of the simplicity of the technique. In particular, these good results might be regarded too rapidly as validation of the classical least squares weighted by use of the inverse of a covariance matrix $\mathbf{Q}$ of the measurement errors commonly assumed to be proportional to $\mathbf{I}$ by arguing that:

1. the errors made by distinct detectors are certainly uncorrelated; and
2. detectors of the same type make errors of similar magnitude.

These arguments, acceptable for the errors resulting from the detectors, are not relevant for the prevailing errors associated with the model. We argue that the success of the ordinary least squares arises not from proportionality of $\mathbf{Q}$ and $\mathbf{I}$, but from the approximate proportionality of $\mathbf{H}_\varphi$ and $\mathbf{I}$ when a monitoring network consists of point detectors at different locations, as occurs most often in practice.

In this case, the main difference between renormalised and ordinary least squares is their behaviour near the detectors. The locally high values of φ indicate that a release there will be retrieved from renormalised least squares with much better resolution than anywhere else. Owing to the small extent of the area where the renormalised resolution becomes very high, this property is probably not very useful.

However, it helps in understanding the following important result.

Proposition 2 *A point release is well retrieved from least squares weighted by the identity matrix if the measurements correspond to point detectors at different locations subjected to similar weather conditions.*

The proposition is true away from the detectors where classical and renormalised least squares are mostly similar. It is also true close to them because the high resolution there of the renormalised least squares is generally in excess of the requirements. The diagonal terms of $\mathbf{H}_\varphi$ are mostly determined by the behaviour of the a_i near the detectors. If these are subject to significantly different meteorological conditions, the identity matrix will be replaced in the proposition by a diagonal matrix. In this, the form of the covariance matrix of detector errors does not matter if these are small compared with model uncertainties. Finally, in example 1, the main benefit of renormalisation, which is to widely distribute $\hat{a}_\varphi(\vec{x}) = \sqrt{\mathbf{H}_\varphi}^{-1} \vec{a}_\varphi(\vec{x})$ through the unit sphere, is mostly absorbed in the immediate vicinity of the detectors where the resolution becomes very good on a very limited domain. In most parts of the ground, $\hat{a}_\varphi(\vec{x})$ and $\vec{a}(\vec{x})$ remain almost parallel. The other two examples highlight the fact that, theoretically, the coincidence $\mathbf{H}_\varphi \simeq c\,\mathbf{I}$ is accidental despite its frequency and practical importance (proposition 2). In particular in the second example, $\mathbf{H}_\varphi$ is far from proportional to identity, and renormalised and ordinary least squares are very different. The advantage of renormalisation is substantial for two reasons. First, owing to the similar magnitude of the adjoint functions, the direction of the standard adjoint vectors $\vec{a}(\vec{x})$ is confined not only in the positive octant $\mathbb{S}_+^{n-1}$ (Sect. 3.1), but in a small part of it, close to the axis of symmetry in the direction $(1, 1, \ldots, 1)$. This situation may be compared with that illustrated by Fig. 3. The advantage of renormalisation for spreading $\hat{a}_\varphi(\vec{x}) = \sqrt{\mathbf{H}_\varphi}^{-1} \vec{a}_\varphi(\vec{x})$ widely through the unit sphere is all the larger. Second, this advantage is well distributed throughout the emission domain Σ and not absorbed, as in example 1, by the immediate vicinity of the detectors. This example makes it easy to understand that, even though the eight detectors are distinct, owing to the wide overlapping of the adjoint functions, the eight measurements should be similarly affected by the imperfections of the dispersion model. Unlike detector errors, representativity errors should be strongly correlated.

5. Discussion

Two least-squares approaches for identifying the point of release from a set of concentration measurements are described in the Sect. 2. They differ by the norm used to compare the measurements observed with those ideally modelled for all possible emissions. Classically, in order to privilege the most accurate measurements, a norm is defined on the basis of the covariance matrix of the errors. The alternative approach relies on the existence of a norm which is shown to optimally discriminate all possible emissions. The criteria utilised in each case are convincing but different, which means, when dealing with optimization, incompatible. Therefore, the grounds and relevance of both approaches must be carefully discussed. The primary basis of the comparison is the probability distribution of the measurement errors, especially those of the model. The classical approach relies on the knowledge of such probability distribution and this is its Achille's heel.

5.1. Advantages and Disadvantages of the BLUE Approach

The BLUE property of the classical approach recalled in the Sect. 2.3, at least in the linear approximation for small measurement errors, is a serious advantage. If a probability law is known for the measurement errors $\vec{\varepsilon}$ (Eq. 11), there is no doubt that the estimate obtained from the classical least squares is the best. In practice, such a law is not known. Measurement errors may be decomposed as $\vec{\varepsilon} = \vec{\varepsilon}_d + \vec{\varepsilon}_r$ into detector and representativity errors. The instrumental error $\vec{\varepsilon}_d$, on one hand, may be tested technically and, given the accuracy of the current technology, it is very limited. On the other hand, the representativity errors $\vec{\varepsilon}_r$ is certainly much larger and has not been explored fully (DALEY 1992; HANSEN

2002; ZUPANSKI and ZUPANSKI 2006; LI et al. 2009; CARASSI and VANNITSEM 2010). It is generally assumed that $\vec{\varepsilon}, \vec{\varepsilon}_d$ and $\vec{\varepsilon}_r$ all follow normal laws. The assumption of a non-normal law has recently been considered by PIRES et al. (2010) on the basis of phenomenological considerations about dispersion processes. However, proposition A1 proved in Appendix, implies that a normal law is not logically acceptable for $\vec{\varepsilon}_r$, irrespective of the physics. Fundamentally, the normal assumption propagates statistical bonds between measurement vectors of different norms that are not compatible with their linearity as functions of tracer emissions. Indeed, an observation vector $\vec{\mu}_o$, under the assumption of normal errors, may correspond with equal probability density to ideal model vectors $\vec{\mu}_o + \vec{\varepsilon}$ or $\vec{\mu}_o - \vec{\varepsilon}$ though their norms are generally different.

5.2. Advantages and Disadvantages of the Alternative Approach

Utilization of the renormalised geometry for the identification of point sources has been already illustrated by SHARAN et al. (2009). The function φ subject to Eq. 8 may not be obtained analytically but is computed rapidly by use of an iterative algorithm. The domain Σ must be discretised even when the adjoint functions are determined from an analytical dispersion model. On the basis of a vector $\vec{\mu}_o$ of observations, a function $s_{\|\varphi}$ is computed (Eq. 17) and the point emission is estimated at the location $\vec{x}_{\max}$ where $s_{\|\varphi}^2$ becomes maximum (Eq. 16). Estimation quality is better if the release is located in a region where φ is relatively high, because this is related to the sharpness of the maximum. This approach to point source identification is simple but unusual in several respects.

The classical theory transforms an input law of the measurement errors into an output law of the errors of estimation. In the approach of renormalised least squares, this does not make sense. If a law was known for the input error with a covariance matrix $\mathbf{Q}$, the classical approach should be utilised, owing to its BLUE property for minimising the output errors. The point is that, as explained in the Sect. 5.1, no such input law is available for model representativity errors prevailing in the measurement errors. The best way of estimating release location from observations of unknown quality is renormalised inversion, because it best discriminates all possible locations (Sect. 3.2.2). Naturally, output quality is as unknown as input quality, or in other words, output and input are similarly reliable. This simple point is very common in other domains of mathematics. For instance, when dealing with linear systems, the requirement that a matrix be well conditioned to avoid amplification of the input noise does not depend on any probability law of these. It is possible to say that the least squares weighted by $\mathbf{H}_\varphi^{-1}$ are well conditioned.

Use of $\mathbf{H}_\varphi$ has other advantages relatively more formal in nature.

1. A logical continuity is established between the overdetermined inverse problem of identifying a point emission and the underdetermined inverse problem of estimating distributed emissions (Sect. 3.2.1).

2. $\mathbf{H}_\varphi$ is related to the weight function φ which can be given the meaning of a visibility function for characterizing the regions well seen or badly seen. The function φ is relatively higher in the regions close to or upwind of the monitoring network and thus, these are magnified in the weighted geometry of Σ. The geometry weighted by φ has been interpreted as the geometry apparent to the monitoring network and the positive function $\frac{\varphi}{n}$ as the apparent probability distribution of the emissions (Sect. 3.2.3). This is different from a real probability distribution, because the emissions in a region well seen are observed with an enhanced apparent probability.

3. The apparent probability distribution of the emissions is independent of any real prior probability distribution (proposition 1). The independence stated here is an original result not recognized in earlier work by ISSARTEL et al. (2007) and SHARAN et al. (2009). It means that, depending on the arrangement of the monitoring network and meteorological conditions, there exists a probability distribution of the emissions intrinsically defined through the possible emission domain Σ. This result should be helpful in the investigations, so far unsuccessful, that have been conducted in

recent decades for elicitation of Bayesian priors in the assimilative inverse problem of estimating distributed emissions.

6. Conclusions

This paper introduced and justified a new criterion for weighting a least-squares cost function when the covariance matrix of the measurement errors is not known. This point of view arises from a difficulty with the prevailing part of measurement errors because of the imperfect model representativity. In practice, no statistical description is available for this. This fact is most often minimised, ignored, or bypassed with ad hoc assumptions. Here, its importance is emphasized, in particular by showing the illogical nature of the commonly used Gaussian assumption. This leads to weighing of the least squares by means of a matrix $\mathbf{H}_\varphi$ introduced by ISSARTEL et al. (2007) in their assimilative framework of renormalisation theory. This matrix is related to a function φ defined in the space-time domain of possible emissions that can be interpreted as the apparent probability distribution of the emissions.

The renormalised least squares are not characterised by the usual BLUE property for minimising the errors between the real and estimated data of a point release. They are characterised differently, by global optimization of the resolution, i.e. the ability of the inversion technique to discriminate two points of the possible emission domain. This best guarantees the consistency of the output identification with the input data without knowing the quality of these. A similar situation is encountered in the field of linear systems, where the requirement that a matrix be well conditioned does not require a statistical description of the noise.

The theory has been illustrated by several examples. The examples clarify the interpretation of φ as a visibility function and point out a difficulty arising with monitoring networks consisting of point detectors at different locations: the visibility is uselessly focussed at the detectors and $\mathbf{H}_\varphi$ is close to the identity so that renormalised least squares almost degenerate to ordinary ones. This degeneracy can be exploited. It is, nevertheless, essential to understand it well, because it adds much confusion to this complex field of source identification by incorrectly substantiating the utilization of classical least squares. In practice, point detectors are the most common.

Statistically, the results in this paper raise two points. First, it has been seen that the renormalising function, which can also be interpreted as the apparent probability distribution of the emissions, is independent of any prior statistics. This means that a monitoring network, under given meteorological conditions, itself contains natural statistics about the emissions. This is certainly of interest for the still unsolved Bayesian debate about the elicitation of a background error covariance matrix. Second, the utilisation of $\mathbf{H}_\varphi$ in this study suggests that this matrix is itself the covariance matrix of the representativity errors. This idea is valuable but cannot be developed as is because the representativity error may not follow a normal law. It is possible to show that this error is subject to other constraints. These together lead to unusual statistics in which $\mathbf{H}_\varphi$ is indeed central. However, the volume of the explanations required in this development puts them clearly beyond the scope of this paper.

Appendix

This appendix explores the statistical behaviour of the representativity error $\vec{\varepsilon}_r$ of a dispersion model (Eq. 11). Here is described, apparently for the first time, a surprising though simple constraint in the statistics of $\vec{\varepsilon}_r$. This implies that a normal law is not compatible, for $\vec{\varepsilon}_r$, with the linearity of the modelled and observed measurements with respect to the emissions of tracer. For the sake of simplicity, detector errors are ignored so that $\vec{\varepsilon} = \vec{\varepsilon}_r$. Let us define the following probability densities and suppose that they are all continuously differentiable:

$p(\vec{\mu})$ the probability density of the model measurement vector $\vec{\mu}_m$ such that $p(\vec{\mu})d\vec{\mu}$ is the probability that $\vec{\mu}_m$ falls within $d\vec{\mu}$ around $\vec{\mu}$

$p'(\vec{\varepsilon} \, / \, \vec{\mu})$ the probability density of the representativity errors $\vec{\varepsilon}$ when the model vector is $\vec{\mu}$, $q(\vec{\mu})$ the probability density of the observed measurement vector $\vec{\mu}_o$ such that $q(\vec{\mu})d\vec{\mu}$ is the probability that $\vec{\mu}_o$ falls within $d\vec{\mu}$ around $\vec{\mu}$,

$q'(\vec{\varepsilon} \, / \, \vec{\mu})$ the probability density of representativity errors $\vec{\varepsilon}$ when the observed vector is $\vec{\mu}$.

Both model and observed measurements are linear with regard to the emissions. It is consistent to argue, for any $l > 0$ and $\vec{\mu} \in \mathbb{R}^n$, that the probabilities are same for the representativity error to fall:

1. within $d\vec{\varepsilon}$ around $\vec{\varepsilon}$ when the model vector is $\vec{\mu}_m = \vec{\mu}$; or
2. within $l^n d\vec{\varepsilon}$ around $l\vec{\varepsilon}$ when the model vector is $\vec{\mu}_m = l\vec{\mu}$.

This implies for p' and similarly for q' :

$$p'(l\vec{\varepsilon} \, / \, l\vec{\mu}) = l^{-n}p'(\vec{\varepsilon} \, / \, \vec{\mu}), q'(l\vec{\varepsilon} \, / \, l\vec{\mu}) = l^{-n}q'(\vec{\varepsilon} \, / \, \vec{\mu}) \tag{20}$$

The above scaling deduced from linearity restricts the acceptable forms of the conditional probabilities. A normal q' would imply that an observed vector $\vec{\mu}_o = \vec{\mu}$ could be associated equally likely with errors $\vec{\varepsilon}$ or $-\vec{\varepsilon}$. In particular, if $\alpha > 0$ is a real number, the observed vector $\vec{\mu}$ may equally likely correspond to model vector $(1 + \alpha)\vec{\mu}$ or $(1 - \alpha)\vec{\mu}$. As shown below, this is not compatible with the linear scaling: p' and q' may not be normal.

Proposition A1 *The probability density functions p, p', q, q' may not fulfil all of the following conditions,*

1. *p, p', q, q' are continuously differentiable*
2. *p', q' are both invariant, for all $\vec{\mu}$, when the error is changed from $\vec{\varepsilon}$ to $-\vec{\varepsilon}$,*
3. *p', q' both satisfy the linear scaling (Eq. 20).*

Proof Suppose that conditions (1)–(3) are all fulfilled. Let $\vec{\mu}_o = \mu\vec{\omega}$ be the vector observed in which μ is a real number and $\vec{\omega}$ a vector of norm unity. In view of property (2) of q', errors $\vec{\varepsilon}$ and $-\vec{\varepsilon}$ equally likely imply that model vectors $\vec{\mu} + \vec{\varepsilon}$ and $\vec{\mu} - \vec{\varepsilon}$ are equally likely. This is true, in particular, for $\vec{\varepsilon} = \alpha\vec{\mu}$ where α is a real number. In this case, the equal likelihood is written: $p'[\alpha\mu\vec{\omega} \, / \, (1 - \alpha)\mu\vec{\omega}] \, p[(1 - \alpha)\mu\vec{\omega}] = p'[-\alpha\mu\vec{\omega}/ (1 + \alpha)\mu\vec{\omega}] \, p[(1 + \alpha)\mu\vec{\omega}]$. This can be transformed by using the conditions (2) and (3), both for p':

$$\frac{p[(1 + \alpha)\mu\vec{\omega}]}{p[(1 - \alpha)\mu\vec{\omega}]} = \left(\frac{1 + \alpha}{1 - \alpha}\right)^n \frac{p'[\frac{\alpha}{1-\alpha}\mu\vec{\omega}/\mu\vec{\omega}]}{p'[\frac{\alpha}{1+\alpha}\mu\vec{\omega}/\mu\vec{\omega}]}$$

Let us consider the logarithm of the three terms in the left or right hand side of this equation and divide by

$2\alpha\mu$. When α tends to zero, the terms have the following limits:

$$\frac{\log p[(1+\alpha)\mu\vec{\omega}] - \log p[(1-\alpha)\mu\vec{\omega}]}{2\alpha\mu} \longrightarrow \frac{d \log p(\mu\vec{\omega})}{d\mu}$$

$$\frac{n\log(1+\alpha) - n\log(1-\alpha)}{2\alpha\mu} \longrightarrow \frac{n}{\mu}$$

$$\frac{\log p'[\frac{\alpha}{1-\alpha}\mu\vec{\omega} \, / \, \mu\vec{\omega}] - \log p'[\frac{\alpha}{1+\alpha}\mu\vec{\omega} \, / \, \mu\vec{\omega}]}{2\alpha\mu} \longrightarrow 0$$

The limit of the last term is obtained by means of the first order development, allowed by the condition (1), of the function $\xi \longrightarrow \log p'[\xi\mu\vec{\omega} \, / \, \mu\vec{\omega}]$ around 0 for both $\xi = \frac{\alpha}{1-\alpha} = \alpha + \alpha^2 + o(\alpha^2)$ and $\xi = \frac{\alpha}{1+\alpha} = \alpha - \alpha^2 + o(\alpha^2)$. The expression $o(\alpha^2)$ denotes terms negligible compared with α^2. Then, the equation $\frac{d \log p(\mu\vec{\omega})}{d\mu} = \frac{n}{\mu}$ implies $p(\mu\vec{\omega}) = C(\vec{\omega})\mu^n$ where $C(\vec{\omega})$ is independent of μ. In view of condition (1), C is a continuous function of $\vec{\omega}$. Such probability density p may not be integrable over $\mathbb{R}^n$ which is in contradiction with the requirement that $\int p(\vec{\mu})d\vec{\mu} = 1$.

REFERENCES

CARRASSI, A. and VANNITSEM, S. (2010), *Accounting for model error in variational data assimilation: A deterministic formulation*, Monthly Weather Review, *138*, 3369-3386.

DALEY R. (1992), *The effect of serially correlated observation and model error on atmospheric data assimilation*, Monthly Weather Review, *120*, 164-177.

FISHER R. (1953), *Dispersion on a Sphere*, Proc. Mathematical and Physical Sciences, *217*, 295-305

HANSEN J.A. (2002), *Accounting for model error in ensemble-based state estimation and forecasting*, Monthly Weather Review, *130*, 2373-2391.

ISSARTEL, J.P (2005), *Emergence of a tracer source from air concentration measurements: a new strategy for linear assimilation*, Atmospheric Chemistry and Physics, *5*, 249-273

ISSARTEL J.P, SHARAN M, MODANI M. (2007), *An inversion technique to retrieve the source of a tracer with an application to synthetic satellite measurements*, Proceedings of the Royal Society A, *463*, 2863-2886

KRYSTA, M., BOCQUET, M., SPORTISSE, B. and ISNARD, O. (2006), *Data assimilation for short-range dispersion of radionuclides: An application to wind tunnel data*, Atmospheric Environment *40*, 7267-7279.

LI, H., KALNAY, E., MIYOSHI, T. and DANFORTH C.M. (2009), *Accounting for model errors in ensemble data assimilation*, Monthly Weather Review, *137*, 3407-3419.

MARDIA, K. V., Characterizations of directional distributions, (in G. P. Patil et al. editors, Statistical Distributions in Scientific Work, D. Reidel Publishing Company, Dordrecht Holland 1975), 365-385.

MARDIA, K. V. and JUPP, P. E., Directional statistics, (Wiley Series in Probability and Statistics, Wiley, 2000).

MATTHES, J., GRÖLL, L. and KELLER, H.B. (2005), *Source localization by spatially distributed electronic noses for advection and diffusion*, IEEE Trans. Signal Process *53*, 1711-1719.

MILLIKEN, A. and ALBOHALI, M. (1984), *On Necessary And Sufficient Conditions For Ordinary Least Squares Estimators To Be Best Linear Unbiased Estimator*, The Arnericati Statistician *38*, 298-299.

MULHOLLAND, M. and SEINFELD, J. H. (1994), *Inverse air pollution modelling of urban-scale carbon monoxide emissions*. Atmospheric Environment *29*, 497-516.

NORLEN, U. (1975), *The Covariance Matrices for Which Least Squares Is Best Linear Unbiased*, Scandinavian Journal of Statistics, *2*, 85-90.

PIRES, C., TALAGRAND, O., BOCQUET, M. (2010), *Diagonis and impacts of non-Gaussianity of innovations in data assiliimation*, Physica D, Nonlinear Phenomena, *239*, 1701-1717.

PUDYKIEWICZ, J. A (1998), *Application of adjoint tracer transport equations for evaluating source parameters*, Atmospheric Environment, *32*, 3039-3050.

RAO, K. S. (2007), *Source estimation methods for atmospheric dispersion*, Atmospheric Environment, *41*, 6964–6973.

SHARAN, M., SINGH, M.P., YADAV, A.K., AGGARWAL, P., NIGAM, S. (1996), *A mathematical model for the dispersion of pollutants in low wind conditions*, Atmospheric Environment, 30, 1209–1220.

SHARAN, M., ISSARTEL, J.-P., SINGH, S.K., KUMAR, P. (2009), *An inversion technique for the retrieval of single point emissions from atmospheric concentration measurements*, Proceedings of the Royal Society A, *465*, 2069–2088

ZUPANSKI D. AND ZUPANSKI M. (2006), *Model error estimation employing an ensemble data assimilation approach*, Monthly Weather Review, *134*, 1337-1354.

(Received December 1, 2010, accepted April 28, 2011, Published online August 30, 2011)

Pure Appl. Geophys. 169 (2012), 483–497
© 2011 Springer Basel AG
DOI 10.1007/s00024-011-0382-3

Least Square Data Assimilation for Identification of the Point Source Emissions

MAITHILI SHARAN,[1] SARVESH KUMAR SINGH,[1] and J. P. ISSARTEL[2]

Abstract—The identification of single and multiple-point emission sources from limited number of atmospheric concentration measurements is addressed using least square data assimilation technique. During the process, a new two-step algorithm is proposed for optimization, free from initialization and filtering singular regions in a natural way. Source intensities are expressed in terms of their locations reducing the degree of freedom of unknowns to be estimated. In addition, a strategy is suggested for reducing the computational time associated with the multiple-point source identification. The methodology is evaluated with the synthetic, pseudo-real and noisy set of measurements for two and three simultaneous point emissions. With the synthetic data, algorithm estimates the source parameters exactly same as the prescribed in all the cases. With the pseudo-real data, two and three point release locations are retrieved with an average error of 17 m and intensities are estimated on an average within a factor of 2. Finally, the advantages and limitations of the proposed methodology are discussed.

Key words: Inverse modeling, multiple-point source identification, least square, data assimilation.

1. Introduction

The rapid urbanization, industrialization and Chemical Biological and Radiological (CBR) releases raise an issue of major concern around the world regarding the contamination of environment, public health and national security. Rapid detection and an early response can reduce the extent of subsequent contamination and associated mortality of the incident. This assessment requires information about number of sources, their locations, emission rates, time, and duration of releases. However, in most of the realistic events, the difficulty increases when distributed contaminant sensor networks detect concentrations over threshold value but have no particular idea about the releases including their origin (LIU and ZHAI, 2007). This necessitates the development of a methodology that can help in identifying potential contaminant sources from limited concentration measurements.

Several researchers have focused on the problem of recovering the parameters of the unknown sources on the basis of available concentration measurements. Notable efforts in this direction are from PUDYKIEWICZ (1998), ROBERTSON and LANGNER (1998), PENENKO *et al.* (2002), BOCQUET (2005a, b), YEE (2005, 2006), YEE *et al.* (2006), KEATS *et al.* (2007a, b), ELBERN *et al.* (2007), ISSARTEL *et al.* (2007), SHARAN *et al.* (2009), etc. In all these studies, the problem of identification of source parameters was primarily restricted to single-point source. The identification of the multiple-point simultaneous releases is a challenging task and assumes significant importance in real world applications. In an extensive study, HAUPT (2005) emphasized that identification of the multiple-point sources becomes difficult when (1) two sources are seen at the same angle from the receptor but at different distances, (2) sources are located far away from the receptors and (3) meteorological conditions are not variable then distinguishing the contributions from different sources become difficult. However, the inverse problem of separating several influences merged into a set of concentration measurements has not been explored completely (ISSARTEL *et al.* 2011).

In recent years, few studies addressed the inverse problem of multiple-point source identification from finite number of noisy concentration measurements. MATTHES *et al.* (2005) attempted the identification of source locations using the least square by spatially

---

[1] Centre for Atmospheric Sciences, Indian Institute of Technology Delhi, Hauz Khas, New Delhi 110016, India. E-mail: mathilis@cas.iitd.ac.in
[2] Centre d'Etudes du Bouchet, 5, rue Lavoisier, BP 3, 91710 Vert le Petit Cedex, France.

distributed electronic noises in an indoor release experiment for industrial storage of toxic chemicals conducted at the Sigma-Aldrich Company (Germany). Later, the identification of the multiple sources in the atmosphere was briefly addressed by YEE (2007), but the approach was based on the assumption that the number of sources is known a priori. Later, YEE (2008) extended the theory for reconstruction of an unknown number of contaminant sources using probabilistic inference in conjunction with Metropolis-coupled reversible-jump Markov chain Monte Carlo (MCMC) method. However, the implementation of the theory was computationally expensive and limited to noisy synthetic data only. Recently, LUSHI and STOCKIE (2010) described an inverse Gaussian plume approach for estimating atmospheric pollutant emissions from four-point sources in a large lead–zinc smelting operation in Trail, British Columbia. The study was performed only for the estimation of emission rates but did not include the identification of locations of point sources. Recently, ISSARTEL et al. (2011) proposed a renormalization algorithm for identification of multiple-point sources using the concepts from quantum mechanics and differential geometry.

Least square technique is often used in parametric estimation as well as in the data assimilation problems in various geophysical applications (LEWIS et al. 2006). KRYSTA et al. (2006) have used a least square approach as a tool in the inverse modelling to estimate the source parameters from the concentration measurements. This method is not fully explored in the source identification as (1) the method becomes computationally expensive in the source-oriented modeling involving forward computation of the advection diffusion equation (RAO, 2007), (2) singularity arises at the point of receptors in receptor-oriented modeling (RAO, 2007) and (3) classical optimization methods may not be adequate as they require a priori knowledge of the initial guess of unknown parameters, which is not feasible in reality. In view of these, an attempt has been made here to overcome some of these inherent problems in the source identification.

In the present study, an approach based on least square technique is described to retrieve the multiple-point simultaneous releases from the atmospheric concentration measurements. As a part of the study, an algorithm is proposed for minimizing the sum of square of residuals. The proposed algorithm is advantageous as (1) it does not require any prior knowledge regarding the location of the sources and (2) takes care of singularities in a natural way. In addition, an approach is coupled with the inversion algorithm to reduce the computational time taken by estimation procedure. The study is evaluated with the pseudo-real data generated from the diffusion experiment conducted at Indian Institute of Technology (IIT) Delhi, India.

2. Methodology

In this study, identification of several simultaneous point sources is explored using a finite set of concentration measurements μ_1, μ_2,..., μ_n within the framework of least square theory (LEWIS et al. 2006). It is primarily assumed that the emission is continuously distributed through out the domain and no particular region is taken as a prior release location. The number of simultaneous releases is assumed to be known a priori. The emissions are taken linear with respect to the measurements. Continuous emission is considered from the ground level sources. The methodology begins with the identification of single-point emission for clarity and then it is extended to two and multiple-point simultaneous emissions.

2.1. Least Square Formulation

The least square theory is based on minimizing the error cost function of sum of squares of residuals between the receptor's measured and model predicted concentrations. The initial requirement of the formulation is a source-receptor relationship which describes mapping between source and receptors. Here, we followed a receptor oriented approach to avoid the unnecessary sampling of the source parameters and forward computation of the advection diffusion equation. The sensitivity of the potential source with respect to each sampled measurement is described by introducing the adjoint functions (MARCHUK, 1995) as:

$$\mu_i = q\, a_i \quad \text{for } i = 1, 2, \ldots, n, \qquad (1)$$

where n is the number of receptors, a_i is the adjoint function corresponding to the ith receptor and q is an unknown source strength. Adjoint functions are estimated from adjoint of the dispersion model assuming the unit release at the receptor's location (SHARAN et al, 2009). The adjoint function essentially describes backward transport of the pollutant's concentration from the receptors. Mathematically, if σ_i is the source and L is a linear operator, a direct source-receptor relationship is $L(\sigma_i) = \mu_i$. Using the properties of an inner product ($\langle ., . \rangle$), a fundamental relationship can be described as (ISSARTEL and BAVEREL, 2002), $\langle L(\sigma_i), \mu_i \rangle = \langle \sigma_i, L^*(\mu_i) \rangle = \langle \sigma_i, a_i \rangle$ in which L^* is the adjoint of the linear operator.

Generally, the concentrations measured by receptors will not agree with those predicted by the model owing to noise imposed on the concentration data and turbulence parameters, which by its varying nature is expected to have a complicated structure. For this purpose, the model predicted adjoint functions and the receptor measured concentrations are related as:

$$\mu_i = q\, a_i + \varepsilon_i \quad \text{for } i = 1, 2, \ldots, n, \qquad (2)$$

where ε_i is an additive noise associated with i^{th} concentration measurement.

The least square method for estimation of source parameters (location and intensity) of a single-point source from n observed concentration measurements μ_i's is based on minimizing the sum of square of residuals represented by the function J as (LEWIS et al. 2006):

$$J(\mathbf{x}, q) = \sum_{i=1}^{n} \varepsilon_i^2 = \sum_{i=1}^{n} [\mu_i - q a_i(\mathbf{x})]^2, \qquad (3)$$

subject to the constraints $q > 0$ and $\mathbf{x}_\ell \leqslant \mathbf{x} \leqslant \mathbf{x}_u$ where $\mathbf{x} = (x, y)$ is a position vector. The vectors $\mathbf{x}_\ell$ and $\mathbf{x}_u$ are, respectively, lower and upper limits of the computational domain containing the monitoring network. Measurement errors are assumed uncorrelated with equal variance (LEWIS et al. 2006).

The conditions for minimization of J lead to a system of non-linear algebraic equations in terms of parameters to be estimated. The complexity grows with increasing number of degrees of freedom or parameters. A number of optimization algorithms such as Newton–Raphson (BEYER, 1964), Steepest–Descent (DEBYE, 1909), Levenberg–Marquardt (LEVENBERG, 1944; MARQUARDT, 1963), conjugate gradient (HESTENES and STIEFEL, 1952) exist in the literature for the estimation of source parameters. These algorithms differ from the use of iterative techniques and on the appropriate choice of the initial values of the unknown parameters with respect to iterations. Since the locations of the sources are not known a priori and it is not feasible to prescribe their initial values, existing optimization methods fail for such estimation problems. In view of this, an alternative algorithm is proposed here.

The algorithm proposed here, is essentially a two step minimization process, in which as a first step, for a fixed location, the function J is minimized with respect to q to obtain its estimate and then fixed location along with estimated $\widehat{q}$ is used to compute the value of function $\widehat{J}$. This process is repeated for all the grid points of the domain and the corresponding values of $\widehat{J}$ are stored. In the second step, a sequential search algorithm is applied to look for the global minimum among the stored values of $\widehat{J}$. The parameters corresponding to the global minimum of $\widehat{J}$ will be the estimation of the source intensity and its location.

Incidentally, this algorithm can be represented mathematically for the single-point source as follows: The function J (Eq. 3) is rewritten in the matrix notation as:

$$J(\mathbf{x}, q) = \frac{1}{2}[\boldsymbol{\mu} - q\mathbf{a}(\mathbf{x})]^{\mathrm{T}}[\boldsymbol{\mu} - q\mathbf{a}(\mathbf{x})], \qquad (4)$$

where $\boldsymbol{\mu} = (\mu_1, \mu_2, \ldots, \mu_n)^{\mathrm{T}}$ and $\mathbf{a}(\mathbf{x}) = (a_1(\mathbf{x}), a_2(\mathbf{x}), \ldots, a_n(\mathbf{x}))^{\mathrm{T}}$ are the vector of measurements and adjoint functions, respectively. The superscript 'T' denotes the transpose.

For a fixed $\mathbf{x}$, first order derivative of J with respect to q is given by:

$$\frac{\partial J}{\partial q} = q\mathbf{a}^{\mathrm{T}}(\mathbf{x})\mathbf{a}(\mathbf{x}) - \boldsymbol{\mu}^{\mathrm{T}}\mathbf{a}(\mathbf{x}). \qquad (5)$$

The condition ($\partial J / \partial q = 0$) leads to an estimate $\widehat{q}$ as:

$$\widehat{q} = \frac{\boldsymbol{\mu}^{\mathrm{T}}\mathbf{a}(\mathbf{x})}{\mathbf{a}^{\mathrm{T}}(\mathbf{x})\mathbf{a}(\mathbf{x})}. \qquad (6)$$

177

Notice that the second order derivative of J, $\partial^2 J/\partial q^2 = \mathbf{a}^T(\mathbf{x})\mathbf{a}(\mathbf{x})$ (square of adjoint function at fixed $\mathbf{x}$) is positive which implies that estimated $\widehat{q}$ (Eq. 6) minimizes the function J for a fixed $\mathbf{x}$. The minimum value of J corresponding to estimate $\widehat{q}$ (Eq. 6) for a fixed $\mathbf{x}$ is obtained from simplifying Eq. (4) as:

$$\widehat{J}\left(\mathbf{x}, \widehat{q}\right) = \frac{1}{2}\left[\boldsymbol{\mu}^T\boldsymbol{\mu} - \frac{(\boldsymbol{\mu}^T\mathbf{a}(\mathbf{x}))^2}{\mathbf{a}^T(\mathbf{x})\mathbf{a}(\mathbf{x})}\right], \quad (7)$$

Notice that, in Eq. (7) the first term on RHS (Right Hand Side) is a square of measurement vector and is independent of $\mathbf{x}$ and the second term on RHS is also a square term with negative sign. Hence, the minimization of $\widehat{J}$ with respect to $\mathbf{x}$ in Eq. (7) is equivalent to the maximization of the function

$$S(\mathbf{x}) = \frac{(\boldsymbol{\mu}^T\mathbf{a}(\mathbf{x}))^2}{\mathbf{a}^T(\mathbf{x})\mathbf{a}(\mathbf{x})}. \quad (8)$$

The point at which $S(\mathbf{x})$ becomes maximum in the domain will be the estimate of location of the source. Once the location of the source is identified, its intensity is computed from Eq. (6) at the estimated location $\mathbf{x}_e = (x_e, y_e)$.

For single-point emission, the estimate $\widehat{q}$ (Eq. 6) turns out to be an explicit function of $\mathbf{x}$. In addition, $\widehat{J}$, the minimum value of J estimated at $\widehat{q}$ for fixed $\mathbf{x}$, becomes an explicit function of $\mathbf{x}$ which is relatively easier to optimize with respect to $\mathbf{x}$. In view of these facts, the proposed algorithm becomes simpler for single-point emission. Now, we describe the source estimation for two simultaneous releases.

2.2. Two-Point Sources

For two simultaneous point releases, the error function J is written as:

$$J(\mathbf{x}_1, \mathbf{x}_2, \mathbf{q}) = \frac{1}{2}[\boldsymbol{\mu} - q_1\mathbf{a}(\mathbf{x}_1) - q_2\mathbf{a}(\mathbf{x}_2)]^T$$
$$\times [\boldsymbol{\mu} - q_1\mathbf{a}(\mathbf{x}_1) - q_2\mathbf{a}(\mathbf{x}_2)], \quad (9)$$

in which the components of $\mathbf{q} = (q_1, q_2)^T$ are the intensities of two simultaneous point releases corresponding to the locations $\mathbf{x}_1$ and $\mathbf{x}_2$ respectively. Henceforth, the subscripts '1' and '2' will correspond to first and second source.

For fixed $\mathbf{x}_1$ and $\mathbf{x}_2$, conditions ($\partial J/\partial q_1 = 0$ and $\partial J/\partial q_2 = 0$) for obtaining the critical points lead to a system of equations, written in matrix form, as:

$$\mathbf{Aq} = \mathbf{B}, \quad (10)$$

where:

$$\mathbf{A} = \begin{bmatrix} \mathbf{a}_1^T\mathbf{a}_1 & \mathbf{a}_1^T\mathbf{a}_2 \\ \mathbf{a}_2^T\mathbf{a}_1 & \mathbf{a}_2^T\mathbf{a}_2 \end{bmatrix}; \mathbf{q} = \begin{bmatrix} q_1 \\ q_2 \end{bmatrix} \text{ and } \mathbf{B} = \begin{bmatrix} \mathbf{a}_1^T\boldsymbol{\mu} \\ \mathbf{a}_2^T\boldsymbol{\mu} \end{bmatrix} \quad (11)$$

Here $\mathbf{a}_1 = \mathbf{a}(\mathbf{x}_1)$ and $\mathbf{a}_2 = \mathbf{a}(\mathbf{x}_2)$ are the vectors of adjoint functions evaluated at the points $\mathbf{x}_1$ and $\mathbf{x}_2$, respectively. Notice that $\mathbf{A}$ is the Gram matrix of $\mathbf{a}_1$ and $\mathbf{a}_2$ which implies that it is a real symmetric and positive. It is definite, i.e. det $\mathbf{A} \neq 0$, if and only if these vectors are linearly independent. Then, the system of equations can be solved for the estimated value of $\mathbf{q}$ denoted as $\widehat{\mathbf{q}}$

$$\widehat{q}_1 = \frac{(\mathbf{a}_1^T\boldsymbol{\mu})(\mathbf{a}_2^T\mathbf{a}_2) - (\mathbf{a}_2^T\boldsymbol{\mu})(\mathbf{a}_1^T\mathbf{a}_2)}{\det \mathbf{A}} \text{ and}$$
$$\widehat{q}_2 = \frac{(\mathbf{a}_2^T\boldsymbol{\mu})(\mathbf{a}_1^T\mathbf{a}_1) - (\mathbf{a}_1^T\boldsymbol{\mu})(\mathbf{a}_1^T\mathbf{a}_2)}{\det \mathbf{A}}. \quad (12)$$

The "Hessian matrix" of $J(\mathbf{x}_1, \mathbf{x}_2, \mathbf{q})$ with respect to q_1 and q_2 (square matrix of second order partial derivatives) is simply $\mathbf{A}$. Since this matrix is positive definite implying all its eigenvalues are real and positive, it is non-singular and J attains a local minimum at the critical point $\widehat{\mathbf{q}} = \left(\widehat{q}_1 \ \widehat{q}_2\right)^T$ (Eq. 12) (AYRES, 1962; GOLUB and VANLOAN, 1996). Thus, the pairs of $\mathbf{x}_1$ and $\mathbf{x}_2$ in the domain ensuring the linear independence of $\mathbf{a}_1$ and $\mathbf{a}_2$ are considered.

The minimum value of J is obtained from Eq. (9) for $\widehat{\mathbf{q}}$ and $\mathbf{x}_1$ and $\mathbf{x}_2$ and the resulting value is denoted as $\widehat{J}$. In fact, this $\widehat{J}$ is a function of $\mathbf{x}_1$ and $\mathbf{x}_2$. Since the minimization of $\widehat{J}$ with respect to $\mathbf{x}_1$ and $\mathbf{x}_2$ is not obvious, it is minimized numerically using a sequential algorithm.

Over all algorithm for the source identification is summarized as: (1) choose the pair of $\mathbf{x}_1$ and $\mathbf{x}_2(\mathbf{x}_1 \neq \mathbf{x}_2)$ in the domain such that Hessian matrix is positive definite, (2) estimate $\widehat{\mathbf{q}}$ and $\widehat{J}$ for the pair of $\mathbf{x}_1$ and $\mathbf{x}_2$, (3) store the set $\{\widehat{J}, \widehat{\mathbf{q}}, \mathbf{x}_1, \mathbf{x}_2\}$ of values, (4) repeat the steps (1–3) for all possible pairs of $\mathbf{x}_1$ and $\mathbf{x}_2$ in the domain, (5) employ a sequential algorithm to find the minimum $\widehat{J}$ among its stored values, (6) values of $\widehat{\mathbf{q}}$, $\mathbf{x}_1$ and $\mathbf{x}_2$ corresponding to the minimum $\widehat{J}$ in (5) will be the desired estimation of source parameters.

Now, we generalize this approach for the retrieval of m-simultaneous releases.

2.3. Multiple-Point Sources

For m unknown simultaneous point sources, the function J is written as:

$$J(\mathbf{x}_1, \mathbf{x}_2, \ldots, \mathbf{x}_m, \mathbf{q}) = \frac{1}{2} \left[\boldsymbol{\mu} - \sum_{i=1}^{m} q_i \mathbf{a}(\mathbf{x}_i) \right]^{\mathrm{T}}$$
$$\times \left[\boldsymbol{\mu} - \sum_{i=1}^{m} q_i \mathbf{a}(\mathbf{x}_i) \right] : \mathbf{x}_i \neq \mathbf{x}_j,$$
$$\tag{13}$$

where q_i is the intensity of the ith source and the condition $\mathbf{x}_i \neq \mathbf{x}_j$ indicates that the locations of the unknown sources are distinct. For a fixed set of $\mathbf{x}_1$, $\mathbf{x}_2, \ldots, \mathbf{x}_m$, the condition $(\partial J / \partial q_i = 0$ for $i = 1, 2, \ldots, m)$ for estimation of critical points will lead to a system of m equations, written in matrix notation as:

$$\mathbf{A}\mathbf{q} = \mathbf{b}, \tag{14}$$

in which:

$$\mathbf{A} = \begin{bmatrix} \mathbf{a}_1^{\mathrm{T}}\mathbf{a}_1 & \mathbf{a}_1^{\mathrm{T}}\mathbf{a}_2 & \cdot & \cdot & \mathbf{a}_1^{\mathrm{T}}\mathbf{a}_m \\ \mathbf{a}_2^{\mathrm{T}}\mathbf{a}_1 & \mathbf{a}_2^{\mathrm{T}}\mathbf{a}_2 & \cdot & \cdot & \mathbf{a}_2^{\mathrm{T}}\mathbf{a}_m \\ \cdot & \cdot & \cdot & \cdot & \cdot \\ \cdot & \cdot & \cdot & \cdot & \cdot \\ \cdot & \cdot & \cdot & \cdot & \cdot \\ \mathbf{a}_m^{\mathrm{T}}\mathbf{a}_1 & \mathbf{a}_m^{\mathrm{T}}\mathbf{a}_2 & \cdot & \cdot & \mathbf{a}_m^{\mathrm{T}}\mathbf{a}_m \end{bmatrix}_{m \times m} ;$$

$$\mathbf{q} = \begin{bmatrix} q_1 \\ q_2 \\ \cdot \\ \cdot \\ \cdot \\ q_m \end{bmatrix}_{m \times 1} \text{and } \mathbf{b} = \begin{bmatrix} \mathbf{a}_1^{\mathrm{T}}\boldsymbol{\mu} \\ \mathbf{a}_2^{\mathrm{T}}\boldsymbol{\mu} \\ \cdot \\ \cdot \\ \cdot \\ \mathbf{a}_m^{\mathrm{T}}\boldsymbol{\mu} \end{bmatrix}_{m \times 1}.$$

Now, the outlines given for the retrieval of two-point emissions in Sect. 2.2 are followed for the identification of m-point sources.

The Hessian matrix generated from second order partial derivatives of J with respect to $q_1, q_2, \ldots, q_m$ for a given set $\mathbf{x}_1, \mathbf{x}_2, \ldots, \mathbf{x}_m$ is found to be the same as

the coefficient matrix $\mathbf{A}$. This matrix is examined for its positive definiteness by showing that all its computed eigenvalues are real and positive. In case any of the eigenvalue failed to be real and positive, another combination of $\mathbf{x}_1, \mathbf{x}_2, \ldots, \mathbf{x}_m$ is chosen. For the chosen set of $\mathbf{x}_1, \mathbf{x}_2, \ldots, \mathbf{x}_m$ the coefficient matrix $\mathbf{A}$ is non-singular and then the system of m equations in m unknowns (Eq. 14) is solved numerically using Gauss elimination method to compute $\widehat{\mathbf{q}}$

Now the set of $\mathbf{x}_1, \mathbf{x}_2, \ldots, \mathbf{x}_m$ along with the computed $\widehat{\mathbf{q}}$ is used to estimate the minimum value of J denoted as $\widehat{J}$ (Fig. 1). These set of values $\{\widehat{J}, \mathbf{x}_1, \mathbf{x}_2, \ldots, \mathbf{x}_m, \widehat{\mathbf{q}}\}$ are stored. This process is repeated for all the possible combination of $\mathbf{x}_1$, $\mathbf{x}_2, \ldots, \mathbf{x}_m$ in the domain. A sequential algorithm is utilized for searching the minimum of $\widehat{J}$ among all its stored values. The values of $\mathbf{x}_1, \mathbf{x}_2, \ldots, \mathbf{x}_m$ and $\widehat{\mathbf{q}}$ corresponding to the min $(\widehat{J})$ are identified as the source parameters. These outlines are given in a flow diagram (Fig. 1).

3. Diffusion Data

For evaluation of the proposed technique for the source retrieval, a diffusion data is required. The measurements utilized for identification of single as well as multiple-point emissions are described here briefly.

3.1. Measurements for Single-Point Emission

For the identification of single-point emission, data from IIT diffusion experiment conducted at Delhi ($28°52'$ N, $77°18'$ E) for surface release of tracer SF_6 in low-wind conditions is considered. Details of the experiment including atmospheric stability are given in SINGH et al. (1991) and SHARAN et al. (1996). The tracer was released at a height of 1 m above the ground and the receptors were also placed approximately at the same height. For the computations, source and receptors are assumed at ground level.

In all, 14 test runs were conducted. Only seven (runs 1, 6, 7, 8, 11, 12 and 13) corresponding to the unstable steady conditions are chosen for the analysis (SHARAN et al., 2009). Run 2 was ignored due to a relatively large variability in wind direction and the

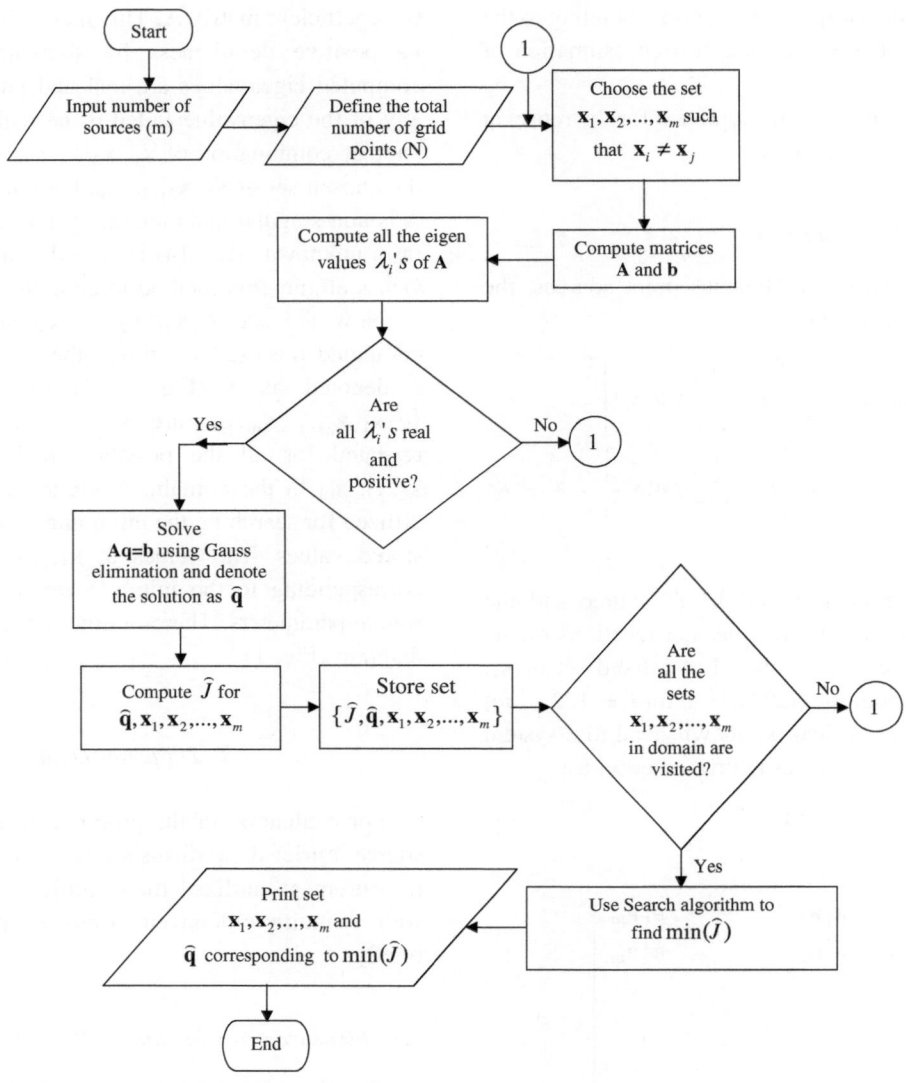

Figure 1
Flow-diagram for the steps of proposed retrieval algorithm

remaining runs because of neutral and stable conditions. In runs 1, 6, 7 and 11, the release point was located at the centre of the circular arcs, whereas in the remaining runs (8, 12 and 13), the release point was shifted 100 m towards northwest. In runs 1, 8, 12 and 13, the release rate was 5,000 µg/s while in the remaining runs (6, 7 and 11), it was 3,000 µg/s. The monitoring network involved 20 receptors on 50, 100 and 150 m, and in some cases on 200 m circular arcs with 45° angular spacing between them (Fig. 2). When the source is shifted from the centre onto the arc at 100 m, the corresponding receptor is moved to the centre. In some of the runs, a few measurements are discarded (SHARAN *et al.*, 1996), thus reducing the effective number of receptors in the monitoring network.

Wind and temperature measurements were obtained at four levels (2, 4, 15 and 30 m) from a 30 m micrometeorological tower. The values of wind speed, wind direction, atmospheric stability and mixing height required in the dispersion model are taken from SHARAN *et al.* (1996).

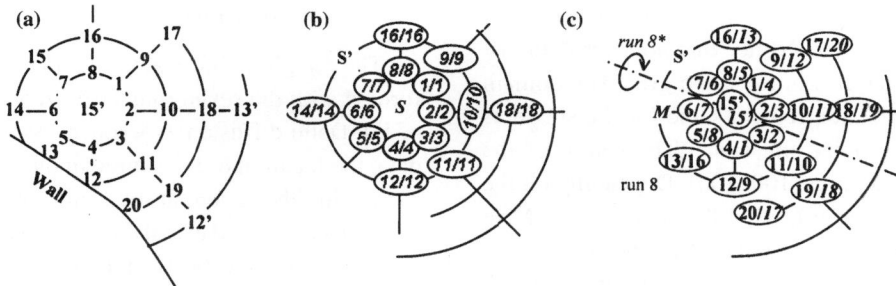

Figure 2
Layout of the site for IIT Delhi tracer diffusion experiment (**a**). The samplers 1–20, 12′, 13′ and 15′ are essentially arranged on circles of radius 50, 100 and 150 m, with regular angular spacing of 45°. In case of shifted source runs, source at the centre S and receptor 15 are interchanged with each other corresponding to positions S′ and 15′ (adopted from SHARAN *et al.* 2009). **b** Representation of pseudo-runs 11–13 and 11–8 obtained by combining the run 11 with central source at S and runs 13 and 8 with shifted source at S′. Within each ellipse the right and left *italic numbers* indicate a sampler respectively in run 11 and 13. **c** Schematic similar to **b** for pseudo-run 8–8* in which run 8 is combined with itself after a mirror symmetry sending the source to M (adopted from ISSARTEL *et al.* 2011)

3.2. Pseudo Measurements for Multiple-point Emissions

Diffusion data is very scarce. Mostly available diffusion data in the literature pertains to the single-point emission. For the evaluation purpose, a pseudo data set is generated from the IIT diffusion experiment by combining the single release with different source locations under similar meteorological conditions.

Notice that, in IIT data, the monitoring network remains same for all the runs only the release location and wind direction vary. Thus, it allows us, with minor modifications, to combine the runs with similar meteorological conditions and different release locations, just as if the two releases had occurred simultaneously. In view of this, four pseudo runs have been prepared by combining the runs with same wind conditions and different location of the source.

Winds are almost same in runs 11 (central source), 8 and 13 (shifted source). Runs 11 (central source), 8 and 13 (shifted source) have almost same wind speed (1.1, 0.9, 1.1 m/s respectively) and direction (125°, 121°, 141°). The pseudo-runs 11–8 and 11–13 are thus prepared by adding the measurements at the receptors common to the combined runs.

A third pseudo-run is prepared based on the fact that wind direction in run eight (shifted source) is close to a symmetry axis of the monitoring network in the direction of 112.5°. Under this mirror symmetry, the source 100 m northwest of the centre of the monitoring network becomes 100 m west of it. Most receptors coincide with the mirror image of another receptor (Fig. 2). The pseudo-run 8–8* is obtained by adding the concentrations observed at the common receptors; the mirror symmetric run is indicated by a star (*).

Finally, a next pseudo-run 11–8–8* is prepared by combining the runs 11, 8 and mirror symmetric run 8* to simulate a triple-point release. For the pseudo run 11–8–8*, wind direction is taken in the direction of 112.5° to follow the mirror symmetry of runs 8 and 8*, and wind speed is averaged between runs 8 and 11.

4. Numerical Computations

The numerical computations are performed for retrieval of both single as well as multiple-point emissions. In both the cases, the computations are performed for (1) the complete set of concentrations observed or pseudo generated and (2) the analogous set of synthetic measurements generated using the dispersion model given in SHARAN *et al.* (1996). A primary step of the inversion algorithm is the generation of adjoint functions. The adjoint functions are generated from the dispersion model (SHARAN *et al.*, 1996) in the backward mode assuming unit release at the receptor's locations and rotating the wind direction by 180° (SHARAN *et al.*, 2009). The implementation of the inversion algorithm requires a discretization of the domain. For this purpose, a square domain of size 995 m × 995 m is chosen and discretized into 199 × 199 grid points. Each mesh is a square of 5 m × 5 m. Notice that no more than one receptor or source lies in the same grid. The domain and grid discretization remain same for the single as well as multiple-point emissions. The centre of the arcs is at the grid point (100, 100). Depending on the run, source is either at (100, 100) or (86, 114). Note that adjoint function is singular at the position of the receptor. In order to resolve this singularity, the mesh containing the receptor and the neighboring meshes are further subdivided into 99 × 99 grid points and an average value of the adjoint function is computed for the receptor's mesh. The steps for the inversion algorithm for single as well as multiple-point emissions are discussed in Sect. 2. The computations are performed on an Intel® Core™ 2 Duo E8135 @ 2.66 GHz desktop machine.

5. Results and Discussion

Source estimation has been carried out for single as well as two- and three-point emissions using two types of data: (1) synthetic and (2) real. Synthetic data for each run is the concentration at the receptors generated from the dispersion model, described as noise free and minimizes the errors associated with the model. Therefore, it is used to verify the mathematical consistency of the inversion algorithm. However, real data corresponds to concentration measurements sampled during the experiments. Accordingly, the results are presented in the following subsections.

5.1. Single-Point Emission

The single-point source reconstruction is performed with the seven runs (1, 6, 7, 8, 11, 12 and 13) of IIT-Delhi diffusion experiment. With the synthetic data, the location of the source in the runs (1, 6, 7 and 11) having the source at the centre of the monitoring network is retrieved exactly at the grid point (100, 100) whereas in the runs (8, 12 and 13) with shifted release, the retrieved location (86, 114) coincides with that prescribed (Table 1). The intensity in the runs (1, 8, 12 and 13) is retrieved approximately 2,999.8 μg/s in lieu of 3,000 μg/s. Similarly an approximate 4,999.7 μg/s intensity is retrieved in place of prescribed 5,000 μg/s in runs (6, 7 and 11). The exact retrieval of the source parameters with the synthetic data provides a mathematical consistency of the technique used here.

With real data, source is reconstructed in all the runs with an average error of 20.5 m from the original release locations (Table 1). The location is estimated in run six at the grid point (100, 100) with a negligible error and at point (92, 102) in run seven with a maximum error of 41.2 m. The intensity of the source is estimated within a factor two in all the runs (Table 1). ·

5.2. Errors in the Retrieval

With the real data, the errors in the retrieval are described in terms of the departure of the real concentration measurements from their ideal synthetic values. This departure is traditionally called

Table 1

Reconstruction results for single-point release

Run	1	6	7	8	11	12	13
Experimental release							
Location	(100, 100)	(100, 100)	(100, 100)	(86, 114)	(100, 100)	(86, 114)	(86, 114)
Intensity	5,000	3,000	3,000	5,000	3,000	5,000	5,000
Least square estimate							
Location	(99, 103)	(100, 100)	(92, 102)	(91, 111)	(97, 103)	(82, 117)	(88, 115)
Intensity	9,529	5,967	4,928	4,155	5,279	9,544	3,664
Error							
E_L	15.8	0	41.2	29.2	21.2	25	11.2
θ_e^0	17.3	25.2	20	4.7	13.7	10.1	6.1

The experimental point releases (first and second rows) with least square estimates (third and fourth rows) are indicated in terms of location in grid and intensity in μg/s. On the fifth and sixth rows, inversion error is presented in terms of departure of retrieved location from experimental release location grid coordinate (represented as E_L in meters) and the angular deviation (θ_e^0 in degrees) measurement errors (COHN, 1997) though it is due to not only instrumental inaccuracy but also to lack of representativity of the dispersion model. For the ideal measurements, the noise free measurement vector always lies along or parallel to the adjoint vector (Eq. 1). However, for a noisy measurement, the equality relation (Eq. 1) will not be true. Thus, the measurement vector will lie along an angle to the adjoint vector. This angular deviation indicates the error incurred in the retrieval of intensity.

As explained in Sect. 2, the location of the source is estimated from the observed noisy measurements μ_r at a location $\mathbf{x}_e$ such that $S(\mathbf{x}_e) = \frac{\left(\mu_r^T \mathbf{a}(\mathbf{x}_e)\right)^2}{\mathbf{a}^T(\mathbf{x}_e)\mathbf{a}(\mathbf{x}_e)}$ is maximum. This is equivalent to maximize $\frac{S(\mathbf{x}_e)}{\|\mu_r\|^2} = \left[\left(\frac{\mu_r}{\|\mu_r\|}\right)^T \frac{\mathbf{a}(\mathbf{x}_e)}{\|\mathbf{a}(\mathbf{x}_e)\|}\right]^2$, where $\|\mathbf{p}\|^2 = \mathbf{p}^T\mathbf{p}$ for any vector $\mathbf{p}$. Since vectors $\frac{\mu_r}{\|\mu_r\|}$ and $\frac{\mathbf{a}(\mathbf{x}_e)}{\|\mathbf{a}(\mathbf{x}_e)\|}$ both have norm 1, the estimated location $\mathbf{x}_e$ of the release is the one such that the vector $\frac{\mathbf{a}(\mathbf{x}_e)}{\|\mathbf{a}(\mathbf{x}_e)\|}$ minimizes the angular distance $\theta_e = \arccos\left(\left(\frac{\mu_r}{\|\mu_r\|}\right)^T \frac{\mathbf{a}(\mathbf{x}_e)}{\|\mathbf{a}(\mathbf{x}_e)\|}\right)$ from the observations μ_r. Thus, this angular departure θ_e provides an indication of the intensity of the noise contained in the data. If there is no noise in the data, i.e. $\mu_r = \mu = q_0\mathbf{a}(\mathbf{x}_0)$, one would obtain $\mathbf{x}_e = \mathbf{x}_0$ with $\theta_e = 0$. This is verified in the computations with the synthetic data, where the estimates of intensities are almost exactly same to those prescribed.

Similarly, error estimates are computed for real data in all the runs (Table 1) and it is observed that the adjoint vector does not coincide with the

measurement vector exactly and depart by an average angular distance of $13.8°$ in all the runs. A relatively higher value of angular departure θ_e indicates a relatively large deviation in the estimated intensity from that prescribed.

5.3. Two and Three Simultaneous Point Sources

The proposed algorithm is used for the identification of multiple-point sources with the pseudo-real as well as corresponding model generated synthetic data for (1) two unknown sources (runs 11–13, 11–8 and 8–8*) and (2) three unknown sources (run 11–8–8*). With the synthetic data, the grid points [(100, 100), (86, 114)] are estimated as two point release locations in case of runs (11–13 and 11–8) and grid points [(86, 114), (80, 100)] in run 8–8* which are exactly same as prescribed. Similarly, in run 11–8–8*, the points [(100, 100), (86, 114), (80, 100)] are estimated as three source locations exactly similar to those prescribed. In all these four runs, the strength at each location is estimated same as prescribed within a maximum round-off error of 0.02%.

With the pseudo-real data, in average, the releases are retrieved 17 m away from their true locations (Table 2). In two-point sources (runs 11–13, 11–8 and 8–8*), the locations are retrieved at a minimum error of 5 m in run 11–13 and maximum 40 m in run 8–8* from their original release locations. In runs (11–13 and 11–8), the corresponding intensities are retrieved within a factor of 2. However, in run 8–8*, the error in the retrieval is relatively large and increases up to a factor 3.8. Similarly, in three-point sources (run 11–8–8*), all the three source locations

Table 2

Reconstruction results for two and three point simultaneous releases with pseudo-real data

Pseudo run	11–13 SS′		11–8 SS′		8–8* SM		11–8–8* SS′M		
Experimental release									
Location	(100, 100)	(86, 114)	(100, 100)	(86, 114)	(86, 114)	(80, 100)	(100, 100)	(86, 114)	(80, 100)
Intensity	3,000	5,000	3,000	5,000	5,000	5,000	3,000	5,000	5,000
Least square estimate									
Location	(103,104)	(88,116)	(99,102)	(86,115)	(78,114)	(80,101)	(97,101)	(82,115)	(82,100)
Intensity	3,403	4,357	3,970	7,115	7,449	18,914	4,715	9,623	5,159
Error									
E_L	25	14	11	5	40	5	16	21	10

Notations and legends are same as in Table 1

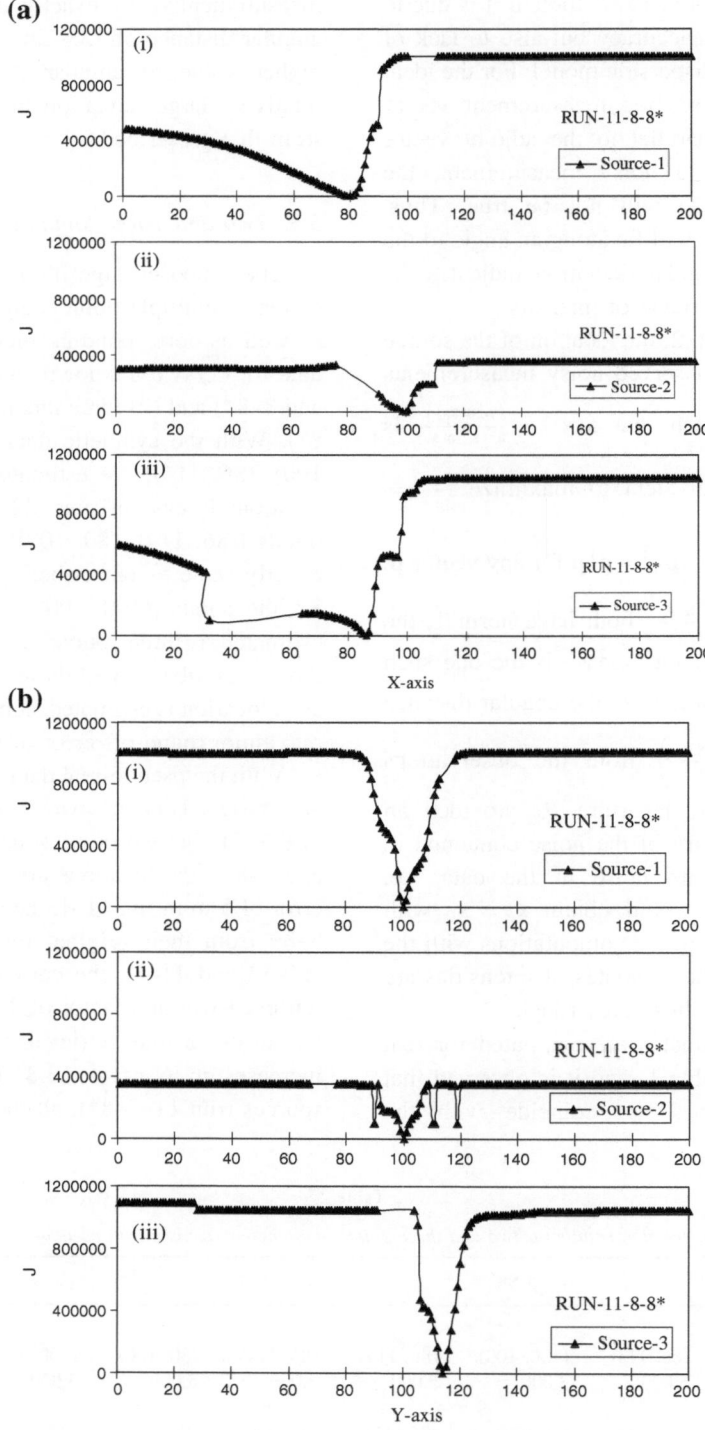

are retrieved within 21 m and intensities are retrieved within a factor of 2. A slight discrepancy may be explained in view of the facts that (1) in average, the real runs contain 19 samplers but only 16 can be combined in the pseudo-runs and (2) the pseudo-runs are not natural runs and the averaging of the winds

Figure 3

a The variation of function J along x-axis in three-point sources (run 11–8–8*) for (1) fixed y-coordinate (grid point 100) of *source-1* and other two fixed sources at grid point (100,100) (*source-2*) and (86, 114) (*source-3*) (2) fixed y-coordinate (100) of source-2 and other two fixed sources at grid point (80,100) (*source-1*) and (86, 114) (*source-3*) and (3) fixed y-coordinate (114) of *source-3* and other two fixed sources at grid point (80,100) (*source-1*) and (100, 100) (*source-2*). **b** The variation of function J along x-axis in three-point sources (run 11–8–8*) for (1) fixed x-coordinate (grid point 80) of *source-1* and other two fixed sources at grid point (100,100) (*source-2*) and (86, 114) (*source-3*), (2) fixed x-coordinate (100) of *source-2* and other two fixed sources at grid point (80,100) (*source-1*) and (86, 114) (*source-3*) and (3) fixed x-coordinate (86) of *source-3* and other two fixed sources at grid point (80,100) (*source-1*) and (100, 100) (*source-2*)

from two real runs is not fully consistent with the addition of the corresponding concentration measurements.

In case of single-point emission, an expression is obtained to quantify the error incurred in the retrieval of intensity in terms of angular departure of the measurement vector from the adjoint vector. However, such an expression is not feasible to quantify the errors in case of more than one-point sources. In addition, it is difficult with the real data to quantify the measurement errors in which representativity term is likely to be dominant.

In the present study, the identification of two-point and three-point sources includes estimation of six and nine unknown parameters respectively. Since intensities are estimated as a function of locations of the sources, the degree of freedom of source identification of the two- and three-point simultaneous releases reduces to four and six respectively. Even in this case, the visualization of all these estimated parameters for two- as well as three-point sources is not possible collectively on a paper or in two-dimension. To overcome this, a representation is drawn in Fig. 3 showing the variation of the function J in case of three-point sources (run 11–8–8*) with respect to each estimated parameters keeping the other source parameters constant. It clearly shows the occurrence of global minimum of J at the estimated parameters.

5.4. Reduction in Computational Time

The general methodology for minimization of function J for a number of simultaneous point emissions is based on visiting all the doublets or triplets or m-set of grid points in the domain for two-, three- and m-point sources, respectively. This approach is time consuming as the number of pairs $N(N-1)/2 \approx 7.8 \times 10^8$ ($N =$ total number of grid points in the domain) for two-point sources or triplets $N(N-1)(N-2)/6 \approx 10^{13}$ for three-point sources or m-set $N(N-1)\ldots(N-m+1)/(m!) \approx 10^{4m+1}$ for m-point sources involved in the optimization are very large. The computational time taken in the estimation of source parameters for simultaneous three-point releases is approximately 90 h. This is expected to increase further with the increase in the number of sources.

In order to minimize the computational time, an alternative approach is adopted here. In this approach, the source parameters are estimated in two steps: (1) as a first step, a gross estimation of the source parameters is carried out by visiting one point out of five grid points in each direction using the algorithm described in Sect. 2 and then (2) these estimated set of source locations are refined by using the same retrieval algorithm for all the grossly estimated locations in their 11-grid point neighborhood. This reduces roughly the visit of number of pairs by a factor of 5^4, number of triplets by 5^6 and m-set by 5^{2m} for two-, three- and m-sources respectively. This approach reduces the computational time to 2 min for two-point and 12 min for three-point simultaneous sources. However, both the generalized and modified approaches lead to the similar results.

5.5. Noisy Measurements

Generally, the concentration measurements are not available for the multiple-point emissions for evaluating the inversion approach for the retrieval of more than one-point emission sources. In such a situation, researchers (YEE, 2008; LUSHI and STOCKIE, 2010) have generated the concentration measurements for evaluation purposes by adding a proportion of noise in the model generated concentrations. The numerical simulations with such noisy measurements are also important for understanding the effect of possible errors in concentration measurements on emission estimates (LUSHI and STOCKIE, 2010).

The model generated concentration measurements are scaled by a normally distributed random number chosen from the interval $(1-\alpha, 1 + \alpha)$ for values of $\alpha = 0.1$, 0.2 and 0.3. These values of α correspond to 10, 20 and 30% noise. The error in the location as well as in the intensity is found to increase as the percentage of noise increases in the concentration measurements (Table 3). In all four pseudo-runs, the maximum error in the estimation of source location increases up to 16 m from the original release location and departure in the retrieval of intensities increases up to 30% in comparison to the prescribed with the increase in random noise (10–30%).

6. Advantages and Limitations

The present study is focused for the identification of multiple-point surface releases in steady state conditions using a least square approach. The algorithm is described within a general framework for the retrieval of m-point emission sources. This is applied (1) with the real data for single-point source and (2) with pseudo-real corresponding to two and three-point emission sources. The source parameters are retrieved almost exactly with the model generated synthetic data. These parameters are also retrieved with real and pseudo-real data. In the following, we discuss the advantages/limitations with the approach used here for the identification of sources.

6.1. Advantages

The proposed inversion algorithm is based on simple concepts of least square theory. This algorithm is advantageous in comparison to other methods as:

1. Minimization of the sum of the squares of residuals (Eq. 3) leads to a system of non-linear algebraic equations which needs an iterative algorithm to determine an approximate solution. The existing algorithms have limited applicability as they require the initial value of the parameters to begin the iterative process. However, such information is, in fact, not needed in the proposed algorithm as in reality there is no idea from where plume has originated.
2. The set of locations for which system of equations becomes ill-posed or ill-conditioned are discarded

in a natural way and computations are performed only for those set of locations where a local minimum of the function J exists.
3. Source intensities are expressed in terms of their locations reducing the degree of freedom of unknowns to be estimated.
4. The proposed algorithm allows visiting a relatively less number of set of grid points in order to search for the location of the sources, resulting in a significant reduction in the computational time.

6.2. Assumptions/Limitations

The proposed inversion algorithm is subject to the following assumptions: (1) emissions and the concentration measurements are related linearly, (2) the observation error covariance matrix is taken as an identity matrix, while in reality observations may have small order of covariance, (3) the domain is discretized into grids in such a way that no more than one source or receptor lies in the same grid and (4) the dispersion model used here describes perfect relationship between the emission and the corresponding measurements implying the model representativity errors, if any, are negligibly small. However, in reality, models can not be exactly representative of observations in all the atmospheric conditions. The uncertainty in meteorological variables influences the representation of the concentration measurements by a dispersion model which eventually affects the source retrieval. Recently, we are able to retrieve the source reasonably well (Sharan et al., 2009) with IIT diffusion data in convective conditions in spite of a sparse monitoring network whereas the sources are severely under-estimated with the Idaho low wind diffusion experiment in stable conditions even with a dense monitoring network because of the model representativity errors arising due to large variability in the wind direction (Sharan et al., 2011). This aspect is being investigated further.

The algorithm is limited in its applicability as (1) the number of sources is known a priori, (2) adjoint functions are not assigned any weights according to the visibility of the region from the monitoring network, however, in a recent study, an improvement

Table 3

Reconstruction results for two and three point simultaneous releases with the noisy synthetic data

Pseudo run	Noise (%)	Source parameters	11–13 SS'	11–8 SS'	8–8* SM	11–8–8* SS'M
Experimental release	0	Location	(100, 100) (86, 114)	(100, 100) (86, 114)	(86, 114) (80, 100)	(100, 100) (86, 114) (80, 100)
		Intensity	3,000 5,000	3,000 5,000	5,000 5,000	3,000 5,000 5,000
Least square estimate	10	Location	(100, 99) (86, 115)	(102, 99) (86, 114)	(86, 114) (81, 100)	(97, 101) (84, 115) (80, 100)
		E_L	5 5	11.2 0	0 5	15.8 11.2 0
		Intensity	2,984 5,243	2,439 5,421	5,329 4,351	3,518 6,037 5,198
		E_I	0.53 4.9	18.7 8.4	6.6 13	17.3 20.7 4
	20	Location	(100, 99) (86, 113)	(102, 99) (87, 114)	(85, 115) (79, 100)	(98, 101) (83, 115) (80, 100)
		E_L	5 5	11.2 5	7 7	11.2 15.8 0
		Intensity	2,860 4,859	2,857 4,539	5,830 5,043	2,641 6,512 5,448
		E_I	4.7 2.8	4.8 9.2	16.6 0.86	12 30.2 9
	30	Location	(100, 99) (87, 114)	(103, 99) (86, 113)	(88, 112) (81, 100)	(100, 99) (88, 113) (80, 100)
		E_L	5 5	15.8 5	14.1 5	5 11.2 0
		Intensity	2,987 4,093	2,168 4,535	5,393 5,195	2,873 4,360 4,633
		E_I	0.43 18.1	27.7 9.3	7.9 3.9	4.2 12.8 7.3

The experimental point releases (first and second rows) are shown in terms of location in grid and intensity in ug/s. The least square estimates are presented with 10%, 20% and 30% noises in third, seventh, and eleventh rows in terms of location and in fifth, ninth and thirteenth rows in terms of intensity. On the fourth, eighth and twelfth rows, error E_L in meter is indicated corresponding to noises

The departure in the retrieved intensity from the prescribed release is shown as E_I (in %)

in this direction is proposed by ISSARTEL *et al.* (2007) by renormalizing the domain with the weights according to available visibility from the direction of monitoring network, (3) computational complexity of the method increases in proportion to the number of unknown sources and number of grids in the computational domain and (4) the quantification of errors individually related to each source in the multiple-point source identification is not feasible at this moment, however, a combined estimate of error is reflected by the value of the function J.

6.3. Data Limitation

The proposed algorithm for the identification of sources is evaluated with the (1) model generated synthetic data, (2) noisy data obtained from synthetic measurements after adding a random noise and (3) real data in single-point emission and pseudo real data for two and three point emission sources. The data from IIT diffusion experiment used here corresponds to primarily the single-point emission in convective conditions. Further, the monitoring network is very sparse in this data set. In general, the diffusion data is very limited even for single-point emission sources. Existing studies on the retrieval of multiple-point emissions utilize the noisy synthetic data. Recently, an attempt has been made in the literature (LUSHI and STOCKIE, 2010) to design an experiment with four-point emission sources. Thus, the technique proposed here needs to be evaluated further with the availability of the concentration measurements not only with the single-point emissions in different atmospheric stability conditions but as well as from the simultaneous releases from more than one-point emission sources.

7. Conclusions

In this paper, we have presented an inversion approach based on least square technique for single- and multiple-point source estimation from limited number of atmospheric concentration measurements. The source estimation method is based on two-step minimization of sum of square of residuals between the receptor measured and the model predicted concentrations. The novelty of our approach stems from its simplicity and advantages in comparison to the other classical optimization methods.

The proposed algorithm has been successfully applied to identify the single as well as two and three simultaneous point emissions from synthetic, noisy and with real or pseudo-real concentration measurements from IIT diffusion experiment. With the synthetic measurements, release locations and intensities are retrieved exactly in all the runs for single as well as two and three simultaneous point releases. For the noisy measurements, it is observed that the retrieval error grows significantly as the noise in the concentration measurement increases. The retrieval of source is presented with noisy measurements obtained by adding 10, 20 and 30% random noise to the model generated concentrations.

In case of single-point release with real data, the source is identified with an average error of 20.5 m from the original source location. The corresponding intensity is retrieved within a factor of two in all the runs. An error estimate for the departure of intensities is also given in terms of angular departure of measurement vector from the corresponding model generated vector.

With the pseudo-real measurements, two and three point release locations are retrieved with an average error of 17 m and intensities are estimated on an average within a factor of two. The incurred errors in the retrieval are correlated with the errors involved in the observations and dispersion model. In addition, an alternative simplified approach is proposed in order to reduce the computational time required in the estimation of source parameters.

REFERENCES

AYRES, F. JR., Schaum's Outline of Theory and Problems of Matrices, (Schaum, New York 1962), 219 pp.
BEYER, W.A. (1964), A Note on Starting the Newton-Raphson Method, Communications of the ACM (CACM) 7, 442 pp.
BOCQUET, M. (2005a), *Reconstruction of an atmospheric tracer source using the principle of maximum entropy-I: Theory*, Q. J. R. Meteorol. Soc. *131*, 2191–2208.
BOCQUET, M. (2005b), *Reconstruction of an atmospheric tracer source using the principle of maximum entropy-II: Applications*, Q. J. R. Meteorol. Soc. *131*, 2209–2223.
COHN, S.E. (1997), *An introduction to estimation theory*, J. of the Met. Soc. of Japan *75*, 1B, 257–288.

DEBYE, P. (1909), *Näherungsformeln für die Zylinderfunktionen für große Werte des Arguments und unbeschränkt veränderliche Werte des Index*, Mathematische Annalen *67*(4), 535–558.

ELBERN H., STRUNK A. SCHMIDT H., TALAGRAND O. (2007), *Emission rate and chemical state estimation by 4-dimensional variational inversion*, Atmos. Chem. Phys. *7*, 3749–3769.

GOLUB, G.H. and VANLOAN, C.F., Matrix computations, (Johns Hopkins University Press, Baltimore 1996), 694 pp.

HAUPT, S.E. (2005), A demonstration of coupled receptor/dispersion modelling with a genetic algorithm, Atmos. Environ. *39*, 7181–7189.

HESTENES, M.R. and STIEFEL, E. (1952), *Methods of Conjugate Gradients for Solving Linear Systems*, Journal of Research of the National Bureau of Standards *49*, 409–436.

ISSARTEL, J.-P., BAVEREL, J. (2002), *Adjoint backtracking for the verification of the Comprehensive Nuclear Test Ban Treaty*, Atmospheric Chemistry and Physics Discussions *2*, 2133–2150.

ISSARTEL, J.P., SHARAN, M. and MODANI, M. (2007), *An inversion technique to retrieve the source of a tracer with an application to synthetic satellite measurements*, Proc. R. Soc. A *463*, 2863–2886.

ISSARTEL, J.P., SHARAN, M. and SINGH S.K. (2011), A retrieval technique for multiple point emissions from atmospheric concentration measurements, Submitted for publication.

KEATS, A., YEE, E. and LIEN, F.S. (2007a), *Bayesian inference for source determination with applications to a complex urban environment*, Atmos. Environ. *41*, 465–479.

KEATS, A., YEE, E. and LIEN, F.S. (2007b), *Efficiently characterizing the origin and decay rate of a nonconservative scalar using probability theory*, Ecological Modelling *205*, 437–452.

KRYSTA, M., BOCQUET, M., SPORTISSE, B. and ISNARD, O. (2006), *Data assimilation for short-range dispersion of radionuclides: An application to wind tunnel data*, Atmos. Environ. *40*, 7267–7279.

LEVENBERG, K. (1944), *A Method for the Solution of Certain Non-Linear Problems in Least Squares*, The Quarterly of Applied Mathematics *2*, 164–168.

LEWIS, J.M., LAKSHMIVARAHAN, S. and DHALL S.K., Dynamic Data Assimilation: A Least Square Approach, (Cambridge University Press, 2006), 654 pp.

LIU, X. and ZHAI, Z. (2007), *Inverse modeling methods for indoor airborne pollutant tracking: literature review and fundamentals*, Indoor Air *17*, 419–438.

LUSHI, E. and STOCKIE, J.M. (2010), *An inverse Gaussian plume approach for estimating atmospheric pollutant emissions from multiple point sources*, Atmos. Environ. *44*, 1097–1107.

MARCHUK, G.I., *Adjoint equations and analysis of complex systems*, (Kluver Academic Publishers, Dordrecht, 1995).

MARQUARDT, D. (1963), *An Algorithm for Least-Squares Estimation of Nonlinear Parameters*, SIAM Journal on Applied Mathematics *11*, 431–441.

MATTHES, J., GRÖLL, L. and KELLER, H.B. (2005), *Source localization by spatially distributed electronic noses for advection and diffusion*, IEEE Trans. Signal Process *53*, 1711–1719.

PENENKO, V., BAKLANOV, A. and TSVETOVA, E. (2002*), Methods of sensitivity theory and inverse modelling for estimation of source term*. Future Generation Computer Systems *18*, 661–671.

PUDYKIEWICZ, J.A. (1998), *Application of adjoint tracer transport equations for evaluating source parameters*, Atmos. Environ. *32*, 3039–3050.

RAO, K.S. (2007), *Source estimation methods for atmospheric dispersion*, Atmos. Environ. *41*, 6964–6973.

ROBERTSON, L. and LANGNER, J. (1998), *Source function estimate by means of variational data assimilation applied to the ETEX-I tracer experiment*, Atmos. Environ. *32*, 4219–4225.

SHARAN, M., ISSARTEL, J.P., SINGH, S. K. and KUMAR, P. (2009), *An inversion technique for the retrieval of single-point emissions from atmospheric concentration measurements*, Proc. R. Soc. A *465*, 2069–2088.

SHARAN, M., ISSARTEL, J.P. and SINGH, S. K. (2011), A point-source reconstruction from concentration measurements in low wind stable conditions, Submitted for publication.

SHARAN, M., SINGH, M.P., YADAV, A.K., AGGARWAL, P. and NIGAM, S. (1996), *A mathematical model for the dispersion of pollutants in low wind conditions*, Atmos. Environ. *30*, 1209–1220.

SINGH, M.P., AGARWAL, P., NIGAM, S. and GULATI, A. (1991), Tracer experiments—a report. Technical report, Centre for Atmospheric Science, IIT Delhi.

YEE, E. (2005), Probabilistic inference: an application to the inverse problem of source function estimation, The Technical Cooperation Program (TTCP) Chemical and Biological Defence (CBD) Group Technical Panel 9 (TP-9) Annual Meeting, Defence Science and Technology Organization, Melbourne, Australia.

YEE, E. (2006), A Bayesian approach for reconstruction of the characteristics of a localized pollutant source from a small number of concentration measurements obtained by spatially distributed "electronic noses", Russian-Canadian Workshop on Modeling of Atmospheric Dispersion of Weapon Agents, Karpov Institute of Physical Chemistry, Moscow, Russia.

YEE, E. (2007), Bayesian probabilistic approach for inverse source determination from limited and noisy chemical or biological sensor concentration measurements, Chemical and Biological Sensing VIII (Augustus W. Fountain III, ed), Proc of SPIE *6554*, 12 pp.

YEE, E. (2008), *Theory for Reconstruction of an Unknown Number of Contaminant Sources using Probabilistic Inference*, Boundary-Layer Meteorology *127*, 359–394.

YEE, E., Lien, F.S., Keats, A., Hseih, K.J. and D'Amours, R. (2006), Validation of Bayesian inference for emission source distribution using the Joint Urban 2003 and European Tracer Experimnts, Fourth International Symposium on Computational Wind Engineering (CWE2006), Yokohama, Japan, 4 pp.

(Received December 19, 2010, accepted April 20, 2011, Published online August 4, 2011)

Pure Appl. Geophys. 169 (2012), 499–517
© 2011 Her Majesty the Queen in Right of Canada
DOI 10.1007/s00024-011-0384-1

Probability Theory as Logic: Data Assimilation for Multiple Source Reconstruction

EUGENE YEE[1]

Abstract—Probability theory as logic (or Bayesian probability theory) is a rational inferential methodology that provides a natural and logically consistent framework for source reconstruction. This methodology fully utilizes the information provided by a limited number of noisy concentration data obtained from a network of sensors and combines it in a consistent manner with the available prior knowledge (mathematical representation of relevant physical laws), hence providing a rigorous basis for the assimilation of this data into models of atmospheric dispersion for the purpose of contaminant source reconstruction. This paper addresses the application of this framework to the reconstruction of contaminant source distributions consisting of an unknown number of localized sources, using concentration measurements obtained from a sensor array. To this purpose, Bayesian probability theory is used to formulate the full joint posterior probability density function for the parameters of the unknown source distribution. A simulated annealing algorithm, applied in conjunction with a reversible-jump Markov chain Monte Carlo technique, is used to draw random samples of source distribution models from the posterior probability density function. The methodology is validated against a real (full-scale) atmospheric dispersion experiment involving a multiple point source release.

Key words: Bayesian inference, data assimilation, inverse dispersion, Markov chain Monte Carlo, sensor/model data fusion, source reconstruction.

1. Introduction

A considerable research effort has been focussed on the problem of the forward prediction of the concentration resulting from the turbulent diffusion of a contaminant released into the environment. In contrast, much less work has been directed toward the related problem of the inverse prediction of the source characteristics (e.g., location, emission rate) from the concentration measured by an array of sensors. Nevertheless, a solution of the inverse source characterization problem is important for a number of applications of scientific, engineering, and practical interest. These include localization and characterization of pollutant gas emissions into the atmosphere, applications to groundwater remediation, and source attribution for the assignment of litigation and remediation costs to organizations responsible for hazardous gas releases.

From a purely mathematical viewpoint, the source reconstruction problem is an inverse problem. Let $c(\mathbf{x}, t)$ denote the instantaneous concentration at location $\mathbf{x}$ and time t, resulting from the release of a contaminant into the turbulent atmosphere. Because atmospheric turbulence consists of random fluctuations of various flow properties, the instantaneous concentration will be expected to fluctuate on a range of spatial-temporal scales. Models of atmospheric dispersion predict only the ensemble-mean concentration $C(\mathbf{x}, t) \equiv \langle c(\mathbf{x}, t) \rangle$ ($\langle \cdot \rangle$ denotes an ensemble-averaging operation where the ensemble average is taken over a large number of realizations of the atmospheric turbulence) which is related to the source density function $S(\mathbf{x}, t)$ through the following integral equation:

$$C(\mathbf{x}, t) = \int_{t_0}^{t} \int_{\mathbb{R}^3} p(\mathbf{x}, t | \mathbf{x}', t') S(\mathbf{x}', t') d\mathbf{x}' dt', \quad (1)$$

where for simplicity it is assumed implicitly that at the (arbitrary) initial time t_0, $C(\mathbf{x}, t_0) = 0$.

The right-hand-side (RHS) of Eq. 1 defines a formal operator G as follows:

$$(GS)(\mathbf{x}, t) \equiv \int_{t_0}^{t} \int_{\mathbb{R}^3} p(\mathbf{x}, t | \mathbf{x}', t') S(\mathbf{x}', t') d\mathbf{x}' dt', \quad (2)$$

which allows us to write Eq. 1 in the symbolic form

$$C = GS. \quad (3)$$

[1] Defence R&D Canada-Suffield, P.O. Box 4000 Stn Main, Medicine Hat, AB T1A 8K6, Canada. E-mail: eugene.yee@drdc-rddc.gc.ca

Formally, the source reconstruction problem can be solved by constructing the inverse operator G^{-1}. Unfortunately, this direct mathematical inversion is not possible for a number of reasons. Firstly, as a pure integral equation the problem may have no solution (singular operator G), and when it does the solution may not be uniquely determined. Secondly, $C(\mathbf{x}, t)$ cannot be specified completely and precisely—measurements of C are available only at a finite number of space-time points and these measurements are usually noisy suggesting that even if G^{-1} can be constructed exactly, the presence of noise in the concentration data may introduce instabilities into the solution. For these reasons, the source reconstruction problem using incomplete and noisy concentration data is an ill-posed problem. The reader is referred to Hanson (1998) for a more detailed description of the main features of the underlying ill-posed and inverse problems in the form of integral equations similar to Eq. 1 that arise in the physical sciences.

A mathematical tool that has been developed for treating instabilities that occur in inverse problems is regularization (TIKHONOV and ARSENIN, 1977; HANSON, 1998). In this approach, the class of admissible solutions for S is restricted by imposing further constraints on the solution (although frequently these constraints are somewhat arbitrary and ad-hoc). More specifically, regularization of the source reconstruction problem involves finding the best result from among all those source distributions that agree with the incomplete and noisy concentration data, by minimizing a cost functional of the form:

$$J(S) = ||\tilde{C} - GS||^2 + \lambda \Phi(S), \qquad (4)$$

where the first term on the RHS of Eq. 4 is a measure of the misfit between the measured (noisy) concentration $\tilde{C}$ and the model concentration GS ($|| \cdot ||$ is some appropriately defined norm), $\Phi(S)$ is the regularization functional that is used to impose some constraint on the solution, and λ is a regularizing parameter that imposes a relative weight (or, importance) between the data and the constraint.

ROBERTSON and PERSSON (1993) and ROBERTSON and LANGNER (1998) minimized Eq. 4 with $\Phi(S) = 0$, using a four-dimensional (space and time) variational data assimilation method with an adjoint transport equation, to recover emission rate profiles for an experiment with synthetic data and for the European Tracer Experiment (ETEX) (VAN DOP et al., 1998), respectively. SEIBERT and STOHL (2000) and SEIBERT (2000) used a regularization functional of the form $\Phi(S) = ||S||^2$ (which imposes an upper bound on the energy of the source distribution S) to reconstruct the distribution of emission rates for application to the radionuclide monitoring system implemented under the Comprehensive Test Ban Treaty (CTBT) (HOURDIN and ISSARTEL, 2000; ISSARTEL and BAVEREL, 2003). THOMSON et al. (2007) investigated the use of three different regularization functionals $\Phi(S)$ for source reconstruction, with the minimum of the cost functional $J(S)$ obtained using a random search algorithm with simulated annealing. BOCQUET (2005) and KRYSTA and BOCQUET (2007) used entropy as a regularizer (viz., applied the principle of maximum entropy on the mean) and solved the resulting constrained minimization of the cost functional by using duality theory to convert the problem to a simpler unconstrained minimization problem for the Lagrangian multipliers. Finally, ALLEN et al. (2007) minimized a misfit measure defined as the squared difference between the logarithmic measured and modeled concentrations with $\Phi(S) = 0$. In effect, these investigators assumed that it was known a priori that the unknown source distribution is a single continuous point source with unknown source location (x, y) and source strength Q, and used a genetic algorithm to select the source location and strength so as to minimize the misfit measure.

The regularization approach selects a particular source distribution S by minimizing, maximizing, or optimizing some form of cost functional $J(S)$ (viz, seeks or selects a single optimal source distribution from the entire ensemble of acceptable distributions that are consistent with the finite number of noisy concentration data). Unfortunately, the reliability of the inferred S cannot be obtained from any single selection of a source distribution, no matter what optimal properties in terms of quality and utility this single selection of S purportedly embodies. To deal with uncertainty (which reflects also the non-uniqueness of solutions for S using incomplete and noisy concentration data), it is necessary to apply Bayesian probability theory (or probability theory as logic) to the problem, rather than simply apply regularization. To this purpose, a probabilistic approach using a Bayesian inferential scheme that allowed the uncertainty in the inference for S to be determined was developed by YEE (2005) and demonstrated using Project Prairie Grass data for dispersion over open terrain. This methodology was further developed, refined, and generalized in subsequent work: (1) application of the methodology to complex environments (dispersion in built-up or urban environments) by YEE (2006), KEATS et al. (2007a) and CHOW et al. (2008); (2) generalization of the methodology to deal with a non-conservative scalar by KEATS et al. (2007b); (3) Bayesian experimental design for receptor placement in order to maximize the expected information in the measured concentration data for improving estimates of the source location and strength (KEATS et al., 2010); and, (4) application of the methodology to source reconstruction for long-range dispersion on continental scales by YEE et al. (2008).

YEE (2007) generalized the methodology to the reconstruction of multiple sources when the number of sources was known a priori. FLESCH et al. (2009) investigated the estimation of emission rates from multiple area sources on the ground surface, assuming that both the number and the location of the area sources were known a priori. However, these investigators used a regularized least-squares approach for the estimation of the emission rates, rather than a Bayesian approach. YEE and FLESCH (2010) applied Bayesian probability theory to the inference of emission rates from multiple ground area sources and demonstrated the significant advantages of applying an inference procedure relative to that of applying an inversion procedure using a regularized least-squares methodology. Finally, YEE (2008a, b) developed the theory underlying the application of a Bayesian probabilistic inferential framework for addressing the problem of source reconstruction for the difficult case of multiple point sources when the number of sources is unknown a priori.

In this paper, we extend the Bayesian probabilistic inferential methodology for multiple source reconstruction in the following manner. Firstly, the computational framework used by YEE (2008b) for sampling from the posterior distribution of the source parameters is significantly improved in the current paper. Secondly, the methodology proposed herein for multiple source reconstruction is validated using

some new concentration data obtained from a full-scale field experiment involving a multiple continuous point source release.

2. Bayesian Inference for Source Reconstruction

2.1. Structure-Based Representation for Source Distribution

In this paper, we focus on the reconstruction of a source distribution S consisting of an unknown number of continuously emitting localized sources. To this purpose, we will utilize a structure-based representation for S in which the source distribution is constructed from a number of (elemental) source atoms that are positioned at locations $\mathbf{x}$ over the domain of interest. In principle, the source atoms can be chosen to be as simple or as complex as one wishes (e.g., the source atom might represent a point source, a finite line source, an area source with a specified shape, etc.).

Because we are interested in the reconstruction of a unknown number of continuously emitting localized sources, we choose as the source atom a point source located at a vector position $\mathbf{x}_s$ that is emitting contaminant at a constant emission rate Q. The number of source atoms N_s required to provide a structure-based description of the source distribution (or source molecule) is allowed to vary. This number will be determined from the concentration data by the source reconstruction algorithm. As such, the source distribution has the following structure-based representation:

$$S(\mathbf{x}, t) = \sum_{k=1}^{N_s} Q_k \delta(\mathbf{x} - \mathbf{x}_{s,k}), \tag{5}$$

where $\delta(\cdot)$ is the Dirac delta function and the subscript k (on $\mathbf{x}_s$ and Q) are used to label the parameters associated with the k-th source atom in this representation.[1]

---

[1] To allow for the presence of localized sources with time-varying emission rates, it is straightforward to deal with this case by simply selecting the source atoms in the structure-based representation of Eq. 5 to have the following form: $Q_k \delta(\mathbf{x} - \mathbf{x}_{s,k})$ $\left[\mathcal{H}(t - T_b^k) - \mathcal{H}(t - T_e^k) \right]$ ($k = 1, 2, \ldots, N_s$) where $\mathcal{H}(\cdot)$ denotes the Heaviside unit step function; and, T_b and T_e are the activation and deactivation times of the source atom, respectively.

Now, let us assemble the parameters for the source distribution of Eq. 5 (consisting of N_s point sources) into the following source parameter vector:

$$\theta_{N_s} \equiv (\mathbf{x}_{s,1}, Q_1, \ldots, \mathbf{x}_{s,N_s}, Q_{N_s}) \in \mathbb{R}^{4N_s}. \quad (6)$$

Furthermore, let $\Theta \equiv (N_s, \theta_{N_s})$ be the parameter vector that encodes the structural information about the source molecule S (cf. Eq. 5).

2.2. Measurement Model

The measured concentration d_J obtained by a sensor at receptor location $\mathbf{x}_{d_J}$ is assumed be the sum of a modeled concentration signal $\overline{C}(\mathbf{x}_{d_J}; \Theta)$ and noise e_J:

$$d_J = \overline{C}(\mathbf{x}_{d_J}; \Theta) + e_J \equiv \overline{C}_J(\Theta) + e_J, \quad J = 1, 2, \ldots, N, \quad (7)$$

where N is the number of sensors. Furthermore,

$$\overline{C}(\mathbf{x}_{d_J}; \Theta) = \int_{\mathcal{D} \subset \mathbb{R}^3} d\mathbf{x} C(\mathbf{x}) h(\mathbf{x}|\mathbf{x}_{d_J}) \equiv \langle\langle C|h \rangle\rangle(\mathbf{x}_{d_J}), \quad (8)$$

is the expected concentration seen by the sensor at location $\mathbf{x}_{d_J}$, where $\mathcal{D}$ is the volume in space enclosing the source distribution S and the sensors in the array (recall $C(\mathbf{x})$ is the model concentration determined by Eq. 1^2 with S represented in accordance to Eq. 5). In Eq. 8, $h(\mathbf{x}|\mathbf{x}_d)$ is the spatial filtering function (of $\mathbf{x}$) for the concentration measurement made by the sensor at location $\mathbf{x}_d$. Note from Eq. 8 that $\overline{C}$ can be determined as the inner (or scalar) product $\langle\langle C|h \rangle\rangle$ of the concentration C and the detector spatial filtering function h.

With this background, the problem of reconstruction of the source distribution S (or source molecule) reduces to the following: estimate N_s (number of source atoms) and θ_{N_s} (parameters for each source atom) or, equivalently, estimate Θ given the noisy concentration data $\mathbf{D} \equiv (d_1, d_2, \ldots, d_N)$.

---

² Actually, the source–receptor relationship considered explicitly here is a special case of Eq. 1 involving the steady continuous emission of a contaminant from a sustained source into a statistically stationary atmosphere.

2.3. Bayesian Probability Theory

To determine Θ given the concentration data $\mathbf{D}$, we apply probability theory as logic (or Bayesian probability theory). The components of the Bayesian inference scheme for source reconstruction are shown in Fig. 1. Bayesian probability theory can be derived from more fundamental principles starting with the formulation of a small number of requirements that any theory of plausibility (or inference) ought to verify. These requirements were first provided by Cox (1946), with an eloquent description of the complete development described by JAYNES (2003). The theory can be interpreted as *the* unique extension of Aristotelian deductive logic (which deals only with propositions that are either certainly true or certainly false) to cases where there is uncertainty so that propositions need to be assessed in terms of their plausibilities (inductive logic). In view of this, probability theory when interpreted as logic is the quantitative theory of inference and, in consequence, is the unique consistent tool for dealing with uncertainty.

Within the context of the source reconstruction problem formulated above, the application of Bayesian probability theory yields the following result:

$$p(\Theta|\mathbf{D}, I) = \frac{p(\Theta|I)p(\mathbf{D}|\Theta, I)}{p(\mathbf{D}|I)}, \quad (9)$$

where I is the background (contextual) information available in the problem (e.g., model that defines the mapping from a source distribution S to the concentration C, background meteorology). The various factors that appear in Eq. 9 have the following interpretation. Firstly, $p(\Theta|I)$ is the prior probability density function (PDF) for a proposition Θ about the source distribution, predicated on the contextual information specified by I, with "|" denoting "conditional upon". The prior PDF encodes all the prior information about the source distribution before receipt of the concentration data $\mathbf{D}$. Secondly, $p(\mathbf{D}|\Theta, I)$ is the likelihood function and is identical to the direct PDF for observing the concentration data $\mathbf{D}$, when Θ is known exactly (viz, the source distribution is known). Thirdly, $p(\mathbf{D}|I)$ is referred to as the evidence and, in our case here, is simply a normalization constant. Finally, $p(\Theta|\mathbf{D}, I)$ is the posterior PDF for the proposition Θ about the source, in light

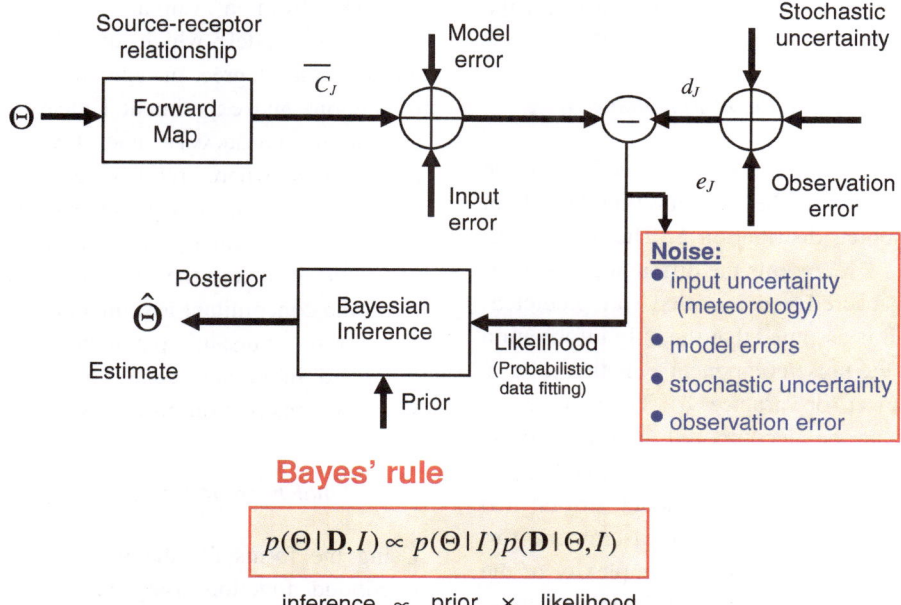

Figure 1
The components of the Bayesian inference scheme for source reconstruction

of the new information introduced through the newly acquired concentration data $\mathbf{D}$ (cf. Fig. 1).

We are seeking the posterior PDF $p(\Theta|\mathbf{D}, I)$, which encodes our inferences about Θ. Because $p(\mathbf{D}|I)$ is simply a normalization constant, the posterior PDF of interest in Eq. 9 can be specified to within a normalization constant as

$$p(\Theta|\mathbf{D}, I) \propto p(\Theta|I)p(\mathbf{D}|\Theta, I). \qquad (10)$$

The problem now reduces to the assignment of $p(\Theta|I)$ (prior distribution) and $p(\mathbf{D}|\Theta, I)$ (likelihood function).

2.3.1 Assignment of Prior

For the prior distribution, the logical independence of the source parameters is assumed, in which case $p(\Theta|I)$ factorizes as follows:

$$p(\Theta|I) \equiv p(N_s, \theta_{N_s}|I) = p(N_s|I)\prod_{k=1}^{N_s} p(Q_k|I)p(\mathbf{x}_{s,k}|I). \qquad (11)$$

The prior for the number of source atoms $p(N_s|I)$ is chosen to be a binomial distribution with parameter $p^* \in (0, 1)$ (binomial rate), and with a domain of definition between $N_{s,\min}$ (minimum number of source

atoms) and $N_{s,\max}$ (maximum number of source atoms); viz., $N_s \sim B(p^*; N_{s,\min}, N_{s,\max})$[3] where the symbol $\sim$ denotes "is distributed as". The prior for the emission rate $p(Q_k|I)$ for $k = 1, 2, \ldots, N_s$ is chosen to be a Bernoulli-uniform mixture: $Q_k \sim \mathcal{BU}(\gamma; Q_{\max})$ where γ is the probability that the source atom is turned on (viz., $\Pr\{Q_k > 0\} = \gamma$) and $Q_{\max}$ is an *a priori* upper bound on the expected emission rate.[4] The prior for the source atom location $p(\mathbf{x}_{s,k}|I)$ for

---

[3] The binomial distribution $B(p^*; N_{s,\min}, N_{s,\max})$ has a probability distribution function defined as follows:

$$p(N_s|I) = \frac{(N_{s,\max} - N_{s,\min})!}{(N_s - N_{s,\min})!(N_{s,\max} - N_s)!}p^{*(N_s - N_{s,\min})}\left(1 - p^*\right)^{N_{s,\max} - N_s},$$

for $N_s = N_{s,\min}, N_{s,\min} + 1, \ldots, N_{s,\max}$. Note that in this definition, the standard form of the binomial distribution has been offset by the minimum number of source atoms $N_{s,\min}$. If the expected degree of complexity in the source molecule is unknown a priori, then the assignment of the maximum number of source atoms can be omitted if the binomial distribution for $p(N_s|I)$ is replaced by a discrete probability distribution having support over the natural numbers $\mathbb{N}$ (e.g., Poisson distribution, geometric distribution).

[4] More specifically, a Bernoulli-uniform mixture model for Q_k has the following form:

$$p(Q_k|I) = (1 - \gamma)\delta(Q_k) + \gamma\mathbb{I}_{(0, Q_{\max})}(Q_k)/Q_{\max},$$

$(k = 1, 2, \ldots, N_s)$ and $\mathbb{I}_A(x)$ is the indicator function for set A, with $\mathbb{I}_A(x) = 1$ if $x \in A$ and $\mathbb{I}_A(x) = 0$ if $x \notin A$.

$k = 1, 2..., N_s$ is chosen to be uniform (flat) over the some spatial region $\mathcal{D} \subset \mathbb{R}^3$, so $\mathbf{x}_{s,k} \sim \mathcal{U}(\mathcal{D})$.

2.3.2 Assignment of Likelihood Function

The likelihood function $p(\mathbf{D}|\Theta, I)$ encodes all the information provided by the concentration data about the unknown source distribution S. The noise component e_J in Eq. 7 represents the difference (residual) between the measured and modeled (or, predicted) concentration. It is assumed that the only information we have about the noise component e_J is that it has a known noise power (or variance) σ_J^2. The variance σ_J^2 of this noise can be decomposed into four basic contributions as discussed by RAO (2005) (see Fig. 1): namely, (1) input errors; (2) model errors; (3) stochastic uncertainty; and, (4) measurement (or, observation) noise. In spite of the complexity of the noise structure, application of the principle of maximum entropy (see JAYNES, 2003) to our state of knowledge concerning the noise, results in the following Gaussian form for the likelihood function:

$$p(\mathbf{D}|\Theta, I) = \frac{1}{\prod_{J=1}^{N} \sqrt{2\pi}\sigma_J} \exp\left(-\frac{1}{2}\chi^2(\Theta)\right), \quad (12)$$

where

$$\chi^2(\Theta) \equiv \sum_{J=1}^{N} \left(\frac{d_J - \overline{C}_J(\Theta)}{\sigma_J}\right)^2. \quad (13)$$

In essence, the likelihood function is simply a probabilistic data fitting function that fuses the measured concentration data d_J with the predicted concentration data $\overline{C}_J(\Theta)$, and uses this information within the posterior distribution to assess the plausibility of various hypotheses about the unknown source distribution (encoded in the parameter vector Θ).

For applications, the noise power σ_J^2 in Eq. 13 is assumed to be known. The main contributions to σ_J, reflecting the scale of the discrepancy between the observed and predicted values of the concentration, are the measurement noise (arising from the sensor) and the model error (arising from the atmospheric dispersion model). The contribution from the former can be obtained from a noise model for the sensor, whereas the contribution from the latter can be

obtained from a validation of the atmospheric dispersion model, which quantifies using various metrics, the degree of agreement between model predictions and experimental measurements of concentration (OBERKAMPF and BARONE, 2006). For applications where reliable estimates of σ_J are difficult to obtain, it is possible within the Bayesian framework to treat σ_J as a nuisance parameter and eliminate it using the process of marginalization (with the concomitant assignment of a prior distribution for this nuisance parameter). This approach is described in greater detail and applied to various source reconstruction problems in YEE (2010).

2.3.3 Final Form for Posterior Distribution

Using the forms for the prior distribution and the likelihood function assigned above, the posterior distribution of Eq. 10 can be written explicitly as follows:

$$p(\Theta|\mathbf{D}, I) \equiv p(N_s, \theta_{N_s}|\mathbf{D}, I)$$

$$\propto \frac{1}{\prod_{J=1}^{N} \sqrt{2\pi}\sigma_J} \exp\left(-\frac{1}{2}\sum_{J=1}^{N} \left(\frac{d_J - \overline{C}_J(\Theta)}{\sigma_J}\right)^2\right)$$

$$\times \frac{(N_{s,max} - N_{s,min})!}{(N_s - N_{s,min})!(N_{s,max} - N_s)!} p^{*(N_s - N_{s,min})}$$

$$\times (1 - p^*)^{N_{s,max} - N_s}$$

$$\times \prod_{k=1}^{N_s} \left[(1-\gamma)\delta(Q_k) + \gamma \mathbb{I}_{(0, Q_{max})}(Q_k)/Q_{max}\right].$$

$$(14)$$

The posterior PDF $p(\Theta|\mathbf{D}, I)$ embodies the state of knowledge about the source parameters given the prior information encoded in $p(\Theta|I)$ and the newly acquired concentration data $\mathbf{D}$, the latter of which modulates our prior belief about Θ through the likelihood function $p(\mathbf{D}|\Theta, I)$. In consequence, $p(\Theta|\mathbf{D}, I)$ allows us to estimate all the interesting statistics about the source parameter Θ. More specifically, appraisal of the ensemble of source molecules drawn from $p(\Theta|\mathbf{D}, I)$ will allow one to make quantitative inferences about various features of the unknown source distribution, as well as perform an error analysis to quantify the uncertainty in the reconstruction of these source features.

3. Computational Framework

This section describes the computational procedures which are used for extracting the source parameter estimates required for stochastic event reconstruction. In this paper, the description of the computational methodology is necessarily brief. The reader is referred to YEE *et al.* (2008) and YEE (2008b) for a more complete description of various aspects of this topic. There are two major issues in the computational framework applied to Bayesian inference for source reconstruction that need to be addressed; namely, (1) a computationally efficient methodology for the computation of the source–receptor relationship required in the determination of the likelihood function, and (2) a methodology for sampling from the posterior distribution of the source parameters.

3.1. Fast Computation of Source–Receptor Relationship

To apply the Bayesian inference methodology to source reconstruction, we need to relate the hypotheses of interest about the unknown source distribution S (encoded in Θ) to the available concentration data d_J measured by the network of sensors. This requires the computation of a modeled (or predicted) concentration $\overline{C}_J$ as prescribed by Eqs. 8 and 1. This is the source–receptor relationship which is computed using an atmospheric dispersion model.

The likelihood function given in Eqs. 12 and 13 is not a closed-form expression and its evaluation is computationally expensive owing to the fact that $\overline{C}_J$ ($J = 1, 2, \ldots, N$) needs to be determined for a given source distribution Θ (using Eqs. 8 and 1). Moreover, a simulation-based posterior inference using Markov chain Monte Carlo sampling requires a large number of computations of the source–receptor relationship to be undertaken. In consequence, a fast and efficient technique for performing computations of the source–receptor relationship (for any given source distribution Θ) is required for the rapid sampling of the posterior distribution. To this purpose, KEATS *et al.* (2007a) and YEE *et al.* (2008) described a computationally efficient methodology for determination of the source–receptor relationship using an adjoint

representation for this relationship (see, also ISSARTEL and BAVEREL 2003; HOURDIN and TALAGRAND 2006).

3.1.1 Duality

To facilitate the fast computation of the source–receptor relationship given in Eq. 8, we utilize the following dual (or adjoint) representation of this relationship:

$$
\begin{aligned}
\overline{C}(\mathbf{x}_{d_J}; \Theta) &\equiv \int_{\mathcal{D}} d\mathbf{x}\, C(\mathbf{x}) h(\mathbf{x}|\mathbf{x}_{d_J}) \equiv \langle\!\langle C|h \rangle\!\rangle (\mathbf{x}_{d_J}) \\
&= \int_{\mathcal{D}} d\mathbf{x}'\, C^*(\mathbf{x}'|\mathbf{x}_{d_J}) S(\mathbf{x}') \\
&\equiv \langle\!\langle C^*|S \rangle\!\rangle (\mathbf{x}_{d_J}),
\end{aligned}
$$

$$(15)$$

where $C^*(\mathbf{x}'|\mathbf{x}_{d_J})$ is an adjunct (dual) concentration at point $\mathbf{x}'$ associated with the sensor concentration data at location $\mathbf{x}_{d_J}$. Equation 15 provides the receptor-oriented representation of the source–receptor relationship. This representation is mathematically equivalent to the source-oriented representation of the source–receptor relationship summarized in Eq. 8.

3.1.2 Forward and Backward Lagrangian Stochastic Models

In the source-oriented approach, the computation of $\overline{C}_J$ requires the determination of $C(\mathbf{x})$ which, in turn, requires the specification of the kernel function $p(\mathbf{x}|\mathbf{x}')$ (see Eq. 1 applied to the special case of a steady continuous emitting source in a statistically stationary atmosphere). There are no known analytical solutions for $p(\mathbf{x}|\mathbf{x}')$ for the case of complex turbulent flows (e.g., atmospheric flows). Approximations for the kernel function $p(\mathbf{x}|\mathbf{x}')$ can be obtained using either an Eulerian or Lagrangian description of atmospheric diffusion. Backtracking in an Eulerian framework using the adjoint of the direct transport equation has been described in HOURDIN and TALAGRAND (2006).

In this paper, we will use a first-order Lagrangian stochastic (LS) trajectory simulation method to approximate $p(\mathbf{x}|\mathbf{x}')$ (and subsequently, $C(\mathbf{x})$). To this purpose, we consider THOMSON'S (1987) model for inhomogeneous Gaussian turbulence, where the forward increments of the velocity $\mathbf{U}(t) \equiv (U_i(t))$ and position $\mathbf{X}(t) \equiv (X_i(t))(i = 1, 2, 3)$ of a marked fluid element (particle) are given by the following stochastic differential equation:[5]

$$dX(t) = U(t)dt,$$

$$dU(t) = \mathbf{a}(\mathbf{X}(t), \mathbf{U}(t), t)dt + (C_0\epsilon(\mathbf{X}(t), t))^{1/2}d\mathbf{W}(t),$$
$$(16)$$

where ϵ is the turbulence kinetic energy (TKE) dissipation rate, C_0 is the Kolmogorov universal constant, $d\mathbf{W}(t) \equiv (dW_i(t))$ are the increments of a vector-valued (three-dimensional) Wiener process[6], and $\mathbf{a} \equiv (a_i)$ is the drift coefficient vector which assumes the following form:

$$a_i = -\frac{1}{2}(C_0\epsilon)\Gamma_{ik}^{-1}(u_k - \overline{U}_k) + \frac{\phi_i}{P_E}, \quad (17)$$

where

$$\frac{\phi_i}{P_E} \equiv \frac{1}{2}\frac{\partial \Gamma_{il}}{\partial x_l} + \frac{\partial \overline{U}_i}{\partial t} + \overline{U}_l\frac{\partial \overline{U}_i}{\partial x_l}$$
$$+ \left(\frac{1}{2}\Gamma_{lj}^{-1}\left[\frac{\partial \Gamma_{il}}{\partial t} + \overline{U}_m\frac{\partial \Gamma_{il}}{\partial x_m}\right] + \frac{\partial \overline{U}_i}{\partial x_j}\right)(u_j - \overline{U}_j)$$
$$+ \frac{1}{2}\Gamma_{lj}^{-1}\frac{\partial \Gamma_{il}}{\partial x_k}(u_j - \overline{U}_j)(u_k - \overline{U}_k).$$
$$(18)$$

Here, $\overline{U}_i$ is the mean Eulerian velocity, $\Gamma_{ij} \equiv \overline{(u_i - \overline{U}_i)(u_j - \overline{U}_j)}$ is the Reynolds stress tensor (where an overline is used to denote an ensemble average), and P_E is the background (Eulerian) velocity PDF (which is implicitly assumed here to possess a Gaussian form). In this source-oriented approach, marked particles with *initial* spatial coordinates $\mathbf{x}_i$ are sampled from a spatial density function that is proportional to the source distribution $S(\mathbf{x})$, and the forward-time trajectories of these tagged fluid particles are computed using Eqs. 16–18. The displacement statistics of these particles are used to determine $C(\mathbf{x})$.

---

[5] The forward increments of the marked particle position and velocity are defined as $dX(t) \equiv X(t + dt) - X(t)$ and $dU(t) \equiv U(t + dt) - U(t)$, respectively, with $dt > 0$.

[6] A Wiener process W_i is a stochastic process which has independent increments with $dW_i(t + dt) \equiv W_i(t + dt) - W_i(t) \sim \mathcal{N}(0, dt)$ where $\mathcal{N}(\mu, \sigma^2)$ denotes a Gaussian distribution with mean μ and variance σ^2.

For source reconstruction, we use the receptor-oriented approach described by Eq. 15 for the efficient calculation of $\overline{C}_J$, which requires the determination of the adjunct (or dual) concentration $C^*(\mathbf{x}'|\mathbf{x}_{d_J})$. To this purpose, it was shown by THOMSON (1987) and FLESCH et al. (1995) that the following backward-time Lagrangian trajectory simulation model is the dual to the forward-time Lagrangian trajectory simulation model given by Eqs. 16–18:

$$d\mathbf{X}^b(t') = \mathbf{U}^b(t')dt',$$

$$d\mathbf{U}^b(t') = \mathbf{a}^b(\mathbf{X}^b(t'), \mathbf{U}^b(t'), t')dt' \quad (19)$$
$$+ (C_0\epsilon(\mathbf{X}^b(t'), t'))^{1/2}d\mathbf{W}(t'),$$

with

$$U_i^b = U_i, \quad a_i^b = \frac{1}{2}(C_0\epsilon)\Gamma_{ik}^{-1}(u_k - \overline{U}_k) + \frac{\phi_i}{P_E}. \quad (20)$$

In Eq. 19, $d\mathbf{X}^b(t') \equiv \mathbf{X}^b(t') - \mathbf{X}^b(t' - dt')$ and $d\mathbf{U}^b(t') \equiv \mathbf{U}^b(t') - \mathbf{U}^b(t' - dt')(dt' > 0)$ are the backward increments of the position $\mathbf{X}^b(t')$ and velocity $\mathbf{U}^b(t')$ of a marked particle at time t' along a backward-time trajectory. In this receptor-oriented (dual) approach, marked particles with *final* spatial coordinates $\mathbf{x}_f$ are sampled from a spatial density function that is proportional to the sensor response function $h(\mathbf{x}'|\mathbf{x}_d)$ (at the sensor spatial location $\mathbf{x}_d$), and the backward-time trajectories of these tagged fluid particles are computed using Eqs. 19 and 20. The displacement statistics of these particles are used to determine $C^*(\mathbf{x}'|\mathbf{x}_d)$.

Substituting Eq. 5 into Eq. 15, the model concentration seen by the sensor at spatial point $\mathbf{x}_{d_J}$ is given explicitly as follows:

$$\overline{C}_J(\Theta) \equiv \overline{C}(\mathbf{x}_{d_J}; \Theta) = \sum_{k=1}^{N_s} Q_k C^*(\mathbf{x}_{s,k}|\mathbf{x}_{d_J}). \quad (21)$$

It should be noted that the computation of $\overline{C}_J(\Theta)$ can be obtained for any source distribution S (encoded as the parameter vector Θ), without having to re-compute C^*. Indeed, it should be emphasized that because C^* does not depend on the source distribution (viz., is independent of Θ), it can be pre-calculated using the backward-time LS trajectory simulation model for each available detector space-time location, and this pre-calculated C^* can be used in Eq. 21 for a computationally efficient determination of $\overline{C}_J(\Theta)$ $(J = 1, 2, \ldots, N)$ required for the rapid evaluation of the likelihood function $p(\mathbf{D}|\Theta, I)$.

3.2. Markov Chain Monte Carlo Sampling

All the information arising from the application of Bayesian probability theory to the problem of source reconstruction is embodied in the posterior PDF of the parameters $\Theta \equiv (N_s, \theta_{N_s}) \in \mathbb{R}^{4N_s+1}$ (see Eq. 14) that define the source distribution S. Estimates of the features (number of component sources, component emission rates, component source locations) of the unknown source distribution, along with the uncertainties in these estimates, can be obtained from samples[7] of source molecules (encoded by Θ) drawn from the posterior PDF. To this purpose, we apply a Markov chain Monte Carlo (MCMC) algorithm for posterior sampling (see GILKS *et al.*, 1996; GELMAN *et al.*, 2003). YEE (2008b) described the formulation of a reversible-jump MCMC (RJMCMC) algorithm applied with parallel tempering for generating samples from the posterior distribution $p(\Theta|\mathbf{D}, I) \equiv p(N_s, \theta_{N_s}|\mathbf{D}, I)$. For brevity, only the important details of the algorithm that are required to understand the following Bayesian analysis of simulated and real concentration data, as well as the significant modifications to the algorithm as used in the current paper (in order to increase the computational efficiency), will be described here.

The objective of MCMC sampling is to construct an auxiliary Markov chain[8] whose stationary (or, invariant) distribution is the posterior distribution $p(\Theta|\mathbf{D}, I)$. The difficulty in the construction of the Markov chain in the current application resides in the fact that the number of source atoms N_s required to represent the source distribution is unknown a priori, so the dimension of the hypothesis (or, parameter) space is unknown (viz, $\theta_{N_s} \in \mathbb{R}^{4N_s}$ with N_s unknown). More specifically, with reference to the posterior distribution $p(\Theta|\mathbf{D}, I)$ given by Eq. 14, it is necessary to consider

$(N_{s,\max} - N_{s,\min} + 1)$ candidate model structures for the source distribution S and associated with each of these candidate model structures is a posterior distribution $p(N_s, \theta_{N_s}|\mathbf{D}, I)$ [for *fixed* N_s] depending upon an unknown parameter vector $\theta_{N_s} \in \mathbb{R}^{4N_s}$ where $N_s \in \{N_{s,\min}, N_{s,\min} + 1, \ldots, N_{s,\max}\}$ is a model indicator that defines the dimension ($4N_s$) of the hypothesis (parameter) space. In order to allow changes in the dimensionality of the model (or, equivalently, changes in the model structure), a reversible-jump MCMC algorithm is used to construct the Markov chain for Θ. The formalization of RJMCMC algorithms for dealing with variable dimension models was described initially by GREEN (1995), and since then has been applied to a wide range of problems (among many others, RICH-ARDSON and GREEN, 1997; HUERTA and WEST, 1999).

3.2.1 Probabilistic Exploration

Consider a Markov chain with state vector $\{\Theta^{(t)}\} \equiv \{N_s^{(t)}, \theta_{N_s^{(t)}}\}$ ($t = 0, 1, 2, \ldots$) that is constructed so that its stationary distribution coincides with $p(\Theta|\mathbf{D}, I)$ given by Eq. 14. After convergence of the Markov chain to its stationary distribution, the samples drawn from this chain can be used to estimate any posterior statistic of interest (viz, any feature of the source distribution). The construction of the Markov chain uses a RJMCMC algorithm in which the MCMC moves are separated into two distinct categories; namely, propagation moves which do not change the dimensionality of the source distribution and trans-dimensional jump moves which change the source distribution by ± 1 discrete source atom. In the current application, the trans-dimensional jump move changes the dimensionality of the hypothesis space by ± 4 (as each source atom in Eq. 5 involves four degrees of freedom).

For the dimension-conserving moves, we partition the parameter vector θ_{Ns} (N_s fixed) as follows: $\theta_{Ns} = (\theta^1, \theta^2)$ where $\theta^1 \equiv (Q_1, Q_2, \ldots, Q_{N_s}) \in \mathbb{R}^{N_s}$ and $\theta^2 \equiv (\mathbf{x}_{s,1}, \ldots, \mathbf{x}_{s,N_s}) \in \mathbb{R}^{3N_s}$. The parameters associated with θ^1 are *linearly* related to the model concentration data, as is evident from Eq. 21. For these parameters, a Gibbs sampler is used for the update (propagation) move. More specifically, the Gibbs sampler updates Q_k as a direct draw from the univariate full conditional posterior distribution

---

[7] In this paper, we use the term "sampler" to refer to any algorithm that has been designed to draw samples from the posterior PDF of the source parameters.

[8] A Markov chain is a discrete-time random process with the Markov property (viz, the next state in the chain depends only on the current state and not on any of the past states).

$p(Q_k|\theta^1_{-k}, \theta^2, \mathbf{D}, I)$ where θ^1_{-k} is the vector θ^1 with its k-th component removed $(k = 1, 2, \ldots, N_s)$. From Eq. 14, it can be shown that the full conditional posterior distribution for Q_k assumes a simple Bernoulli-Gaussian (truncated) distribution which can be sampled from directly (YEE, 2008b).

The remaining parameters in θ^2 (e.g., locations of the source atoms) are related *non-linearly* to the model concentration. In consequence, the Gibbs sampler cannot be used to update these parameters owing to the fact that the full conditional posterior distribution for these parameters cannot be determined analytically. For this reason, the Metropolis–Hastings (M–H) sampler is used to update the parameters in θ^2. To this purpose, we update $\theta^2_l \in \theta^2 (l = 1, 2, \ldots, 3N_s)$ by randomly sampling a new candidate $\theta^{2'}_l$ from a proposal distribution that is taken to be a mixture of Gaussian distributions, each with mean θ^2_l and different variances β^2_l. The variances in this mixture of Gaussian distributions are chosen typically to cover several orders of magnitude across the domain of definition for θ^2_l.

The dimension-changing moves that modify the source distribution by ± 1 source atom (and the dimensionality of the hypothesis space by ± 4) are provided by: (1) a creation move $\mathcal{C}$ that results in the addition of a single source atom, so $\Theta' = (N_s + 1, \theta_{N_s+1}) \equiv \mathcal{C}(\Theta) = \mathcal{C}((N_s, \theta_{N_s}))$ where $\Theta' \in \mathbb{R}^{1+4N_s+4}$; and, (2) an annihilation move $\mathcal{C}^\dagger$ that involves the removal of a single existing source atom from the current source distribution, so $\Theta' = (N_s - 1, \theta_{N_s-1}) \equiv \mathcal{C}^\dagger(\Theta) = \mathcal{C}^\dagger((N_s, \theta_{N_s}))$ where $\Theta' \in \mathbb{R}^{1+4N_s-4}$. If a creation move $\mathcal{C}$ is selected, the coordinates of the new source atom are obtained by drawing random samples from a proposal density that is chosen to be the prior density for each coordinate. On the other hand, if an annihilation (reverse) move $\mathcal{C}^\dagger$ is selected, a source atom in the current source distribution is randomly picked and removed.

To summarize, the Markov chain consists of a sequence of states $\Theta^{(t)} (t = 0, 1, 2, \ldots)$ resulting from individual updates consisting of three basic moves: (1) $\mathcal{M}_{\mathcal{C},\mathcal{C}^\dagger}$ involving either the creation of a source atom at a random location or the annihilation of an existing source atom; (2) $\mathcal{M}_1$ involving updates of the emission rates of the source atoms using Gibbs

sampling; and, (3) $\mathcal{M}_2$ involving updates of the locations of the source atoms using M–H sampling. The state vector $\Theta^{(t-1)}$ of the Markov chain at iteration $t - 1$ is updated to the state vector $\Theta^{(t)}$ at iteration t using the following procedure:

1. Specify values for $(p^*, N_{s,min}, N_{s,max}, \gamma, Q_{max}, \mathcal{D})$ which define $p(\Theta|I)$;
2. Choose an initial state $\Theta^{(0)}$ for the Markov chain by sampling from $p(\Theta|I)$;
3. For $t \in \{1, 2, \ldots, t_{upper}\}$, conduct the following sequence of moves:

$$\Theta^{(t-1)} \overset{\mathcal{M}_{\mathcal{C},\mathcal{C}^\dagger}}{\rightarrow} \Theta_\star \overset{\mathcal{M}_1}{\rightarrow} \Theta_{\star\star} \overset{\mathcal{M}_2}{\rightarrow} \Theta^{(t)}, \tag{22}$$

where $\Theta_\star$ and $\Theta_{\star\star}$ denote some intermediate transition states between iterations $(t - 1)$ and t.

3.2.2 Simulated Annealing

To improve the speed with which a Markov chain traverses the hypothesis space (or, to increase the mixing rate of the chain in the hypothesis space), YEE (2008b) implemented a form of parallel tempering based on a Metropolis-coupled MCMC algorithm described by GEYER (1991). In this study, rather than use a parallel tempering scheme, we employ instead a related (but simpler and computationally more efficient) simulated annealing scheme to facilitate chain mobility in the hypothesis space. In this scheme, we consider an ensemble of N_{mem} (typically between 50 and 200) different source distributions (or, source molecules) that have been randomly drawn from a modified posterior having the following form: $p_\lambda(\Theta|\mathbf{D}, I) \propto p(\Theta|I)p^\lambda(\mathbf{D}|\Theta, I)$, where $\lambda \in [0, 1]$ is a coolness parameter. These samples will be labelled $\Theta_k(\lambda)(k = 1, 2, \ldots, N_{mem})$. Note that $p_0(\Theta|\mathbf{D}, I) = p(\Theta|I)$ (prior distribution of Θ) and $p_1(\Theta|\mathbf{D}, I) = p(\Theta|\mathbf{D}, I)$ (posterior distribution of Θ). In this framework, it is useful to interpret the parameter λ as an inverse temperature parameter T (so, $\lambda = 1/T$), with $\lambda \in [0, 1]$ implying $T \in [1, \infty]$. The posterior distribution $p_1(\Theta|\mathbf{D}, I)$ corresponds to the temperature $T = 1$, whereas the modified $p_\lambda(\Theta|\mathbf{D}, I)(\lambda \in [0, 1))$ corresponds to heating the posterior distribution to a temperature $T = 1/\lambda > 1$ which results in a flattening of the distribution.

When the stochastic sampling scheme begins and $\lambda = 0$ (infinite temperature), we randomly draw N_{mem} source molecules $\Theta_k(0)(k = 1, 2, \ldots, N_{\text{mem}})$ from $p_0(\Theta|\mathbf{D}, I)$ (prior distribution); viz, $\Theta_k(0) \sim p(\Theta|I).$[9] Given an ensemble of N_{mem} source molecules $\Theta_k(\lambda)$ that has achieved equilibrium (at temperature $T = 1/\lambda$) with respect to the modified posterior $p_\lambda(\Theta|\mathbf{D}, I),$[10] an ensemble of N_{mem} source molecules $\Theta_k(\lambda + \delta\lambda)$ that is consistent with $p_{\lambda+\delta\lambda}(\Theta|\mathbf{D}, I)$ (at the reduced temperature $T = 1/(\lambda + \delta\lambda)$, $\delta\lambda > 0$) can be obtained by using the weighted resampling method (GAMERMAN and LOPES, 2006) applied to $\Theta_k(\lambda)(k = 1, 2, \ldots, N_{\text{mem}})$. To this purpose, each source molecule $\Theta_k(\lambda)$ is assigned a normalized importance weight w_k as follows:

$$w_k = p^{\delta\lambda}(\mathbf{D}|\Theta_k(\lambda), I) \times \left(\sum_{j=1}^{N_{\text{mem}}} w_j\right)^{-1}, \quad (23)$$

for $k = 1, 2, \ldots, N_{\text{mem}}$. Next, a resample is drawn from the discrete distribution concentrated at the sample of source molecules $\{\Theta_k(\lambda)\}_{k=1}^{N_{\text{mem}}}$ from $p_\lambda(\Theta|\mathbf{D}, I)$ with respective weights $\{w_k\}_{k=1}^{N_{\text{mem}}}$. In other words, the resampling step here involves generating a new set $\{\Theta_{k*}\}_{k*=1}^{N_{\text{mem}}}$ by resampling (with replacement) N_{mem} times from the set $\{\Theta_k(\lambda)\}_{k=1}^{N_{\text{mem}}}$ so that $\Pr\{\Theta_{k*} = \Theta_k(\lambda)\} = w_k$. This resample can be interpreted as an ensemble of source molecules Θ_{k*} $(k* = 1, 2, \ldots, N_{\text{mem}})$ that is drawn from the modified posterior $p_{\lambda+\delta\lambda}(\Theta|\mathbf{D}, I)$. There are many ways to implement this resampling (with replacement) from $\{\Theta_k(\lambda)\}_{k=1}^{N_{\text{mem}}}$, but in this paper we use the systematic resampling scheme described by KITAGAWA (1996). This scheme requires $O(N_{\text{mem}})$ time to execute and minimizes the Monte Carlo variation in the resample.

An annealing schedule for $\lambda \in [0, 1]$ is required for the simulated annealing. In this paper, we applied simulated annealing with 200 values of λ uniformly spaced in the interval [0,0.05] and 400 values of λ geometrically spaced in the interval (0.05,1]. This gentle annealing schedule allows the ensemble of N_{mem} source molecules to transition slowly through a series of quasi-equilibrium states from the prior

distribution ($\lambda = 0$, or infinite temperature) at one end of the annealing schedule to the posterior distribution ($\lambda = 1$, or unit temperature) at the other end of the schedule. When $\lambda = 1$, the annealing phase is complete and probabilistic exploration of the hypothesis space proceeds (for each of the N_{mem} source molecules in the ensemble) in accordance to the scheme summarized in Eq. 22. The annealing phase of the scheme, corresponding to values of $\lambda \in [0, 1)$, is associated with the burn-in phase of the algorithm. When $\lambda = 1$, the MCMC algorithm has reached an equilibrium, at which point the probabilistic exploration phase corresponding to the sampling from the posterior distribution $p(\Theta|\mathbf{D}, I)$ begins. These samples of source molecules drawn from the posterior distribution can be used to make inferences about the various characteristics of the source distribution.

3.2.3 Evidence and Information Gain

Interestingly, the simulated annealing phase of the proposed scheme can be used to estimate the normalization constant (or, evidence) $Z \equiv p(\mathbf{D}|I)$ for the posterior distribution $p(\Theta|\mathbf{D}, I)$ (see Eq. 9). To accomplish this objective, we use an approach referred to as thermodynamic integration (VON DER LINDEN et al., 1996), which has its origins in the problem of the evaluation of partition functions in statistical mechanics familiar to physicists. Briefly, thermodynamic integration focusses on the modified evidence $Z(\lambda)$ defined as follows:

$$Z(\lambda) \equiv \int p(\Theta|I)p^\lambda(\mathbf{D}|\Theta, I)\mathrm{d}\Theta, \quad (24)$$

which can be seen to be simply the normalization constant (evidence) corresponding to the modified posterior $p_\lambda(\Theta|\mathbf{D}, I)$. Obviously, $Z(0) = 1$ because the prior $p(\Theta|I)$ is normalized and $Z(1) = Z$ is the desired normalization constant (evidence) for the posterior $p(\Theta|\mathbf{D}, I)$. Taking the logarithmic derivative of Eq. 24 with respect to λ, and re-arranging yields

$$\ln[Z(1)] - \ln[Z(0)] = \ln[Z(1)]$$
$$= \int_0^1 \langle \ln[p(\mathbf{D}|\Theta, I)]\rangle_\lambda \mathrm{d}\lambda, \quad (25)$$

---

[9] The prior distribution $p(\Theta|I)$ is composed of standard probability distributions for which independent sampling is easy.

[10] This simply implies that $\{\Theta_k(\lambda)\}_{k=1}^{N_{\text{mem}}}$ can be interpreted as an ensemble of source molecules drawn from $p_\lambda(\Theta|\mathbf{D}, I)$.

where $\langle \cdot \rangle_\lambda$ denotes the mathematical expectation operation with respect to the modified posterior $p_\lambda(\Theta|\mathbf{D}, I)$.

Additionally, the computation of $Z(\lambda)$ allows the determination of the information (gain) corresponding to the receipt of the concentration data $\mathbf{D}$ and the updating of our state of knowledge concerning the (unknown) source S (encoded as Θ) from the prior distribution $p(\Theta|I)$ to the posterior distribution $p(\Theta|\mathbf{D}, I)$. This information gain (amount of useful information about Θ embodied in $\mathbf{D}$) is given by the Kullback-Leibler divergence $D_{KL}(\lambda)$ at $\lambda = 1$ [viz., by $D_{KL}(1)$] where $D_{KL}(\lambda)$ is defined as follows (COVER and THOMAS, 1991):

$$
\begin{aligned}
D_{KL}(\lambda) &\equiv \int \ln\left(\frac{p_\lambda(\Theta|\mathbf{D}, I)}{p(\Theta|I)}\right) p_\lambda(\Theta|\mathbf{D}, I) \mathrm{d}\Theta \\
&= \lambda \int \ln[p(\mathbf{D}|\Theta, I)] p_\lambda(\Theta|\mathbf{D}, I) \mathrm{d}\Theta \\
&\quad - \ln[Z(\lambda)] \int p_\lambda(\Theta|\mathbf{D}, I) \mathrm{d}\Theta \\
&= \lambda \langle \ln[p(\mathbf{D}|\Theta, I)] \rangle_\lambda - \ln[Z(\lambda)],
\end{aligned} \tag{26}
$$

on using the fact that $p_\lambda(\Theta|\mathbf{D}, I) = p(\Theta|I) p^\lambda(\mathbf{D}|\Theta, I)/Z(\lambda)$. The Kullback-Leibler divergence defined in Eq. 26 for $\lambda = 1$ is simply the negative of the entropy (negentropy) of the posterior relative to the prior, and as such, is the information gain provided by the receipt of the concentration data $\mathbf{D}$. More specifically, the information gain compresses the posterior relative to the prior, so that $D_{KL}(1)$ can simply be interpreted as the logarithm of the volumetric factor (in hypothesis space) by which the prior has been compressed to become the posterior. The greater this compression, the greater is the information gain provided by the measured concentration data.

4. Application to FFT-07 Experimental Data

4.1. Experiment

In this section, we apply the source reconstruction algorithm described above to a real dispersion data set obtained from an experiment (involving a multiple source release) conducted under a multinational field campaign known as **F**using **S**ensor **I**nformation from **O**bserving **N**etworks (FUSION) Field Trial 2007 (FFT-07). The experiments in FFT-07 were carried out in September 2007 at Tower Grid on US Army Dugway Proving Ground, Utah. The terrain was flat, uniform, and homogeneous with short grass interspersed with low shrubs that are between 0.25 to 0.75 m in height, providing an upwind fetch that is uniform and unobstructed for 5 km or more in a wide sector. In all the experiments, the tracer gas used was propylene (C_3H_6). The concentration sensors used were fast-response digital photo-ionization detectors (dPIDs).

The experiment used to test the source reconstruction methodology involved four continuously emitting sources. The network (or array) of concentration sensors used in FFT-07 is shown in Fig. 2. A total of 100 dPIDs were arranged in a staggered configuration consisting of 10 rows of 10 sensors. The rows of sensors were spaced 50 m apart. The spacing between sensors along each row was 50 m. The concentration sensors along the ten sampling lines in the array were placed at a height, z_d, of 2.0 m. The (x, y) local coordinate system that will be used here is shown in Fig. 2. We used 62 sensors in the array (indicated by the filled blue squares in Fig. 2) for the source reconstruction. All the functional detectors in the array that measured a significantly non-zero concentration were used for the reconstruction, as well as a number of detectors for which the measured concentration was nominally zero (viz., over the sampling time for the experiment, the instantaneous concentration did not exceed the detection threshold concentration).

In this experiment, the mean wind direction was normally incident to the sensor array. An estimate of the momentum roughness length, $z_0 = 1.3 \pm 0.2$ cm was obtained from a three-dimensional sonic anemometer (R. M. Young, Model 81000) mounted on a lattice tower at a height of 2 m above ground level. Furthermore, the sonic anemometer was used to characterize the background micro-meteorological state of the atmospheric surface layer. For this experiment, the horizontal mean wind speed S_2 at the 2-m level, the friction velocity (u_*), the atmospheric stability (Obukhov length L), and the standard deviations of the velocity fluctuations in the alongwind (σ_u), crosswind (σ_v), and vertical (σ_w) directions

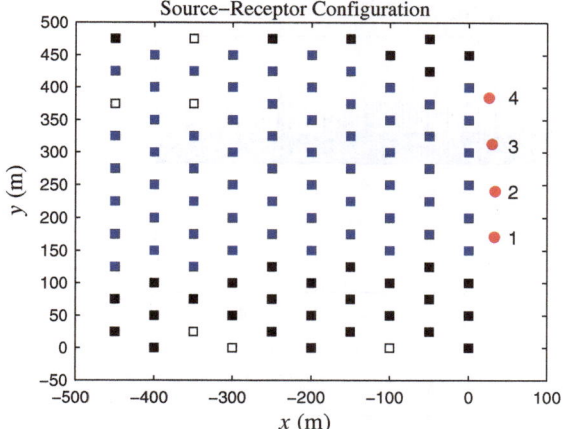

Figure 2

The *solid dots* show the locations of the four sources. The sources are labelled from 1 to 4 with increasing values for the *y*-coordinate of the source. *Squares* show the location of the sensors in the array: *open and filled squares* indicate that the sensor at the given location is missing and present (for the experiment from FFT-07 used herein), respectively, in the array. A *filled blue square* marks the functional sensors that were used for the source reconstruction. The local coordinate system used is indicated, with the *x*- and *y*-directions referring to the alongwind and crosswind directions, respectively (viz., the mean wind direction for the experiment was normally incident to the sampling lines of detectors in the sensor array)

were: $S_2 = 3.61$ m s^{-1}, $u_* = 0.282$ m s^{-1}, $L = -27.3$ m, $\sigma_u/u_* = 2.33$, $\sigma_v/u_* = 1.86$, and $\sigma_w/u_* = 1.10$.

The backward-time LS model given by Eqs. 19 and 20 was used to determine $\overline{C}_J(\Theta)$ from Eq. 21. This model was applied to short-range dispersion in the atmospheric surface layer over a level and unobstructed terrain. The mean wind and turbulence statistics for the prediction of $\overline{C}_J$ were assumed to be horizontally homogeneous and stationary, with the relevant flow statistics for the background micrometeorological atmospheric state of the surface layer summarized above. The choice of the Kolmogorov constant $C_0 = 4.8$ is used for our simulations, the latter value having been obtained by calibrating the LS model against concentration data measured during a benchmark field experiment Project Prairie Grass (WILSON *et al.*, 2001). The vertical profiles of mean wind and turbulence required by the LS model are prescribed in accordance with well-known surface-layer relations based on Monin–Obukhov theory (STULL, 1988).

4.2. Reconstruction

4.2.1 Initialization of Algorithm

The experiment used here to test the source reconstruction methodology involved continuously emitting sources (each with a constant emission rate). As a consequence, the relevant source parameters are as follows: $\Theta = (N_s, \mathbf{x}_{s1}, Q_1, \ldots, \mathbf{x}_{sN_s}, Q_{N_s})$. Furthermore, it is assumed that the height of the sources above ground level ($z_s = 2.0$ m) is known a priori, so the only unknown location parameters are (x_s, y_s) of the sources in the horizontal plane. In view of these assumptions, the adjunct concentration $C^*(\mathbf{x}_s|\mathbf{x}_d)$ in Eq. 21 can be pre-calculated for one detector position $\mathbf{x}_d$ (for the known source height z_s), with the adjunct concentration C^* (considered as a function of $\mathbf{x}_s$) at all other detector positions obtained by a simple translation of $C^*(\mathbf{x}_s|\mathbf{x}_d)$ in the horizontal (x, y) plane.

The proposed stochastic sampling algorithm randomly initializes all unknown source parameters in accordance to the prior distribution $p(\Theta|\mathbf{D})$. In the application of the algorithm here, we choose $N_{s,min} = 1$ and $N_{s,\,max} = 8$, with $p^* = 1/7$ in the specification of $p(N_s|I)$.[11] It is noted that our initial specification for the prior distribution of N_s favors the wrong choice for the actual number of sources. The remaining hyperparameters defining $p(\Theta|\mathbf{D})$ are chosen as follows: $\gamma = 0.25$ and $Q_{max} = 100$ g s^{-1} for specification of the prior for the emission rate Q; and, $\mathcal{D} = [0, 100]$ m $\times$ $[0, 500]$ m which is used to define the prior bounds for the location (x_s, y_s) of any source atom.

In addition to the measured d_J and predicted $\overline{C}_J(\Theta)$ concentrations, it is also necessary to specify the noise error variance σ_J^2. The noise error variance includes the sensor sampling error variance, $\sigma_{s,J}^2$, in the measurement of d_J and the model error variance, $\sigma_{m,J}^2$, in the prediction of $\overline{C}_J(\Theta)$. These two contributions to the noise error variance have been included by adding

---

[11] The finite value of $N_{s,max} = 8$ chosen here assumes that the user knows *a priori* that the unknown source distribution consists of a (small) number of discrete point sources. Furthermore, the choice of $p^* = 1/7$ for $p(N_s|I)$ implies that the expected *a priori* number of sources in the domain is $\langle N_s \rangle = N_{s,min} + (N_{s,max} - N_{s,min})p^* = 2$.

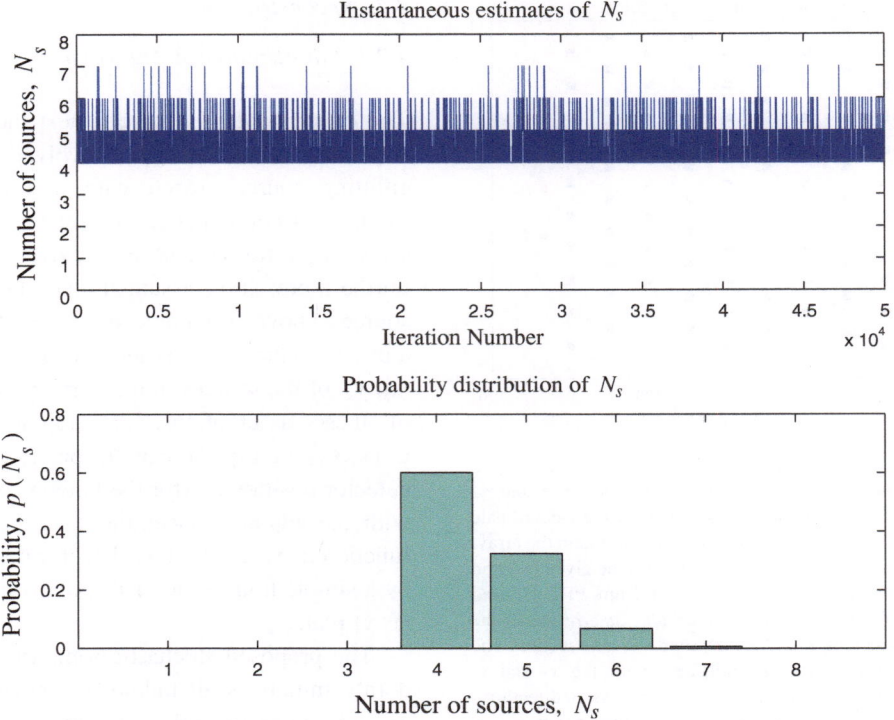

Figure 3

Trace plot (*top*) of the number of discrete sources N_s in the source distribution model samples drawn from $p(\Theta|\mathbf{D}, I)$ during the probabilistic exploration phase of the stochastic sampling algorithm (62 sensors used for source reconstruction), and the corresponding posterior distribution for the number of sources, $p(N_s) \equiv p(N_s|\mathbf{D}, I)$ (*bottom*). In the trace plot, the number of samples displayed has been decimated by a factor of 10

them in quadrature, so $\sigma_J^2 = \sigma_{s,J}^2 + \sigma_{m,J}^2$. The sensor sampling error standard deviation (or square root of the variance) was specified as $\sigma_{s,J} = \max(0.05, 0.02\ d_J)$ ppm$_v$ (parts-per-million by volume), where the lower limit of 0.05 ppm$_v$ represents the precision of the measurements of the concentration using the dPIDs. The model error standard deviation was assumed to be 20% of the predicted value of concentration $\bar{C}_J(\Theta)$ (viz., $\sigma_{m,J} = 0.20\bar{C}_J(\Theta)$ ppm$_v$).

An ensemble of $N_{\mathrm{mem}} = 50$ members of source distribution models Θ were drawn from $p(\Theta|\mathbf{D})$ and used for the simulated annealing phase of the MCMC algorithm. After $\lambda = 1$ was achieved, 1000 further iterations of the RJMCMC algorithm were applied to each of these source distribution model members during the probabilistic exploration phase of the algorithm to give 50,000 samples of source distribution models drawn from the posterior distribution $p(\Theta|\mathbf{D}, I)$.

4.2.2 Results

Figure 3 (top) shows a trace plot for the number of discrete sources in a source distribution model against the sample (or iteration) number for the source molecules drawn from the posterior distribution during the probabilistic exploration phase. The samples of source distribution models drawn from $p(\Theta|\mathbf{D}, I)$ during the probabilistic exploration phase generally mixes well over N_s. In particular, annihilation moves for models from $N_s = 4$ to 3 do not occur. However, dimension-changing moves involving transitions from $N_s = 4$ to 5 (and, vice-versa), as well as higher-order transitions (e.g., from $N_s = 6$ to 7 and its reverse) are seen to occur. Even so, it is seen that the transitions to large values of N_s (e.g., $N_s \approx N_{s,\mathrm{max}}$) are rare and short-lived.

Figure 3 (bottom) displays the marginal posterior distribution $p(N_s) \equiv p(N_s|\mathbf{D}, I)$ for the number of sources.[12] Note that the most probable number of sources for this example is 4 ($\widehat{N}_s \equiv \text{argmax}_{N_s}$ $p(N_s|\mathbf{D}, I) = 4$), which is favoured with a probability of about 0.6. The most probable value for N_s in this case coincides with the correct number of sources $N_s = 4$. This result is obtained in spite of the fact that the prior distribution for N_s was initialized with $p^* = 1/7$ (with an a priori expected number of sources $\langle N_s \rangle = 2$), implying that the algorithm is not sensitive to this hyperparameter.[13] The information embodied in the concentration data was sufficient to move the stochastic simulations toward the more complex model with $N_s = 4$ source atoms.

Figure 4a displays a sample density plot of the locations of the source atoms for all source distribution models drawn from the posterior distribution $p(\Theta|\mathbf{D}, I)$ with $N_s = 4$ (most probable number of sources). Note that there are four clusters of points, with the centroids of each of these clusters coinciding (approximately or better) with the true location of the four sources (see Fig. 2). It is evident that the concentration data $\mathbf{D}$ strongly constrain the y_s-locations of the four sources, but the x_s-locations of these sources are subject to greater uncertainty.

An examination of Fig. 4a suggests that identifiability of the sources can be obtained by ordering the discrete sources of each source distribution model on the y_s-coordinate. More specifically, the labels k for the source atoms in a source distribution model sample are reordered so that the $y_{s,k}$-locations of the atoms verify $y_{s,1} < y_{s,2} < y_{s,3} < y_{s,4}$. With this relabelling of the source atoms in a source molecule, the index k for a source atom follows the labelling of the actual sources shown in Fig. 2.

Figure 4b shows traces of the source parameters associated with each source atom of the source distribution model samples (with $N_s = 4$), after the relabelling of atoms. As can be seen from this figure, the y_s-locations of the source atoms are well separated, but the x_s-locations and emission rates $Q \equiv q_s$ of the atoms are very similar and significantly overlap one another.

Figure 5 shows the marginal posterior distribution (histogram) of the parameters x_s, y_s, and q_s for each of the four source atoms identified in Fig. 4. The posterior mean and standard deviation, as well as the lower and upper bounds for the 95% highest posterior density (HPD) interval,[14] for each of these four identified source atoms are summarized in Table 1. For this example, it is seen that generally the estimates for the source parameters are quite good and, certainly, the true values of the parameters (when these are known) lie within the stated errors.

The mass flow controllers for sources 1, 2, and 4 failed to properly regulate the flow owing to the fact that impurities in the low-grade of propylene used for the experiment contaminated the sensor and electronic circuitry in these controllers. Consequently, the control signals that were used to set the value to regulate the flow rates in these controllers failed to function correctly (and, as a consequence, the actual emission rates for sources 1, 2, and 4 were not known). Nevertheless, an examination of Table 1 shows that the emission rates for the three unregulated sources (viz, sources 1, 2, and 4) (estimated using the source reconstruction algorithm) generally are larger than that for the regulated source (viz, source 3). Finally, measurements of the emission rate from source 3 yielded $Q = 3.8$ g s^{-1}. The estimated value given in Table 1 for this source of $\widehat{q}_s = 4.1 \pm 0.4$ g s^{-1} agrees with the measured emission rate.

As mentioned above, the localization of the source in the x_s-direction (alongwind) is generally poorer than that in the y_s-direction (crosswind). It can be seen that the crosswind locations of the sources

---

[12] The marginal posterior distribution $p(N_s|\mathbf{D}, I)$, which is required for the inference of the number of sources, N_s, is estimated as follows:

$$\hat{p}(N_s|\mathbf{D}, I) = \frac{1}{N_*} \# \left\{ t : N_s^{(t)} = N_s \right\} = \frac{1}{N_*} \sum_{t=1}^{N_*} \mathbb{I}_{N_s}(N_s^{(t)}),$$

where N_* is the number of samples (source molecules) drawn from the Markov chain. Note that this estimate counts the number of source molecules with exactly N_s source atoms and normalizes the count by N_*.

[13] The algorithm was applied with other values for $p^* \in (0, 1)$ with essentially no change in the results with respect to both the best estimates of the source parameters and the uncertainty quantification for these estimates.

[14] A $p\%$ highest posterior density (HPD) interval encloses a source parameter with $p\%$ probability, and is constructed so that the lower and upper bounds of the specified interval are such that the probability density function within the interval is everywhere larger than outside it. This interval can be used as a uncertainty specification for a source parameter.

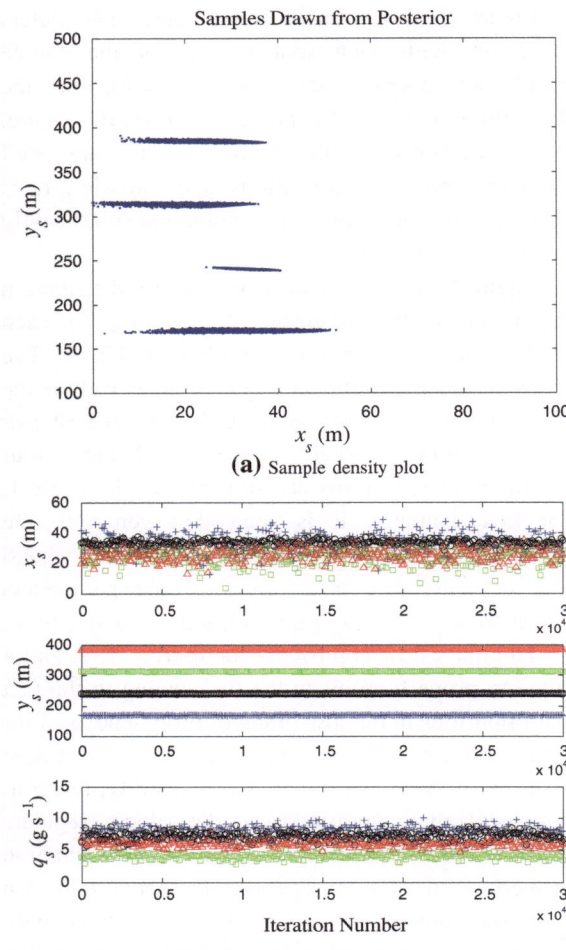

(a) Sample density plot

(b) Trace plots of source parameters

Figure 4

a Density plot consisting of samples of source distribution models obtained for $N_s = 4$ projected onto the (x_s, y_s) subspace. **b** Trace plots of source parameter estimates against sample (or, iteration) number, after relabelling of the source atoms [*plus symbols* (blue): $k = 1$; *circles* (black): $k = 2$; *squares* (green): $k = 3$; *triangles* (red): $k = 4$]. In the trace plots, the number of samples has been decimated by a factor of 100

can be determined with an accuracy of about ± 1 m (standard deviation), but the alongwind locations of the sources can only be determined with an accuracy of about ± 5 m (standard deviation).

Finally, for this example, the information gain obtained from the concentration data **D** was found to be $D_{KL}(1) = 52.5$ natural units (nits), implying that the information contained in the concentration data allowed the posterior volume of the hypothesis space

(volume of hypothesis space of reasonably large plausibility after receipt of the concentration data) to decrease by a factor of $\exp(D_{KL}(1)) \approx 6.3 \times 10^{22}$ relative to the prior volume of the hypothesis space (volume of hypothesis space of reasonably large plausibility before the receipt of the concentration data).

5. Conclusions

In this paper, we have developed and tested an innovative Bayesian method for reconstruction of a source distribution consisting of an unknown number of localized sources. More specifically, Bayesian probability theory was applied to formulate a posterior distribution for the parameters used in the structure-based representation for this source distribution. The evaluation of this posterior distribution in order to extract its features of interest is realized using an efficient computational technology. The computational algorithm uses simulated annealing for the burn-in phase and a reversible-jump MCMC method for the probabilistic exploration phase.

The source reconstruction methodology has been applied successfully to a real dispersion experiment involving a multiple continuous point source release (FFT-07), with measurements of the resulting concentration field obtained from an array of sensors. For this case (involving a small unknown number of localized sources), it has been demonstrated that by calculating the marginal posterior probability distribution of the number of sources from the random samples of source distribution models drawn from the posterior probability density function, a maximum *a posteriori* estimate $\widehat{N}_s$ for the number of sources can be obtained, and samples of source distribution models with exactly $\widehat{N}_s$ discrete sources can then be used to provide best estimates for the source parameters.

As a possible future avenue for investigation, the source reconstruction methodology described herein can be potentially generalized to the reconstruction of an arbitrary (general) source distribution. Towards this objective, it is noted that Eq. 5 generalized to include also time dependence of emission rates can in principle provide a structure-based representation for

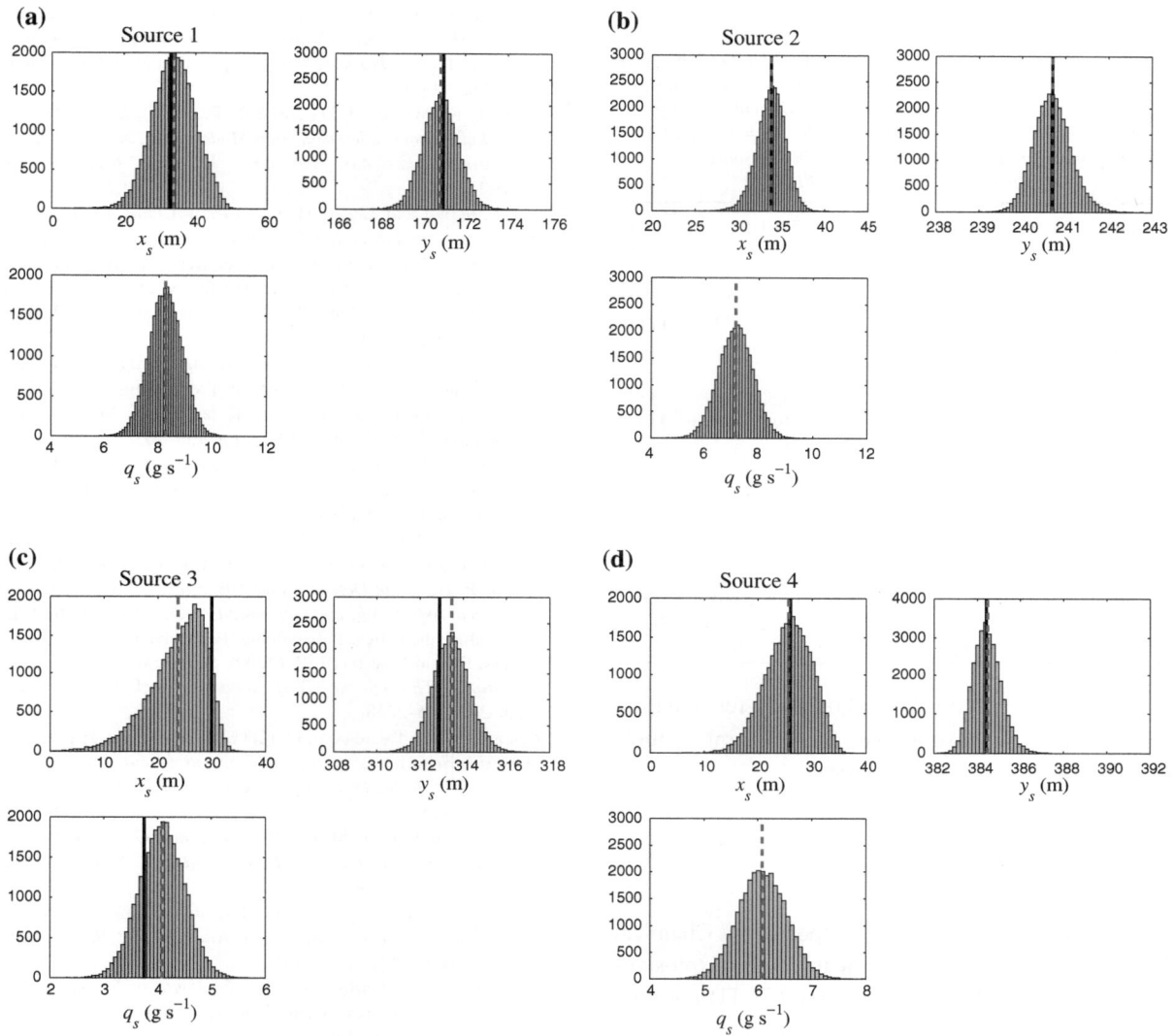

Figure 5

Inference of the discrete source parameters obtained from samples drawn from the posterior distribution $p(\Theta|\mathbf{D}, I)$ having exactly four source atoms. **a, b, c, d** Histograms for the three parameters, namely alongwind location x_s (m), crosswind location y_s (m), and emission rate q_s (g s^{-1}) that characterize atoms 1, 2, 3, and 4 (cf. Fig. 2). In each frame, the *solid vertical line* indicates the true value of the parameter (if known) and the *dashed vertical line* corresponds to the best estimate of the parameter obtained as the posterior mean of the marginal posterior distribution for the parameter

an arbitrary source distribution, whereby the amount of structure required to approximate the geometrical and physical characteristics of the source can be accommodated to any degree of accuracy by varying the number of source atoms used in this representation. More specifically, an arbitrary source distribution of unknown geometry (e.g., area or volume source whose geometry and release history are unknown a priori) can be represented as accurately as

desired by simply allowing N_s to be very large, or even infinite. Consequently, the use of probability theory as logic (Bayesian probability theory) can potentially offer a coherent and rational approach for fusing measured concentration data with model (predicted) concentration data for the reconstruction of arbitrary (general) source distributions. However, much more validation with more complex source distributions will need to be undertaken before the

Table 1

The posterior mean, posterior standard deviation, and lower and upper bounds of the 95% HPD interval of the parameters $x_{s,k}$ (m), $y_{s,k}$ (m), and $q_{s,k}$ (g s^{-1}) for k = 1, 2, 3, and 4 calculated from samples of source distribution models with $N_s = 4$ (the latter corresponding to the most probable number of sources in the domain as inferred from Fig. 3)

Parameter	Mean	Standard deviation	95% HPD	Actual
k = 1				
x_s (m)	34.0	6.0	(22.8, 45.9)	33.0
y_s (m)	170.9	0.8	(169.3, 172.5)	171.0
q_s (g s^{-1})	8.2	0.6	(7.0, 9.4)	–
k = 2				
x_s (m)	33.8	1.6	(30.5, 37.0)	33.8
y_s (m)	240.5	0.4	(239.8, 241.5)	240.7
q_s (g s^{-1})	7.1	0.6	(5.9, 8.4)	–
k = 3				
x_s (m)	23.7	5.2	(13.1, 32.1)	30.0
y_s (m)	313.4	0.9	(311.8, 315.2)	312.9
q_s (g s^{-1})	4.1	0.4	(3.3, 4.9)	3.8
k = 4				
x_s (m)	25.6	4.3	(17.5, 33.8)	26.0
y_s (m)	384.5	0.7	(383.2, 385.8)	384.4
q_s (g s^{-1})	6.1	0.5	(5.2, 7.0)	–

potential of the proposed methodology for quantitative source reconstruction for general source distributions can be fully assessed.

Acknowledgments

This work has been partially supported by Chemical Biological Radiological-Nuclear and Explosives Research and Technology Initiative (CRTI) program under project number CRTI-07-0196TD.

REFERENCES

ALLEN, C. T., YOUNG, G., and HAUPT, S. E. (2007), *Improving Pollutant Source Characterization by Better Estimating Wind Direction with a Genetic Algorithm*, Atmos. Environ. *41*, 2283–2289.

BOCQUET, M. (2005), *Reconstruction of an Atmospheric Tracer Source Using the Principle of Maximum Entropy. I: Theory*, Quart. J. Roy. Meteorol. Soc. *131*, 2191–2208.

CHOW, F. K., KOSOVIC, B., and CHAN, S. (2008), *Source Inversion for Contaminant Plume Dispersion in Urban Environments Using Building-Resolving Simulations*, J. Appl. Meteorol. Climatol. *47*, 1553–1572.

COVER, T. M., and THOMAS, J. A., *Elements of Information Theory* (John Wiley & Sons, Inc., New York 1991).

COX, R. T. (1946), *Probability, Frequency, and Reasonable Expectation*, Am. J. Phys. *14*, 1–13.

FLESCH, T. K., LOWRY, L. A., DESJARDINS, R. L., GAO, Z., and CRENNA, B. P. (2009), *Multi-Source Emission Determination Using an Inverse-Dispersion Technique*, Boundary-Layer Meteorol. *132*, 11–30.

FLESCH, T. K., WILSON, J. D., and YEE, E. (1995), *Backward-Time Lagrangian Stochastic Dispersion Models and Their Application to Estimate Gaseous Emissions*, J. Appl. Meteorol. *34*, 1320–1332.

GAMERMAN, D., and LOPES, H. F., *Markov Chain Monte Carlo: Stochastic Simulation for Bayesian Inference*, Second edition (CRC Press, Chapman and Hall, Boca Raton, Florida 2006).

GELMAN, A., CARLIN, J., STERN, H., and RUBIN, D., *Bayesian Data Analysis*, Second edition (CRC Press, Chapman and Hall, Boca Raton, Florida 2003).

GEYER, C. J., *Markov Chain Monte Carlo Maximum Likelihood*, In Computing Science and Statistics: Proceedings of the 23rd Symposium on the Interface (ed. Keramidas E. M.) (Interface Foundation, Fairfax Station 1991), pp. 156–163.

GILKS, W. R., RICHARDSON, S., and SPIEGELHALTER, D. J., *Markov Chain Monte Carlo In Practice* (CRC Press, Chapman and Hall, Boca Raton, Florida 1996).

GREEN, P. J. (1995), *Reversible Jump MCMC Computation and Bayesian Model Determination*, Biometrika *82*, 711–732.

HANSON, P. R., *Rank-Deficient and Discrete Ill-Posed Problems: Numerical Aspects of Linear Inversion* (Society for Industrial and Applied Mathematics, Philadelphia, Pennsylvania 1998).

HOURDIN, F., and ISSARTEL, J.-P. (2000), *Sub-Surface Nuclear Tests Monitoring Through the CTBT Xenon Network*, Geophys. Res. Lett. *27*, 2245–2248.

HOURDIN, F., and TALAGRAND, O. (2006), *Eulerian Backtracking of Atmospheric Tracers. I: Adjoint Derivation and Parameterization of Subgrid-Scale Transport*, Quart. J. Roy. Meteorol. Soc. *132*, 567–583.

HUERTA, G., and WEST, M. (1999), *Priors and Component Structures in Autoregressive Time Series Models*, J. Roy. Stat. Soc. (Series B) *51*, 881–899.

ISSARTEL, J.-P., and BAVEREL, J. (2003), *Inverse Transport for the Verification of the Comprehensive Nuclear Test Band Treaty*, Atmos. Chem. Phys. *3*, 475–486.

JAYNES, E. T., *Probability Theory: The Logic of Science* (Cambridge University Press, Cambridge, UK 2003).

KEATS, A., YEE, E., and LIEN, F.-S. (2007a), *Bayesian Inference for Source Determination With Applications to a Complex Environment*, Atmos. Environ. *41*, 465–479.

KEATS, A., YEE, E., and LIEN, F.-S. (2007b), *Efficiently Characterizing the Origin and Decay rate of a Nonconservative Scalar Using Probability Theory*, Ecol. Model. *205*, 437–452.

KEATS, A., YEE, E., and LIEN, F.-S. (2010), *Information-Driven Receptor Placement for Contaminant Source Determination*, Environ. Model. Software *25*, 1000–1013.

KITAGAWA, G. (1996), *Monte Carlo Filter and Smoother for Non-Gaussian Nonlinear State Space Models*, J. Comput. Graph. Statist. *5*, 1–25.

KRYSTA, M., and BOCQUET, M. (2007), *Source Reconstruction of an Accidental Radionuclide Release at European Scale*, Quart. J. Roy. Meteorol. Soc. *133*, 529–544.

OBERKAMPF, W. L., and BARONE, M. F. (2006), *Measures of Agreement Between Computation and Experiment: Validation Metrics*, J. Comput. Phys. *217*, 5–36.

RAO, K. S. (2005), *Uncertainty Analysis in Atmospheric Dispersion Modeling*, Pure Appl. Geophys. *162*, 1893–1917.

RICHARDSON, S., and GREEN, P. (1997), *Reversible Jump Markov Chain Monte Carlo Computation and Bayesian Model Determination (With Discussion)*, J. Roy. Stat. Soc. (Series B) *59*, 731–758.

ROBERTSON, L., and LANGNER, J. (1998), *Source Function Estimate by Means of Variational Data Assimilation Applied to the ETEX-1 Tracer Experiment*, Atmos. Environ. *32*, 4219–4225.

ROBERTSON, L., and PERSSON, C. (1993), *Attempts to Apply Four-Dimensional Data Assimilation of Radiological Data Using the Adjoint Technique*, Rad. Protect. Dos. *50*, 333–337.

SEIBERT, P., *Methods for Source Determination in the Context of the CTBT Radionuclide Monitoring System*, In Informal Workshop on Meteorological Modeling in Support of CTBT Verification, 2000, 6 pp.

SEIBERT, P., and STOHL, A., *Inverse Modeling of the ETEX-1 Release With a Lagrangian Particle Model*, In Proceedings of the Third GLOREAM Workshop (eds. Barone G., Builtjes P., and Giunta, G.) (2000) pp. 95–105.

STULL, R. B., *An Introduction to Boundary-Layer Meteorology* (Kluwer Academic Publishers, Dordrecht, The Netherlands, 1988).

THOMSON, D. J. (1987), *Criteria for the Selection of Stochastic Models of Particle Trajectories in Turbulent Flows*, J. Fluid Mech. *180*, 529–556.

THOMSON, L. C., HIRST, B., GIBSON, G., GILLESPIE, S., JONATHAN, P., Skeldon, K. D., and Padgett, M. J. (2007), *An Improved Algorithm for Locating a Gas Source Using Inverse Methods*, Atmos. Environ. *41*, 1128–1134.

TIKHONOV, A. N., and ARSENIN, V. Y., *Solutions of Ill-Posed Problems* (Wiley, New York, 1977).

VAN DOP, H., ADDIS, R., FRASER, G., GIRARDI, F., GRAZIANI, G., INOUE, Y., KELLY, N., KLUG, W., KULMALA, A., NODOP, K., and PRETEL, J. (1998), *ETEX: A European Tracer Experiment; Observations, Dispersion Modeling and Emergency Response*, Atmos. Environ. *32*, 4089–4094.

VON DER LINDEN, W., FISCHER, R., and DOSE, V., *Evidence Integrals*, In Maximum Entropy and Bayesian Methods, (eds. Hanson K. M., and Silver R. N.) (Kluwer Academic Publishers, Dordrecht, The Netherlands 1996), pp. 443–450.

WILSON, J. D., FLESCH, T. K., and HARPER, L. A. (2001), *Micro-Meteorological Methods for Estimating Surface Exchange With a Disturbed Wind Flow*, Agric. Forest Meteorol. *107*, 207–225.

YEE, E., *Probabilistic Inference: An Application to the Inverse Problem of Source Function Estimation*, The Technical Cooperation Program (TTCP) Chemical and Biological Defence (CBD) Group Technical Panel 9 (TP-9) Annual Meeting, Defence Science and Technology Organization, Melbourne, Australia, 2005.

YEE, E., *A Bayesian Approach for Reconstruction of the Characteristics of a Localized Pollutant Source from a Small Number of Concentration Measurements Obtained by Spatially Distributed "Electronic Noses"*, Russian-Canadian Workshop on Modeling of Atmospheric Dispersion of Weapon Agents, Karpov Institute of Physical Chemistry, Moscow, Russia, 2006.

YEE, E., *Bayesian Probabilistic Approach for Inverse Source Determination from Limited and Noisy Chemical or Biological Sensor Concentration Measurements*, In Proceedings of SPIE, Chemical and Biological Sensing VIII, **6554** (ed. Augustus W. Fountain III) (2007) 12 pp.

YEE, E., *Inverse Dispersion of an Unknown Number of Contaminant Sources*, In 15th Joint Conference on the Applications of Air Pollution Meteorology with the A&WMA, New Orleans, LA, Paper 7.1 (2008a) 17 pp.

YEE, E. (2008b), *Theory for Reconstruction of an Unknown Number of Contaminant Sources Using Probabilistic Inference*, Boundary-Layer Meteorol. *127*, 359–394.

YEE, E. (2010), *An Operational Implementation of a CBRN Sensor-Driven Modeling Paradigm for Stochastic Event Reconstruction*, DRDC Suffield TR 2010–070, Defence R&D Canada – Suffield, 68 pp.

YEE, E. and FLESCH, T. K. (2010), *Inference of Emission Rates from Multiple Sources using Bayesian Probability Theory*, J. Environ. Monit. *12*, 622–634.

YEE, E., LIEN, F.-S., KEATS, A., and D'AMOURS, R. (2008), *Bayesian Inversion of Concentration Data: Source Reconstruction in the Adjoint Representation of Atmospheric Diffusion*, J. Wind Eng. Ind. Aerodyn. *96*, 1805–1816.

(Received November 1, 2010, accepted February 26, 2011, Published online August 5, 2011)

Pure Appl. Geophys. 169 (2012), 519–537
© 2011 Springer Basel AG
DOI 10.1007/s00024-011-0385-0

A Genetic Algorithm Variational Approach to Data Assimilation and Application to Volcanic Emissions

KERRIE J. SCHMEHL,[1] SUE ELLEN HAUPT,[2] and MICHAEL J. PAVOLONIS[3]

Abstract—Variational data assimilation methods optimize the match between an observed and a predicted field. These methods normally require information on error variances of both the analysis and the observations, which are sometimes difficult to obtain for transport and dispersion problems. Here, the variational problem is set up as a minimization problem that directly minimizes the root mean squared error of the difference between the observations and the prediction. In the context of atmospheric transport and dispersion, the solution of this optimization problem requires a robust technique. A genetic algorithm (GA) is used here for that solution, forming the GA-Variational (GA-Var) technique. The philosophy and formulation of the technique is described here. An advantage of the technique includes that it does not require observation or analysis error covariances nor information about any variables that are not directly assimilated. It can be employed in the context of either a forward assimilation problem or used to retrieve unknown source or meteorological information by solving the inverse problem. The details of the method are reviewed. As an example application, GA-Var is demonstrated for predicting the plume from a volcanic eruption. First the technique is employed to retrieve the unknown emission rate and the steering winds of the volcanic plume. Then that information is assimilated into a forward prediction of its transport and dispersion. Concentration data are derived from satellite data to determine the observed ash concentrations. A case study is made of the March 2009 eruption of Mount Redoubt in Alaska. The GA-Var technique is able to determine a wind speed and direction that matches the observations well and a reasonable emission rate.

Key words: GA-Var, data assimilation, source term estimation, genetic algorithm, volcanic eruption.

1. Introduction

The field of meteorology has been a breeding ground for data assimilation techniques because of the pressing need to provide accurate forecasts to the public. The atmosphere, however, is a nonlinear dissipative system with chaotic behavior that displays sensitivity to initial conditions (LORENZ, 1963). Thus, robust assimilation methods are critical in order to incorporate available observational data to nudge the solution toward the correct portion of the dynamic attractor. Meteorologists have developed multiple techniques to assimilate data into models, such as those reviewed by DALEY (1991), KALNAY (2003), IDE et al. (1997), and LEWIS et al. (2006), among others. One technique that has become popular is the variational approach (SASAKI, 1970). It has been widely applied in both its three dimensional (3D-Var) and four dimensional (4D-Var) incarnations (LEWIS and DERBER, 1985; TALAGRAND and COURTIER, 1987; RABIER and COURTIER, 1992; ERRICO et al., 1993; COURTIER, 1997; LEWIS et al., 2006). The theoretical derivation is elegant, although difficult to tailor to specific equations. It is considered state-of-the-art, and is used operationally by the European Center for Medium Range Forecasting (ECMWF) (http://www.ecmwf.int/products/). A disadvantage, however, is that it requires accurate estimates of both the analysis and observational error covariances. These covariance matrices are often difficult to specify with the required degree of accuracy, in particular for problems that involve short-term releases of contaminants.

Problems in atmospheric transport and dispersion are also amenable to improved solutions via data assimilation (DENG et al., 2004; KRYSTA et al., 2006; HAUPT et al., 2009b; among others). For instance,

---
[1] Applied Research Laboratory, The Pennsylvania State University, P.O. Box 30, State College, PA 16804-0030, USA.
[2] Research Applications Laboratory, National Center for Atmospheric Research, 3450 Mitchell Lane, Boulder, CO 80301, USA. E-mail: haupts2@asme.org
[3] National Oceanic and Atmospheric Administration, National Environmental Satellite, Data, and Information Service, 1225 W. Dayton St., Madison, WI 53706, USA.

measurements of contaminant concentration may be available that can be used to improve subsequent predictions. The problem, however, is that the physics is governed by a one-way coupled system: the flow field is governed by the Navier–Stokes equations, which are self contained, while the concentration is modeled by the advective dispersion equation, which includes the wind variables. A simplified version of this system is:

$$\frac{\partial \vec{v}}{\partial t} = \mathbf{M}_v(\vec{v})\vec{v} + \eta_v + G_v(\vec{v}^o, \vec{v}^f, C^o, C^f) \qquad (1a)$$

$$\frac{\partial C}{\partial t} = \mathbf{M}_C(\vec{v})C + \eta_C + G_C(\vec{v}^o, \vec{v}^f, C^o, C^f) \qquad (1b)$$

where $\vec{v}$ denotes a continuous two- or three-dimensional wind field and C is the two- or three-dimensional concentration field. Subscripts v and C on the two dynamics operators, $\mathbf{M}_v$ and $\mathbf{M}_C$, denote separation into a wind operator and a concentration operator. G is an assimilation function that is added to include the most recent observations. The wind equation (1a) depends on the previous state of the wind field and the concentration equation (1b) similarly depends on the previous state of the concentration field. Although both operators are functions of the wind, there is no direct influence of the concentration on the wind field through the dynamics operator. We do have an indirect impact, however, through the assimilation function $G_C(\vec{v}^o, \vec{v}^f, C^o, C^f)$. These assimilation functions depend on both the forecast (superscript f) and the observed (superscript o) fields.

Thus, although the wind field governed by the momentum and continuity equations transports the contaminant in the dispersion equation, the concentration does not feed back into the wind equations. Therefore, if only concentration observations are available, they can be used directly only in the dispersion equation. If one wishes to retrieve the wind field from concentration observations only (assuming that no wind observations are available) then the wind field must be inferred by the changes of concentration pattern in time. Such a procedure requires a new approach. The Genetic Algorithm-Variational (GA-Var) technique was developed and has evolved into a technique to address such atmospheric transport and dispersion problems.

The purpose of this paper is to review the GA-Var technique and its use in atmospheric transport and dispersion problems (Sect. 2). Both forward assimilation problems and inverse or retrieval problems (source term estimation) are discussed. A source term estimation problem and its solution via GA-Var as well as its subsequent assimilation into the dispersion model are presented here. Specifically, GA-Var is applied to estimate the emission rate of a volcanic event and the transporting wind speed and direction of the resulting ash cloud given satellite observations. The problem and the procedure for the solution are presented in Sect. 3, followed by the results of a case study in Sects. 4 and 5. These sections present a progression of the modeling study that begins with a basic application, sets up a synthetic environment to analyze system behavior, and then presents a more refined approach to the volcanic emission retrieval problem. Section 6 summarizes the conclusions, evaluates the use of GA-Var, and discusses the prospects for future applications.

2. The Genetic Algorithm Variational Approach

2.1. Technique Description

The central issue of the GA-Var technique is to compare a predicted concentration with an observed concentration by evaluating a cost function. Typically, each of these concentrations are field variables and the cost function sums the differences between the two over all locations in the field. The observed concentrations are derived from sensor field measurements. The predicted concentrations are those forecast by a transport and dispersion model. That model requires input data, namely information regarding the source of the contaminant (source location, effective source height, rate of contaminant release, etc.) and meteorological information (such as wind speed, wind direction, indication of atmospheric stability, depth of the boundary layer). The problem is that some of these data may not be available, which is the rationale for using an optimization method to retrieve, or estimate, the unknown variables. Thus, one must use some method to generate guesses to those data to feed into the dispersion model, and then iteratively improve those estimates. The problem is nonlinear in some of the variables, and thus requires a very robust optimization method to update the estimates. The method chosen here is a genetic algorithm, which has proven its worth on both these and other difficult optimization problems.

The GA-Var technique can be applied to either the forward atmospheric transport and dispersion problem, or the inversion or backward problem, that is, estimating the source characteristics. The forward problem applies when one knows the information about the source and uses observations of concentration and wind variables to improve the subsequent calculation of transport and dispersion. The forward problems are often applied dynamically; that is, one estimates the variables at the current time then uses them to calculate the next step in the transport and dispersion. For the source term estimation problem, however, one may not know the source information in advance. For the source apportionment subclass of these problems, one knows the location of the source but not the emission rate. There are more complex situations where the location of the source may not even be known, such as in the case of an intentional release by a terrorist. Such cases require solution techniques to retrieve all source term information as well as any necessary meteorological variables. It is typical for the source term estimation problem to use all available temporal data summed together to estimate the best source variables.

2.2. The Solution Technique: A Genetic Algorithm

The GA is an artificial intelligence optimization method inspired by the biological processes of genetic recombination and evolution. It begins with a population of potential solutions and evolves them closer to the correct solution through implementing two genetic operations: mating and mutation. Although we briefly review the GA procedure, we refer the reader to the GA literature (HOLLAND, 1975; DE JONG, 1975; GOLDBERG, 1989; DAVIS, 1991; GORDON and WITLEY, 1993; MITCHELL, 1996; HAUPT and HAUPT, 2004; among others) for more details and flavors of evolutionary algorithms in general. The genetic algorithm flavor that we have found useful for variational assimilation is the continuous parameter

genetic algorithm, which is described briefly here and in more detail by HAUPT and HAUPT (2004).

A first requirement in any optimization technique is to define the quantity to be minimized,[1] known as the *cost function* or *objective function*. Here we choose to minimize the difference between the observed concentration and the modeled concentration, summed over the entire field. This difference is often called the *innovation* in the context of assimilation. A formulation in which this innovation is normalized is:

$$\text{cost} = \frac{\sqrt{\sum_{s=1}^{\text{TS}} [O_s - C_s]^2}}{\sqrt{\sum_{s=1}^{\text{TS}} [O_s] \sum_{s=1}^{\text{TS}} [C_s]}} \qquad (2)$$

where C_s is the forecast concentration at sensor, s; O_s is the observed concentration at sensor, s; and TS is the total number of sensors

Note that in some cases it is more efficient to take the logarithm of the observed and predicted concentrations so as to better measure the spread of the contaminant.

We wish to minimize the cost in (2) using a continuous parameter GA; that is, one in which the parameters are real numbers. Figure 1 flowcharts the GA solution process. The genetic algorithm begins with a population of randomly initialized solution vectors[2] called *chromosomes*. The chromosomes comprises *genes*, each of which represents one variable to be optimized. Each gene is one real-valued variable that is sought, such as the emission rate of a contaminant or its location, which would be represented as coordinates in a pre-defined Cartesian grid space. Figure 2a depicts the way the genes are assembled into a chromosome. The next step of the GA (Figs. 1, 2b) is to evaluate the quality or *fitness* of each member of the population of chromosomes. This requires applying the dispersion model using the guessed variables that have been coded into the genes that form the chromosome. The field of concentration values produced by the dispersion model can then be compared to the field of observed concentration

---

[1] One can just as easily maximize a quantity, but the problem here is cast as one in minimization.
[2] These solution vectors are guesses, within predefined bounds, to the unknown variables.

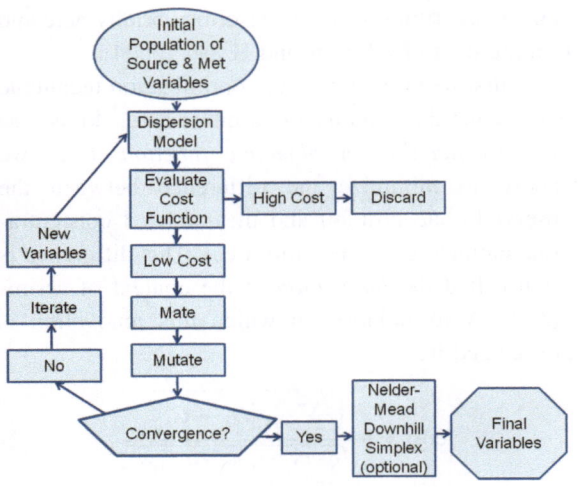

Figure 1
Flowchart of genetic algorithm as used in the GA-Var model

values via the cost function (2) (Figs. 1, 2b). The value of each chromosome is evaluated and ranked to produce a fitness. The ranked chromosomes are then sorted from best to worst for evaluation; that is, from the *most fit* that will progress to the next step, to the *least fit*. The least fit chromosomes are discarded from the population (typically 50%) while the most fit ones continue in the evolution process.

The GA replaces those discarded chromosomes with newly generated chromosomes that blend the information from two highly ranked chromosomes using an operator called *mating*. First the parents must be selected, usually by a scheme that exploits the cost information on their relative fitness (Fig. 2c). Then the information in the parent chromosomes is combined and swapped to produce offspring chromosomes.

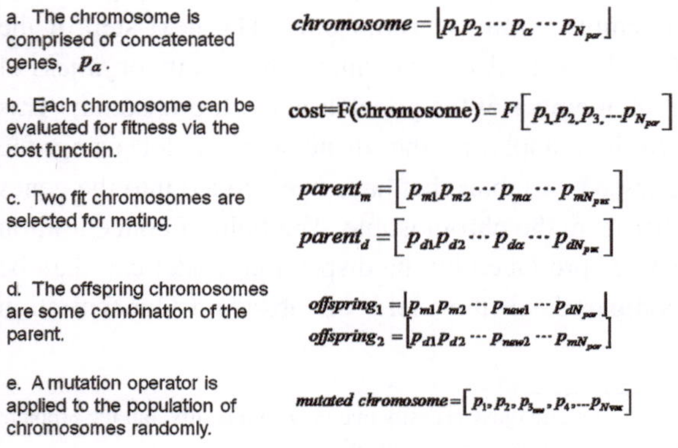

The second GA operator is *mutation*, which replaces individual gene values with new random values (Fig. 2e). The mutation process enables the algorithm to continue to search the entire solution space rather than converge to a local minimum. We employ *elitism* to retain the best solution in the population.

Each round of evaluating the cost function, discarding unfit chromosomes, selecting parent chromosomes, mating, and mutating constitutes one GA *generation*. We run the GA for a pre-determined number of iterations (or generations), or until convergence has occurred. The number of generations, mutation rate, selection rate, and population size are all parameters to be tuned and depend on the problem. This issue is an active topic of research (HAUPT and HAUPT, 2000, 2004) and has been fine-tuned for this problem (ALLEN et al., 2007a; KUROKI et al., 2010).

One way to efficiently fine-tune a solution is to use a hybrid GA, which uses the GA to find the correct solution basin, then applies some gradient descent method, typically the Nelder–Mead Downhill Simplex (NMDS) method, to complete finding the minimum point of that basin. The rationale for this combination is that the GA is sufficiently robust to usually find the basin of the global minima in a reasonable number of iterations. Once that basin is identified, however, the NMDS finds the bottom of that basin more rapidly. As demonstrated in HAUPT et al. (2007), the NMDS method alone is not reliable for finding the global minimum.

2.3. Review of the uses of the GA-Var Approach

The GA-Var technique was originally formulated to apportion source emission rates to a number of sources whose location is known (HAUPT, 2005). The general class of models typically used for this procedure is the receptor model. Receptor models are formulated to begin with contaminant concentration data from one or more receptors and project that information backward to retrieve the source information. This is in contrast to forward transport and dispersion models, which start with the source

a. The chromosome is comprised of concatenated genes, p_α.

$$chromosome = \left[p_1 p_2 \cdots p_\alpha \cdots p_{N_{par}} \right]$$

b. Each chromosome can be evaluated for fitness via the cost function.

$$cost = F(chromosome) = F\left[p_1 p_2 p_3 \cdots p_{N_{par}} \right]$$

c. Two fit chromosomes are selected for mating.

$$parent_m = \left[p_{m1} p_{m2} \cdots p_{m\alpha} \cdots p_{mN_{par}} \right]$$
$$parent_d = \left[p_{d1} p_{d2} \cdots p_{d\alpha} \cdots p_{dN_{par}} \right]$$

d. The offspring chromosomes are some combination of the parent.

$$offspring_1 = \left[p_{m1} p_{m2} \cdots p_{mnew1} \cdots p_{dN_{par}} \right]$$
$$offspring_2 = \left[p_{d1} p_{d2} \cdots p_{mnew2} \cdots p_{mN_{par}} \right]$$

e. A mutation operator is applied to the population of chromosomes randomly.

$$mutated\ chromosome = \left[p_1 p_2 p_{3new} p_4 \cdots p_{Nvar} \right]$$

Figure 2
A chromosome level view of the GA process

characteristics and meteorological conditions then use physical, mathematical, and chemical calculations to predict contaminant concentration at some distance downwind from the source. Important input for these dispersion models includes information about the emissions from the source, the local atmospheric conditions, and the geographical characterization. Both types of models are highly developed and versions of them are widely used for diagnosis and prediction of atmospheric contaminant transport events (EPA, 2003). Combining the technology of the forward-looking dispersion models and backward-looking receptor models enables using monitored concentrations to estimate unknown source and meteorological variables. This coupled model approach integrates the physical basis of the dispersion calculations with the ground truth of the actual monitored pollutant concentrations.

It has been demonstrated that coupling receptor models with dispersion models using a GA is an effective tool for attributing concentration contribution at a receptor to each of a specified number of sources (HAUPT, 2005; HAUPT et al., 2006, 2009a). GA-Var has also been applied with more sophisticated dispersion models such as the operational puff model, SCIPUFF (ALLEN et al., 2007a) and tested on field trial data (ALLEN et al., 2007b).

Previous work has indicated that the measured wind direction is not always representative of the transporting wind or is not available at all (KRYSTA et al., 2006; ALLEN et al., 2007b). Therefore, this variable has been added to the list of unknown variables to optimize. This method was systematically analyzed for sensitivity to formulation (LONG et al. 2010a).

The GA-Var technique is sufficiently general to apply to a wide range of problems on different temporal and spatial scales (HAUPT et al., 2009a, b, 2010; RODRIGUEZ et al., 2010a; KUROKI et al., 2010; ANNUNZIO et al., 2011). It has been tested and used to study details of small-scale emission scenarios such as for the 0.5 × 0.5 km sensor array of the Fusion Field Trial 2007 (FFT07) field experiment (RODRIGUEZ et al., 2010b). It was applied at local scales (10 s of km) when testing on the Dipole Pride data (ALLEN et al., 2007a). Other contemporary demonstrations of using evolutionary strategies to optimize the source

term and meteorological data have been accomplished by CERVONE and FRANZESE (2010) and CERVONE et al. (2010). Here we demonstrate that the technique is also applicable to regional and larger scales by analyzing volcanic emissions as observed from a satellite (LONG et al., 2010b).

3. Example Application of GA-Var to Satellite Data for a Volcanic Eruptive Event

3.1. Background

The 1989–1990 eruption of Mount Redoubt and the resulting damage to aircraft prompted the formation of a multidisciplinary, international group to address the effects of volcanic activity on aviation safety (CASADEVALL, 1994). The findings of that group highlighted the importance of the timely and accurate exchange of information between international agencies and the pilots (CASADEVALL, 1994). In addition, the group stressed the importance of understanding volcanoes, the airspace near the volcanoes, and the detection and tracking of the ash clouds (CASADEVALL, 1994). In 1998, nine Volcanic Ash Advisory Centers (VAACs) were established to facilitate the exchange of information between volcano observatories, meteorological agencies and air traffic control centers (OFCM, 2004). While the monitoring and tracking of volcanic ash clouds has improved, the safest mitigation strategy is for aircraft to avoid volcanic ash clouds entirely (CASADEVALL, 1994; OFCM, 2004; FOX, 2009; CHIVERS, 2010; LANGSTON, 2010). Thus, there is a need to predict the atmospheric transport and dispersion of these ash clouds. A number of dispersion models, such as the Numerical Atmospheric-dispersion Modelling Environment (NAME) (RYALL and MARYON, 1998), a Lagrangian Trajectory Volcanic Ash Tacking Model (PUFF) (SEARCY et al., 1998), the Hybrid Single Particle Lagrangian Integrated Trajectory Model (HYSPLIT) (DRAXLER and HESS, 1998), the Canadian Emergency Response Model (CANERM) (PUDYKIEWICZ, 1989; D'ARMOURS, 1998), and other models are in use by Volcanic Ash Advisory Centers to accomplish these predictions. Atmospheric transport and dispersion models used for predicting ash cloud movement rely on accurate

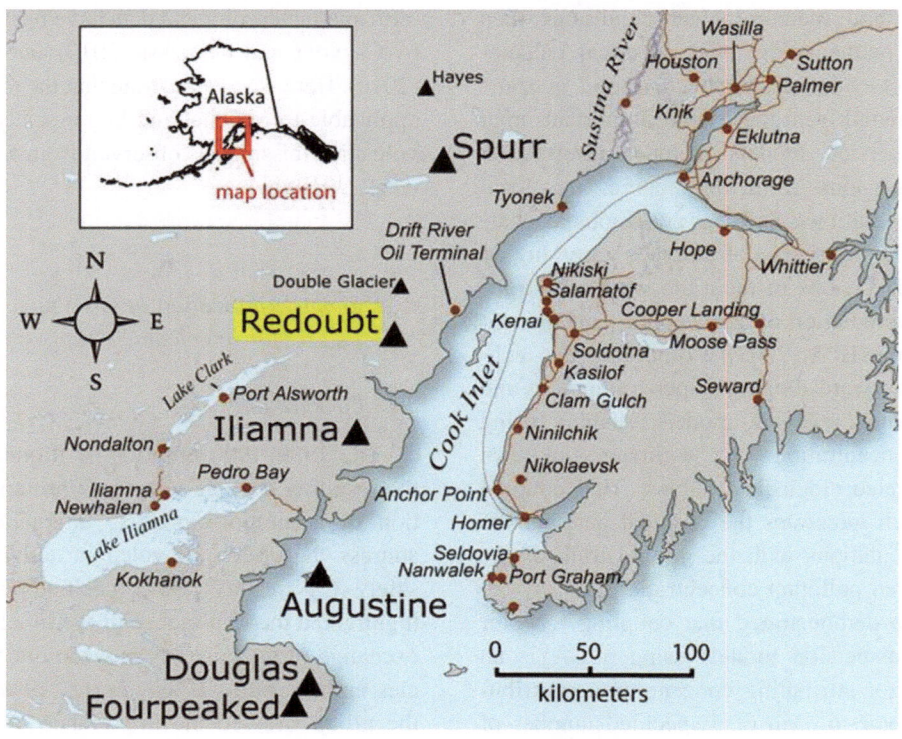

Figure 3
Location of Mount Redoubt in relation to other Alaskan towns. Image courtesy of the AVO/ADGGS. Prepared by Janet Schaefer of the Alaska Volcano Observatory/Alaska Division of Geological & Geophysical Surveys, September 26, 2008

knowledge of the source term and meteorology variables, particularly the emission rate and the transporting wind, in order to make a prediction about the future state of the cloud.

One of the difficulties, however, is estimating a good emission rate to use for that modeling (Michael Richards, personal communication). In this study we focus specifically on retrieving the emission rate that represents the amount of aerosols being pumped into the atmosphere. We have also found that it is necessary to back-calculate the wind speed and wind direction governing the representative ash cloud transport. A method is presented that applies the GA-Var technique introduced in Sect. 2 to observational data in order to retrieve the source term parameters and representative wind information.

The general approach for volcanic source term estimation and the subsequent prediction of ash cloud transport uses the GA-Var algorithm and satellite observations. Following a volcanic eruption a satellite detects the ash cloud. That satellite data is

processed and used as observed concentration data in our model. We then apply our GA-Var technique to retrieve the pertinent information regarding the emission rate of the volcano as well as the wind speed and direction. With that information we can refine the forecast movement of the ash cloud and provide a more accurate forecast to warn and reroute aircraft in the region.

3.2. Case Study

A case study is made of events 1–5 of the March 2009 eruption of Mount Redoubt in Alaska. Located in southern Alaska along the Cook Inlet (Fig. 3), Mount Redoubt is a stratovolcano that rises to 3,108 m above sea level (SIEBERT and SIMKIN, 2002). Event 1 began at approximately 0634 UTC on Monday, 23 March (2234 AKDT on Sunday, 22 March). There were a series of five events that each lasted from 2 min to nearly 40 min as detailed in Table 1. The fifth event ended at approximately 1300

UTC (0500 AKDT) on Monday, 23 March (AVO/USGS 2010). Although the volcano did not emit a continuous rate of ash into the atmosphere, we begin by exploring modeling it as a single, continuous, uniform eruption that lasted for 6.5 h in Sect. 4. In Sect. 5 we refine the approach by modeling the five events separately. Transport of the ash cloud occurs over the entire forecast period. The bulk of the ash was estimated to lie between 7.6 and 9.1 km above sea level (AVO/USGS 2010).

3.3. Satellite Data

The satellite data used in this study are from the Advanced Very High Resolution Radiometer (AVHRR), which is a five-channel (0.65, 0.86, 3.75, 11, and 12 μm) instrument on-board a low-earth orbit satellite. The AVHRR data are used to identify volcanic ash clouds and estimate ash cloud height, mass loading (mass per unit area), and effective particle radius. Since the methodology used to identify volcanic ash is described in PAVOLONIS et al. (2006) and PAVOLONIS (2010), and the cloud property retrieval algorithm is described in HEIDINGER and PAVOLONIS (2009), only a brief summary is given here. Satellite pixels that contain volcanic ash are first identified using the ratio of effective absorption optical depth at 11 and 12 μm (see PAVOLONIS 2010)

and near-infrared measurements (see PAVOLONIS et al., 2006). For each pixel where ash is detected, a bi-spectral (11 and 12 μm) optimal estimation technique is used to retrieve the ash cloud temperature, cloud emissivity, and microphysical parameters (see HEIDINGER and PAVOLONIS, 2009 for details). The ash cloud temperature is converted into an ash cloud height using atmospheric profiles of temperature and height from the Global Forecast Model (GFS) and the cloud emissivity and microphysical parameter are used to determine the volcanic ash mass loading and effective particle radius by applying the same microphysical assumptions described in WEN and ROSE (1994). WEN and ROSE (1994) and PRATA and GRANT (2001) showed that the relative error in the volcanic ash mass loading is about 50%. Figure 4 illustrates the estimated mass loading (a), the height of the cloud ash (b), and the effective particle radius (c). The location of Mount Redoubt is indicated by the inverted triangle in the lower left of Fig. 4. The mass loading provided by the retrieval is given in Ton km^{-2}. We convert this value into concentration using an assumption for cloud thickness found in PRATA and GRANT (2001) where the vertical thickness, dz, is approximated as 0.4 times the height of cloud top, which is determined from Fig. 4b. PRATA and GRANT use this relationship based on work done by BRIGGS (1975) and MANINS (1985) relating smoke

Table 1

Source parameter information provided by the Alaska Division of Geological & Geophysical Surveys Report of Investigations 2011

Event number[a]	Date (UTC)	Time official[b] (UTC)	Duration (seismic SPU[c], min)	Plume height[d] (ft)
Event 1	3/23/2009	0634	2	18,000
Event 2	3/23/2009	0702	7	44,000
Event 3	3/23/2009	0814	20	48,000
Event 4	3/23/2009	0938	38	43,000
Reanalysis	3/23/2009	0948	Undefined	45,000
Reanalysis	3/23/2009	1052	8	Undefined
Event 5	3/23/2009	1230	20	60,000
Reanalysis	3/23/2009	1258	3	Undefined

Adapted from the 2009 Eruption of Redoubt Volcano, Alaska (SCHAEFER 2011)

[a] Event numbers are defined by explosions where a VAN/VONA was issued; "reanalysis" refers to explosions that were interpreted by post-event reanalysis of seismic data; some reanalysis events may be considered pulses of the prior event, but others are unique events between larger signals that were buried in the data and not recognized at the time of initial analysis

[b] Official onset times were derived from seismic signal analysis

[c] Duration reflects the time period at distal station SPU when the signal is twice the background and is rounded to the nearest minute. This is the same reference as used in 1989–1990 eruption

[d] Plume heights vary slightly depending on data source; only maximum plume heights are listed here

Figure 4 ▶

The mass loading in Ton km^{-2} (**a**), height of the ash cloud in km (**b**), and radius of the ash in μm (**c**) derived from the satellite retrieval. The location of Mount Redoubt is indicated by an *inverted triangle*

plume rise and vertical extent. The authors acknowledge that this is a broad assumption considering the highly variable nature of the ash cloud both temporally and spatially. Further study is needed to determine how applicable this assumption is for volcanic ash plumes and to fully quantify the uncertainties inherent in such an assumption but it appears reasonable for the preliminary application here. For this study, we focus on concentration values derived at a height of 6 km, which is where the bulk of the plume lies as indicated by Fig. 4b.

3.4. Dispersion Model

The atmospheric transport and dispersion of the ash cloud is predicted using the Second-Order Closure Integrated PUFF (SCIPUFF) model. SCIPUFF is a sophisticated puff-based transport and dispersion model that accounts for turbulence, terrain, and weather effects in its calculations (SYKES, 2004). SCIPUFF tracks individual puffs, evolves the dispersion coefficients, splits and merges the puffs, and incorporates advanced methods to assess turbulence levels. SCIPUFF allows the user to specify the type of material being released and then define general characteristics such as density and particle size. For this case, we model the transport and dispersion of the ash particles ranging in size from 0.10 to 100 μm (commensurate with the values indicated in Fig. 4c) and having a density of 2,600 kg m^{-3} (Scott and McGIMSEY, 1994). Note that we do not include effects resulting from chemical reactions.

3.5. Application Details

We wish to apply GA-Var to a single level of data. We have shown that even when a problem is treated as two dimensional, we can retrieve information regarding the vertical dispersion (HAUPT *et al.*, 2010). Thus, because we wish to directly match the predicted concentrations with the satellite data (Fig. 4b) which estimates the effective level of

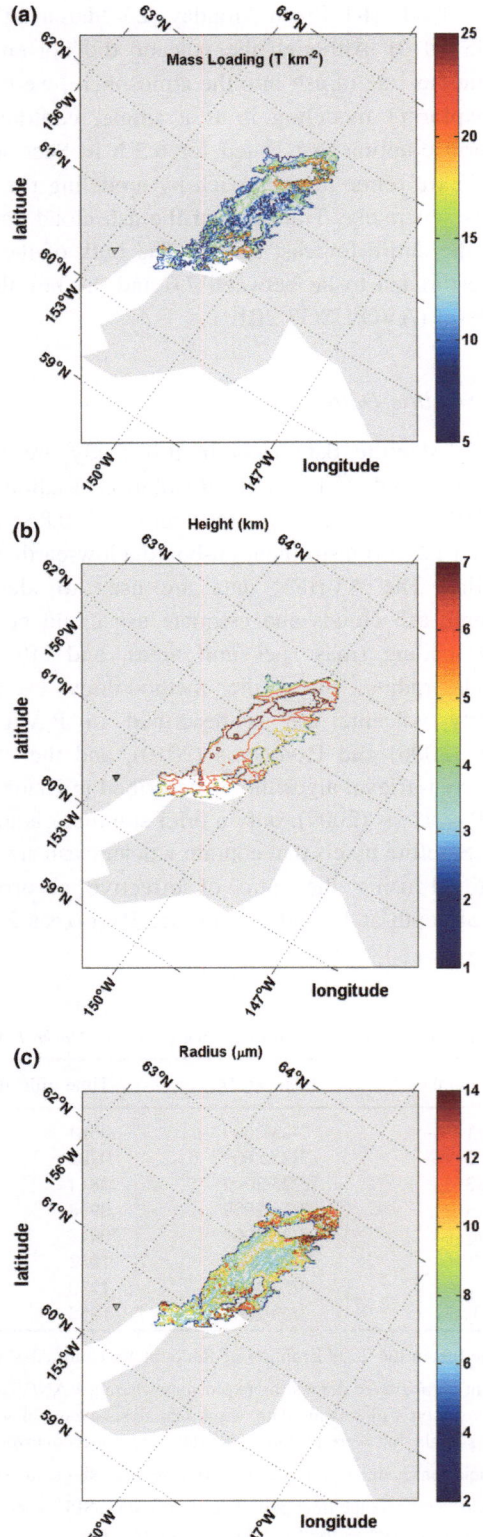

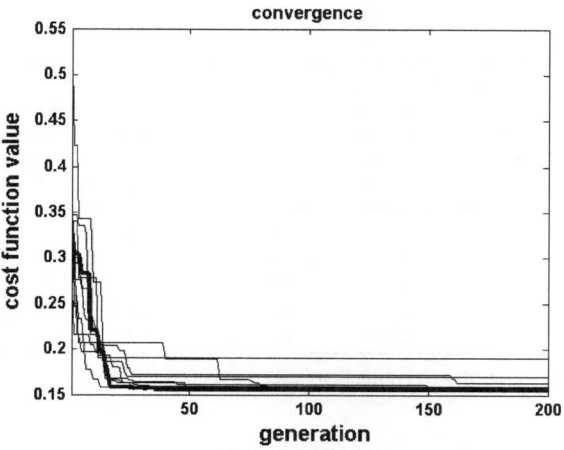

Figure 5

GA convergence for each of the ten runs. The best solution or the run that ultimately yielded the lowest cost function value is indicated by the *heavy black line*

4. Results

4.1. Retrieval Problem

The GA is set up with a population of 64 chromosomes and is run for 200 iterations. Half of the population is selected to participate in mating while the other half is discarded. We use a mutation rate of 20%. Because we are considering a single integrated level of concentration data, it is sufficient to consider a single representative wind speed and direction for the purpose of the retrieval. The transporting wind (speed and direction), along with the emission rate, are the unknown variables, or genes, in the source term description. The GA searches a range of potential values for each of the genes. The emission rate can vary from 1×10^3 to 1×10^7 kg s^{-1}, the wind direction can span the entire range of values, 0°–360°, and the wind speed can vary from 0 to 50 m s^{-1}.

The GA-Var method is initialized with random guesses for the variable values. Thus, every time the algorithm is run, it finds a different path to the optimal solution. Therefore, we run the algorithm ten times in order to gain statistics regarding the algorithm's performance. Figure 5 plots the cost function value for each of the runs as a function of generation. The best solution is indicated by the heavy black line. There is considerable spread in the cost function values during the first 20 generations as is typical. Notice that the best solution begins to converge in

transport for those observations, it is appropriate to compare the satellite-derived concentration data observed at ∼6 km with predicted concentration values at the same level. Therefore, we set the effective height of release for the SCIPUFF predictions at 6 km where the plume has come into equilibrium with the environment. We make no attempt to model the vertical transport because we do not have sufficient data regarding release details and plume rise, nor does the GA-Var algorithm require that information. The wind field of interest is thus also at this same elevation.

Table 2

The wind direction, wind speed, emission rate and cost function value along with the mean and standard deviation for ten model runs

	Wind direction (°)	Wind speed (m s^{-1})	Emission rate (kg s^{-1})	Cost function value
Run 1	215.7	16.4	6.8×10^4	0.1631
Run 2	213.0	23.8	1.1×10^5	0.1589
Run 3	214.0	18.0	6.4×10^4	0.1592
Run 4	*213.2*	*23.7*	*9.7×10^4*	*0.1593*
Run 5	**213.3**	**19.3**	**8.0×10^4**	**0.1555**
Run 6	212.4	21.2	1.0×10^5	0.1594
Run 7	213.5	21.8	9.1×10^4	0.1567
Run 8	213.0	29.3	1.5×10^5	0.1701
Run 9	211.8	34.7	2.3×10^5	0.1903
Run 10	213.7	24.0	9.9×10^4	0.1587
Mean	213.4	23.2	1.1×10^5	0.1631
STD	1.0	5.4	4.8×10^4	0.0104

The best solution (run 5) is indicated with bold and the solution corresponding to the median cost function value (run 4) is indicated with italics

Table 3

Source term parameter values used to determine the relative importance of each event to the total downwind concentration at the time the satellite observations were taken

Event number	Time adjusted UTC	Plume height, observed (km)	Plume height, H^a (km)	Volumetric flow rate DRE[b] ($m^3 s^{-1}$)	Duration (min)	Mass[c] (kg)	Average emission rate over the event period (kg s^{-1})
Event 1	0630	5.5	2.4	2×10^0	2	6.6×10^5	5.5×10^3
Event 2	0700	13.4	10.3	9×10^2	7	9.8×10^8	2.3×10^6
Event 3	0815	14.6	11.5	1×10^3	20	4.4×10^9	3.7×10^6
Event 4	0930	13.1	10.0	8×10^2	38	4.7×10^9	2.1×10^6
Event 5	1230	18.3	15.2	5×10^3	20	1.4×10^{10}	1.2×10^7

Since either plume height of duration values were missing in the reanalysis events those contributions were neglected

[a] Plume height, H, is height of the plume above the vent

[b] DRE (dense rock equivalent) is also known as erupted volume, V. The values for DRE were calculated using equation (1) of MASTIN et al. (2009)

[c] Mass is calculated using a density of 2,600 kg m^{-3} which was not measured but is the same number reported in SCOTT and McGIMSEY (1994) (Redoubt 1989 eruption)

fewer than 40 generations. Several other runs converge to a similar value as the best solution but require more than 40 generations.

The results of ten runs of the GA are listed in Table 2. The mean emission rate of the ten runs is $1.1 \times 10^5 \pm 4.8 \times 10^4$ kg s^{-1}. This value represents an average release rate spanning a period of 6.5 h. Because we assume a continuous, uniform emission in this scenario, this translates into an estimated 2.6×10^9 kg of ash being pumped into the atmosphere throughout the entire period. This value agrees well with previous work where only the emission rate was sought and the total amount of mass emitted was found to be $\sim 1.7 \times 10^9$ kg (LONG et al., 2010b). Sensitivity runs indicate that the cost function is very flat in the region of the best solution, indicating that the large error bars are to be expected.

Verifying the value for emission rate is challenging given the nature of a volcanic release. Using equation (1) of MASTIN et al. (2009), we can estimate the volumetric flow rate of dense-rock equivalent (DRE) for each event based on the plume height. Table 3 lists the duration and plume height of each event. The total volume of DRE is found by multiplying the duration of each event by the volumetric flow rate. Then the total mass is found by multiplying the volumetric flow rate by the density. The density used here is 2,600 kg m^{-3} (SCOTT and McGIMSEY, 1994). The total mass from each event is then summed, yielding a mass for

events 1–5 of 2×10^{10} kg. This value is approximately an order of magnitude higher than our estimate. As we will show later, we believe that the satellite data do not actually represent the entire release, the reason for the refinement to follow. We seek to refine this estimate in Sect. 5.1.

In addition to evaluating the merit of the retrieved emission rate, we can assess the wind speed and direction retrieved by GA-Var against local upper air soundings. Note that we expect a fair amount of wind variability over the 7 h spanning these events as well as over the vertical extent of the plume rise. What we wish to retrieve is a transporting wind that is representative of the actual observed transport. The GA-Var method retrieved a mean wind direction of $213.4° \pm 1.0°$ and a mean wind speed of 23.2 ± 5.4 m s^{-1}. We compare the retrieved values for wind speed and direction with upper air soundings from Anchorage, Alaska, which is about 110 miles away from Mount Redoubt (University of Wyoming Atmospheric Soundings). The first sounding was taken approximately 6 h prior to the first event and the second was taken approximately 30 min before the fifth event ends. The first sounding indicates winds at 6 km of ~ 22 m s^{-1} and wind directions of 225°. The second sounding shows winds at 6 km of ~ 16 m s^{-1} and wind directions of 205°. The GA-Var best solution determined for wind speed and direction of 19.3 m s^{-1} and 213.3° agree well with these observed values. We note that the most

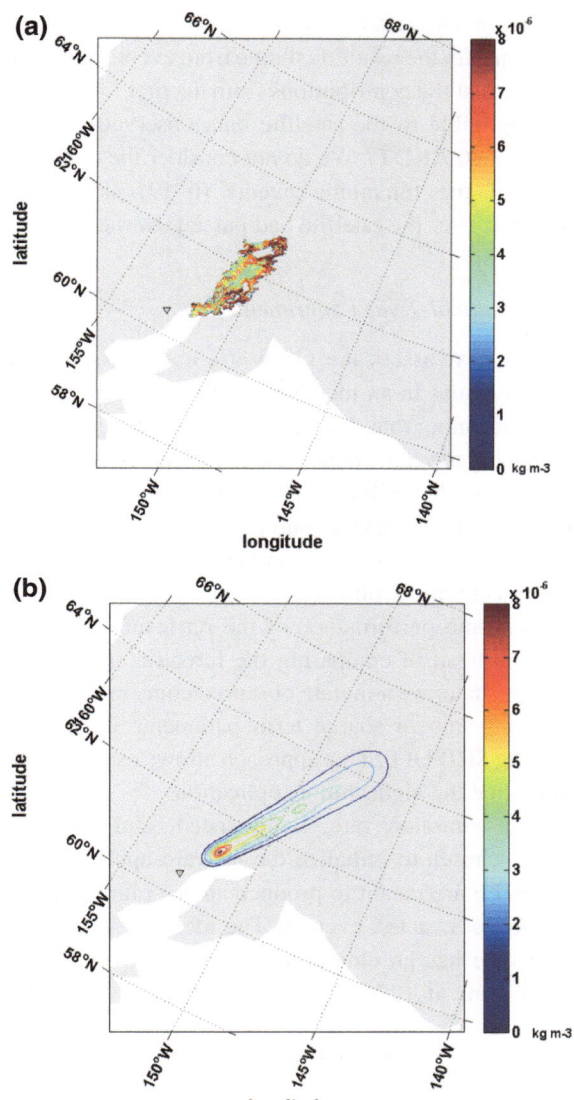

Figure 6
Derived concentration field at 1430 UTC (0630 AKDT) (**a**) and the SCIPUFF forecast with best GA-Var solution for emission rate, wind direction and wind speed (**b**). The location of Mount Redoubt is indicated by an *inverted triangle*

period. Thus, the wind that is back-calculated via the GA would provide additional guidance for the prediction of the future state of the cloud. Note that one could also perform the subsequent prediction with model generated three dimensional winds.

4.2. Forward Assimilation

We compare the observed concentration field with the forecast field predicted using SCIPUFF with our GA-Var retrieved values for emission rate, wind direction, and wind speed in Fig. 6. The best solution yields a wind direction of 213.3°, wind speed of 19.3 m s^{-1}, and an emission rate of 8.0×10^4 kg s^{-1}. Figure 6 indicates that the GA-Var's best solution produces a footprint that is generally similar to the satellite derived concentration data. Specifically, the maximum concentration value in the forward assimilation is 6.4×10^{-6} kg m^{-3}, which is of the same order of magnitude as the observed data (8.2×10^{-6} kg m^{-3}). The match is not exact, as one would expect when comparing the results of an ensemble averaged model (SCIPUFF) to a single realization (the satellite data). The GA-Var retrieved variables that, when inserted into SCIPUFF for subsequent prediction, predicts a plume that extends further north and east than the observed plume. The resulting best GA-Var solution plume in Fig. 6b serves as our first guess for the initial conditions for the subsequent transport and dispersion forecast.

5. Refinements to the Application

Although the approach to back-calculating the emission rate of Mount Redoubt shown in the previous section produces an order of magnitude agreement with other best estimates, we wish to refine that estimate by carefully analyzing the events of that day and using that analysis to refine our approach. To that end, we first choose to model the events in a forward application of SCIPUFF. Then we analyze the behavior of GA-Var in an identical twin application where we can compare our results with known prescribed emission rates. Finally, we construct a more refined approach to the source term estimation problem using this information gained for

representative wind speed and direction are not the surface level winds. In fact, for this particular case the surface level winds at Kenai (65 miles from Mount Redoubt) were 22.5°—nearly opposite the upper level winds that transported the bulk of the cloud (http://wunderground.com). In this case, modelers would need a wind profile that may not be available or they need to capture the mean value over the time

221

such an analysis. The goal of this exercise is to demonstrate the process for optimizing the application of GA-Var in a new scenario.

5.1. Modeling Separate Events

In an effort to obtain a more precise estimation of the total amount of mass emitted from the five events of the eruption, we next model the five events individually rather than assuming a single, continuous emission. The start time, duration and plume height of each event as determined by the Alaska Volcano Observatory (AVO) are listed in Table 1. Table 3 lists the source term parameters used in the modeling of the five separate events. The plume height above the vent, or H, is calculated by subtracting the height of Mount Redoubt (3,108 m) from the observed plume height (MASTIN et al., 2009). From H, we determine the volumetric flow rate in $m^3 s^{-1}$ using equation (1) of MASTIN et al. (2009). The volumetric flow rate is then multiplied by the event duration in order to determine the total volume. The total volume is multiplied by a density of 2,600 kg m^{-3} in order to calculate the total mass. An average value for the emission rate is calculated by taking the total mass and dividing by the duration. Using the approximate start times and emission rates, we run SCIPUFF for each event and calculate the total concentration at time 1430 UTC when the satellite would pass overhead. For events 1–4 we assume a uniform wind speed (16 m s^{-1}) and wind direction (205°) based on the first sounding taken at 0000 UTC 3/23/2009 (University of Wyoming). For event 5, we use a uniform wind speed (22 m s^{-1}) and wind direction (225°) based on the second sounding taken at 1200 UTC 3/23/2009 (University of Wyoming). We set up a fixed sensor network that spans approximately 60.5° to 63.0°N and −152° to −149°E, which covers the entire area of the ash plume sensed by the satellite. Based on these SCIPUFF calculations, we observe that the ash plumes resulting from events 1–4 did not produce measurable concentrations at our sensor network. The plumes had moved beyond our network by the time the satellite would have passed overhead as simple spatial advection arguments might suggest. Event 5, however, did produce measurable concentrations at

the sensor network. For the next phase of the study, we will consider the resulting plume from event 5 only and assume that the contributions from the first four events are negligible to the satellite data observed at 1430 UTC (0630 AKDT). We do not consider the contributions of the remaining events (6–19) since they occurred after the satellite had passed overhead.

5.2. Identical Twin Experiment

We next assess the GA-Var's ability to retrieve known values in an identical twin experiment framework, meaning that we create synthetic concentration data using the same atmospheric transport and dispersion model that will be used for the retrieval. Such an approach eliminates the sources of error in the retrieval related to the observations and produces a "clean" set of concentration observations that can be used to evaluate the performance of the retrieval algorithm. Thus, instead of comparing the forecasts against the satellite data, we simulate observed concentrations by inputting known source term parameter values and running SCIPUFF. This approach allows us to analyze and refine the algorithm's application.

The emission rate, wind speed, and direction values chosen to initialize the forward application of SCIPUFF are meant to produce an ash plume similar to, but not exactly, event 5. The identical twin setup used here has an emission rate of 7.8×10^5 kg s^{-1} that begins at 1230 UTC and lasts for 20 min. The

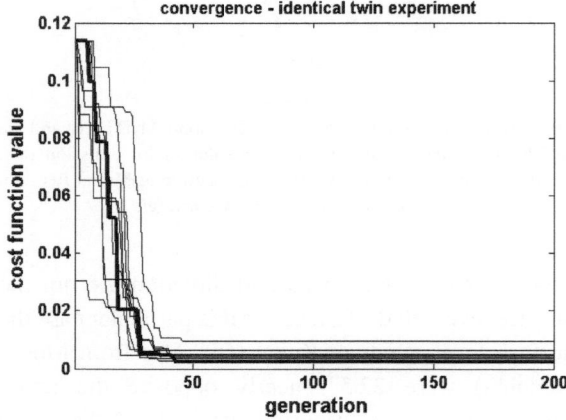

Figure 7
GA convergence for the identical twin experiment runs. The best solution or the run that ultimately yielded the lowest cost function value is indicated by the *heavy black line*

Table 4

Results of the identical twin experiment

	Wind direction (°)	Wind speed (m s^{-1})	Emission rate (kg s^{-1})	Cost function value
Run 1	214.7	16.1	6.8×10^5	0.0065
Run 2	215.0	16.1	6.7×10^5	0.0036
Run 3	215.1	15.9	5.2×10^5	0.0043
Run 4	215.3	16.0	7.1×10^5	0.0047
Run 5	214.7	16.0	8.3×10^5	0.0050
Run 6	215.1	16.3	9.9×10^5	0.0096
Run 7	214.8	16.0	1.0×10^6	0.0036
Run 8	215.0	16.0	6.1×10^5	0.0026
Run 9	**214.9**	**15.9**	$\mathbf{7.9 \times 10^5}$	**0.0023**
Run 10	214.8	15.9	8.5×10^5	0.0046
Truth	*215.0*	*16.0*	7.8×10^5	*0*
Mean	214.9	16.0	7.7×10^5	0.0047
STD	0.2	0.1	1.6×10^5	0.0021

The wind direction, wind speed, emission rate and cost function values along with the mean and standard deviation for ten model runs. The best solution (run 9) is indicated with bold. The source term parameters used to create the observed data are intended to be similar to those describing event 5 of the eruption

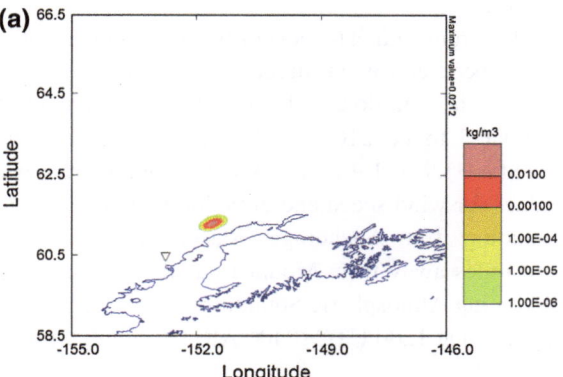

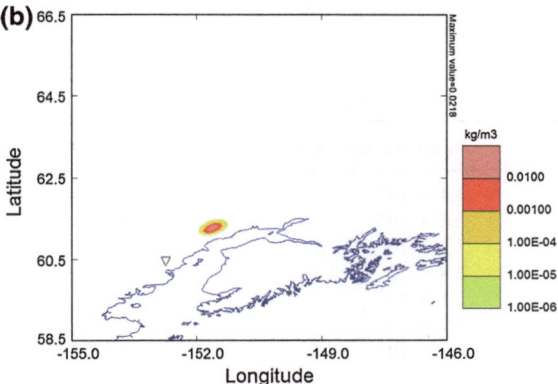

Figure 8

The truth forecast at 1430 UTC (0630 AKDT) used to create the identical twin observations (**a**) and the SCIPUFF forecast initialized with the best GA-Var solution for emission rate, wind direction and wind speed (**b**). The location of Mount Redoubt is indicated by an *inverted triangle*

wind speed and direction are set to be uniform throughout the domain at 16 m s^{-1} and 215°, respectively. The computed concentrations become the new concentration observations that will be used in the retrieval algorithm.

For the GA-Var application we use a cost function that is logarithmic in concentration:

$$\text{cost} = \frac{\sqrt{\sum_{s=1}^{\text{TS}} \left[\log(C_s + \varepsilon) - \log(O_s + \varepsilon)\right]^2}}{\sqrt{\sum_{s=1}^{\text{TS}} \left[\log(O_s + \varepsilon)\right]^2}} \quad (3)$$

where all the variables are the same as defined previously and ε is a small constant used to threshold the data.

With the exception of the observed data now being our synthetically created data, the retrieval problem remains the same as described in Sect. 4.1. The population, number of iterations, and mutation rate are the same. In addition, the range of potential values for each of the genes is the same. We run GA-Var ten times in order to gain statistics regarding its performance. The convergence rates of the ten runs are plotted in Fig. 7. The results of this identical twin retrieval experiment are listed in Table 4. For the best run (9), the percent error in wind direction is 0.04%, in wind speed it is 0.2% and in emission rate it is 1.6%. SCIPUFF is initialized with the best GA-Var found solution and that forecast can be compared with the identical twin data forecast in Fig. 8. The

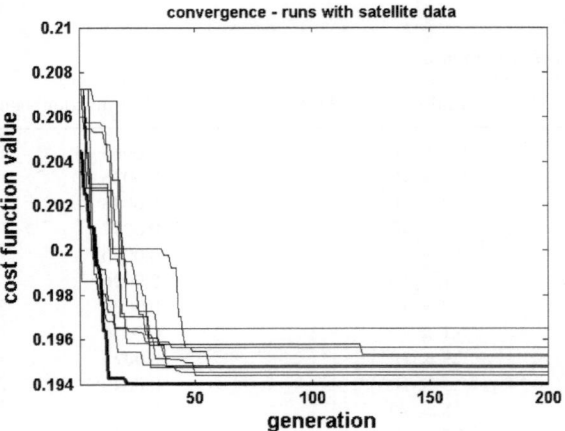

Figure 9
GA convergence for the runs with refined modeling techniques. The best solution or the run that ultimately yielded the lowest cost function value is indicated by the *heavy black line*

performance of the GA with the identical twin data thus gives us confidence that the technique is applicable to this type of problem.

5.3. Application to Mount Redoubt Case Study

Taking into account the contribution of each event discussed in Sect. 5.1, a number of refinements are made to the Mount Redoubt case study using satellite data. First, the modeling now reflects the effects of event 5 alone because the contributions from the earlier events would have moved off our sensor

domain by the time the satellite observed the ash cloud. The simulation period begins with event 5 at 1230 UTC and ends when the satellite would have observed the ash cloud at 1430 UTC. The release is no longer continuous over a period of several hours, but rather lasts 20 min of that 2-h period. As before, a single representative wind speed and wind direction are sought, along with the representative emission rate that would produce the observed concentration values.

The GA parameters of population size, mutation rate and total number of generations are the same as in Sects. 4.1 and 5.2. Ten runs are completed in order to quantify variability in the results. Figure 9 illustrates the convergence of those ten runs. The run corresponding to the best solution (indicated by the lowest cost function value) is indicated by the heavy black line. The specific results for wind direction, wind speed and emission rate are presented in Table 5.

The runs indicate generally good agreement in wind speed and wind direction with little variability among the solutions. The mean wind direction is estimated to be $216.1° \pm 1.4°$ and the mean wind speed is 31.9 ± 1.9 m s^{-1}. We compare the retrieved values for wind speed and direction with an upper air sounding from Anchorage, Alaska, which is about 110 miles away from Mount Redoubt (University of Wyoming Atmospheric Soundings 2009). The sounding taken at 1200 UTC (0400 AKDT) shows winds at

Table 5

Results of the event 5 release using observed satellite data

	Wind direction (°)	Wind speed (m s^{-1})	Emission rate (kg s^{-1})	Cost function value
Run 1	214.0	31.5	4.4×10^5	0.1965
Run 2	214.6	34.6	6.6×10^4	0.1948
Run 3	217.1	32.7	1.2×10^5	0.1945
Run 4	214.5	32.0	6.8×10^4	0.1948
Run 5	216.8	28.6	7.5×10^4	0.1953
Run 6	217.1	33.2	5.7×10^4	0.1948
Run 7	217.3	28.7	4.9×10^4	0.1957
Run 8	217.0	32.8	6.2×10^4	0.1944
Run 9	214.9	32.4	6.4×10^4	0.1953
Run 10	**217.3**	**32.8**	**7.1×10^4**	**0.1940**
Mean	216.1	31.9	1.1×10^5	0.1950
STD	1.4	1.9	1.2×10^5	0.0007

The wind direction, wind speed, emission rate and cost function values along with the mean and standard deviation for ten model runs. The best solution (run 10) is indicated with bold

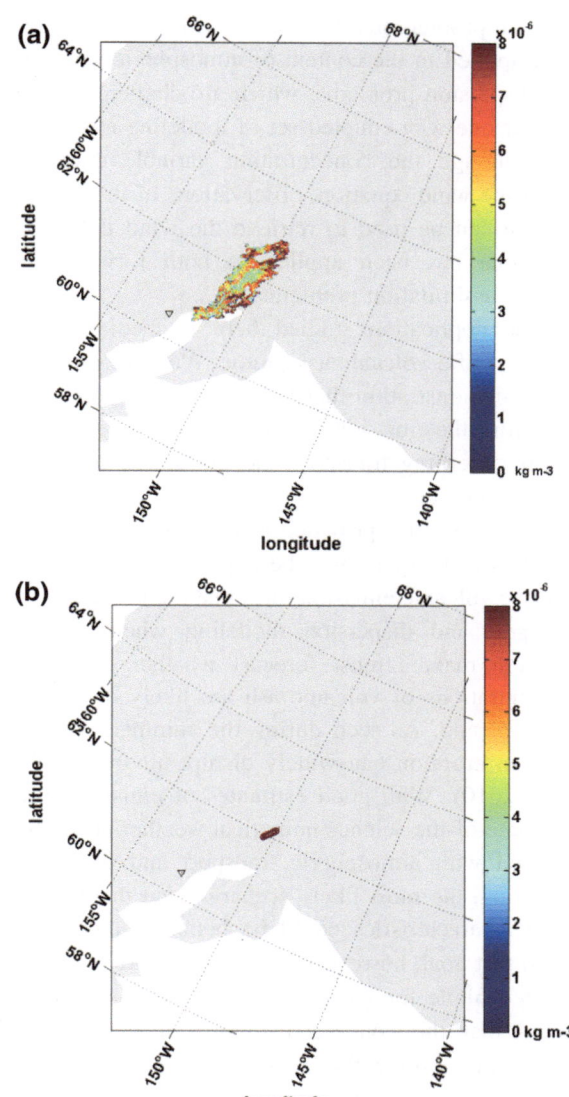

Figure 10
Derived concentration field at 1430 UTC (0630 AKDT) (**a**) and the refined SCIPUFF forecast with best GA-Var solution for emission rate, wind direction and wind speed (**b**). The location of Mount Redoubt is indicated by an *inverted triangle*

6 km of ~ 16 m s^{-1} and wind direction of 205°. In addition, we compare the GA-Var determined wind data to NWP runs using WRFv3.1 (SKAMAROCK *et al.*, 2008) that were run over Alaska at 12 km resolution. The WRF data indicate that in the vicinity of Mount Redoubt at 1200 UTC (0400 AKDT) and 6 km, the wind speed was 22.7 m s^{-1} and the direction was 201°. Downwind of Mount Redoubt and near the edge of the plume pictured in Fig. 10b (62°N

and −150°W) at 1500 UTC (0700 AKDT) and 6 km, the wind speed was 11.3 m s^{-1} and the wind direction was 218°. While the best GA-Var retrieved wind direction of 217.3° compares well with the observed values, the wind speed is much higher than the observed wind speed of 16 m s^{-1}. Note that although the WRF runs were initialized with boundary conditions for the initialization time from the GFS, it did not incorporate data assimilation, and thus is not necessarily tied to the ground truth.

The solutions for emission rate exhibit more variability than the values for wind speed and direction. This was also the case with the identical twin experiment. As with that part of the study, the solution with the lowest cost is considered the best solution. The best solution corresponds to a wind direction of 217.3°, a wind speed of 32.8 m s^{-1} and emission rate of 7.1×10^4 kg s^{-1}. Since this event lasts for a 20 min period, that translates to a total mass of 8.5×10^7 kg of ash.

The best GA computed solution (run 10 of Table 5) was then used to initialize a forward SCIPUFF simulation. Figure 10 compares that SCIPUFF forecast with the observed concentration field. The GA-Var predicted a relatively high wind speed and thus matched a portion of the satellite that was further downwind. Had the wind speed been closer to the values predicted by the NWP model or the observations, then the GA-Var would have matched a portion of the cloud closer to Mount Redoubt. The GA-Var solution captures the relevant trajectory of the cloud well but under-predicts the size of the cloud. Note that one could use NWP model data to drive the dispersion model using the GA-Var-retrieved emission rate.

Using equation (1) of MASTIN *et al.* (2009) and the plume height reports from SCHAEFER (2011) listed in Table 3, the total mass of ash from event 5 alone is 1.4×10^{10} kg. In SCHAEFER (2011), the author estimates the amount of mass emitted from event 5 to be only 4.5×10^9 kg. Both of these estimates are considerably higher than the estimate presented here of 8.5×10^7 kg. Figure 10 confirms that the best GA estimate produces a forecast that only captures a portion of the ash cloud.

The large discrepancy between the estimate presented here and the value reported in SCHAEFER

(2011) is likely due to the fact that we do not take into account the fact that the larger ash particles have likely settled out and are not at transported to the height of the observed satellite data. The fine ash fraction is the fraction of the total emitted mass that is transported long distances from the volcano (DACRE 2011). While the larger particles tend to settle out, these smaller fine ash particles can stay in the ash cloud for hours or days and fall out at rates that are not well understood (MASTIN et al. 2009). MASTIN et al. (2009) lists the mass fraction of particles smaller than 63 µm for a number of eruptions. This value can vary considerably from 0.01 to 0.7 depending on the size of the eruption. In addition, DACRE et al. (2011) modeled the 2010 Eyjafjallajökull eruption using lidar observations and found that only between 3 and 4% of the total emitted mass flux was transported by ash particles smaller than 100 µm in diameter. Here we have attempted to match the concentration values derived from the satellite data without estimating the fine ash fraction. In the case of Mount Redoubt, if the mass of 4.5×10^9 kg (determined by SCHAEFER 2011) is taken to be the total emitted mass, then the GA-determined emission rate represents $\sim 2\%$ of that total emitted mass, which lies within the range listed in MASTIN et al. (2009). Thus, we must recognize that the emission rate computed by GA-Var has essentially matched the fine ash fraction that is appropriate for producing a cloud that matches the satellite data concentration levels. If the purpose of the emission rate retrieval is to obtain the appropriate emission rate to use in subsequent atmospheric transport and dispersion modeling to forecast downwind concentrations, this GA-Var derived value is the appropriate rate that has the fine ash fraction built into it.

6. Conclusions

A genetic algorithm variational approach to data assimilation has been presented here. This technique uses an optimization approach to data assimilation where we seek to minimize the difference between the predicted and the observed concentration fields. The robust optimization technique that solves that retrieval problem is a genetic algorithm. GA-Var has been applied in the context of atmospheric transport and dispersion problems, which are challenging due to their one-way coupled set of modeling equations. Even though the concentration variable does not affect the wind equation, observations of its changes in time can be used to retrieve the wind field. The technique has been applied to both forward and inverse assimilation problems.

The application studied here is medium-range transport of a volcanic ash plume. We have provided a first demonstration that GA-Var can be applied to determine the emission rate, wind direction, and wind speed governing transport and dispersion of a volcanic ash cloud.

This work has provided a preliminary indication that this technique could be applied in real time to retrieve volcanic emission rates that are necessary for transport and dispersion modeling when aviation decision makers must forecast whether dangerous concentrations of volcanic ash are likely downwind of a volcano. As seen during the summer of 2010, such an eruption can widely disrupt air travel (CHIVERS, 2010). With good estimates of emission rates and state-of-the-science numerical weather prediction coupled with atmospheric transport and dispersion modeling, the most likely scenarios and their uncertainty characteristics could be better predicted. To reach that goal, however, will require more testing of this technique for volcano applications. Because we found that the convergence rates of the GA-Var algorithm are quite similar between the various runs, a single run is quite likely to provide a good estimate with a sufficient number of generations. Ways could be found to speed the algorithm and to optimize its performance. For instance, some previous work has followed the GA with a gradient descent method to speed convergence to the global minimum once the basin of attraction has been found. Another way to fine-tune could use a multi-stage approach that uses different cost function formulations at different points in the problem.

Prior to operational use, this method should be tested on more historical cases to determine performance characteristics. The work here, as well as prior studies, has indicated, however, that prospects are promising for advancing the GA-Var method toward usefulness on volcano problems as well as those in homeland and defense security.

Acknowledgments

The GA-Var technique has been developed jointly with Randy L. Haupt, George S. Young, Christopher T. Allen, Yuki Kuroki, Luna M. Rodriguez, and Andrew J. Annunzio. Dustin Truesdell assisted with the first application of GA-Var to the volcano source term estimation problem. The authors would like to thank Michael Richards for suggesting the application of the GA-Var method to the volcano problem. The Anchorage soundings and the Kenai observation data are taken from Wunderground.com. We additionally thank two reviewers for their perceptive comments that led to modifications that have improved the manuscript.

REFERENCES

The Alaska Volcano Observatory U.S. Geological Survey (AVO/USGS) Volcanic Activity Notice for Redoubt. March 23, 2009-12:20 PM Report. Accessed: 29 January 2010. http://www.avo.alaska.edu/activity/avoreport.php

ALLEN, C.T., S.E. HAUPT, and G.S. YOUNG, 2007a: *Source Characterization with a Receptor/Dispersion Model Coupled With A Genetic Algorithm*, Journal of Applied Meteorology and Climatology, **46**, 273-287.

ALLEN, C.T., G.S. YOUNG, and S.E. HAUPT, 2007b: *Improving Pollutant Source Characterization by Optimizing Meteorological Data with a Genetic Algorithm*, Atmospheric Environment, **41**, 2283-2289.

ANNUNZIO, A.J., S.E. HAUPT, and G.S. YOUNG, 2011: *Determining Relevant Turbulent Length Scales and Source Characteristics from Contaminant Concentration Observations*, submitted to Atmospheric Environment.

BRIGGS, G. A., 1975: Plume rise predictions. Pp. 59-1 11 in *Lectures on air pollution and environmental impact analyses*. American Meteorological Society, Boston, USA.

Casadevall TJ (eds), 1994: Volcanic ash and aviation safety. In: Proceedings of the first international symposium, Seattle, Washington, July 1991, U.S. Geological Survey Bulletin B 2047, pp 450. http://www.dggs.dnr.state.ak.us/pubs/pubs?reqtype=citation&ID=376

CERVONE, G. and P. FRANZESE, 2010: *Monte Carlo Source Detection of Atmospheric Emissions and Error Function Analysis*, Computers & Geosciences, **36**, 902-909.

CERVONE G., FRANZESE P., GRADJEANU, A., 2010: *Characterization of atmospheric contaminant sources using adaptive evolutionary algorithms*, Atmospheric Environment, **44**. 3787-3796.

CHIVERS, H., 2010: *Dark Cloud: VAAC and Predicting the Movement of Volcanic Ash*. Meteorological Technical International. June, 60-63.

COURTIER, P., 1997: *Dual formulation of four-dimensional variational assimilation*. Quart. J. Roy. Meteor. Soc., **123**, 2449-2461.

DACRE, H.F., A.L.M. GRANT, R.J. HOGAN, S.E. BELCHER, D.J. THOMSON, B. DEVENISH, F. MARENCO, J.M. HAYWOOD, A. ANSMANN, I. MATTIS, L. CLARISSE, 2011: *Evaluating the structure and magnitude of the ash plume during the initial phase of the 2010 Eyjafjallajökull eruption using lidar observations and NAME simulations*. J. of Geophysical Research, DOI:10-1029, in press.

DALEY, R., 1991. Atmospheric Data Assimilation. Cambridge University Press, Cambridge, 457 pp.

D'AMOURS R. 1998. *Modeling the ETEX plume dispersion with the Canadian emergency response model*. Atmospheric Environment **32**(24): 4335–4341.

Davis, L. ed.: 1991, *Handbook of Genetic Algorithms*, New York, Van Nostrand Reinhold.

De JONG, K. A.: 1975, Analysis of the behavior of a class of genetic adaptive systems, Ph.D. Dissertation, The University of Michigan, Ann Arbor, MI.

DENG A., N.L. SEMAN, G.K. HUNTER, and D.R. STAUFFER, 2004: *Evaluation of interregional transport using the MM5-SCIPUFF system*. Journal of Applied Meteorology, **43**, 1864-1886.

Draxler, R.R., and G.D. Hess, 1998: *An overview of the HY-SPLIT_4 modeling system of trajectories, dispersion, and deposition*. Aust. Meteor. Mag., **47**, 295-308.

EPA 2003 EPA, *Revision to the Guidelines on Air Quality Models: Adoption of a Preferred Long Range Transport Model and Other Revisions*. Federal Register, vol. 68, (72), 40 CFR Part 51, 2003.

ERRICO, R., T. VUKICEVIC and K. RAEDER, 1993: *Examination of the accuracy of a tangent linear model*. Tellus, **45A**, 462-477.

Fox, D., 2009: *Volcano Watchers*, Popular Mechanics, **186**, 56-59.

GOLDBERG, D.E., 1989: *Genetic Algorithms in Search, Optimization, and Machine Learning*, New York: Addison-Wesley.

GORDON, V.S. and D. WHITLEY: 1993, Serial and parallel genetic algorithms as function of optimizers. In S. FORREST (ed.), ICGA-90: 5th Int. Conf. on Genetic Algorithms. Los Altos, CA: Morgan Kaufmann, pp. 177-183. Haupt, R.L., and S.E. Haupt, 1998: *Practical Genetic Algorithms*, John Wiley and Sons, New York, NY, 177 pp.

HAUPT, R.L. and S.E. HAUPT, 2000: *Optimum Population Size and Mutation Rate for a Simple Real Genetic Algorithm that Optimizes Array Factors*, Applied Computational Electromagnetics Society Journal, **15**, No. 2, pp. 94-102.

HAUPT, R.L., and S.E. HAUPT, 2004: *Practical Genetic Algorithms, Second Edition with CD*, John Wiley and Sons, New York, NY, 255 pp.

HAUPT, S.E., R.L. HAUPT, and G.S. YOUNG, 2010: *A Mixed Integer Genetic Algorithm used in Chem-Bio Defense Applications*, Journal of Soft Computing, doi:10.1007/s00500-009-0516-z.

HAUPT, S.E., C.T. ALLEN, and G.S. YOUNG, 2009a: Environmental Science Models and Artificial Intelligence, in *Artificial Intelligence Methods in the Environmental Sciences*, HAUPT, S.E., A. PASINI, and C. MARZBAN, Eds., Springer, 424 pp.

HAUPT, S.E., A. BEYER-LOUT, K.J. LONG, and G.S. YOUNG, 2009b: *Assimilating Concentration Observations for Transport and Dispersion Modeling in a Meandering Wind Field*, Atmospheric Environment, **43**, 1329-1338.

HAUPT, S.E., 2005: *A Demonstration of Coupled Receptor/Dispersion Modeling with a Genetic Algorithm*, Atmospheric Environment, **39**, 7181-7189.

HAUPT, S.E., G.S. YOUNG, and C.T. ALLEN, 2007: A Genetic Algorithm Method to Assimilate Sensor Data for a Toxic Contaminant Release, Journal of Computers, **2**, 85-93.

HAUPT, S.E. G.S. YOUNG, and C.T. ALLEN, 2006: Validation Of A Receptor/Dispersion Model Coupled With A Genetic Algorithm, Journal of Applied Meteorology, **45**, 476–490.

HEIDINGER, A. K. and M. J. PAVOLONIS, 2009: Nearly 30 years of gazing at cirrus clouds through a split-window. Part I: Methodology. J. Appl. Meteorol. and Climatology, **48**(6), 110-1116.

HOLLAND, J.H., 1975: Adaptation in Natural and Artificial Systems, Ann Arbor: The University of Michigan Press.

IDE K., P. COURTIER, M. GHIL, and A.C. LORENC, 1997: Unified Notation for Data Assimilation: Operational, Sequential and Variational. Journal of the Meteorological Society of Japan, **75**, 181-189.

KALNAY E., 2003: Atmospheric Modeling, Data Assimilation and Predictability. Cambridge University Press, Cambridge, 136-204.

KRYSTA M., M. BOCQUET, B. SPORTISSE, and O. ISNARD, 2006: Data assimilation for short-range dispersion of radionuclides: An application to wind tunnel data, Atmospheric Environment, **40**, 7267-7279.

KUROKI, Y., G.S. YOUNG, and S.E. HAUPT, 2010: UAV Navigation by an Expert System for Contaminant Mapping with a Genetic Algorithm, Expert Systems with Applications, **37**, 4687-4697. (doi:10.1016/j.eswa.2009.12.039)

LANGSTON, LEE S., 2010: Asking for Trouble. Mechanical Engineering: The Magazine of ASME. July. 28-31.

LEWIS, J.M. and J.C. DERBER, 1985: The use of adjoint equations to solve a variational adjustment problem with advective constraints. Tellus, **37**, 309-327.

LEWIS, J.M., S. LAKSHMIVAHARHAN, and S.K. DHALL, 2006: Dynamic Data Assimilation: A Least Squares Approach, Cambridge University Press, Cambridge, UK, 654 pp.

LONG, K.J., S.E. HAUPT, and G.S. YOUNG, 2010a: Assessing Sensitivity of Source Term Estimation. Atmospheric Environment, **44**, 1558-1567.

LONG, K.J., D. TRUESDELL, S.E. HAUPT, G.S. YOUNG, 2010b: Using a Genetic Algorithm to Estimate Source Term Parameters of Volcanic Ash Clouds, 8th Conference on Artificial Intelligence Applications to Environmental Science, January 20.

LORENZ, E.N., 1963: Deterministic Nonperiodic Flow, J. Atmos. Sci., **20**, 130-141.

Manins, P. C., 1985: Cloud heights and stratospheric injections resulting from a thermo-nuclear war. Atmos. Environ., 19(8), 1245-1255.

MASTIN, L.G., GUFFANTI, M., SERVRANCKX, R., WEBLEY, P.W., BARSOTTI, S., DEAN, K., DENLINGER, R., DURANT, A., EWERT, J.W., GARDNER, C.A., HOLLIDAY, A.C., NERI, A., ROSE,W.I., SCHNEIDER, D., SIEBERT, L., STUNDER, B., SWANSON, G., TUPPER, A., VOLENTIK, A., WAYTHOMAS, C.F., 2009: A multidisciplinary effort to assign realistic source parameters to model of volcanic ash-cloud transport and dispersion during eruptions. In: MASTIN, LARRY, WEBLEY, PETER (Eds.), Journal of Volcanology and Geothermal Research: Special Issue on Volcanic Ash Clouds, 186, pp. 10–21.

MITCHELL M., 1996: An Introduction to Genetic Algorithms, The MIT Press, Cambridge, MA.

OFCM (2004) Proceedings of 2nd international conference on volcanic ash and aviation safety, Alexandria, Virginia (USA), 21–24 June 2004. U.S. Department of Commerce/National Oceanic and Atmospheric Administration. http://www.ofcm.gov/ICVAAS/Proceedings2004/ICVAAS2004-Proceedings.htm

PAVOLONIS, M. J., W. F. FELTZ, A. K. HEIDINGER, and G. M. GALLINA, 2006: A daytime complement to the reverse absorption technique for improved automated detection of volcanic ash. J Atmos Ocean Technol., 23, 1422-1444.

PAVOLONIS, M. J., 2010: Advances in extracting cloud composition information from spaceborne infrared radiances: A robust alternative to brightness temperatures. Part I: Theory. J. Applied Meteorology and Climatology, 49, 1992-2012.

PRATA, A.J., GRANT, I.F, 2001: Retrieval of microphysical and morphological properties of volcanic ash plumes from satellite data: application to Mt. Ruapehu, New Zealand. Quarterly Journal of the Royal Meteorological Society, Vol. 122, pp. 1–25.

PUDYKIEWICZ, J. (1989) Simulation of the Chernobyl dispersion with a 3-D hemispheric tracer model. Tellus 41B, 391-412.

RABIER, F., and P. COURTIER, 1992: Four-dimensional assimilation in the presence of baroclinic instability. Quart. J. Roy. Meteor. Soc., 118, 649-672.

RODRIGUEZ, L.M., S.E. HAUPT, and G.S. YOUNG, 2010a: Impact of Sensor Characteristics on Source Characterization for Dispersion Modeling, submitted to Measurement.

RODRIGUEZ, L.M., S.E. HAUPT, G.S. YOUNG, A.J. ANNUNZIO, and K.J. SCHMEHL, 2010b: Reanalysis of FFT07 Phase I using a Genetic Algorithm Coupled with Dispersion Models, 14th Annual George Mason University Conference on Atmospheric Transport and Dispersion Modeling, Fairfax, VA, July 13-15.

RYALL, D. B., and MARYON, R. H., 1998, Validation of the UK Met. Office's NAME model against the ETEX dataset, Atmospheric Environment 32:4265-4276.

SASAKI, Y., 1970: Some basic formalisms in numerical variational analysis. Monthly Weather Review, **98**, 875-883.

SCHAEFER, JANET R., 2011, The 2009 Eruption of Redoubt Volcano, Alaska: Alaska Division of Geological & Geophysical Surveys Report of Investigations 2011-XXX, in press.

Scott, W. E., and McGimsey, R. G., 1994, Character, mass, distribution, and origin of tephra-fall deposits of the 1989-1990 eruption of Redoubt volcano, south-central Alaska: in MILLER, T. P. and CHOUET, B. A., (eds.), The 1989-1990 eruptions of Redoubt Volcano, Alaska, Journal of Volcanology and Geothermal Research, v. 62, n. 1, p. 251-272.

SEARCY, C., DEAN, K. and STRINGER, W. (1998). PUFF: A high-resolution volcanic ash tracking model. Journal of Volcanology and Geothermal Research. **80**. p. 1-16.

SIEBERT L, and SIMKIN T (2002). Volcanoes of the World: an Illustrated Catalog of Holocene Volcanoes and their Eruptions. Smithsonian Institution. Global Volcanism Program Digital Information Series, GVP-3, (http://www.volcano.si.edu/world/) (http://www.volcano.si.edu/world/volcano.cfm?vnum=1103-03-&volpage=erupt)

SKAMAROCK, W.C., J.B. KLEMP, J. DUDHIA, D.O. GILL, D.M. BARKER, M.G. DUDA, X-Y. HUANG, W. WANG, and J.G. POWERS, 2008: A description of the Advanced Research WRF Version 3. NCAR Technical Note NCAR/TN-475+STR. 113 pp.

SYKES R.I., et al., 2004: SCIPUFF Version 2.0, Technical Documentation; A.R.A.P. Report no. 727, Titan Corp. Princeton, NJ.

TALAGRAND, O. and P. COURTIER, 1987: Variational assimilation of meteorological observations with the adjoint vorticity equation. I: Theory. Quart. J. Roy. Meteor. Soc., **113**, 1311-1328.

University of Wyoming College of Engineering Department of Atmospheric Science. Atmospheric Soundings from Anchorage, Alaska taken March 23, 2009 0000 UTC and 1200 UTC. http://weather.uwyo.edu/upperair/sounding.html

WEN, S. and W. I. ROSE, 1994: *Retrieval of sizes and total masses of particles in volcanic ash clouds using AVHRR bands 4 and 5.* J. Geophys. Res., **99**, 5421-5431.

(Received October 19, 2010, accepted April 29, 2011, Published online July 23, 2011)

Pure Appl. Geophys. 169 (2012), 539–554
© 2011 Springer Basel AG
DOI 10.1007/s00024-011-0386-z

Ensemble-Based Observation Targeting for Improving Ozone Prediction in Houston and the Surrounding Area

NAIFANG BEI,[1] FUQING ZHANG,[2] and JOHN W. NIELSEN-GAMMON[3]

Abstract—This study examines the effectiveness of targeted meteorological observations for improving ozone prediction in Houston and the surrounding area based on perfect-model simulation experiments. Supplementary observations are targeted for the location that has the highest impact factor (maximum Kalman gain) estimated from an ensemble and is expected to minimize ozone forecast uncertainty at the verification time. It is found that the observational impact factor field varies with time and is sensitive to ensemble resolutions and physics parameterizations. The efficiency of observation targeting is further examined through assimilating observations in areas with different impact factors using an ensemble Kalman filter. It is found that the ensemble sensitivity analysis is capable of locating supplementary observations that may reduce meteorological and ozone forecast error, but not as effectively as expected.

Key words: Ozone prediction, data assimilation, observation targeting, observational impact factor.

1. Introduction

The severity of air pollution situations is determined by a complicated interaction among three factors: the emissions to the atmosphere, chemical reactions, and meteorology (BANTA *et al.*, 2005). However, except for accidental releases or spills, whether high-pollution concentrations form on a given day is dominated principally by meteorological processes, which determine the dilution or accumulation of the pollutant emissions and can also impact other key processes, such as chemical reaction rates.

Air-quality forecasters depend upon numerical weather prediction (NWP) model output for guidance in formulating their forecasts, so the reliability of air-quality forecasts is related to the accuracy of the NWP models. It has been known that the quality of a numerical weather forecast is related to the quality of its initial condition. If the initial condition has large errors, or if it has moderate errors in regions where forecast errors grow fast, the relevant numerical weather forecast may not be accurate enough. The importance of the accurate representation of meteorological conditions for ozone predictions in Houston and the surrounding area has been demonstrated by ZHANG *et al.* (2007) recently.

One possible approach for minimizing forecast errors from day to day or in particular situations is the use of an adaptive observation network (EMANUEL *et al.*, 1995). The existing methods for adaptive observations include the singular vector technique (PALMER *et al.*, 1998; BUIZZA and MONTANI, 1999; BERGOT *et al.*, 1999; GELARO *et al.* 1999, 2000; BERGOT, 2001), the quasilinear inverse approach (PU *et al.*, 1997; PU and KALNAY, 1999), gradient and sensitivity approaches (BERGOT *et al.*, 1999; LANGLAND *et al.*, 1999; BAKER and DALEY, 2000), ensemble spread techniques (LORENZ and EMANUEL, 1998; HANSEN and SMITH, 2000; MORSS, 1998; MORSS *et al.*, 2001), the ensemble transform technique (BISHOP and TOTH, 1999; SZUNYOGH *et al.*, 1999), and the ensemble transform Kalman filter (BISHOP *et al.*, 2001; MAJUMDAR *et al.*, 2001; HAMILL and SNYDER, 2002; LIU and KALNAY, 2008; LIU *et al.*, 2009). KANG (2009) and KANG *et al.* (2011) applied the LETKF for estimating surface CO_2 fluxes and obtained promising results. HAMILL and SNYDER (2002) demonstrated the application of an algorithm to select the optimal adaptive observation location using the background-error

---
[1] Molina Center for Energy and the Environment, La Jolla, CA, USA.
[2] Department of Meteorology, The Pennsylvanian State University, University Park, PA 16802, USA. E-mail: fzhang@psu.edu
[3] Department of Atmospheric Sciences, Texas A&M University, College Station, TX, USA.

statistics from an ensemble Kalman filter coupled to a quasigeostrophic model. They underscored the importance of accurate estimates of the background-error covariance matrix through using different data assimilation schemes. Their work focused on testing adaptive observation strategies for improving analyses but not forecasts. MAJUMDAR et al. (2006) evaluated the similarities and differences among five types of adaptive sampling guidance for tropical cyclones that occurred during the Atlantic hurricane season of 2004. They found that the guidance using the same adaptive sampling technique with different numerical models was often similar while the guidance using the two main techniques usually differed significantly.

Data assimilation has been used operationally for meteorological modeling and prediction. It is also useful as an inverse modeling technique for diagnosing pollutant emission source locations and strengths (i.e. parameter estimation) (CHANG et al., 1997; ELBERN et al., 2000; MENDOZA-DOMINGUEZ and RUSSELL, 2001) and for identifying locations (in time and space) for field observation networks and adaptive observations (DAESCU and CARMICHAEL, 2003). However, much of these data assimilation works have focused on variational data assimilation techniques.

The ensemble-based Kalman filter (EnKF) is an alternative data assimilation approach which has been applied in a number of studies (EVENSEN, 2003) since it was first introduced by EVENSEN (1994). It has also been used to improve air quality modeling (such as VAN LOON et al., 2000; HANEA et al., 2004; HEEMINK and SEGERS, 2002). Its advantages include flow-dependent background error covariance, ease of implementation, and its use of a fully nonlinear model. The EnKF system used here is the same as that employed in ZHANG et al. (2006) and MENG and ZHANG (2007). It is a square root EnKF with 20 ensemble members that uses covariance relaxation (ZHANG et al., 2004, their Eq. 5 where $\alpha = 0.5$) to inflate the background error covariance. More recent development and applications of this system can be found in MENG and ZHANG (2008a, b), ZHANG et al. (2009, 2011), HU et al. (2010), and WENG et al. (2011) while a comprehensive review of mesoscale applications of the EnKF is presented in MENG and ZHANG (2011). STUART et al. (2007) studied the use of ensemble-based Kalman filtering of chemical

observations for constraining meteorological uncertainties and for selecting target observation locations following the method proposed by HAMILL and SNYDER (2002). They further demonstrated the potential usefulness of the above method for locating promising adaptive observations in a predictive model. However, they did not test the strategy through assimilating the additional observations at the selected locations. LIU and KALNAY (2008) proposed an ensemble sensitivity method, which was conducted within an ensemble Kalman filter, to measure observation impact on the reduction of forecast errors due to assimilation of observations. LIU et al. (2009) further investigated the analysis sensitivity, which is proportional to the analysis error and anti-correlated with the observation error. The purpose of this study is to apply and extend the method developed by HAMILL and SNYDER (2002) to determine the meteorological targeted observation locations for improving ozone prediction in Houston and the surrounding area using a real-world NWP model (MM5) and photochemical model (CMAQ/Models3). In addition, the efficiency of the targeted observations is investigated using EnKF data assimilation under the perfect-model assumption.

A high-ozone event that occurred on 30 August during the Texas Air Quality Study of 2000 (TexAQS2000) is chosen for study. An analysis of the meteorological and ozone conditions on this day is provided by BANTA et al. (2005), and mesoscale simulations of the meteorology and photochemistry on this day have been reported by BAO et al. (2005), ZHANG et al. (2007), and CHENG and BYUN (2008a, b). The day featured westerly and northwesterly winds in the morning that died down by early afternoon. During mid- to late-afternoon, bay and sea breezes developed along the immediate coastline, and hourly-averaged ozone levels of nearly 200 ppbv were recorded between Galveston Bay and downtown Houston. Airborne lidar observations and model simulations indicate that very high concentrations of ozone were present over Galveston Bay, and photochemical model simulations also produce high ozone concentrations over the nearshore Gulf of Mexico east and southeast of the Houston–Galveston area. Skies were generally clear, and no deep convection or precipitation was present.

The initial and boundary conditions used in the meteorological model are interpolated from the Eta model's 3-hourly gridded analysis for the Global Energy and Water Cycle Experiment (GEMEX) Continental-Scale International Project (GCIP). Since the GCIP analysis for the initial conditions has already incorporated the conventional observation data from the current fixed network, the targeted observation locations we mention here and after are actually single sounding observations embedded within a generic synoptic-scale observing network. In addition, the method for selecting targeted observation locations is not specific to Houston and its surrounding area.

The methodology for choosing targeted observation locations is given in Sect. 2. Section 3 describes observation targeting results and their sensitivity to the data treatment, resolutions and planetary boundary layer (PBL) parameterization schemes. Section 4 investigates the effectiveness of targeted observations, and the summary and discussion are provided in Sect. 5.

2. Methodology for Choosing Targeted Observation Locations

In this study, we try to explore the targeted meteorological observation locations for improving ozone prediction through both meteorological and photochemical ensemble forecasts.

2.1. Model Configurations

The Pennsylvania State University-National Center for Atmospheric Research fifth-generation nonhydrostatic mesoscale model (MM5) version 3 (DUDHIA, 1993) is used to run the meteorological simulations. Horizontal grid spacings of 12- and 4-km are used in the coarse domain (D1) and fine domain (D2) (Fig. 1), respectively, with one-way nesting. There are 43 vertical layers in the terrain-following coordinate system, with the model top at 50 hPa and vertical spacing smallest within the boundary layer. The MRF boundary layer parameterization scheme (HONG and PAN, 1996) and the simple ice microphysical scheme (DUDHIA, 1993) are

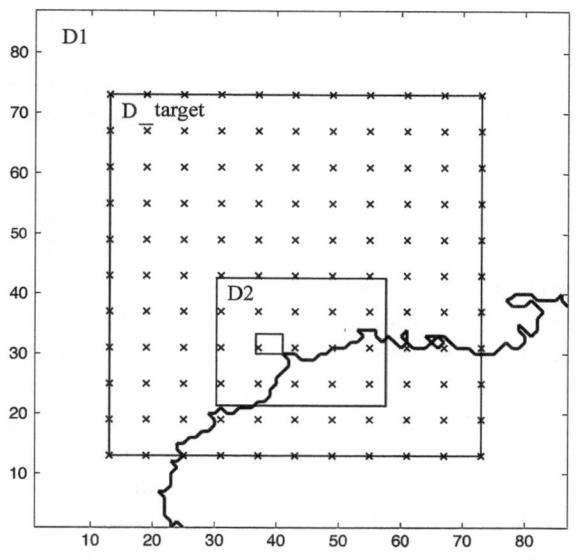

Figure 1

Model domains used in photochemical ensemble simulations. The horizontal grid spacing of domain 1 (D1) and domain 2 (D2) is 12- and 4-km, respectively. The *box* indicated by D_target denotes the domain used in Figs. 2, 3, 7, and 9. The *innermost box* denotes the domain used as the Houston area. *Cross points* indicate positions used to extract the profiles for assimilation. D1 grid points are numbered along the margins

used for both the 12- and 4-km domain. The cumulus scheme of GRELL (1993) with the shallow cumulus option is used only for the 12-km domain while convection in the 4-km domain is fully explicit. The MM5 simulations use the 24 land-use categories created from the 30-s USGS global land cover data, but a separate land-surface model or urban canopy parameterization are not used.

The MM5 simulations consist of 21 ensemble members initialized at 0000 UTC 30 August 2000 (1800 LST 29 August 2000) and integrated for 24 h. The ensemble members are initialized with the climatological ensemble initialization method of AKSOY et al. (2006) and ZHANG et al. (2007) in which dynamically consistent initial and boundary conditions are statistically sampled from the Eta model's 3-hourly gridded (40-km) analyses for the Global Energy and Water Cycle Experiment (GEWEX) Continental-Scale International Project (GCIP). Departures from the initial and boundary condition ensemble means are scaled down by 20% to reduce the ensemble spread over the Houston area to 0.4–0.6 m s^{-1} for horizontal wind components and 0.7–0.8 K for temperature. The scaled perturbations are added to the unperturbed initial and boundary conditions from the GCIP analyses for the model simulation period beginning at 0000 UTC 30 August which are used for the 12-km domain ensemble simulation.

The output from the meteorological ensemble simulations is used to drive a 21-member photochemical ensemble simulation using the EPA photochemical model CMAQ/Model-3 (BYUN and CHING, 1999) with the CBIV gas-phase chemical mechanism (EPA, 2003). The CMAQ model uses the same two horizontal nests as the MM5 simulations but employs just 21 vertical layers with the lowest three levels at approximately 21, 64 and 106 m. Anthropogenic and biogenic emissions are directly downloaded from the online emission inventory of the Texas Commission on Environmental Quality (TCEQ) (ftp.tceq.state.tx.us) and converted to CMAQ-ready emission files. Initial chemistry conditions are obtained from a 24-h spin-up run at 12-km grid spacing initialized at 00Z 29 Aug 2000. Further details regarding the model configurations are found in ZHANG et al. (2007).

2.2. Targeted Observations Sensitivity Assessment

The method adopted for determining targeted observation locations is similar to that developed by HAMILL and SNYDER (2002), which follows closely from the theory of BERLINEAR et al. (1999). Its algorithm is mathematically identical to the ETKF of BISHOP et al. (2001) and is also closely related to the ensemble sensitivity analysis in ZHANG (2005), HAWBLITZEL et al. (2007), TORN and HAKIM (2008), and SIPPEL and ZHANG (2008, 2010). In this method, the norm used for the total decrease in the model uncertainty is the sum over all state variables of the individual differences in variances, or the trace of P^b $- P^a$. $P^b - P^a$ is written as (HAMILL and SNYDER, 2002):

$$P^b - P^a = P^b H^T (H P^b H^T + R)^{-1} H P^b \quad (1)$$

where,

$$P^b = \langle (x^t - x^b)(x^t - x^b)^T \rangle \quad (2)$$

$$P^a = \langle (x^t - x^a)(x^t - x^a)^T \rangle \quad (3)$$

P^a and P^b are the post- and pre-analysis background error covariance matrices, respectively. H is the operator matrix relating the model state variables to the observational variables, superscript T means matrix transpose, and R is the observational error covariance matrix, which includes both the instrument error and the representation error. x^a is the m-dimensional analyzed state vector, x^b is the background state, and x^t is the true state, which is associated with the observations through the following formula:

$$Y_0 = H x^t + \varepsilon, \ \varepsilon \sim N(0, R) \quad (4)$$

Because the true state is not known, here we estimate the background error covariance using the ensemble covariance matrices around the ensemble mean. Thus, P^b in Eq. 2 can be estimated by

$$P^b = \frac{1}{n-1} \sum_{i=1}^{n} \left(x_i^b - \overline{x^b} \right) \left(x_i^b - \overline{x^b} \right)^T = X^b (X^b)^T \quad (5)$$

where the subscript i denotes ensemble members, n represents the ensemble size. Substituting Eq. 5 to 1, we obtain

$$P^b - P^a = X^b (HX^b)^T [HX^b (HX^b)^T + R]^{-1} HX^b (X^b)^T \quad (6)$$

In this study, we intend to choose the locations for additional observations with the goal of minimizing the forecast-error variance. It is necessary to compare the forecasts from the initial condition with and without additional observations. If we mark quantities relating to these two forecasts by superscripts fla and flb, respectively, the change in the forecast-error variance due to the additional observations is tr(P^{fla} − P^{flb}). When the analysis errors are not too large, we have

$$P^{f|b} - P^{f|a} \sim M(P^b - P^a) M^T$$

Here M is the linearization of the nonlinear forecast operator M (BISHOP et al., 2001; MAJUMDAR et al., 2001). From Eq. 6, we can get

$$P^{f|a} - P^{f|b} \approx M X^b (HX^b)^T [HX^b (HX^b) T + R]^{-1} HX^b (MX^b) T \quad (7)$$

Considering the ensemble forecasts from the background state, $x_i^{flb} = M\,(x_i^b)$ for $i = 1,..., n$. MX^b in Eq. 7 can be replaced by X^{flb} with the same accuracy, and HX^b can be replaced by Y, which represents the observation located at the model grid point, so Eq. 7 becomes

$$P^{f|a} - P^{f|b} \approx (X^{f|b} \cdot Y^T)[Y \cdot Y^T + R]^{-1}[Y \cdot (X^{f|b})^T] \tag{8}$$

where X^{flb} represents model state variables and Y represents observation variables. The symbol "$\cdot$" means covariance between two variables. In this particular study, X^{flb} represents photochemical model output variables (such as the surface ozone concentration) or meteorological model output variables (such as u, v, and T), and Y represents meteorological observation variables at model grids (such as u, v and T at all vertical layers). R represents observation error variances of u, v and T. The observational error variances at vertical layers used in this study are interpolated from those provided by PARRISH and DERBER (1992), and are assumed to be uncorrelated between vertical levels and different variables (Table 1).

The trace of $(P^{flb} - P^{fla})$ can be computed for each additional observation location candidate (each H). Like STUART et al. (2007), we define the trace of $(P^{flb} - P^{fla})$ normalized by its domain-wide maximum value as the observational impact factor. The locations with the maximum observational impact factor value are expected to be where additional observations can best improve ozone prediction or

meteorological prediction, depending on which model variables are included in Eq. 8.

If we consider u, v and T in an entire vertical column as one sounding observation and the surface O_3 concentration ($[O_3]$) as the model state variable (or the forecast variable), the above formula can be used to calculate the observational impact factor of sounding observations on the surface $[O_3]$. Figure 1 shows the model domains (D1 and D2) for CMAQ/Models3, the domain (D_target) used for examining the effect of the targeted observations, the Houston area (innermost domain), and the locations (cross points) at which observational sounding data is extracted for assimilation. A total of 121 sensitivity experiments (one for each sounding) are conducted to examine the impact of each individual sounding as compared to the estimated impact derived from the ensemble sensitivity.

With the climatological ensemble initialization scheme employed here, the background error covariance is independent of information regarding previous observation locations. In practice, the background errors would be expected to be larger in places where observations are relatively sparse, such as over the Gulf of Mexico in this case, but such observation-sensitive background error information may not always be available. All other things being equal, the observation sensitivity would be higher than analyzed here where the background error covariance is large and smaller where the background error covariance is small.

3. Diagnosed Targeting Sensitivities

The ensemble mean forecast of ozone, temperature and wind at 21 UTC (15 LST, the approximate time of maximum ozone) is shown in Fig. 2a for the 12-km run over D1 and Fig. 2c for the 4-km run over D2. The highest values of ozone are located over Galveston Bay and the adjacent Gulf of Mexico, but a tongue of high ozone concentrations extends westward from Galveston Bay toward downtown Houston. The winds have weakened and shifted from an offshore land breeze to an onshore bay/sea breeze. The ensemble mean ozone at 21 UTC is 120 ppb in the Houston area and is as high as over 150 ppb over Galveston Bay.

Table 1

Observation error variances for temperature and u and v wind components

Level	Pressure (hPa)	T (K^2)	u (m^{2}s^{-2})	v (m^{2}s^{-2})
1	1,010	3.24	1.96	1.96
2	1,001	3.24	1.96	1.96
3	989	3.16	2.04	2.04
4	967	3.06	2.22	2.22
5	936	2.85	2.53	2.53
6	895	2.65	2.92	2.92
7	830	2.31	3.68	3.68
8	676	1.69	5.95	5.95
9	483	1.85	8.06	8.06
10	291	4.2	11.35	11.35
11	98	9.79	6.3	6.3

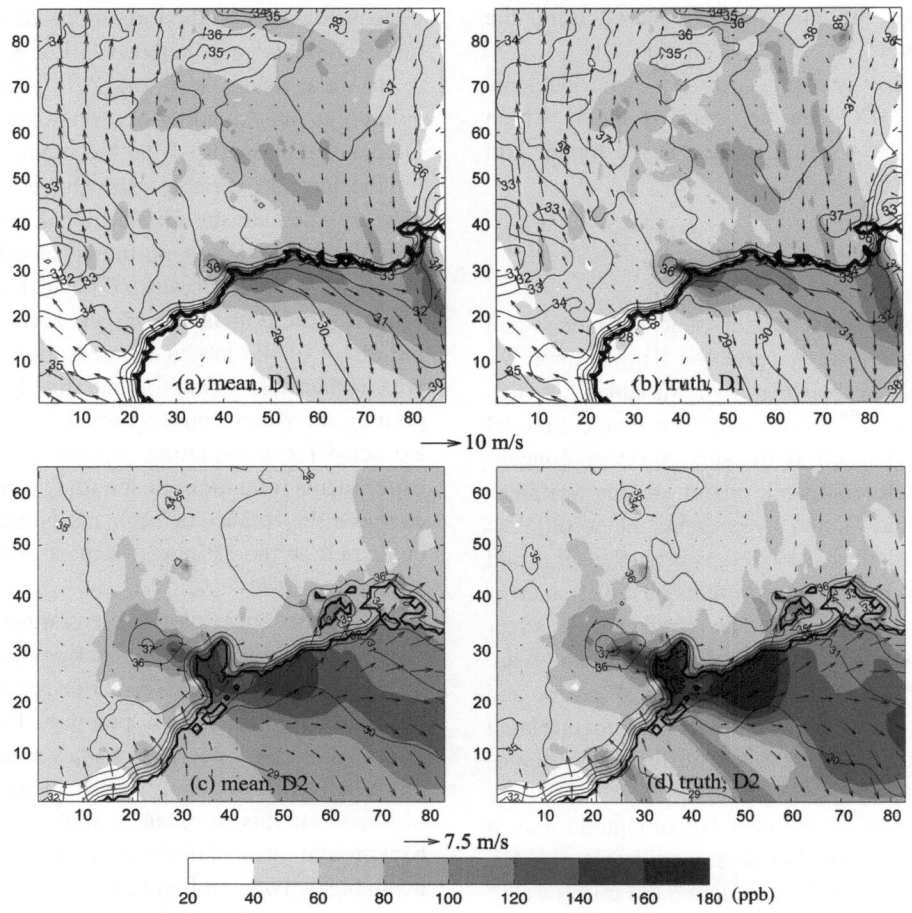

Figure 2
The ensemble mean forecast of ozone (ppb, *shaded*), temperature (Celsius, *contoured*) and wind (*vectors*) at 21 UTC (15 LST) in **a** D1 and **c** D2, and the simulations of ozone, temperature, and wind from the "truth" run in **b** D1 and **d** D2. Domain grid points are numbered along the margins

Figure 2b and d show the simulations of ozone, temperature, and wind from a separate model run (the "truth" run) that is the source of the simulated observations for the experiments described in Sect. 4. Because this is a single deterministic run, the ozone field has a more detailed structure than in the ensemble mean. In the context of this experiment, improvements in the model forecasts are only possible in places where the ensemble mean (Fig. 2a, c) and the truth run (Fig. 2b, d) disagree at 21 UTC.

Using the method described in Sect. 2, we compare the observational impact factors derived from different treatments of forecast variables (Houston urban ozone, domain-wide ozone, and meteorological variables), observation data (sounding vs. single

layer), different model resolutions (coarse vs. fine resolutions), and different PBL schemes (single- vs. multi-scheme). Multi-scheme, hereafter, means ensembles with five different PBL schemes: MRF PBL (HONG and PAN, 1996), High-resolution Blackadar PBL (BLACKADAR, 1979), Burk-Thompson PBL (BURK *et al.*, 1989), Eta PBL (JANJIC, 1990, 1994), and Gayno-Seaman PBL (BALLARD *et al.*, 1991; SHAFRAN *et al.*, 2000).

The spatial and temporal evolution of the observational impact factor for sounding observations on the peak time surface [O$_3$] averaged over Houston area (the inner box) is shown in Fig. 3, along with the ensemble mean of the surface wind and temperature over the simulation period. Initially (00 UTC or 18

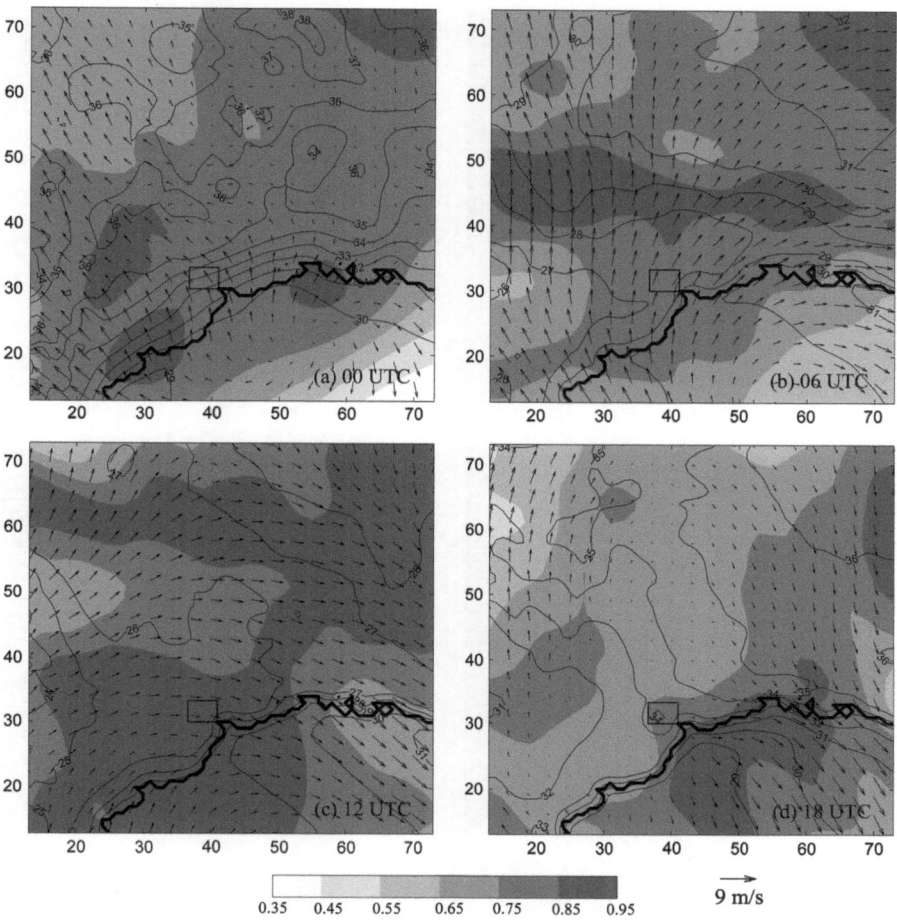

Figure 3
The observational impact factor (*shaded*) for sounding observations on the peak time surface [O₃] (21 UTC) over the Houston area (the innermost domain in Fig. 1) along with the ensemble mean surface temperature (Celsius, *contoured*) and surface wind (*vectors*)

LST, Fig. 3a), the largest observational impact factor is found over land near the coast, where the ensemble mean of the wind direction is almost perpendicular to the coastline and the gradient of the ensemble mean of the temperature has its maximum value. This shows that the location and intensity of the previous day's sea-breeze front has an important correlation in the model simulations with the formation of the next day's urban [O₃]. Another large impact factor area is located in the northeast corner of the domain. At 06 UTC (0 LST, Fig. 3b), the largest observational impact values are farther inland, though still collocated with the stronger winds and remnant temperature gradient of the sea breeze. At 12 UTC (6 LST, Fig. 3c), the greatest impact area is located near and upstream of Houston. Finally, at 18 UTC (12

LST, Fig. 3d), the area of the largest observational impact factors are generally smaller than before, though an area with large impact factor persists along the coast east and southeast of Houston. This area perhaps reflects the importance of the timing and magnitude of the local sea breeze on the ozone concentrations in the Houston area.

Figure 4 shows the observational impact factor for 21 UTC ozone concentrations throughout domain D1. The sensitivities at 00 UTC (Fig. 4a) are quite similar to those for Houston area ozone only (Fig. 3a), suggesting that observational influences on the ozone forecast this far in advance are fairly large-scale and not specific to the ozone in any particular region. As the verification time approaches, differences between the domain-wide impacts and Houston

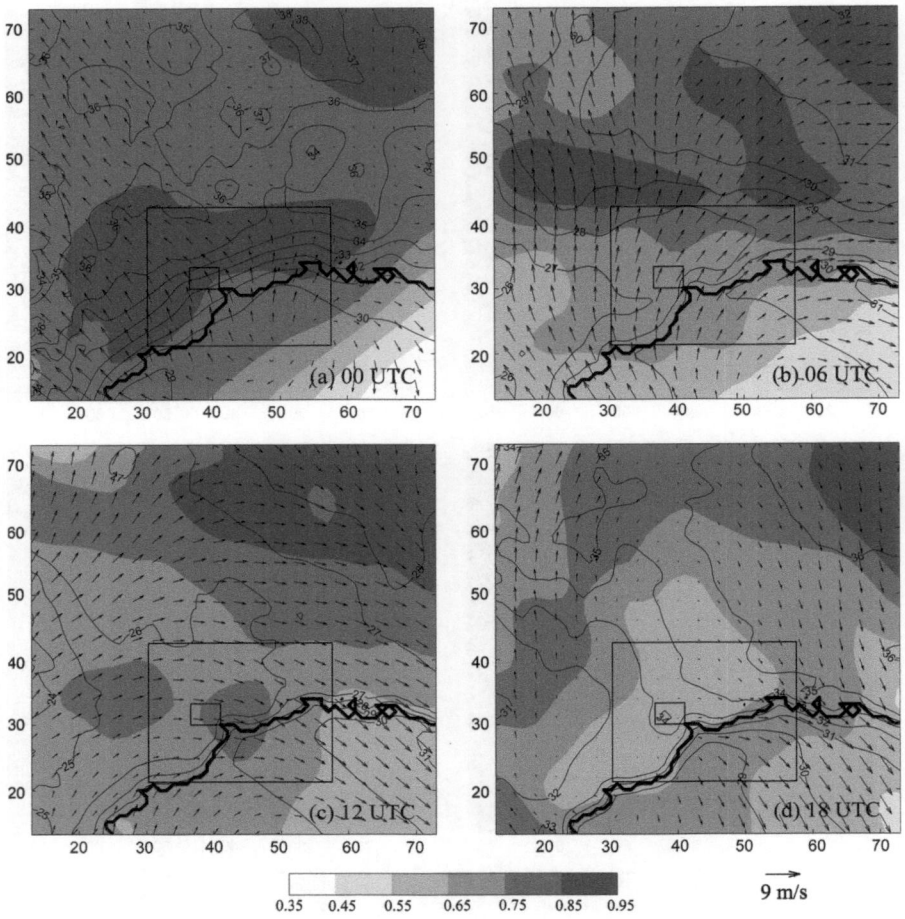

Figure 4
The observational impact factor (*shaded*) for sounding observations on the peak time surface [O₃] (21 UTC) over D1 along with the ensemble mean surface temperature (*contoured*) and surface wind (*vectors*) from 00 to 18 UTC. Inner domain indicates D2 and the innermost domain indicates the Houston area

area impacts become larger. In general, the domain-wide ozone is more sensitive to sounding observations in the northern portion of the domain.

If the targeted sounding is to be assimilated by a sequential data assimilation system that ingests observations of u, v and T at each layer separately, the covariances and correlations between the observations on different layers may not properly be taken into account in the calculation of the observational impact factor. When the multiple observations within a single sounding are redundant for analysis purposes, the impact factor for the multiple single-layer observations can be estimated as the average of the observational impact factor over each layer. Figure 5 shows the average observational impact factor for

single-layer observations on the peak time surface [O₃] over D1. In general, the distribution of the observational impact factor is consistent with that for the sounding observations treated as a whole (Fig. 4), while the impacts themselves are smaller. However, the average single-layer impact factor is much smaller than the full-sounding impact factor in the coastal areas, especially at later assimilation times.

The large difference between single-layer and full-sounding impact factors in coastal areas is probably due to the importance of the sea breeze circulation for determining ozone concentrations. Only a small portion of a coastal or offshore sounding will sample the sea breeze flow or return circulation, and the rest of the observations aloft will be of

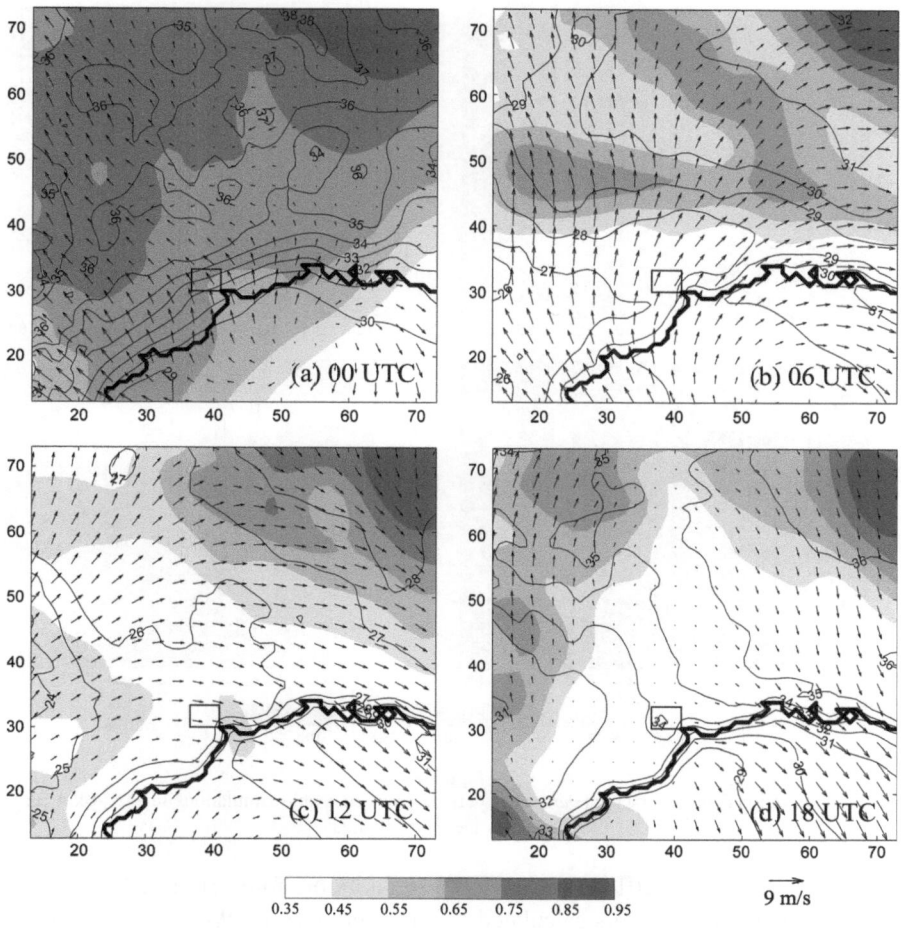

Figure 5
The observational impact factor (*shaded*) for single layer observations on the peak time surface [O_3] (21 UTC) over D1 along with the ensemble mean surface temperature (*contoured*) and surface wind (*vectors*) from 00 to 18 UTC. The inner domain indicates the Houston area

relatively little value. In such a circumstance, averaging of impact factors from the various levels in the sounding would underestimate the impact of the sounding observation. In contrast, farther inland, an observation from anywhere in the lower troposphere would be sufficient to improve the model's analysis of the large-scale flow opposing the sea breeze and controlling its inland penetration. Here, averaging comes closer to the impact factor computed from a full sounding. This demonstrates that the impact of the multiple data points within a sounding should be estimated collectively rather than as a set of independent observations.

Since the model resolution contributes to the quality of numerical weather forecast, we investigate the observational impact factor derived from the high resolution (4-km) ensemble simulations. Figure 6 shows the observational impact factor for sounding observations on the peak time surface [O_3] within domain 2 calculated from 4-km ensemble simulations (D2 in Fig. 1) at different times. The evolutions of the observational impact factor in the 4-km run is chiefly consistent with those from the 12-km run, however, there exist more fine structures along the coastline, especially at 06 and 18 UTC (00 LST and 12 LST), indicating the significant contributions of the local circulation in the coastal area to the formation of the peak time surface [O_3].

Ozone prediction is also sensitive to various PBL schemes used in the meteorological model on an urban

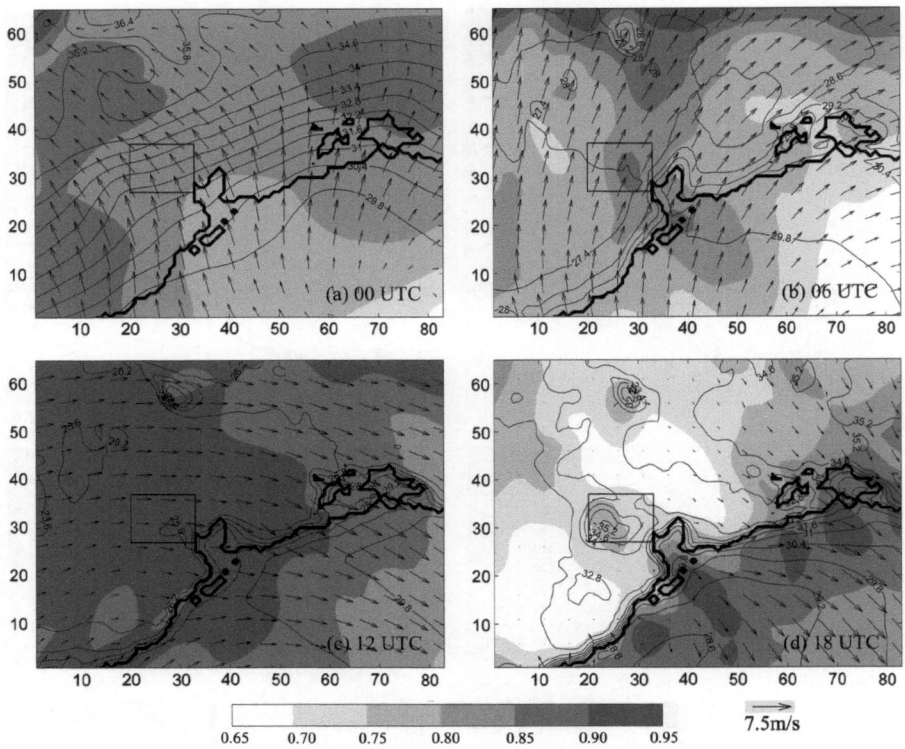

Figure 6
Same as Fig. 4, but the observational impact factor (*shaded*) is derived from ensemble simulations in D2 (4-km run). The inner domain indicates the Houston area

scale (MAO *et al.*, 2006, BEI *et al.*, 2010). We have compared the observational impact factors calculated from the ensemble simulations using a single-PBL scheme versus multi-PBL schemes, respectively (Fig. 7). They exhibit similar basic patterns (compare to Fig. 6), but with large differences in the details of location and magnitude, especially along the coastline. It shows that the targeting depends upon details of the forecast system used in the data assimilation scheme, so we cannot use one technique for identifying target locations and another technique for assimilating the additional data. It also further shows that model errors besides model resolution should also be considered in the targeted observation network design.

4. *Examination of the Improvement from Targeted Observations*

The most interesting question is whether the assimilation of an additional observation actually reduces the forecast error in line with the calculated impact factor. We have conducted a group of experiments in an initial attempt to address this question. In each experiment, a synthetic (fake) observational sounding extracted from the truth run (a perfect-model identical twin experiment) has been assimilated. Random uncorrelated noise of a particular magnitude (see Table 1) is added to each observation. The synthetic soundings are extracted from the places indicated by crosses in Fig. 1. Each synthetic sounding is assimilated using a standard Ensemble Kalman Filter algorithm, without covariance localization, inflation of member deviations, or hybridization. The EnKF used in the current study follows closely that of SNYDER and ZHANG (2003). As in the standard Kalman filter,

$$x^a = x^f + K(\mathbf{y} - Hx^f),$$

where x^f represents the prior estimate or first guess, x^a is the posterior estimate or analysis, $\mathbf{y}$ is the observation vector, H is the observation operator that

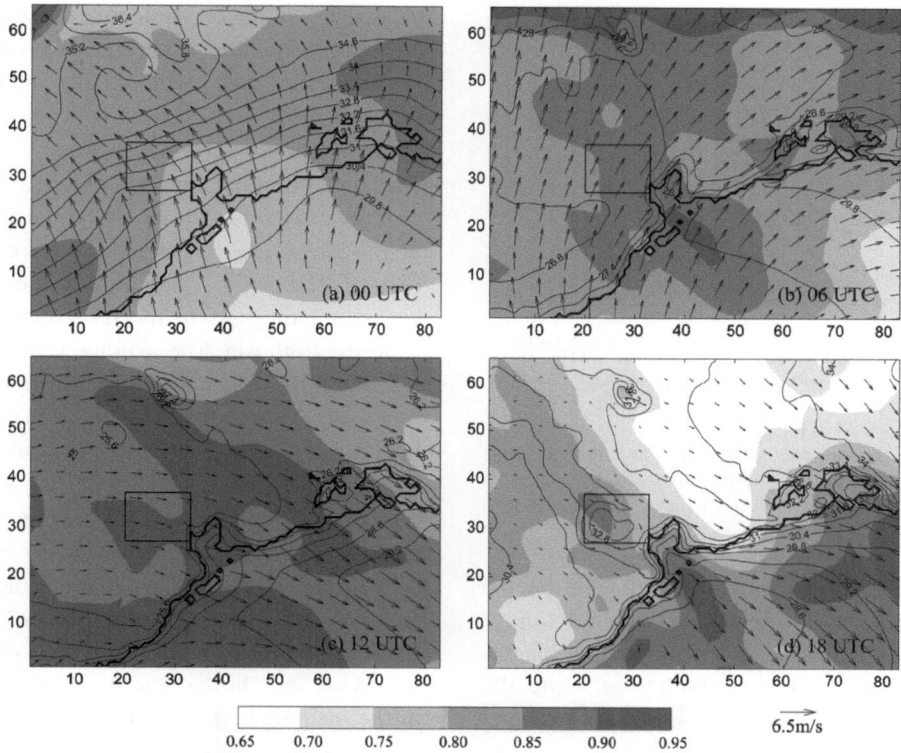

Figure 7
Same as Fig. 6, but the observational impact factor (*shaded*) is derived from ensemble simulations with multi-PBL schemes

returns observed variables given the state, and K is the so-called Kalman gain matrix defined as

$$K = P^f H^T (H P^f H^T + R)^{-1},$$

where P^f and R represent the background and observational error covariance, respectively. In the EnKF, the flow-dependent P^f is estimated through an ensemble of short-range forecasts. Observations are taken sequentially with observations errors assumed to be uncorrelated. Further background on the EnKF refer to SNYDER and ZHANG (2003) and references therein.

The assimilation is performed at 00 h. For each experiment, we calculate the actual reduction in the root mean squared error (with respect to the truth run) of the surface wind speed (hereafter rms-SPD) and the surface [O_3] (hereafter rms-O_3) according to:

$$\text{rms} = \left\| \bar{x}^{f|b} - x^t \right\| - \left\| \bar{x}^{f|a} - x^t \right\| \qquad (9)$$

We also identify the square root of the observational impact factor with the expected reduction in the root mean squared error. The actual reduction in the rms-SPD over D1 and the actual reduction in the rms-O_3 over the Houston area are calculated and compared to their expected reduction.

Figure 8 shows the scatter plot of the expected and actual reduction in the rms-SPD at 21 UTC (15 LST) when the observations were assimilated at 00 UTC (18 LST). The rms-SPD at 21UTC is reduced when any of the synthetic soundings are assimilated at the model initial time. In general, a larger expected reduction in rms-SPD is associated with a larger actual reduction in rms-SPD. However, the expected reduction is always overestimated. The correlation is only 0.13, suggesting the value of the impact factor diagnosis is limited, at least in this case. One possible reason is that the background-error covariances provided to the EnKF may not be perfect since the ensemble size is limited. Also, the initial ensemble perturbations, despite being dynamically consistent, are derived directly from climatological uncertainties in this location, which may not have sufficient flow

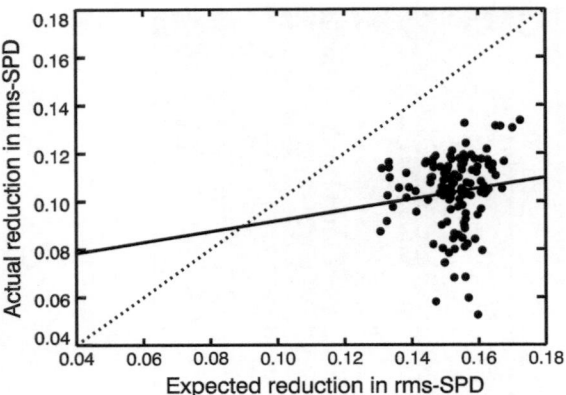

Figure 8

Scatter plot of the expected and actual reduction in the root mean squared error of surface wind speed (rms-SPD, m/s) over D1 at 21 UTC (15 LST). The *solid* (*dotted*) *line* denotes data fitted (ideal) line

dependent statistics at the initial time. In addition, the actual error reduction (through EnKF assimilation of the synthetic soundings) also depends on the magnitude of the error (without assimilation) while the predicted impact factor may not.

Figure 9 shows the horizontal distributions of the expected and actual reduction in the rms-SPD. Both expected and actual reductions in rms-SPD are largest for synthetic soundings ingested from southwest and northeast of Houston. The actual reduction is much smaller than the expected reduction when fake soundings are assimilated in the eastern margin of the domain.

Reductions in meteorological errors should lead to reductions in ozone errors. However, ozone errors,

particularly in the Houston area, should depend partly on the initial configuration and transport of ozone and its precursors and partly on the interaction of the meteorological conditions with emissions during model integration. The latter process is not explicitly considered in the estimation of expected error.

A scatter plot of the expected and actual reduction in the rms-O_3 over the Houston area is shown in Fig. 10. The rms-O_3 over the Houston area is reduced by assimilating any of the fake soundings extracted from the truth run. The average actual ozone impact is similar to the average expected ozone impact. The correlation between the expected and actual reduction in rms-O_3 is 0.23, which is still low but is slightly higher than for rms-SPD. For the low correlation between the expected and actual reduction in rms-O3, one possible reason is the non-linear response of the ozone concentration uncertainty to the meteorological field uncertainty (ZHANG *et al.*, 2007); another is the lack of consideration of ozone emissions mentioned earlier. On the other hand, the better estimate of the magnitude of the impact factor for ozone than for meteorological variables may be due to a greater dependence of O_3 on meteorological conditions throughout the forecast period, including the time at which the impact factor estimate is made, since O_3 at verification time is sensitive to transport and dilution that takes place beginning at analysis time and continuing throughout the forecast period.

The horizontal distributions of the expected and actual reduction in the rms-O_3 (Fig. 11) show that both the expected and actual reduction in rms-O_3 are

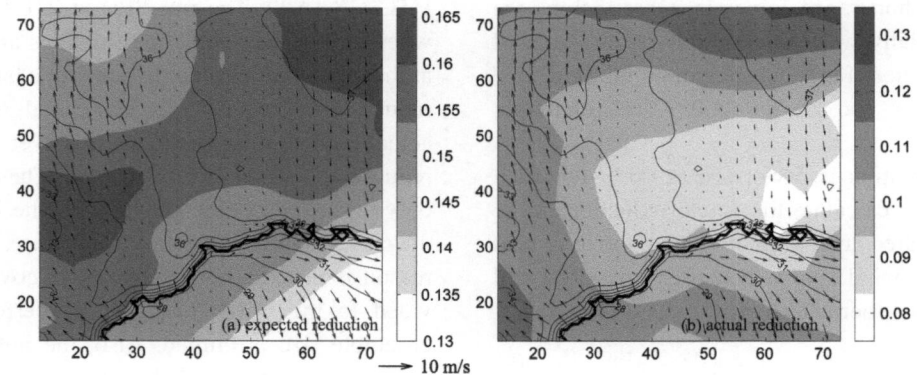

Figure 9

Horizontal distributions of **a** the expected reduction and **b** the actual reduction in rms-SPD (m/s) over D1 at ozone peak time through data assimilation at 00 UTC (6 LST)

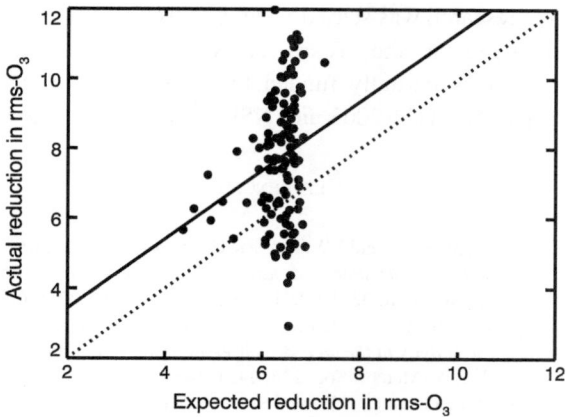

Figure 10
Scatter plot of the expected and actual reduction in the root mean squared error of surface [O_3] (rms-O_3, ppb) over the Houston area at 21 UTC (15 LST). The *solid* (*dotted*) *line* denotes data fitted (ideal) line

largest for synthetic soundings ingested from west and southwest of Houston. The actual reduction tends to be much smaller than the expected reduction when synthetic soundings are assimilated near certain boundaries, particularly the northern, eastern, and southwestern margins of the domain.

5. Summary and Discussion

An ensemble-based method for determining targeted observation locations developed by HAMILL and SNYDER, 2002 has been extended and applied for improving ozone predictions in Houston and the surrounding area. In this method, an observational impact factor has been calculated from an ensemble run and employed to select the targeted observation locations. The locations with larger observational impact factors are expected to be where supplementary observations can better reduce the forecast error. The improvement brought by the targeted observations is further examined through an ensemble Kalman filter assimilation of synthetic observations.

The computational cost of estimating the observational impact factor as proposed here is only slightly more than the cost of performing the set of ensemble forecasts on which the calculation is based. Calculation of actual impact factor through assimilation of test observations requires one EnKF analysis and set of ensemble forecasts for each test observation, which rapidly becomes prohibitively expensive as the number of test observations increases.

The observational impact factor for meteorological observations on the peak time surface [O_3] evolves with time. Initially, the larger impact factor is mainly located at the land side of the coastline and the northeast portion of the domain. The larger impact factor area near the coastline moves to the north and then northeast at later assimilation times with the increase of south and southwest wind. Close to verification time, the observational impact factor along the coastline increases. This shows that the observational impact factor field is changed with the evolution of the wind circulation.

The observational impact factor from considering sounding data for different layers separately on the

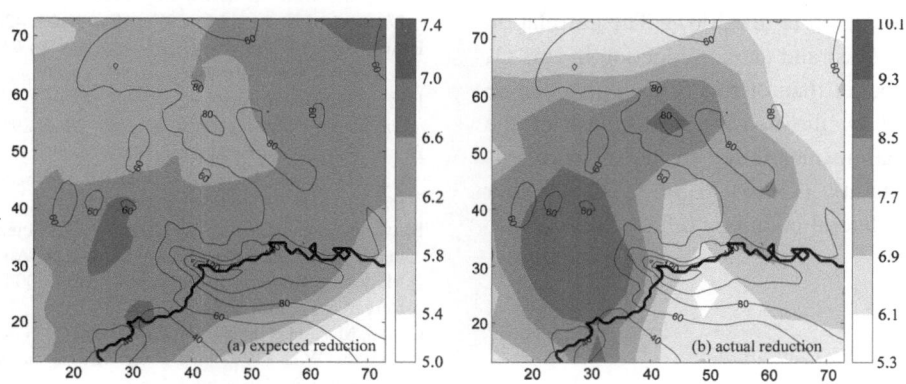

Figure 11
Horizontal distributions of **a** the expected reduction and **b** the actual reduction in rms-O_3 (ppb) over the Houston area at peak ozone time through data assimilation at 00 UTC (6 LST)

peak time surface [O_3] is different from that for the sounding observations over the entire column as a whole, indicating that the targeting strategies should be based on proper consideration of the impact of the sounding in its entirety on the forecast.

The observational impact factor derived from ensemble simulations with different resolution basically has the same pattern except for some minor differences in the distribution. Ensembles created with different PBL schemes can also produce differences in the detailed distribution and magnitude of the calculated observational impact factor fields, especially along the coastline, which indicates the model error due to the PBL scheme should also be included in the observation targeting. It also suggests we should not use one technique for identifying target locations and another technique for assimilating the additional data.

Through assimilating synthetic observational soundings extracted from the truth run using an ensemble Kalman filter, the surface wind speed forecasts are improved in the root mean square sense when any of the synthetic observational soundings is assimilated at the model initial time. In general, larger observational impact factors are associated with larger reductions in rms-SPD. However, at the peak ozone time (21 UTC), the correlation between the expected and actual reduction in the rms-SPD is low (0.13). The lack of agreement between expected and actual observation impacts could be produced by the data assimilation scheme, without using the localization in EnKF, the noise of the observation data, sampling ensemble with a limited ensemble size, and the limit of meteorological predictability (MORSS and EMANUEL, 2002; ZHANG et al., 2007). The correlation between the expected and actual reduction in rms-O_3 is also low but better than of rms-SPD, which can be partly explained by the non-linear response of the ozone concentration uncertainty to the meteorological field uncertainty (ZHANG et al., 2007). Other reasons (such as lateral boundary influence) may also affect the results.

Acknowledgments

The authors are grateful to Yonghui Weng and Zhiyong Meng for help on the EnKF implementation. This research was started when both NB and FZ were employed at the Texas A&M University. This research is partially funded by GTRI/HARC Grant Project No. H24-2003 and NSF grant ATM-084065.

REFERENCES

AKSOY, A., ZHANG, F. and J.W. NIELSEN-GAMMON (2006) Ensemble-based simultaneous state and parameter estimation with MM5, Geophys. Res. Lett. 33, L12801, doi:10.1029/2006GL026186.

BAKER, N.L. and DALEY, R. (2000) Observation and background adjoint sensitivity in the adaptive observation-targeting problem. Quart. J. Roy. Meteor. Soc. 126, 1431–1454.

BALLARD, S.P., GOLDING, B.W. and SMITH, R.N.B. (1991) Mesoscale model experimental forecasts of the Haar of northeast Scotland. Mon. Wea. Rev., 119, 2107–2133.

BANTA, R.M., SENEF, C.J., NIELSEN-GAMMON, J., DARBY, L.S., RYERSON, T.B., ALVAREZ, R.J., SANDBERG, S. P., WILLIAMS, E.J. and TRAINER, M. (2005) A bad air day in Houston. Bull. Amer. Meteor. Soc., 86, 657–659.

BAO, J.-W., MICHELSON, S.A., McKEEN, S.A. and GRELL, G.A. (2005) Meteorological evaluation of a weather-chemistry forecasting model using observations from the TEXas AQS 2000 field experiment. J. Geophys. Res., 110, D21105, doi:10.1029/2004JD005024.

BEI, N., LEI, W., ZAVALA, M., and MOLINA, L.T. (2010) Ozone predictabilities due to meteorological uncertainties in Mexico City Basin using ensemble forecasts, Atmos. Chem. Phys., 10, 6295–6309.

BERGOT, T., (2001) Influence of the assimilation scheme on the efficiency of adaptive observations. Quart. J. Roy. Meteor. Soc., 127, 635–660.

BERGOT, T., HELLO, G., JOLY, A. and MALARDEL, S. (1999) Adaptive observations: a feasibility study. Mon. Wea. Rev., 127, 743–765.

BERLINER, L.M., LU, Z.-Q. and SNYDER, C. (1999) Statistical design for adaptive weather observations. J. Atmos. Sci., 56, 2536–2552.

BISHOP, C.H., and TOTH, Z. (1999) Ensemble transformation and adaptive observations. J. Atmos. Sci., 56, 1748–1765.

BISHOP, C.H., ETHERTON, B.J. and MAJUMJAR, S. (2001) Adaptive sampling with the ensemble transform Kalman filter. Part 1: Theoretical aspects. Mon. Wea. Rev., 129, 420–436.

BLACKADAR, A.K., (1979) High resolution models of the planetary boundary layer, Adv. Environ. Sci. and Eng., 1, 50–85.

BUIZZA, R., and A. MONTANI, (1999) Targeting observations using singular vectors. J. Atmos. Sci., 56, 2965–2985.

BURK, S.D., and THOMPSON, W.T. (1989) A vertically nested regional numerical prediction model with second-order closure physics. Mon. Wea. Rev., 117, 2305–2324.

BYUN, D.W. and CHING, J.K.S. (ed.) (1999) Science Algorithms of the EPA Models-3 Community Multi-scale Air Quality (CMAQ) Modeling System, EPA Report, EPA/600/R-99/030, NERL, Research Triangle Park, NC.

CHANG, M.E., HARTLEY, D.E., CARDELINO, C., HAAS-LAURSEN, D. and CHANGE, W.L. (1997) On using inverse methods for resolving emissions with large spatial in homogeneities. J. Geophys. Res., 102, 16023–16036.

CHENG, F.-Y., and BYUN, D.W. (2008a) Application of high resolution land use and land cover data for atmospheric modeling in the *Houston-Galveston metropolitan area, Part I: Meteorological simulation results.* Atmos. Env., **42**, 7795–7811, doi: 10.1016/j.atmosenv.2008.04.055.

CHENG, F.-Y., and BYUN, D. W. (2008b) *Application of high resolution land use and land cover data for atmospheric modeling in the Houston-Galveston metropolitan area, Part II: Air quality simulation results.* Atmos. Env., **42**, 4853–4869, doi: 10.1016/j.atmosenv.2008.02.059.

DAESCU, D.N., and CARMICHAEL, G.R. (2003) *An adjoint sensitivity method for the adaptive location of the observations in air quality modeling.* J. Atmos. Sci., **60**, 434–450.

DUDHIA, J., (1993) *A nonhydrostatic version of the Penn State-NCAR Mesoscale Model: Validation tests and simulation of an Atlantic cyclone and cold front.* Mon. Wea. Rev., **121**, 1493–1513.

ELBERN, H., SCHMIDT, H., TALAGRAND, O. and EBEL, A. (2000) *4-D variational data assimilation with an adjoint air quality model for emission analysis.* Environmental Modelling & Software, **15**, 539–548.

EMANUEL, K.A., and Coauthors (1995) *Report of the first prospectus development team of the U.S. Weather Research Program to NOAA and the NSF.* Bull. Amer. Meteor. Soc., **76**, 1194–1208.

Environmental Protection Agency (EPA) (2003), Models-3 Community Multiscale Air Quality (CMAQ) model, version 4.3, report, Research Triangle Park, N.C. (Available at http://www.cmascenter.org/modelclear. shtml).

EVENSEN G., (1994) *Sequential data assimilation with a non-linear quasi-geostrophic model using Monte Carlo methods to forecast error statistics.* J. Geophys. Res., **99**, 10143–10162.

EVENSEN G., (2003) *The Ensemble Kalman Filter: theoretical formulation and practical implementation.* Ocean Dynamics, 53, 343–367.

GELARO, R., LANGLAND, R., ROHALY, G. D. and ROSMOND, T.E. (1999) *An assessment of the singular vector approach to target observations using the FASTEX data set.* Quart. J. Roy. Meteor. Soc., **125**, 3299–3327.

GELARO, R., REYNOLDS, C.A., LANGLAND, R.H. and ROHALY, G.D. (2000) *A predictability study using geostationary satellite wind observations during NORPEX.* Mon. Wea. Rev., **128**, 3789–3807.

GRELL, G.A., (1993) *Prognostic evaluation of assumptions used by cumulus parameterizations.* Mon. Wea. Rev., **121**, 764–787.

HAMILL, T.M. and SNYDER, C. (2002) *Using improved background-error covariances from an ensemble Kalman filter for adaptive observations.* Mon. Wea. Rev., **130**, 1552–1572.

HANEA, R.G., VELDERS, G.J.M. and HEEMINK, A. (2004) *Data assimilation of groundlevel ozone in Europe with a Kalman filter and chemistry transport model.* J. Geophys. Res., **109**, D10302, doi:10.1029/2003JD004283.

HANSEN, J.A., and SMITH, L.A. (2000) *The role of operational constraints in selecting supplementary observations.* J. Atmos. Sci., **57**, 2859–2871.

HAWBLITZEL, D.P., ZHANG, F., MENG, Z. and DAVIS, C.A. (2007) *Probabilistic evaluation of the dynamics and predictability of the mesoscale convective vortex of 10-13 June 2003.* Mon. Wea. Rev., **135**, 1544–1563.

HEEMINK, A.W., and SEGERS, A.J. (2002) *Modeling and prediction of environmental data in space and time using Kalman filtering.* Stochastic Environmental Research and Risk Assessment, **16**, 255–240.

HONG, S.-Y. and PAN, H.-L. (1996) *Nonlocal boundary layer vertical diffusion in a medium-range. forecast model,* Mon. Wea. Rev., **124**, 2322–2339.

HU, X., ZHANG, F. and NIELSEN-GAMMON, J.W. (2010) *Ensemble-based simultaneous state and parameter estimation for treatment of mesoscale model error: a real-data study.* Geophys. Res. Lett., **37**, L08802.

JANJIC, Z.I., (1990) *The step-mountain coordinate: physical package.* Mon. Wea. Rev., **118**, 1429–1443.

JANJIC, Z.I., (1994) *The step-mountain eta coordinate model: further developments of the convection, viscous sublayer and turbulence closure schemes.* Mon. Wea. Rev., **122**, 927–945.

KANG, J., (2009) Carbon Cycle Data Assimilation Using a Coupled Atmosphere-Vegetation Model and the Local Ensemble Transform Kalman Filter, Ph. D Thesis, University of Maryland.

KANG, J.-S., KALNAY, E., LIU, J., FUNG, I., MIYOSHI, T. and IDE, K. (2011) *"Variable localization" in an Ensemble Kalman Filter: application to the carbon cycle data assimilation.* J. Geophys. Res., doi:10.1029/2010JD014673, (in press).

LANGLAND, R., GELARO, R., ROHALY, G.D. and SHAPIRO, M.A. (1999) *Target observations in FASTEX: Adjoint based targeting procedures and data impact experiments in IOP/7 and IOP/8.* Quart. J. Roy. Meteor. Soc., **125**, 3241–3270.

LIU, J., and E. KALNAY, 2008: *Estimating observation impact without adjoint model in an ensemble Kalman filter. Quart. J. Roy. Meteor. Soc.,* **134**, 1327-1335.

LIU, J., KALNAY, E., MIYOSHI, T. and CARDINALI, C. (2009) *Analysis sensitivity calculation in an ensemble Kalman filter,* Quart. J. Roy. Meteor. Soc., **135**, 1842–1851.

LORENZ, E.N. and EMANUEL, K.A. (1998) *Optimal sites for supplementary observations: Simulation with a small model.* J. Atmos. Sci., **55**, 399–414.

MAJUMDAR, S.J., BISHOP, C.H., SZUNYOGH, I. and TOTH, Z. (2001) *Can an Ensemble Transform Kalman Filter predict the reduction in forecast error variance produced by targeted observations?* Quart. J. Roy. Meteor. Soc., **127**, 2803–2820.

MAJUMDAR, S.J., ABERSON, S.D., BISHOP, C. H., BUIZZA, R., PENG, M.S. and REYNOLDS, C.A. (2006) *A comparison of adaptive oberving guidance for atlantic tropical cyclones.* Mon. Wea. Rev., **134**, 2354–2372.

MAO Q., L.L., GAUTNEY, T.M., COOK, M.E. JACOBS, SMITH, S.N. and KELSOE, J. J. (2006) *Numerical experiments on MM5-CMAQ sensitivity to various PBL schemes.* Atmos. Environ., **40**, 3092–3110.

MENG, Z., and ZHANG, F. (2007) *Tests of an ensemble Kalman filter for mesoscale and regional-scale data assimilation. Part II: Imperfect Model Experiments.* Mon. Wea. Rev., **135**, 1403–1423.

MENG, Z., and ZHANG, F. (2008a) *Tests of an Ensemble Kalman Filter for Mesoscale and Regional-Scale Data Assimilation. Part III: Comparison with 3DVar in a Real-Data Case Study.* Mon. Wea. Rev., **136**, 522–540.

MENG, Z., and ZHANG, F. (2008b) *Tests of an ensemble Kalman Filter for mesoscale and regional-scale data assimilation. Part IV: comparison with 3DVar in a month-long experiment.* Mon. Wea. Rev., **136**, 3671–3682.

MENG, Z., and ZHANG, F. (2011) *Limited-area ensemble-based data assimilation.* Mon. Wea. Rev., (in press).

MENDOZA-DOMINGUEZ, A. and RUSSELL, A.G. (2001) *Estimation of emission adjustments from the application of four-dimensional data assimilation to photochemical air quality modeling.* Atmos. Environ., **35**, 2879–2894.

MORSS, R.E. (1998) *Adaptive observations: Idealized sampling strategies for improving numerical weather prediction.* Ph.D. dissertation, Massachusetts Institute of Technology, 225 pp. [Available from UMI Dissertation Services, P. O. Box 1346, 300 N. Zeeb Rd., Ann Arbor, MI, 48106-1346].

MORSS, R.E., and EMANUEL, K.A. (2002) *Influence of added observations on analysis and forecast errors: Results from idealized systems.* Quart. J. Roy. Meteor. Soc., **128**, 285–322.

MORSS, R. E., EMANUEL, K.A. and SNYDER, C. (2001) *Idealized adaptive observation strategies for improving numerical weather prediction.* J. Atmos. Sci., **58**, 210–234.

PALMER, T.N., GELARO, R., BARKMEIJER, J. and BUIZZA, R. (1998) *Singular vectors, metrics, and adaptive observations.* J. Atmos. Sci., **55**, 633–653.

PARRISH S. and DERBER, J. (1992) *The national meteorological center's spectral statistical-interpolation analysis system,* Mon. Wea. Rev., **120**, 1747–1763.

PU, Z.-X., and KALNAY, E. (1999) *Targeting observations with the quasilinear inverse and adjoint NCEP global models: Performance during FASTEX.* Quart. J. Roy. Meteor. Soc., **125**, 3329–3337.

PU, Z.-X., KALNAY, E., SELA, J. and SZUNYOGH, I. (1997) *Sensitivity of forecast errors to initial conditions with a quasi-inverse linear model.* Mon. Wea. Rev., **125**, 2479–2503.

SIPPEL, J.A., and ZHANG, F. (2008) *A probabilistic analysis of the dynamics and predictability of tropical cyclogenesis.* J. Atmos. Sci., **65**, 3440–3459.

SIPPEL, J.A., and ZHANG, F. (2010) *Factors Affecting the Predictability of Hurricane Humberto (2007).* J. Atmos. Sci., **67**, 1759–1778.

SHAFRAN, P.C., SEAMAN, N.L. and GAYNO, G.N. (2000) *Evaluation of numerical predictions of boundary layer structure during the Lake Michigan ozone study,* J. Appl. Meteor., **39**, 412–426.

SNYDER, C. and ZHANG, F. (2003) *Assimilation of simulated Doppler radar observations with an ensemble Kalman filter.* Mon. Wea. Rev., **131**, 1663–1677.

STUART, A.L., AKSOY, A., ZHANG, F. and NIELSEN-GAMMON, J.W. (2007) *Ensemble-based data assimilation and targeted observation of a chemical tracer in a sea breeze model.* Atmos. Environ., **41**, 3082–3094.

SZUNYOGH, I., TOTH, Z., EMANUEL, K.A., BISHOP, C.H., SNYDER, C.R., MORSS, WOOLEN, E.J. and MARCHOK, T. (1999) *Ensemble–based targeting experiments during FASTEX: the effect of dropsonde data from the Lear jet.* Quart. J. Roy. Meteor. Soc., **125**, 3189–3218.

TORN, R.D., and HAKIM, G.J. (2008) *Ensemble-based Sensitivity Analysis.* Mon. Wea. Rev., **136**, 663–677.

VAN LOON, M., BUILTJES, P.J.H., SEGERS, A.J. (2000) *Data assimilation of ozone in the atmospheric transport chemistry model LOTUS.* Environ Modeling Software, **15**, 603–609.

WENG, Y., ZHANG, M. and ZHANG, F. (2011) *Advanced data assimilation for cloud-resolving hurricane initialization and prediction.* Computing in Science and Engineering, **13**, 40–49.

ZHANG, F., (2005) *Dynamics and structure of mesoscale error covariance of a winter cyclone estimated through short-range ensemble forecasts.* Mon. Wea. Rev., **133**, 2876–2893.

ZHANG, F., SNYDER, C. and SUN, J. (2004) *Impacts of initial estimate and observation availability on convective-scale data assimilation with an ensemble Kalman filter.* Mon. Wea. Rev., **132**, 1238–1253.

ZHANG, F., MENG, Z. and AKSOY, A. (2006) *Tests of an ensemble Kalman filter for mesoscale and regional-scale data assimilation. Part I: Perfect Model Experiments.* Mon. Wea. Rev., **134**, 722–736.

ZHANG, F., BEI, N.J., NIELSEN-GAMMON, W., LI, G., ZHANG, R., STUART, A. and AKSOY, A. (2007) *Impacts of meteorological uncertainties on ozone pollution predictability estimated through meteorological and photochemical ensemble forecasts.* J. Geophys. Res., **112**, D04304, doi:10.1029/2006JD007429.

ZHANG, F., WENG, Y., SIPPEL, J.A., MENG, Z. and BISHOP, C.H. (2009) *Cloud-resolving hurricane initialization and prediction through assimilation of Doppler Radar observations with an ensemble Kalman filter.* Mon. Wea. Rev., **137**, 2105–2125.

ZHANG, F., Y. WENG, J. F. GAMACHE, and F. D. MARKS (2011), *Performance of convection - permitting hurricane initialization and prediction during 2008–2010 with ensemble data assimilation of inner-core airborne Doppler radar observations,* Geophys. Res. Lett., **38**, doi:10.1029/2011GL04846910.1029/2011GL048469.

(Received December 23, 2010, accepted May 12, 2011, Published online August 9, 2011)

Pure Appl. Geophys. 169 (2012), 555–578
© 2011 Springer Basel AG
DOI 10.1007/s00024-011-0372-5

■ **Pure and Applied Geophysics**

Variational Data Assimilation Technique in Mathematical Modeling of Ocean Dynamics

V. I. Agoshkov[1] and V. B. Zalesny[1]

Abstract—Problems of the variational data assimilation for the primitive equation ocean model constructed at the Institute of Numerical Mathematics, Russian Academy of Sciences are considered. The model has a flexible computational structure and consists of two parts: a forward prognostic model, and its adjoint analog. The numerical algorithm for the forward and adjoint models is constructed based on the method of multicomponent splitting. The method includes splitting with respect to physical processes and space coordinates. Numerical experiments are performed with the use of the Indian Ocean and the World Ocean as examples. These numerical examples support the theoretical conclusions and demonstrate the rationality of the approach using an ocean dynamics model with an observed data assimilation procedure.

Key words: Variational data assimilation, ocean dynamics, mathematical models, numerical algorithms, adjoint equations, multicomponent splitting, World ocean circulation.

1. Introduction

The creation of efficient mathematical models of the World Ocean and its individual subareas represents an urgent problem in present-day research, as well as the comprehensive analysis of observational data (Ivchenko, 2006, 2007). Such models are numerical tools used to understand and predict global ocean circulation (Chassignet and Verron, 2005; Sarkisyan and Suendermann, 2009), its interaction with the atmosphere (Marchuk, 1987), and the regional variability of seas and oceans (Nechaev, 2003). Special emphasis is placed on such models

when dealing with assimilation of observed data (Wenzel et al., 2001; Chassignet and Verron, 2005; Nerger, 2006; Blum et al., 2008).

The urgency of modeling is intensified by the fact that collecting full-scale data concerning the ocean and implementation of observation experiments involve great difficulties and expense. To perform reliable analysis and prognosis of oceanic circulation, it is necessary to develop a system of data assimilation, and one of the main components of this system is a physically complete and verified model of ocean circulation.

Various informational and computational systems of data assimilation for oceans and seas are being developed (Wenzel et al., 2001; Agoshkov et al., 2008; Blum et al., 2008; etc.). According to the methods used, these can be divided into two groups: statistical and variational. These systems are intended for improvement of the quality and reliability of prognosis, including the use of new methods for mathematical modeling and computational algorithms. From the practical viewpoint, this allows us to improve the estimation of climatic variability in sea and ocean processes.

The use of the technique of statistical interpolation was an important breakthrough in solving data assimilation problems. This approach dates back to Kolmogorov (1941), and it became known in Earth sciences thanks to the monograph by Gandin (1963). This method made it possible to place data analysis on a proper statistical foundation. Such an approach is usually called optimal interpolation (OI) (Lorenc, 1981). Observations are assigned weights associated with errors of observations. However, the first approximation field is not the initial value for the analysis as it was before, but is rather an additional useful source of information together with its error characteristics. OI has been used in many operational

---
[1] Institute of Numerical Mathematics, RAS, Gubkin str., 8, 199333 Moscow, Russia. E-mail: agoshkov@inm.ras.ru; zalesny@inm.ras.ru

centers since the late 1970s (LORENC, 1981). Later, this method was developed in works by LORENC (1986). To date, the OI method and its modifications have been used for operational data analysis in weather forecasts, as well as in oceanographic data assimilation.

The Kalman filter (KALMAN, 1961) is a generalization of the OI method. A continuous analog of the Kalman method is often called the Kalman–Bucy filter (KALMAN and BUCY, 1961). There are various generalizations of this method to the nonlinear case (JAZWINSKI, 1970). The extended Kalman filter (EKF) method, which uses model linearization near a certain known state, is very popular now (EVENSEN, 2007). The works of NELEPO (1979) and KOROTAEV (2006) considerably contributed to the development of Kalman filter methods and the methods of four-dimensional analysis of hydrophysical fields on the basis of dynamic–stochastic models of the ocean. The ensemble Kalman filter (EnKF) method (EVENSEN, 2007; FERTIG et al., 2007; KALNAY et al., 2007; ZHANG et al., 2009), based on the use of the Monte Carlo method at each time step, has become increasingly popular in recent years.

The use of variational methods, in particular the optimal control method, played an important role in the solution of data assimilation problems. The idea of minimizing a certain cost function associated with observational data along trajectories of the model was very fruitful. In this approach the problem of data assimilation is formulated as a problem of optimal control. The theoretical foundations of investigating and solving such problems are laid out in classical works by BELLMAN (1957), PONTRYAGIN et al. (1962), LIONS (1968), MARCHUK (1975, 1995), and others. Variational formalism was used for the first time in meteorology by SASAKI (1970) and, in problems of dynamic oceanography, by LE DIMET, and TALAGRAND (1986) and LE PROVOST and SALMON (1986).

As is known, when solving minimization problems, it is necessary to calculate the gradient of the cost function. The use of the theory of adjoint equations was an important step in this direction (LIONS, 1968; MARCHUK, 1995, MARCHUK et al. 1996). Beginning with well-known works (PENENKO 1976; MARCHUK and PENENKO 1978; LE DIMET and TALAGRAND, 1986), adjoint equations have been used

extensively by many researchers to investigate and numerically solve data assimilation problems and to calculate the gradient of the functional (AGOSHKOV and MARCHUK, 1993; MARCHUK and ZALESNY, 1993; MARCHUK and SHUTYAEV, 1994; WENZEL et al. 2001; COURTIER et al. 1998; SHUTYAEV, 2001; BENNETT, 2002; AGOSHKOV et al., 2005, 2008, ZALESNY and RUSAKOV, 2007; BLUM et al., 2008; ZALESNY and GUSEV, 2009; NAVON, 2009). Currently, growing interest is being aroused by four-dimensional data assimilation (4DVAR), which uses forward and adjoint models for observational data assimilation not at a concrete moment in time but rather in a specified time interval (BLUM et al., 2008).

The solvability of nonlinear problems of data assimilation and rigorous substantiation of numerical methods for their solution is not a simple task. Results indicating that weakly nonlinear problems of data assimilation are solvable were obtained by AGOSHKOV and MARCHUK (1993) and SHUTYAEV (2001). Further generalizations and new applications have been proposed in recent years (MARCHUK et al., 2001; LE DIMET and SHUTYAEV, 2005; AGOSHKOV, 2007b; AGOSHKOV et al., 2008; IPATOVA, 2008; SHUTYAEV and PARMUZIN, 2009).

Algorithms for four-dimensional data assimilation (BENNETT, 2002; KALNAY, 2003) seem to be quite effective. In recent years, the ensemble Kalman method and variational data assimilation have been compared in many works (CAYA et al., 2005; FERTIG et al., 2007; KALNAY et al., 2007; ZHANG et al., 2009). Additionally, the so-called hybrid approach, combining the ensemble method and variational data assimilation, has come into use (TIAN et al., 2008).

In this work we consider a class of 4DVAR data assimilation problems for satellite observed sea surface temperature and sea level height data as well as temperature and salinity measurements from the Argo data system. At each time step, derivatives with respect to time of the basic mathematical model are substituted by difference approximations, and the assimilation process is implemented successively in all time subintervals. In this case, the assimilation process at each time step becomes locally three dimensional, which makes it possible to substantially decrease the volume of assimilated information and, consequently, computational expenditure. On the other hand, in spite of such a simplification, the assimilation problem remains four dimensional in the entire calculated interval.

It should be particularly noted that, in formulating problems of variational assimilation of measurement data, we introduce only the quantities which can be uniquely defined with the use of the observational data under consideration as additional unknowns.

The water areas of the Indian Ocean or the entire World Ocean were used as test regions for performing numerical experiments with the use of versions of the mathematical model of ocean circulation developed at the Institute of Numerical Mathematics, Russian Academy of Sciences (INM RAS) (DIANSKY, 2002, 2006; SARKISYAN, 2005).

Herein, we present a flexible hierarchical mathematical model using a variational data assimilation procedure that describes the large-scale circulation of the World Ocean. The hierarchical structure of the model is based on the method of its numerical solution, i.e., multicomponent splitting. The method includes splitting with respect to physical processes and space coordinates (MARCHUK, 1982, 1990; MARCHUK and KUZIN, 1982; MARCHUK et al., 2005). The computational implementation of the model has a modular structure: a particular splitting step is implemented by a particular program module. As a result of splitting, the complicated system of ocean dynamics equations is split into a series of separate subsystems, which are modules of a simpler structure. The adjoint model consists of the respective subsystems adjoint to the split subsystems of the forward model.

Since the formulations of the problems considered in this work and the algorithms for their solution are given in (AGOSHKOV, et al., 2005, 2008, 2009), they, as well as the designations used below, are presented here in a brief form.

2. Ocean primitive equation system

Consider the following primitive equation system describing the large-scale ocean circulation in the domain $D(t)$, $t \in (0, \bar{t})$ and written in spherical coordinates (x, y, z) (SARKISYAN and SUENDERMANN, 2009; AGOSHKOV, 2007a, 2010):

$$\frac{d\vec{u}}{dt} + \begin{bmatrix} 0 & -f \\ f & 0 \end{bmatrix} \vec{u} + \frac{1}{\rho_0} \operatorname{grad} P + A_u \vec{u} + (A_k)^2 \vec{u} = \vec{f},$$

$$\frac{\partial P}{\partial z} = g\rho + f_p,$$

$$\operatorname{div} \vec{u} + \frac{1}{r^2} \frac{\partial}{\partial z} r^2 w = 0$$

$$\frac{dT}{dt} + A_T T = f_T$$

$$\frac{dS}{dt} + A_S S = f_S$$

$$\rho = \rho_0 (1 + \beta_T (T - T^{(0)}) + \beta_S (S - S^{(0)}) + \beta_{TS}(T, S)).$$

$$(1)$$

Here we use the following notations: $x \in [0, 2\pi]$ is the longitude, $y \in [-\pi/2, \pi/2]$ is the latitude, $r \equiv r(z) \equiv R_E - z \cong R_E$ is the distance from the center of the Earth with mean value of R_E; $\vec{U} \equiv (\vec{u}, w) \equiv (u, v, w)$ is the velocity vector, P is the pressure, ρ is the potential density anomaly, and T and S are potential temperature and salinity;

$$f(u) = 2\omega \sin y + mu \sin y \equiv l + f_1(u),$$
$$n \equiv 1/r, m \equiv 1/(r\cos y);$$

$$\frac{d}{dt} = \frac{\partial}{\partial t} + (\vec{u}, \operatorname{grad}) + w\frac{\partial}{\partial z}, \quad (\vec{u}, \operatorname{grad})$$
$$= mu\frac{\partial}{\partial x} + nv\frac{\partial}{\partial y},$$

$$\operatorname{grad}\varphi \equiv \left(m\frac{\partial\varphi}{\partial x}, n\frac{\partial\varphi}{\partial y} \right), \quad \operatorname{Grad}\varphi \equiv \left(\operatorname{grad}\varphi, \frac{\partial\varphi}{\partial z} \right),$$

$$\operatorname{div}\vec{u} \equiv m\frac{\partial u}{\partial x} + m\frac{\partial}{\partial y}\left(\frac{n}{m}v \right), \quad \operatorname{Div}\vec{U} \equiv \operatorname{div}\vec{u} + \frac{1}{r^2}\frac{\partial r^2 w}{\partial z},$$

A_u, A_T, and A_S are the second-order differential operators for momentum, temperature, and salinity describing the large-scale turbulent processes (with exchange coefficients $\hat{a}_\varphi$).

$$A_\varphi \equiv -\operatorname{Div}(\hat{a}_\varphi \operatorname{Grad}\varphi), \quad A_u = A_v \equiv A,$$

$$\operatorname{div}(\operatorname{grad}\varphi) \equiv \Delta\varphi \equiv m^2\frac{\partial^2\varphi}{\partial x^2} + mn\frac{\partial}{\partial y}\left(\frac{n}{m}\frac{\partial\varphi}{\partial y} \right),$$

$$\operatorname{Div}(\operatorname{Grad}\varphi) \equiv \operatorname{div}(\operatorname{grad}\varphi) + \frac{1}{r^2}\frac{\partial}{\partial z}r^2\frac{\partial\varphi}{\partial z},$$

$(A_k)^2$ is the fourth-order differential operator (DIANSKY et al., 2002, 2006) which determines by matrix $\hat{k} = \operatorname{diag}\{k_{ii}\}$ with nonnegative diagonal elements k_{ii};

$\vec{f} = (f_1, f_2), f_T, f_S, f_P$ are given functions describing external sources; $g = \text{const.} > 0$;

$\rho_0, T^{(0)}, S^{(0)}$ are reference density, temperature, and salinity;

$\beta_T, \beta_S, \beta_{TS}(T, S)$ are given coefficients determining by state equation $\rho = \rho(T, S)$. Note that, using function f_P, one can describe dependence of density from pressure.

Let us introduce some assumptions to simplify the governing Eq. (1). Note that ocean sea surface is defined by equation $z = \xi(x, y, t)$ or $F_0(x, y, z, t) \equiv \xi(x, y, t) - z = 0$, $t \in [0, \bar{t}](\bar{t} < \infty)$, and bottom relief by $z = H(x, y)$ or $F_H(x, y, z, t) \equiv -H(x, y) + z = 0$, $H(x, y) > 0$.

Generally speaking, sea surface height $\xi = \xi(x, y, t)$ is an upper boundary of the domain $D(t)$ and is an unknown function. Assume now that we consider (1) not in the domain $D(t)$ with moving boundary but in the reduced domain $D = \{(x, y, z) : (x, y, R_E) \in \Omega, 0 < z < H(x, y)\}$. The boundary of the ocean domain $\Gamma \equiv \partial D$ is represented as the sum of four non-crossing parts $\Gamma_S, \Gamma_{w,op}, \Gamma_{w,c}, \Gamma_H$, where $\Gamma_S \equiv \Omega$ is the undisturbed surface, $\Gamma_{w,op}$ is the open boundary, $\Gamma_{w,c}$ is the solid (coastal) boundary, and Γ_H is the bottom. The characteristic functions for the $\Gamma_S, \Gamma_{w,op}, \Gamma_{w,c}, \Gamma_H$, parts of the boundary Γ are defined as $m_S, m_{w,op}, m_{w,c}, m_H$.

We further assume that $\partial\Omega$ and ∂D are piecewise-smooth surfaces that belong to the space $C^{(2)}$ and that satisfy a local Lipschitz condition. $\vec{N} \equiv (N_1, N_2, N_3)$ denotes the external normal unit vector to Γ. Note that $\vec{N} = (0, 0, -1)$ on Γ_S and $\vec{N} = (N_1, N_2, 0)$ on $\Gamma_w = \Gamma_{w,op} \cup \Gamma_{w,c}$, where $\vec{n} \equiv (N_1, N_2) \equiv (n_1, n_2)$, is the external normal unit vector to $\partial\Omega$. We denote by $U_n : U_n = \vec{U} \cdot \vec{N} = uN_1 + vN_2 + wN_3$ the normal component of the velocity vector $\vec{U} = (u, v, w)$ on the boundary Γ and $U_n^{(+)} \equiv (|U_n| + U_n)/2$, $U_n^{(-)} \equiv (|U_n| - U_n)/2$. Note that $U_n = U_n^{(+)} - U_n^{(-)}$ on Γ.

Assume that all functions and coefficients in (1) are defined for $z \in (0, H)$, and $\rho = \rho_0$, $f_P \equiv 0$ for $z \in (\xi, 0)$. Consider Eq. (1) in the domain D for $t \in (0, \bar{t})$, where D does not depend on ξ.

Reformulate problem (1) in evolutionary form for prognostic functions u, v, ξ, T, and S. After some

transformations (AGOSHKOV, 2007a, 2008), we obtain the following system of equations:

$$\begin{cases} \frac{d\vec{u}}{dt} + \begin{bmatrix} 0 & -f \\ f & 0 \end{bmatrix}\vec{u} - g \cdot \text{grad}\xi + A_u\vec{u} + (A_k)^2\vec{u} = \vec{f} - \frac{1}{\rho_0}\text{grad}P_a \\ \qquad\qquad - \frac{g}{\rho_0}\text{grad}\int_0^z \rho_1(T, S)dz', \\ \frac{\partial\xi}{\partial t} - m\frac{\partial}{\partial x}\left(\int_0^H \Theta(z)u\,dz\right) - m\frac{\partial}{\partial y}\left(\int_0^H \Theta(z)\frac{n}{m}v\,dz\right) = f_3, \\ \frac{dT}{dt} + A_T T = f_T, \frac{dS}{dt} + A_S S = f_S, \end{cases}$$
$$(2)$$

where the second equation is considered in Ω and the others in D for $t > 0$. Here and below for convenient operations into curvilinear orthogonal system we use the following weight function: $\Theta(z) \equiv r(z)/R$ (note that, because $H < < R$, we have $\Theta(z) = 1 - z/R \cong 1$). Boundary conditions for system (2) at Γ_S are

$$\begin{cases} \left(\int_0^H \Theta\vec{u}\,dz\right)\vec{n} + \beta_0 m_{op}\sqrt{gH}\xi = m_{op}\sqrt{gH}d_s Ha\partial\Omega, \\ U_n^{(-)}u - v\frac{\partial u}{\partial z} - k_{33}\frac{\partial}{\partial z}A_k u = \tau_x^{(a)}/\rho_0, U_n^{(-)}v - v\frac{\partial v}{\partial z} - k_{33}\frac{\partial}{\partial z}A_k v = \tau_y^{(a)}/\rho_0, \\ \qquad\qquad A_k u = 0, \; A_k v = 0, \\ \qquad\qquad -v_T\frac{\partial T}{\partial z} = Q^{(T)}, \\ \qquad\qquad -v_S\frac{\partial S}{\partial z} = Q^{(S)}, \end{cases}$$
$$(3)$$

where $Q^{(T)} \equiv Q_T + U_n^{(-)}d_T - \gamma_T(T - T_a) - U_n^{(-)}T$, $Q^{(S)} \equiv Q_S + U_n^{(-)}d_S - \gamma_S(S - S_a) - U_n^{(-)}S$.

Boundary conditions in the other parts of the boundary can be found in (AGOSHKOV, 2007a, 2009; AGOSHKOV et al., 2008).

The initial conditions for the prognostic functions are

$$u = u^0, v = v^0, T = T^0, S = S^0, \xi = \xi^0, \quad \text{for } t = 0,$$
$$(4)$$

where u^0, v^0, T^0, S^0, and ξ^0 are known functions.

Note that we exclude from (2) two diagnostic functions $P = P(u, v, \xi, T, S)$ and $w(x, y, z, t)$ using the following formulas

$$P(x, y, z, t) = P_a(x, y, t) + \rho_0 g(z - \xi)$$
$$+ \int_0^z g\rho_1(T, S)dz', \qquad (5)$$

$$\begin{cases} \frac{\partial P}{\partial z} = \rho g + f_P & \text{for } 0 < z < H, \\ P = P_a - \rho_0 g\xi & \text{for } z = 0, (x, y, t) \in \Omega \times (0, \bar{t}). \end{cases}$$
$$(6)$$

$$w(x, y, z, t) = \frac{1}{r} \left(m \frac{\partial}{\partial x} \left(\int\limits_z^H ru \, dz' \right) + m \frac{\partial}{\partial y} \right.$$

$$\left. \left(\frac{n}{m} \int\limits_z^H rv \, dz' \right) \right), \quad (x, y, t) \in \Omega \times (0, \bar{t}). \quad (7)$$

Now the ocean dynamics problem is formulated as follows: Find functions u, v, ξ, T, S, satisfying (2–4). If u, v, ξ, T, and S are defined functions, P and w can be found from Eqs. (5, 7).

3. Weak approximation of the ocean primitive equation model by splitting method

For a numerical solution of (2, 3) it is possible to simultaneously use the splitting method and a σ-coordinate system. Based on this approach, a numerical model of ocean dynamics was developed at INM RAS. This model is used in this work in the numerical experiments (see below). Now, we present the splitting scheme applying for approximation of (2, 3) in the case of $\vec{f} = g \, \mathrm{grad}G$, where G is a given scalar function $G = G(x, y, z)$.

Introduce the grid $0 = t_0 < t_1 < \cdots < t_{J-1} < t_J = \bar{t}$, $\Delta t_j = t_j - t_{j-1}$ in the time interval $[0, \bar{t}]$ and consider Eqs. (2, 3, 4–7) on each time subinterval (t_{j-1}, t_j), assuming that a vector-solution at the previous time steps $\phi_k \equiv (u_k, v_k, \xi_k, T_k, S_k)$, $k = 1, 2, \ldots, j-1$ is known. Applying splitting with respect to physical

processes (MARCHUK, 1990), we represent the numerical solution of our problem in two main steps.

Step 1. Convection–diffusion of temperature and salinity. In this step, the following problem is considered:

$$\begin{cases} T_t + (\bar{U}, \mathrm{Grad})T - \mathrm{Div}(\hat{a}_T \cdot \mathrm{Grad}\, T) = f_T \ B \ D \times (t_{j-1}, t_j), \\ T = T_{j-1} \quad \text{for } t = t_{j-1} \ B \ D, \\ -v_T \frac{\partial T}{\partial z} = Q^{(T)} \quad \text{on } \Gamma_S \times (t_{j-1}, t_j), \\ \frac{\partial T}{\partial N_T} = 0 \quad \text{on } \Gamma_{w,c} \times (t_{j-1}, t_j), \\ \bar{U}_n^{(-)}T + \frac{\partial T}{\partial N_T} = \bar{U}_n^{(-)}d_T + Q_T \quad \text{on } \Gamma_{w,op} \times (t_{j-1}, t_j)(), \\ \frac{\partial T}{\partial N_T} = 0 \text{ on } \Gamma_H \times (t_{j-1}, t_j), \\ T_j \equiv T \text{ on } D \times (t_{j-1}, t_j). \end{cases}$$

$$\quad (8)$$

$$\begin{cases} S_t + (\bar{U}, \mathrm{Grad})S - \mathrm{Div}(\hat{a}_S \cdot \mathrm{Grad}\, S) = f_S \quad \text{in } D \times (t_{j-1}, t_j), \\ S = S_{j-1} \quad \text{for } t = t_{j-1} \ B \ D, \\ -v_S \frac{\partial S}{\partial z} = Q^{(S)} \quad \text{on } \Gamma_S \times (t_{j-1}, t_j), \\ \frac{\partial S}{\partial N_S} = 0 \quad \text{on } \Gamma_{w,c} \times (t_{j-1}, t_j), \\ \bar{U}_n^{(-)}S + \frac{\partial S}{\partial N_S} = \bar{U}_n^{(-)}d_S + Q_S \quad \text{on } \Gamma_{w,op} \times (t_{j-1}, t_j), \\ \frac{\partial S}{\partial N_S} = 0 \quad \text{on } \Gamma_H \times (t_{j-1}, t_j), \\ S_j \equiv S \quad \text{on } D \times (t_{j-1}, t_j). \end{cases}$$

$$\quad (9)$$

After solving (8, 9), we get T_j, S_j which approximation of the exact solution T, S in $D \times (t_{j-1}, t_j)$. Here we assume that the velocity vector $\bar{U} = (\bar{u}, \bar{v}, w(\bar{u}, \bar{v}))$ and the turbulent diffusion coefficients $\hat{a}_T = \hat{a}_T(x, y, z, t)$, $\hat{a}_S = \hat{a}_S(x, y, z, t)$ are known functions.

Step 2. Adaptation of density and velocity fields. This step consists of three substages and includes the solution of three subproblems. The first is

$$\begin{cases} \underline{u}_t^{(1)} + \begin{bmatrix} 0 & -\ell \\ \ell & 0 \end{bmatrix} \underline{u}^{(1)} - g \cdot \mathrm{grad}\xi^{(1)} = g \cdot \mathrm{grad}G - \frac{1}{\rho_0}\mathrm{grad}\left(P_a + g \int\limits_0^z \rho_1(\bar{T}, \bar{S}) dz' \right) \\ \qquad\qquad\qquad \text{in } D \times (t_{j-1}, t_j), \\ \xi_t^{(1)} - \mathrm{div}\left(\int\limits_0^H \Theta \underline{u}^{(1)} dz \right) = f_3 \quad \text{in } \Omega \times (t_{j-1}, t_j), \\ \underline{u}^{(1)} = \underline{u}_{j-1}, \xi^{(1)} = \xi_{j-1} \quad \text{for } t = t_{j-1}, \\ \left(\int\limits_0^H \Theta \underline{u}^{(1)} dz \right) \cdot n + \beta_0 m_{op}\sqrt{gH}\xi^{(1)} = m_{op}\sqrt{gH}d_s \quad \text{on } \partial\Omega \times (t_{j-1}, t_j), \\ \underline{u}_j^{(1)} \equiv \underline{u}^{(1)}(t_j) \quad \text{in } D \end{cases}$$

$$\quad (10)$$

and the function $\xi_j \equiv \xi^{(1)}$ approximates ξ in (t_{j-1}, t_j). The second subproblem is

$$\begin{cases} \underline{u}_t^{(2)} + \begin{bmatrix} 0 & -f_1(\bar{u}) \\ f_1(\bar{u}) & 0 \end{bmatrix} \underline{u}^{(2)} = 0 \quad \text{in } D \times (t_{j-1}, t_j), \\ \quad \underline{u}^{(2)} = \underline{u}_j^{(1)} \quad \text{for } t = t_{j-1} \; B \, D, \\ \quad \underline{u}_j^{(2)} \equiv \underline{u}^{(2)}(t_j) \quad \text{in } D, \end{cases}$$

$$(11)$$

and the third is

$$\underline{u}_t^{(3)} + (\bar{U}, \text{Grad})\,\underline{u}^{(3)} - \text{Div}(\hat{a}_u \cdot \text{Grad}))\,\underline{u}^{(3)}$$
$$+ (A_k)^2\,\underline{u}^{(3)} = 0 \quad \text{in } D \times (t_{j-1}, t_j),$$
$$\underline{u}^{(3)} = \underline{u}_j^{(2)} \quad \text{for } t = t_{j-1} \text{ in } D, \qquad (12)$$
$$\bar{U}_n^{(-)}\,\underline{u}^{(3)} - v_u \frac{\partial \underline{u}^{(3)}}{\partial z} - k_{33} \frac{\partial}{\partial z}(A_k\,\underline{u}^{(3)}) = \frac{\tau^{(a)}}{\rho_0},$$
$$A_k\,\underline{u}^{(3)} = 0 \quad \text{on } \Gamma_S \times (t_{j-1}, t_j),$$

where $u^{(3)} = (u^{(3)}, v^{(3)}), \tau^{(a)} = (\tau_x^{(a)}, \tau_y^{(a)}), \; U^{(3)} = (u^{(3)}, w^{(3)}(u^{(3)}, v^{(3)})), \bar{U}^{(3)} = (u^{(3)}, 0), \tau^{(b)} = (\tau_x^{(b)}, \tau_y^{(b)})$.

Note that, if the finite-difference method is used for a numerical solution of (8–12), it will be possible to assume that $\Theta(z) \cong 1$. If the finite-element method should be used (KUZIN, 1985), it will be necessary first to pass to the weak formulation of (8–12), whereupon it is assumed that $\Theta(z) \cong 1$.

As noted above, the use of a σ-coordinate system is one of the specific features of the numerical solution of the complete problem (2–7) (MARCHUK et al., 2005; ZALESNY and RUSAKOV, 2007). We can switch to a σ-coordinate system at the stage when considering the complete problem before the application of suitable splitting schemes and other numerical procedures. However, switching to σ-systems is also possible after the application of splitting schemes, i.e., in our case, as applied to problems (8–12). We will not discuss the merits and shortcomings of either of the specified approaches to the use of a coordinate system here. Writing problems (8–12) in a σ-coordinate system, we can subsequently use suitable numerical methods for their numerical approximation. Some of these

methods are described in (MARCHUK, 1987, 2005; DIANSKY et al., 2006; ZALESNY and RUSAKOV, 2007; ZALESNY and GUSEV, 2009). We should only note that all of the subproblems in the complete numerical model based on (2–7), which was used for numerical calculations in this work, were solved with the aid of a σ-coordinate system in accordance with (ZALESNY and GUSEV, 2009).

Here we will make some remarks concerning the methods of solving problem (10). Suppose that problem (10) is considered in (t_{j-1}, t_j) and an ordinary implicit scheme of first order is used for its approximation with respect to time (MARCHUK, 1982). Two methods can be used for solving the problem obtained. According to the first method, the velocity components are eliminated and the problem is reduced to the equation for the sea level (AGOSHKOV, 2005). The second method consists of solving the obtained system of algebraic equations, i.e., the problem in the terms $(\bar{u}, \bar{v}, \xi)$. Both of these methods have their own advantages and shortcomings, which are discussed in (AGOSHKOV, 2007a). Like in (AGOSHKOV, 2007a), this work uses the method of solving the system (10). Additionally, this method makes it possible to switch algebraically from the forward problem to the conjugate problem in an exact way, which is especially important for the iterative procedure.

Let us describe the numerical solution of the convection–diffusion equation, for example (8), in more detail. Represent $Div(\bar{U}T)$ in the symmetric form

$$Div(\bar{U}T) = \frac{1}{2}((\bar{u}, \text{grad}T) + (\text{div}(\bar{u}T))$$
$$+ \frac{1}{2}\left(w\frac{\partial T}{\partial z} + \frac{1}{r^2}\frac{\partial}{\partial z}(r^2 wT)\right)$$

and write (8) in the weak form

$$(T_t, \hat{T}) + (LT, \hat{T}) = F(\hat{T}) + B(Q_T, \hat{T}) \qquad (13)$$
$$\text{for } \forall \hat{T} \in W_2^1(D),$$

where $L, F, B(\cdot, \cdot)$ are defined by

$$(LT, \hat{T}) \equiv \int_D (-T \mathrm{Div}(\bar{U}\hat{T})) \mathrm{d}D + \int_{\Gamma_{w,op} \cap \Gamma_S} \bar{U}_n^{(+)} T\hat{T} \mathrm{d}\Gamma$$

$$+ \int_{\Omega} \gamma_T T\hat{T}|_{z=0} \mathrm{d}\Omega$$

$$+ \int_D \hat{a}_D \mathrm{Grad}\,(T) \cdot \mathrm{Grad}\,(\hat{T})\,\mathrm{d}D,$$

$$F(\hat{T}) = \int_{\Gamma_{w,op}} (Q_T + \bar{U}_n^{(-)} d_T)\hat{T}\mathrm{d}T$$

$$+ \int_{\Omega} (\gamma_T T_a + \bar{U}_n^{(-)} d_T)\hat{T}\mathrm{d}\Omega + \int_D f_T\hat{T}\mathrm{d}D,$$

$$(T_t, \hat{T}) = \int_D T_t\hat{T}\mathrm{d}D, \quad B(Q_T, \hat{T}) = \int_{\Omega} Q_T\hat{T}\mathrm{d}\Omega.$$

$$(14)$$

Represent the operator L as the sum of two operators L_1, L_2:

$$L = L_1 + L_2, \qquad (15)$$

where L_1, L_2 are:

$$(L_1 T, \hat{T}) \equiv \int_D \left(-T\frac{1}{2}(\mathrm{div}(\bar{u}\hat{T}) + (\bar{u}, \mathrm{grad})\hat{T}\right)$$

$$- \frac{1}{2}T\frac{1}{r^2}\frac{\partial}{\partial z}(r^2 W\hat{T}) + \mu_T \mathrm{grad}T \cdot \mathrm{grad}\hat{T})\mathrm{d}D$$

$$+ \int_{\Gamma_{w,op}} \bar{U}_n^{(+)} T\hat{T}\mathrm{d}\Gamma,$$

$$(L_2 T, \hat{T}) \equiv \int_D \left(-T\frac{1}{2}\left(w_1\frac{\partial T}{\partial z} + \frac{1}{r^2}\frac{\partial(r^2 w_1 T)}{\partial z}\right)\right.$$

$$\left. + v_T\frac{\partial T}{\partial z}\frac{\partial \hat{T}}{\partial z}\right)\mathrm{d}D + \int_{\Omega} (\bar{U}_n^{(+)} + \gamma_T)T\hat{T}\mathrm{d}\Omega,$$

$$W = \frac{z}{H}\left(mu\frac{\partial H}{\partial \lambda} + nv\frac{\partial H}{\partial \theta}\right),$$

$$w_1 = w - W.$$

Let us also assume that

$$F(\hat{T}) = F_1(\hat{T}) + F_2(\hat{T}),$$

where

$$F_1(\hat{T}) = \int_{\Gamma_{w,op}} (Q_T + \bar{U}_n^{(-)} d_T)\hat{T}\mathrm{d}T,$$

$$F_2(\hat{T}) = \int_{\Omega} (\gamma_T T_a + \bar{U}_n^{(-)} d_T)\hat{T}\mathrm{d}\Omega + \int_D f_T\hat{T}\mathrm{d}D.$$

Equation (13) can be written in the operator form

$$T_t + LT = F + BQ, \quad t \in (t_{j-1}, t_j)$$
$$T = T_{j-1} \quad \text{for } t = t_{j-1}, j = (1, 2, \dots, J),$$
$$(16)$$

where $L = L_1 + L_2$, $F = F_1 + F_2$. To solve (16) we apply a splitting scheme with respect to coordinates:

$$(T_1)_t + L_1 T_1 = 0, \quad t \in (t_{j-1}, t_j),$$
$$T_1 = T_{j-1} \quad \text{for } t = t_{j-1}, \qquad (17)$$

$$(T_2)_t + L_2 T_2 = f_T, \quad t \in (t_{j-1}, t_j)$$
$$T_2 = T_1(t_j), \qquad (18)$$

with the corresponding b.c., where

$$L_1 T_1 = \frac{1}{2}(u, \mathrm{grad})T_1 + \frac{1}{2}\mathrm{div}(uT_1) + \frac{1}{2}W\frac{\partial T_1}{\partial z}$$
$$+ \frac{1}{2r^2}\frac{\partial r^2 WT}{\partial z} - \mathrm{div}\,\mu_T \mathrm{grad}T_1,$$

$$L_2 T_2 = \frac{1}{2}\left(w_1\frac{\partial T_2}{\partial z} + \frac{1}{r^2}\frac{\partial(r^2 w_1 T_2)}{\partial z}\right) - \frac{1}{r^2}\frac{\partial}{\partial z}r^2 v_T\frac{\partial T_2}{\partial z},$$

and μ_T and v_T are the lateral and vertical components of the turbulent diffusion coefficient $\hat{a}_T$. The boundary conditions for T_1 follow from conditions for T. After solving (18) we get the function $T_2(t_j)$ as the approximate solution of (8) for $t = t_j$:

$$T_2(t_j) \equiv T_j \cong T \quad \text{for } t = t_j. \qquad (19)$$

The differential form of (18) is (for $T_2 \equiv T$)

$$T_t + \frac{1}{2}\left(w_1\frac{\partial T}{\partial z} + \frac{1}{r^2}\frac{\partial(r^2 w_1 T)}{\partial z}\right)$$

$$- \frac{1}{r^2}\frac{\partial}{\partial z}r^2 v_T\frac{\partial T}{\partial z} = f_T \quad \text{in } D \quad \text{for } t \in (t_{j-1}, t_j),$$

$$- v_T\frac{\partial T}{\partial z} = Q^{(T)} \quad \text{for } z = 0, \; -v_T\frac{\partial T}{\partial z} = 0$$

$$\text{for } z = H, T = T_1(t_j) \quad \text{for } t = t_{j-1}, \qquad (20)$$

253

where

$$\bar{U}_n^{(-)} = \frac{|\bar{U}_n| - \bar{U}_n}{2} = \frac{1}{2}(|\bar{w}_1| + \bar{w}_1) = \frac{1}{2}(|\bar{w}| + \bar{w})$$
$$\text{for } z = 0,$$

and

$$Q^{(T)} \equiv Q_T - \gamma_T(T - T_a) - \bar{U}_n^{(-)}T + \bar{U}_n^{(-)}d_T \quad (21)$$

is a turbulent heat flux at the sea surface.

As a result of the splitting procedure, the original problem (2–4) at each time interval (t_{j-1}, t_j) is replaced by the system of Eqs. (17, 18, 9–12). This system is a weak approximation of (2–4) and we will call it "problem I."

4. Sea level data assimilation problem

Let us consider the assimilation of satellite altimetric data into a primitive equation ocean model. Assume that satellite observes only the sea surface height of the ocean. With the assimilation of these data, we reconstruct the four-dimensional space–time circulation of the ocean including the vertical motion. The problem is solved by the variational technique and the adjoint method.

Suppose that the only observed function is the function ξ_{obs} on $\overline{\Omega} \equiv \Omega \cup \partial\Omega$ at $t \in (t_{j-1}, t_j)$, $j = 1, 2, \ldots, J$. A case is admitted when ξ_{obs} exists only in some subset of $\Omega \times (0, \bar{t}) : (\Omega \times (0, \bar{t}))_{obs}$ with characteristic function m_0:

$$m_0(x, y, z, t) = \begin{cases} 1, & (x, y, z, t) \in (\Omega \times (0, \bar{t}))_{obs} \\ 0, & (x, y, z, t) \notin (\Omega \times (0, \bar{t}))_{obs} \end{cases}.$$

Suppose that the functions ξ_0, G, and f_3 are also additional unknowns. Introduce the cost function

$$J_\alpha \equiv J_a(\xi_0, G, f_3, \Phi)$$
$$= \frac{1}{2}\left\{ \alpha_0 \bar{t} \left\| \xi_0 - \xi^{(0)} \right\|_{L_2(\bar{g}, \Omega)}^2 + \alpha_f \left\| f_3 - f_3^{(0)} \right\|_{L_2(0, \bar{t}; L_2(\bar{g}, \Omega))}^2 \right.$$
$$\left. + \alpha_G \left\| G - G^{(0)} \right\|_{L_2(0, \bar{t}; L_2(\bar{g}, \Omega))}^2 \right\} \equiv \sum_{j=1}^{J} \int_{t_{j-1}}^{t_j} J_\alpha^{(j)} \mathrm{d}t,$$

$$(22)$$

where

$$J_0(\Phi) \equiv J_0(\xi) = \frac{1}{2} \left\| m_0(\xi - \xi_{obs}) \right\|_{L_2(0, \bar{t}; L_2(\bar{g}; \Omega))}^2,$$

$$J_\alpha^{(j)} = \frac{1}{2}\left\{ \alpha_0 \| \xi_0 - \xi^{(0)} \|_{L_2(\bar{g}; \Omega)}^2 + \alpha_f \| f_3 - f_3^{(0)} \|_{L_2(\bar{g}; \Omega)}^2(t) \right.$$
$$+ \alpha_G \| G - G^{(0)} \|_{L_2(\bar{g}; \Omega)}^2(t) + \| m_0(\xi - \xi_{obs}) \|_{L_2(\bar{g}; \Omega)}^2(t) \right\}.$$

$$(23)$$

Here $\alpha = (\alpha_0, \alpha_f, \alpha_G)$, $\alpha_0 \geq 0, \alpha_f \geq 0, \alpha_G \geq 0$ are the regularization parameters, which can be dimensional values. Additionally, α_f and α_G can be specified as quantities depending on $\alpha_0 \geq 0$, for instance $\alpha_G \geq \alpha_0$, $\alpha_f \geq \alpha_0 \bar{t}^2$, etc.

The variational data assimilation problem is formulated as follows: Find the functions ξ_0, G, f_3 and the solution to problem I so that the functional (22) takes a minimum value.

As is shown in (AGOSHKOV, 2007a), the numerical solution of this variational data assimilation problem can be reduced to the following steps [for each time interval (t_0, t_1)]:

Step 1. We solve the equation for temperature

$$T_t + (\bar{U}, \mathrm{Grad})T - \mathrm{Div}(\hat{\alpha}_T \cdot \mathrm{Grad}T) = f_T \text{ in}$$
$$D \times (t_0, t_1)T = T_0 \quad \text{for } t = 0 \text{ in } D \quad (24)$$

with the corresponding boundary conditions. After that, T is assumed to be an approximation of temperature and $T(t_1)$ is taken as the initial condition for solving the problem in the interval (t_1, t_2).

Step 2. We solve the equation for salinity:

$$S_t + (\bar{U}, \mathrm{Grad})S - \mathrm{Div}(\hat{\alpha}_S \cdot \mathrm{Grad}S) = f_S \quad \text{in}$$
$$D \times (t_0, t_1)S = S_0 \quad \text{for } t = 0 \text{ in } D \quad (25)$$

with the corresponding boundary conditions. After that, S is assumed to be an approximation of salinity, and $S(t_1)$ is taken as the condition for solving the problem in the interval (t_1, t_2).

Step 3. In this step we have to solve the system of variational equations arising in minimizing $J_\alpha^{(1)}$ during the solutions of the splitting equations (AGOSHKOV, 2007a), which consists of (8–12) and the following adjoint problem:

$$\begin{cases} -(\underline{u}_1^*)_t - \begin{bmatrix} 0 & -\ell \\ \ell & 0 \end{bmatrix} \underline{u}_1^* + g\,\mathrm{grad}\,\xi^* = 0 \quad \text{in } D \times (t_0, t_1), \\[2mm] -\xi_t^* + \mathrm{div}\left(\int\limits_0^H \Theta\,\underline{u}_1^*\,\mathrm{d}z \right) = m_0(\xi - \xi_{obs}) \quad \text{in } \Omega \times (t_0, t_1), \\[2mm] -\left(\int\limits_0^H \Theta\,\underline{u}_1^*\,\mathrm{d}z \right)\cdot n + \beta_0 m_{op}\sqrt{gH}\,\xi^* = 0 \quad \text{on } \partial\Omega \times (t_0, t_1), \\[2mm] \xi^* = 0,\,\underline{u}_1^* = 0 \quad \text{for } t = t_1, \end{cases}$$
$$(26)$$

where $\xi^* = \xi_1^*$.

The system of optimality conditions in this subproblem is

$$\begin{cases} t_1 \alpha_0(\xi_0 - \xi^{(0)}) + \xi^*(t_0) = 0 \quad \text{in } \Omega, \\ \alpha_f(f_3 - f_3^{(0)}) + \xi^* = 0 \quad \text{in } \Omega \times (t_0, t_1), \\ \alpha_G(G - G^{(0)}) - \mathrm{div}\left(\int\limits_0^H \Theta\,\underline{u}_1^*\,\mathrm{d}z \right) = 0 \quad \text{in } \Omega \times (t_0, t_1). \end{cases} \quad (27)$$

Note that the solutions of systems (11) and (12) do not participate in the formation of systems (26) and (27). Subproblems (11) and (12) only correct the horizontal velocities u, v, and can be performed only at the final iteration.

Theorems concerning the unique and dense solvability of the formulated problem of the variational sea level data assimilation are presented in (AGOSHKOV, 2007a).

For the numerical realization of the algorithm for the solution to this problem in (t_0, t_1), it is sufficient to solve two initial-boundary-value problem for the parabolic equations [whereupon T, S will be determined in $D \times (t_0, t_1)$] and perform step 3, including the data assimilation block. In this step, the problems are numerically solved with the use of the following iterative algorithm: If $f_3^{(k)}, G^{(k)}, \xi_0^{(k)}$ are determined, we will solve (10) at $\xi_0 = \xi_0^{(k)}$, $f_3 = f_3^{(k)}$, $G = G^{(k)}$, whereupon we solve adjoint problem (26) and calculate the new approximations $f_3^{(k+1)}, G^{(k+1)}, \xi_0^{(k+1)}$:

For a suitable choice of the parameters $\{\gamma_k\}$, the process converges and, with the property of dense solvability, we can suppose that the values

$$\gamma_k = \frac{1}{2} \int\limits_{t_0}^{t_1} \int\limits_\Omega gm_0(\xi^{(k)} - \xi_{obs})^2\,\mathrm{d}\Omega\mathrm{d}t\bigg/$$

$$\left(g\int\limits_\Omega (\xi^*(t_0))^2\mathrm{d}\Omega + \int\limits_{t_0}^{t_1}\int\limits_\Omega g(\xi^*)^2\,\mathrm{d}\Omega\mathrm{d}t \right.$$

$$\left. + \int\limits_{t_0}^{t_1}\int\limits_\Omega g\left(\mathrm{div}\int\limits_0^H \underline{u}_1^{*(k)}\,\Theta\mathrm{d}z \right)^2 \mathrm{d}\Omega\mathrm{d}t \right) \quad (29)$$

can be an effective choice of $\{\gamma_k\}$. After meeting the criterion of the iterative process termination, it will be necessary to solve problems (10) to calculate $f_3^{(k+1)}, G^{(k+1)}, \xi_0^{(k+1)}$ and obtain the approximate solution for the entire problem in $D \times (t_0, t_1)$.

After all of the problems are solved and the iterative process is fulfilled in (t_0, t_1), the variational assimilation problem is solved analogously in the subsequent intervals (t_{j-1}, t_j), $j = 2, 3, \ldots$ Due to the property of the unique and dense solvability of the data assimilation problems under consideration in each time interval, it is possible to state that the system of all approximate solutions $\{\phi_j\}$ yields the minimal value of J_α; i.e., it is the solution of the data assimilation problem under consideration in the entire interval $(0, \bar{t})$.

5. Variational sea surface temperature data assimilation problem and algorithm of its solution

Let us assume that to each $t \in (0, \bar{t})$ corresponds some subset Ω_0 of Ω and that its characteristic

$$\begin{cases} \xi_0^{(k+1)} = \xi_0^{(k)} - \gamma_k(\alpha_0(\xi_0^{(k)} - \xi^{(0)}) + \xi^*(t_0)) \quad \text{in } \Omega \\ f_3^{(k+1)} = f_3^{(k)} - \gamma_k(\alpha_f(f_3^{(k)} - f_3^{(0)}) + \xi^*) \quad \text{in } \Omega \times (t_0, t_1), \\ G^{(k+1)} = G^{(k)} - \gamma_k\left(\alpha_G(G^{(k)} - G^{(0)}) - \mathrm{div}\left(\int\limits_0^H \Theta u_1^*\mathrm{d}z \right) \right) \quad \text{in } \Omega \times (t_0, t_1), \\ \text{provided that } \int\limits_\Omega G^{(k)}\mathrm{d}\Omega = 0 \quad \forall k. \end{cases} \quad (28)$$

function is $m_0 = m_0(t)$. Let also subset $\Omega_0 \equiv \Omega_0^{(j)}$ and characteristic function $m_0 \equiv m_0^{(j)}$ at each time interval $t \in (t_{j-1}, t_j)$, $j = 1, 2, \ldots, J$ be constants which are varied from one interval to another. Rewrite the first boundary condition (8) for temperature T at $z = 0$ in the following form:

$$-v_T \frac{\partial T}{\partial z} = Q \quad \text{for } z = 0 \text{ in } \Omega_0^{(j)} \times (t_{j-1}, t_j),$$

where $Q \equiv Q^{(T)}$ is a turbulent heat flux

$$Q = Q_T - \gamma_T(T - T_a) - \bar{U}_n^{(-)}T + \bar{U}_n^{(-)}d_T,$$

whereas on $(\Omega \backslash \Omega_0^{(j)}) \times (t_{j-1}, t_j)$ the boundary condition has the form

$$\bar{U}_n^{(-)}T - v_T \frac{\partial T}{\partial z} + \gamma_T(T - T_a) = Q_T + \bar{U}_n^{(-)}d_T$$

$$\text{at } z = 0 \text{ in } (\Omega \backslash \Omega_0^{(j)}) \times (t_{j-1}, t_j).$$

Consider function Q in $\Omega_0^{(j)} \times (t_{j-1}, t_j)$ as an additional unknown which should be found together with the other solution's components. It should be noted that, in $(\Omega \backslash \Omega_0^{(j)}) \times (t_{j-1}, t_j)$, all functions except T entering into the boundary condition are given.

Suppose that the function T_{obs} in the subset $\Omega_0^{(j)}$ of Ω is the only function that at $t \in (t_{j-1}, t_j)$ is obtained from the processing of satellite observed data.

Let us now formulate the following inverse problem: Find the solution $\varphi = (u, v, \xi, T, S)$ of problem I and the function Q on $\{\Omega_0^{(j)}\}$, so that the following boundary conditions are satisfied on Ω, $\forall t \in (t_{j-1}, t_j)$:

$$-v_T \frac{\partial T_2}{\partial z} = Q \quad \text{at } z = 0 \text{ on } \Omega_0^{(j)} \times (t_{j-1}, t_j)$$
$$\text{for } j = 1, \ldots, J,$$

$$\bar{U}_n^{(-)}T_2 - v_T \frac{\partial T_2}{\partial z} + \gamma_T(T_2 - T_a) = Q_T + \bar{U}_n^{(-)}d_T$$
$$\text{at } z = 0 \text{ on } (\Omega \backslash \Omega_0^{(j)}) \times (t_{j-1}, t_j),$$

and the additional conditions

$$T_2 = T_{obs}^{(j)} \quad \text{on } \Omega_0^{(j)} \times (t_{j-1}, t_j) \quad \text{for } j = 1, \ldots, J$$

are fulfilled.

The formulated problem can also be considered as the *exact control problem* if Q is considered as a

control function. This problem can be reduced to the variational data assimilation problem.

Introduce the cost function

$$J_\alpha \equiv J_\alpha(Q, \varphi) = \frac{1}{2} \int_0^t \alpha |Q - Q^{(0)}|^2 d\Omega \, dt + J_0(\varphi)$$

$$= \sum_{j=1}^J J_{\alpha,j},$$

(30)

where

$$J_0 = \frac{1}{2} \int_0^t \int_{\Omega_0(t)} m_0 |T - T_{obs}|^2 d\Omega \, dt,$$

$$J_{\alpha,j} = \frac{1}{2} \int_{t_{j-1}}^{t_j} \int_{\Omega_0^{(j)}} \alpha |Q - Q^{(0)}|^2 d\Omega \, dt$$

$$+ \frac{1}{2} \int_{t_{j-1}}^{t_j} \int_\Omega m_0^{(j)} \left|T - T_{obs}^{(j)}\right|^2 d\Omega \, dt.$$

Here, $\alpha \equiv \alpha(x, y, t)$ is the function that plays the role of a regularization or a penalty function and which can be a dimensional quantity, and $Q^{(0)} \equiv Q^{(0)}(x, y, t)$ is the given function.

The variational data assimilation problem is formulated as follows: Find the solution $\varphi = (u, v, \xi, T, S)$ of problem I and the function Q on $Q_0^{(j)}$ that make the functional (30) attain a minimum value.

Following (AGOSHKOV, 2003, AGOSHKOV et al., 2008), we can show that the solution of the optimality system which determines the solution of the formulated variational data assimilation problem reduces to successive solutions of some subproblems for $t \in (t_{j-1}, t_j)$, $j = 1, \ldots, J$. We represent their solutions for the interval (t_0, t_1) as three main steps.

Step 1. At the first step it is necessary to solve the system of adjoint equations for temperature that appears during the minimization of J_α in the solutions of the forward splitting model. We have the following adjoint problems:

$$\begin{cases} -(T_2^*)_t + L_2^* T_2^* = B^* m_0^{(1)}(T - T_{obs}^{(1)}) & \text{in } D \times (t_0, t_1) \\ T_2^* = 0 & \text{for } t = t_1 \end{cases}$$

$$(31)$$

$$\begin{cases} -(T_1^*)_t + L_1^* T_1^* = 0 & \text{in } D \times (t_0, t_1) \\ T_1^* = T_2^*(t_0) & \text{for } t = t_1 \end{cases} \quad (32)$$

and optimality condition

$$\alpha(Q - Q^{(0)}) + T_2^* = 0 \quad \text{in } \mathbf{\Omega}_0^{(1)} \times (t_0, t_1), \quad (33)$$

where L_1^*, L_2^* are the operators adjoint to L_1, L_2, respectively, and T_1^*, T_2^* are the solutions to adjoint problems (31) and (32). After that, the functions T_2 and $Q(t_1)$ are taken to be approximations to the components T and Q of the complete solution of problem I at $t = t_1$, and the function $T_2(t_1) \equiv T(t_1)$ is considered to be the initial condition for T when the problems are solved in (t_1, t_2).

Step 2. In this step, we solve the problem for salinity:

$$S_t + (\bar{U}, \text{Grad})S - \text{Div}(\hat{\alpha}_S \cdot \text{Grad } S) = f_S \quad \text{in } D \times (t_0, t_1) \quad (34)$$

under the corresponding boundary and initial conditions. Then, function S is regarded as an approximate solution, and $S(t_1)$ is used as the initial condition when the problem is solved in (t_1, t_2).

Step 3. In this step, it is necessary to solve the system of Eqs. (10–12). The unique dense solvability of this variational data assimilation problem is considered in (AGOSHKOV et al., 2005, 2008), and the following statements hold:

- Let the functional J be defined by formula (30) at T_{obs}, $Q^{(0)} \in L_2(\mathbf{\Omega} \times (0, \bar{t}))$. Then the variational data assimilation problem of finding the solution to problem I and the function Q so that for them the functional takes the minimal value is solved uniquely at any $\alpha > 0$. At $\alpha = 0$, this problem is uniquely and densely solvable and the sequence of solutions at $\alpha \to +0$ can be chosen as the sequence of regularized solutions minimizing the functional J_0. In this case $\inf J_0 = 0$.
- Provided that the variational data assimilation problem in (t_{j-1}, t_j) is uniquely and densely solvable, the solution to the original data assimilation problem for the total time interval $(0, \bar{t})$ is

reduced to the successive solution of the corresponding problems in the intervals (t_{j-1}, t_j).

Consider the following iterative algorithm for the solution of above-formulated problem and determination of Q based on the observed data T_{obs}. If $Q^{(k)}$ is the already constructed approximation to Q, we solve the subproblem arising after splitting along every vertical line for $Q \equiv Q^{(k)}$ and the adjoint problem corresponding to it, whereupon the next approximation $Q^{(k+1)}$ is determined:

$$Q^{(k+1)} = Q^{(k)} - \gamma_k(\alpha(Q^{(k)} - Q^{(0)}) + T_2^*) \quad \text{on } \mathbf{\Omega}_0^{(1)} \times (t_0, t_1) \quad (35)$$

with the parameter γ_k that is chosen in such a way that the iterative process under consideration converges (AGOSHKOV, 2003). After $Q^{(k+1)}$ is determined, the solution is repeated with the new approximation $Q^{(k+1)}$ to find $Q^{(k+2)}$, and so on. The iterations are repeated until an appropriate criterion of process convergence is met.

After T and Q are found within the necessary accuracy, steps 1–3 are realized in accordance with the splitting scheme considered previously, which completes the determination of the approximate solution to the complete problem in D at $t = t_1$.

The complete problem, with an additional determination of Q for other sets $Q_0^{(j)} \times (t_{j-1}, t_j)$ with $j > 1$, is solved in an analogous way.

Note that, during the iterative procedure described above, the functional J_α is minimized and the parameters γ_k are chosen in an appropriate way to ensure the convergence $Q^{(k)} \to Q$, $T^{(k)} \to T_{obs}$ as $k \to \infty$ and $\alpha \to +0$ in the norm of the space $L_2(\mathbf{\Omega}_0^{(j)}) \times (t_{j-1}, t_j)$.

The variational data assimilation problem considered above belongs to the class of four-dimensional problems of variational assimilation. Here, the splitting method is regarded as a method of approximation of the initial model, and the variational data assimilation problem itself is solved for the set $D \times (t_0, t_1)$, or for $D \times (t_{j-1}, t_j)$, $j > 1$.

6. Variational initialization problem for the temperature and salinity fields

We consider now a numerical algorithm for the variational data initialization problem also based on splitting and adjoint equation methods. The basic assimilated data arrays are climatic or Argo temperature and salinity data, and we should find initial conditions T^0, S^0.

So, we assume that as data measurements we have the functions T_{obs}, S_{obs} denoted in $D \times (t_{j-1}, t_j)$, $j = 1, 2, \ldots, J$. We assume also that T_{obs}, S_{obs} can be done in a subset of $D \times (t_{j-1}, t_j)$ with its characteristic function m_0.

Consider problem I with unknown initial conditions T^0, S^0 and introduce functional J_α

$$J_\alpha = \frac{1}{2 \cdot mes(D)} \int_D \left[\alpha_T^0 (T^0 - \hat{T}^0)^2 + \alpha_S^0 (S^0 - \hat{S}^0)^2 \right] dD$$

$$+ \frac{1}{(t_J - t_0)} \int_{t_0}^{t_J} \int_D m_0 \left[\alpha_T (T - T_{obs})^2 + \alpha_S (S - S_{obs})^2 \right]$$

$$\times dDdt, \tag{36}$$

where α_T^0, α_S^0, α_T, α_S are given nonnegative regularization parameters and coefficients, and $\hat{T}^0$, $\hat{S}^0$ are given functions approximating in some manner T^0, S^0.

Let us formulate the following variational problem: Find functions T^0, S^0 and solution of the problem I that make the functional J_α attain a minimum value.

The following are the steps used to solve this problem (Zalesny and Gusev 2009, AGOSHKOV et al., 2010, see also Appendix).

Step 1. Solve numerically the primitive equations system (1) on the time interval $t \in \Sigma_{j=1}^J (t_{j-1}, t_j)$ with the corresponding boundary conditions and initial conditions

$$u(t_0) = u^0, v(t_0) = v^0, \xi(t_0) = \xi^0, T(t_0) = \hat{T}^0,$$
$$\times S(t_0) = \hat{S}^0, \tag{37}$$

where (t_0, t_J) is an assimilation interval.

Step 2. Solve the optimality system to minimize the functional (36) on the solutions to the temperature and salinity convection–diffusion splitting equations with known coefficients $u = u^j, v = v^j, w = w^j, \xi = \xi^j, \hat{a}_T = \hat{a}_T^j, \hat{a}_S = \hat{a}_S^j$ at the time interval $t \in \Sigma_{j=1}^J (t_{j-1}, t_j)$

$$\begin{cases} (T_1)_t + L_1 T_1 = 0, & T_1(t_0) = T_{iter}^0 \\ (S_1)_t + L_1 S_1 = 0, & S_1(t_0) = S_{iter}^0 \end{cases}$$

$$\begin{cases} (T_2)_t + L_2 T_2 = f_T, & T_2(t_0) = T_1(t_1) \\ (S_2)_t + L_2 S_2 = f_S, S_2(t_0) = S_1(t_1), & t \in (t_0, t_1) \end{cases}$$

$$\begin{cases} (T_1)_t + L_1 T_1 = 0, & T_1(t_{j-1}) = T_2(t_{j-1}) \\ (S_1)_t + L_1 S_1 = 0, & S_1(t_{j-1}) = S_2(t_{j-1}) \end{cases}$$

$$\begin{cases} (T_2)_t + L_2 T_2 = f_T, T_2(t_{j-1}) = T_1(t_j) \\ (S_2)_t + L_2 S_2 = f_S, S_2(t_{j-1}) = S_1(t_j), & t \in \Sigma_{j=2}^J (t_{j-1}, t_j) \end{cases}$$

$$\begin{cases} (S_2^*)_t + L_2^* S_2^* = m_0 \tilde{\alpha}_S (S_2 - S_{obs}), & S_2^*(t_J) = 0 \\ (T_2^*)_t + L_2^* T_2^* = m_0 \tilde{\alpha}_T (T_2 - T_{obs}), & T_2^*(t_J) = 0 \end{cases}$$
$$\tag{38}$$

$$\begin{cases} (S_1^*)_t + L_1^* S_1^* = 0, S_1^*(t_J) = S_2^*(t_{J-1}) \\ (T_1^*)_t + L_1^* T_1^* = 0, T_1^*(t_J) = T_2^*(t_{J-1}), & t \in (t_J, t_{J-1}) \end{cases}$$

$$\begin{cases} (S_2^*)_t + L_2^* S_2^* = m_0 \tilde{\alpha}_S (S_2 - S_{obs}), & S_2^*(t_j) = S_1^*(t_{j-1}) \\ (T_2^*)_t + L_2^* T_2^* = m_0 \tilde{\alpha}_T (T_2 - T_{obs}), & T_2^*(t_j) = T_1^*(t_{j-1}) \end{cases}$$

$$\begin{cases} (S_1^*)_t + L_1^* S_1^* = 0, & S_1^*(t_j) = S_2^*(t_{j-1}) \\ (T_1^*)_t + L_1^* T_1^* = 0, & T_1^*(t_j) = T_2^*(t_{j-1}), & t \in \Sigma_{j=J-1}^1 (t_j, t_{j-1}) \end{cases}$$

Problem (38) is solved iteratively until the optimality conditions

$$\begin{aligned} \alpha_T^0 (T_{iter}^0 - \hat{T}^0) + T_1^*(t_0) &= 0 \\ \alpha_S^0 (S_{iter}^0 - \hat{S}^0) + S_1^*(t_0) &= 0 \end{aligned} \tag{39}$$

are satisfied and new initial temperature and salinity fields T_{opti}^0, S_{opti}^0 are found. So, the solution of the variational initialization problem consists in determination of T_{opti}^0 and S_{opti}^0 so that optimality Eqs. (38–39) hold and the minimum of functional (36) is attained. In numerical experiments the minimization of the functional was performed by the procedure M1QN3 (Gilbert and Lemarechal 1989).

Step 3. Solve the system of the primitive Eq. (1) on the time interval $t \in \Sigma_{j=1}^J (t_{j-1}, t_j)$ with the optimal initial conditions for temperature and salinity

$$u(t_0) = u^0, v(t_0) = v^0, \xi(t_0) = \xi^0,$$
$$T(t_0) = T_{opti}^0, S(t_0) = S_{opti}^0. \tag{40}$$

Note that, if there are no observed data for a time interval $t = (t_0, t_J)$, the optimality system (38–39) gives a trivial solution such that $T_{opti}^0 = \hat{T}^0, S_{opti}^0 = \hat{S}^0$. In this case our procedure is reduced to the solution of the forward problem I with the original initial conditions.

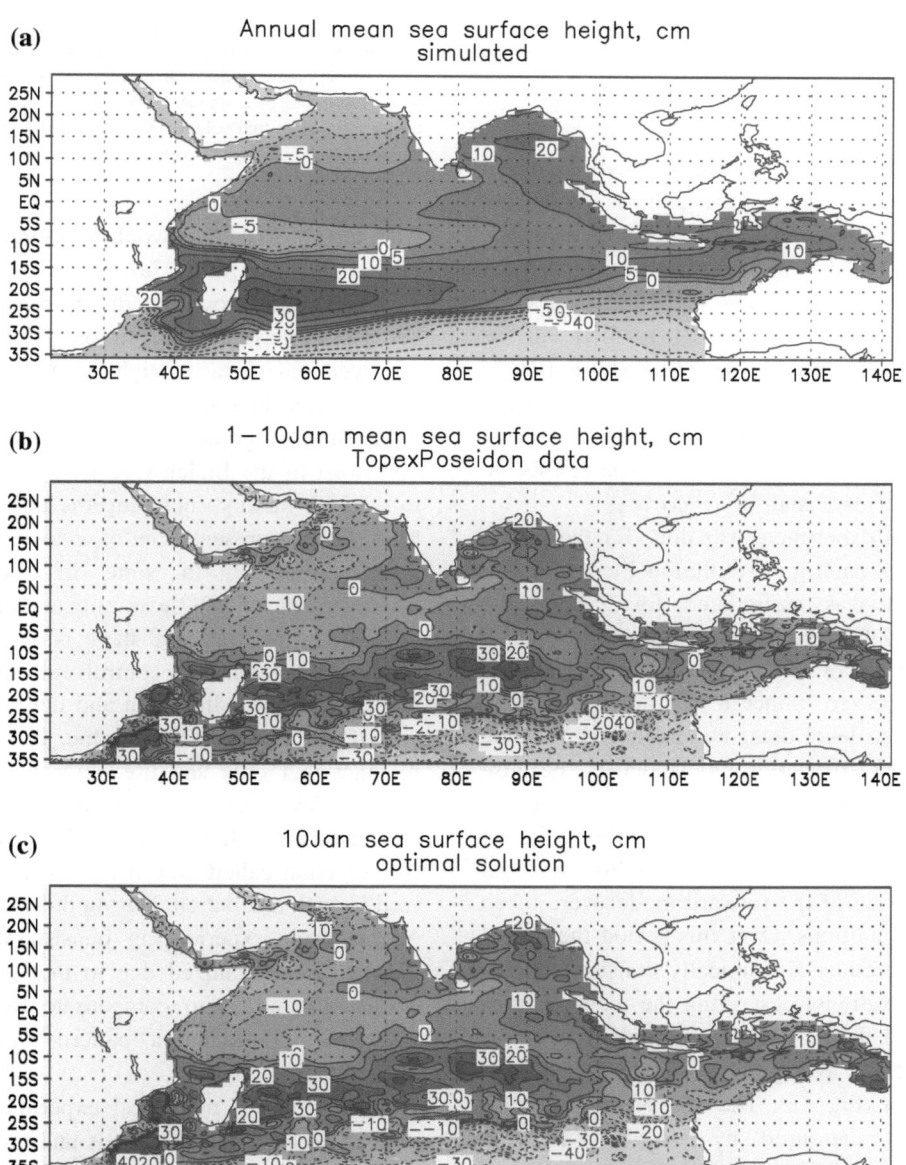

Figure 1

Sea surface height of the Indian Ocean: **a** model annual mean run without assimilation; **b** TOPEX/Poseidon observed data averaged for 1–10 January 2000; **c** optimal solution averaged for 1–10 January 2000

7. *Numerical experiments*

Variational sea level data assimilation. Let us present the result of numerical experiments on the inverse problem solution with respect to functions f_3, G, and ξ_0 through the sea level data assimilation. We chose the following modifications of

the boundary conditions and coefficients in them. The functions of total heat and salinity fluxes Q_T and Q_S are specified at the surface $z = 0$ as

$$Q_T = (Q_0 - 0.4Q_{SW})/(\rho_0 c_0) + \gamma_T(T_0 - T),$$
$$Q_S = -Q_{WS}S + \gamma_S(S_0 - S).$$

The bottom friction is parameterized as

$$\frac{1}{\rho_0}\left(\tau_x^b, \tau_y^b\right) = -C_d\sqrt{\left(u^2 + v^2 + e_b^2\right)}(u, v),$$

where $C_d = 2.5 \times 10^{-3}$ is the coefficient of quadratic friction and $e_b = 5\,\mathrm{cm/s}$ is the empirical addition characterizing the bottom friction at low velocities.

The coefficient k_{33} in the fourth-order vertical viscosity operator is assumed to be zero, and the coefficients k_{11} and k_{22} are chosen to filter out the high-frequency numerical noise. The condition $U_n = 0$ was set at the entire lateral boundary and at the bottom, and Dirichlet conditions $T = T_{obs}$, $S = S_{obs}$ were set at the open boundary. In all experiments we assume that $\Theta(z) \cong 1$ and $\frac{1}{r^2}\frac{\partial}{\partial z}\left(r^2\varphi\right) \cong \frac{\partial\varphi}{\partial z}$.

A numerical experiment was done for the Indian Ocean with horizontal grid resolution $1^0 \times 0.5^0$. The sea level data was assimilated for 2 days; the functions G and ξ_0 were restored in the course of 50 iterations, and all of the regularization parameters were taken to be zero. Since it was only possible to use observed sea level data, we did not correct the initial conditions for barotropic velocities, because in the absence of information about the velocity, the solution of the problem of its initialization is not unique. Therefore, the fields obtained for the 20th year of integration of the forward model were taken as the initial conditions. Over this period the model was found to produce dynamically consistent fields (close to observations) describing the main features of the observed fields. See, for example, the model sea surface height averaged over the final year of the 20-year run (Fig. 1a). The observed data included anomalies measured by satellite (TOPEX/Poseidon) superimposed on the annual mean sea level (Fig. 1b).

The computations showed that correction of the potential G is totally justified. The main advantage of this approach is that observed sea level anomalies one can control circulation without disturbing nondivergence of the velocity field. The results show that the mean quadratic deviation from the observed sea level after 50 iterations is 2.5%. Comparison of the convergences of iterative processes with and without control variable G show that the role of this control function is quite important. If one were to use only initial condition ξ_0 as a control variable, the deviation

Figure 2

a Monthly mean sea surface temperature. Model solution without assimilation, January 2004. **b** Monthly mean sea surface temperature. Observed data, January 2004. **c** Monthly mean sea surface temperature. Optimal solution for assimilation period of 1 month, January 2004

from the observed data would be more than five times greater (AGOSHKOV, 2007a).

The results show that the correlation coefficient between the optimal solution (Fig. 1c) and the observed sea surface height (Fig. 1b) is 0.99. The only significant region of discrepancy between the model and observed sea level can be found in the northwest of the Indian Ocean. This may be caused by various factors, including the coarse space resolution and representation of coastline and bottom topography, etc. In spite of the pronounced improvement of sea level model–data consistency due to data assimilation, there is no significant differences between the control run of the forward model without assimilation and the optimal solution with regard to circulation regime. This may be due to the fact that only satellite observed data averaged over 10 days were available for this water area and these data are stationary in the given experiment. The phenomenon called "the inverse barometer" is realized in such a situation. During this phenomenon, the ocean level, experiencing quasistationary external forcing, is reorganized due to the quasigeostrophic equilibrium so as to compensate for this forcing, whereas the circulation pattern remains virtually unchanged.

The series of numerical experiments conducted with the use of the algorithms and models proposed above showed that the procedure of assimilating sea level observed data, which includes the solution of the forward and adjoint system of shallow-water equations and the correction of the initial conditions and forcing functions, can be fulfilled simultaneously with the assimilation of sea surface temperature and salinity. This procedure is performed within the framework of the iterative process, which is controlled by an optimizer based on the quasi-Newton algorithm and ensures its fastest convergence (AGOSHKOV et al., 2010).

Variational assimilation of sea surface temperature observed data. Consider the problem of

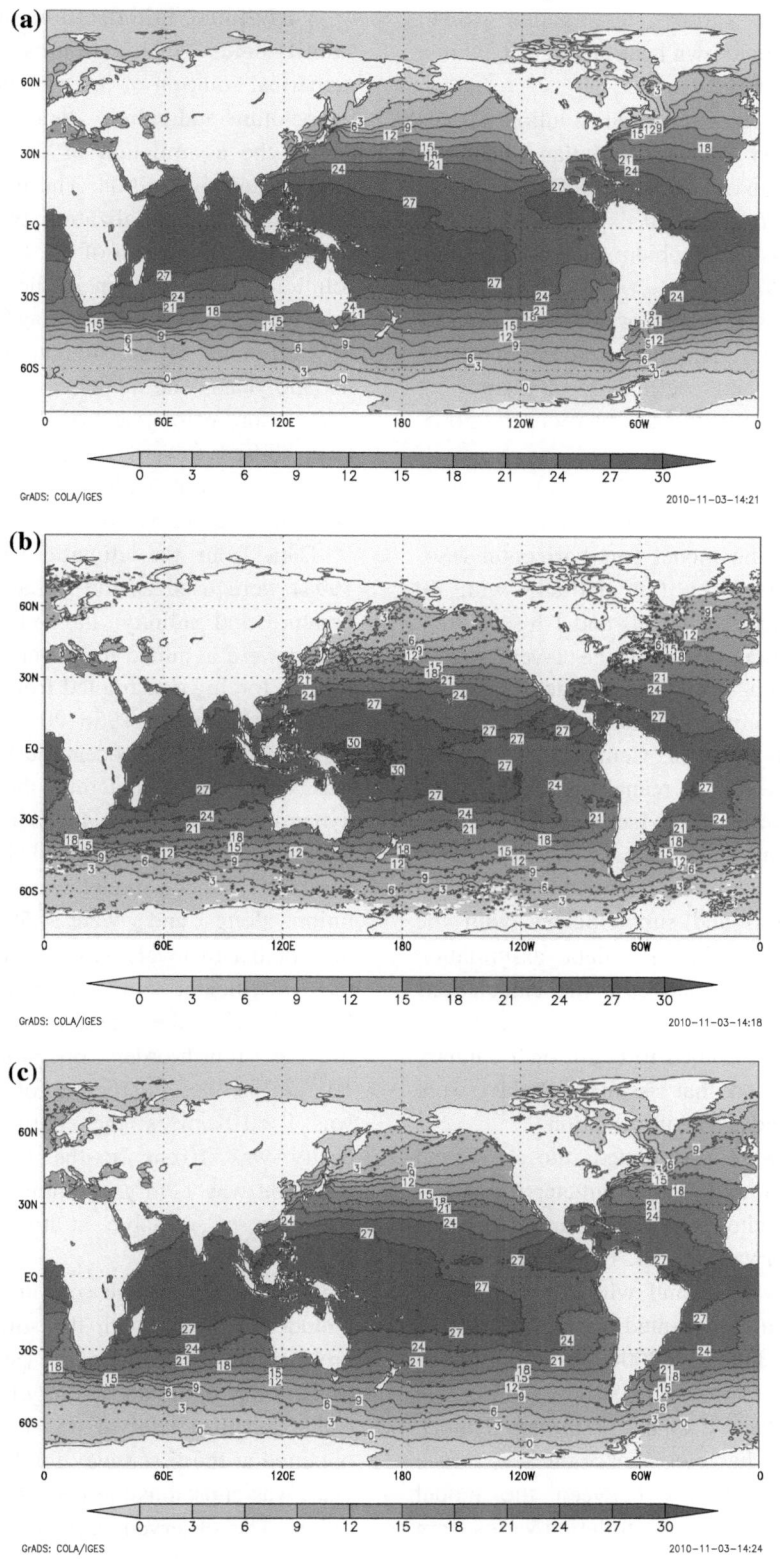

assimilation of sea surface temperature (SST) observed data at the unknown heat flux $Q \equiv Q^{(T)}$ in a part of the ocean surface.

Some results on the numerical solution to this problem for the water area of the Indian Ocean are presented in (AGOSHKOV, *et al.*, 2009). Here we present the results of numerical experiments for the World Ocean basin. The problem of reconstructing the function Q on $\Omega_0^{(j)} \times (t_{j-1}, t_j)$, $j = 1, 2, \ldots, J$ is solved through the variational assimilation of SST measurements. In the numerical experiments, the boundary conditions and coefficients in them were modified in accordance with (AGOSHKOV *et al.*, 2005, AGOSHKOV 2007c).

The global three-dimensional model of ocean dynamics developed at INM RAS was used for these experiments. The model had horizontal resolution of $1°$ longitude by $0.5°$ latitude with 40 unequally spaced vertical σ-levels and 1 h time step. The SST measurement data for the 1-year period 2004 was taken from the World Ocean INM RAS database. The mean climatic heat flux over January obtained from the National Center for Environmental Prediction (NCEP) reanalysis data or the model flux calculated on the basis of the forward model were used as $Q^{(0)}$. The model runs were performed with the use of the forward ocean model without (control run) and supplemented with the block of the SST variational data assimilation (optimal run). In the second case the calculations included the assimilation of T_{obs} for 1 month, and the assimilation was assumed to begin on 1 January 2004. It was supposed that assimilated data were perfect and contained exact information.

The SST data were assimilated into the ocean model with the regularization parameter $\alpha = 10^{-5}$. The heat flux was chosen in accordance with the control run of the forward model. The model runs for a 1-month period without and with the use of the assimilation block are presented in Fig. 2a and c, respectively. The January 2004 monthly mean observed sea surface temperature is shown in Fig. 2b. Results show that the use of the assimilation block improved the model characteristics of the sea surface temperature. The difference between the model results with and without data assimilation varies from $0.5°C$ to $1°C$.

Variational initialization of temperature and salinity fields. Finally, we present some results on the numerical solution of the problem of initializing the temperature and salinity fields in the World Ocean during the assimilation of data from buoys of the Argo system over 2008. The numerical experiments on solving the initialization problem included two stages: the calculation of the forward model and the solution to the problem in the initialization–prediction regime. At the first stage, we calculated the forward model of the circulation of the World Ocean for 350 years. The model equations were formulated in a σ-coordinate system on the sphere with the North Pole shifted toward the continent. The pole was located at a point with coordinates $60°E$ and $60.5°N$ (Zalesny and Gusev 2009).

Data from the climatic array (LEVITUS DATA, 1998) were used as the initial conditions for temperature and salinity; the velocities and sea level height were assumed to be zero. The climatic atmospheric forcing constructed from the averaged CORE data over the period 1958–2004 with 6 h discreteness was specified at the ocean surface. The experiment was aimed at calculating the seasonal cycle of dynamically consistent fields of currents, temperature, and salinity in the World Ocean. The main input parameters of the model were as follows: the resolutions along x and y were 2 .5° and $2°$, respectively; the number of levels was 33; the time step was 6 h; the coefficient of horizontal viscosity was $5 \times 10^4 \, cm^2/s$; the coefficient of horizontal viscosity in the fourth-order operator was about $1 \times 10^{22} \, cm^4/s$; the coefficient of horizontal diffusion was $1 \times 10^7 \, cm^2/s$; the coefficient of vertical viscosity was $10 \, cm^2/s$; the coefficient of vertical diffusion was $1 \, cm^2/s$, jumping to $1 \times 10^3 \, cm^2/s$ in the case of unstable stratification (ZALESNY and GUSEV, 2009).

The second stage of numerical experiments included, together with the solution to the forward problem, the procedure of the variational initialization of the temperature and salinity fields. In this case, the calculated time interval was 1 year and the solution obtained at the first stage, for January 1 of the 351st year, was chosen as the first guess of the initial condition. The observed Argo monthly mean fields of temperature and salinity were used as assimilated

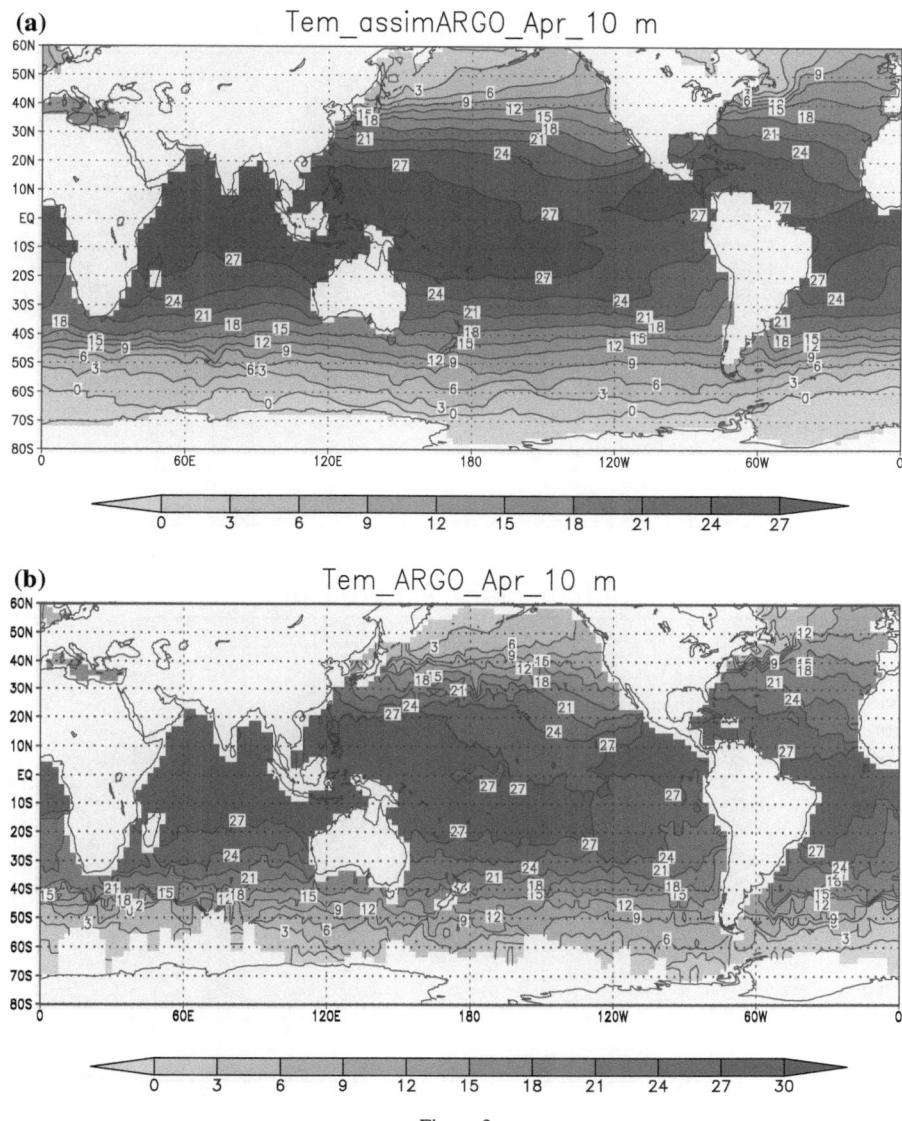

Figure 3
a Temperature at 10 m depth. Model results for the variational initialization–prediction run, 15 April 2008. **b** Temperature at 10 m depth. Argo monthly mean data for April 2008

observational data. The calculation was performed for each month of the year in the variational initialization–prediction regime with recalculation of the initial condition in the period of the first five days of each month. The initialization regime of our experiment consisted of two steps. At the first step, every month we solved the optimality system for temperature and salinity fields (38–39). At the second step we ran the full forward model, also over the period of assimilation (over the first five days of each month) and recalculated sea level, currents, temperature, and salinity. Then we assumed that the new initial conditions for the sea level and horizontal velocities were equal to their values at the end of the assimilated interval. Iteratively repeating this procedure, the initial sea surface height and horizontal velocities fields were adjusted to the optimal solution for temperature and salinity. Finally, we simulated all dynamical

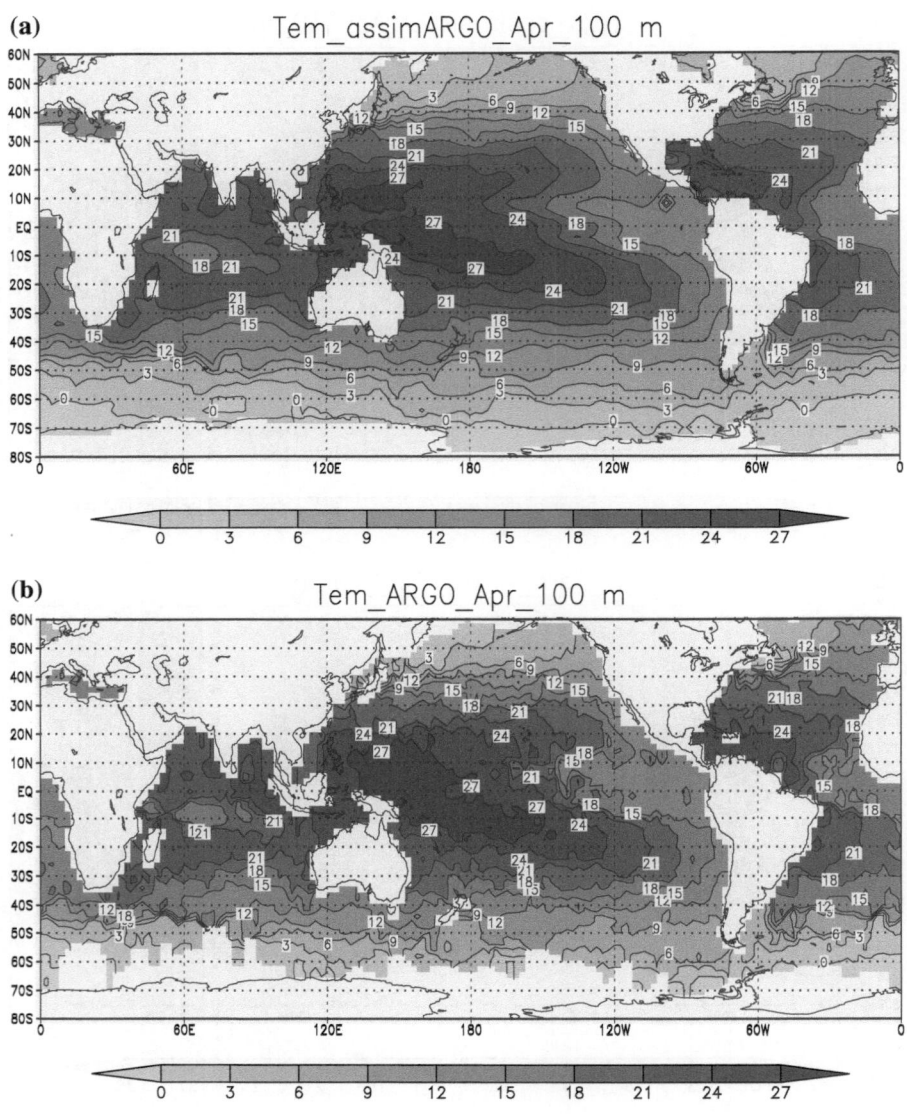

Figure 4

a Temperature at 100 m depth. Model results for the variational initialization–prediction run, 15 April 2008. **b** Temperature at 100 m depth. Argo monthly mean data for April 2008

fields to the end of each month using the forward model (prediction regime).

The results showed that the value of the cost function maximally decreased from 2.3 to 0.013 during the first assimilation interval of 1–5 January. This decrease was caused by the fact that the model solution considerably deviates from the observational data. The deviation from these data is the result of the low spatial resolution of the model, the viscous character of the model resolution, the inaccuracy of the subgrid parameterizations,

etc. Use of the procedure of variational assimilation of observational data immediately reduces the model error by two orders of magnitude. In solving the prognostic problem, the deviation from observational data increased during each calculated month by about 15–20 times until the next application of the initialization procedure. Then, as a result of variational data assimilation, the error decreased again to about 0.02.

An analysis of the spatial distribution of thermohaline fields shows that the assimilation of

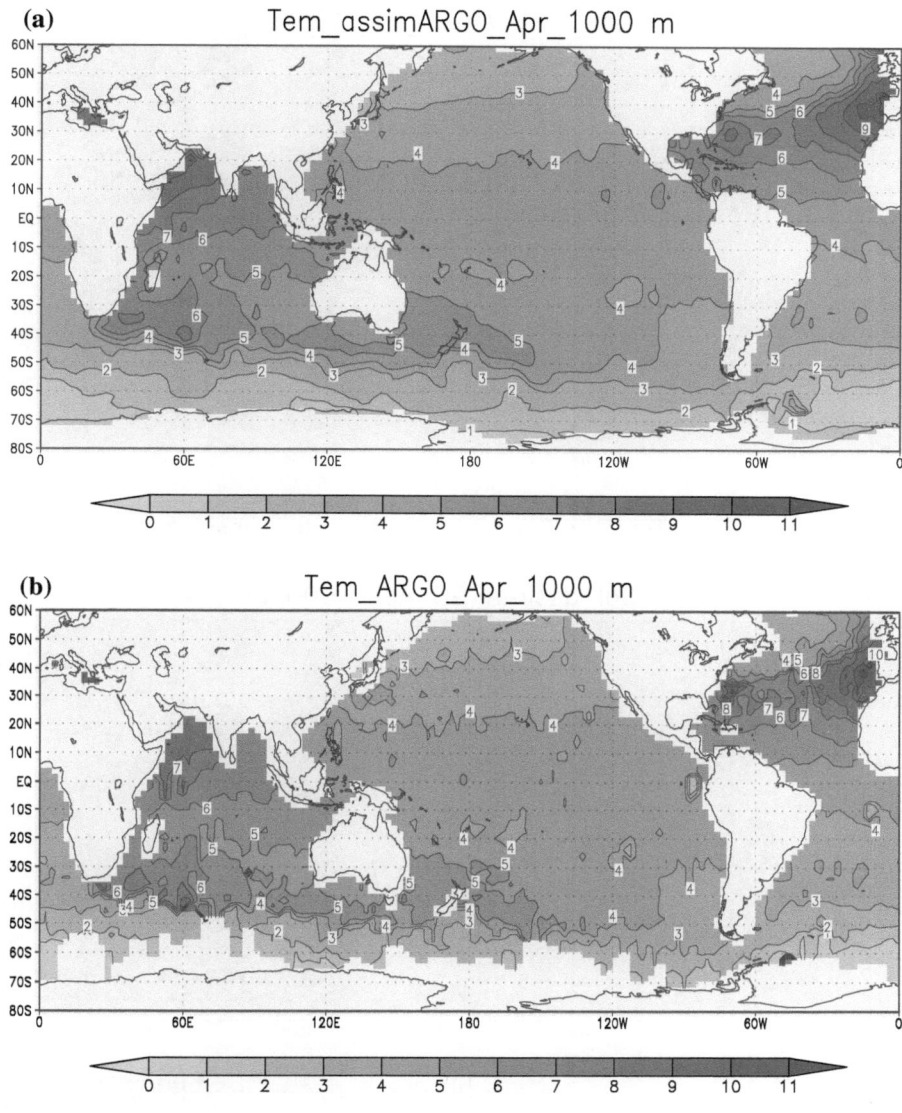

Figure 5

a Temperature at 1,000 m depth. Model results for the variational initialization–prediction run, 15 April 2008. **b** Temperature at 1,000 m depth. Argo monthly mean data for April 2008

observational data noticeably changed the model characteristics. Due to the data assimilation procedure, the simulated temperature and salinity fields better reflect their natural structures (Figs. 3, 4, 5, 6).

8. Conclusions

In this work we formulated problems of the variational assimilation of observed data of sea level height, as well as the temperature and salinity

measurements from the Argo system of buoys, with the use of the global three-dimensional model of ocean dynamics developed at the Institute of Numerical Mathematics, RAS. The basic algorithms for the numerical solution of problems are elaborated, and data assimilation blocks are developed and incorporated into the global three-dimensional model.

There are three main peculiarities of the INM model which distinguish it from other ocean models (MARCHUK *et al.*, 2005; ZALESNY *et al.*, 2010):

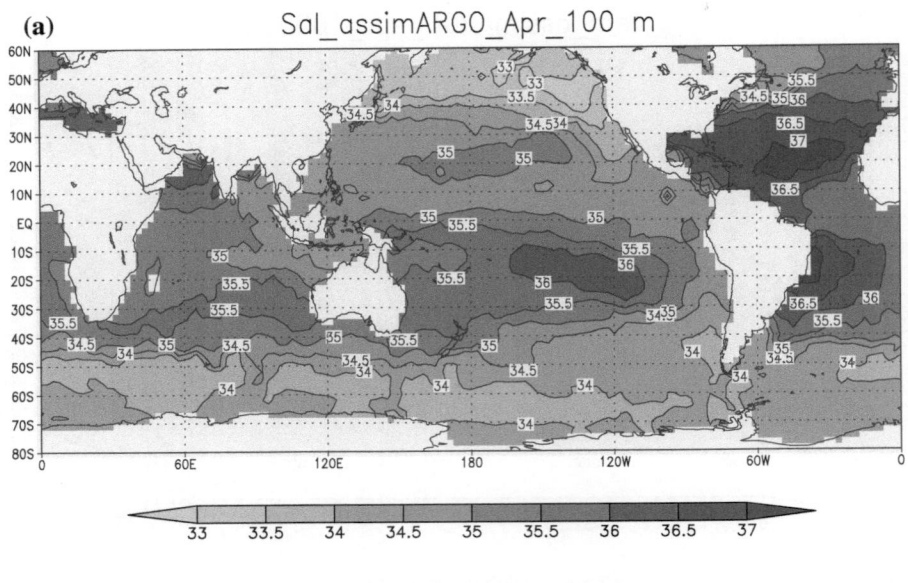

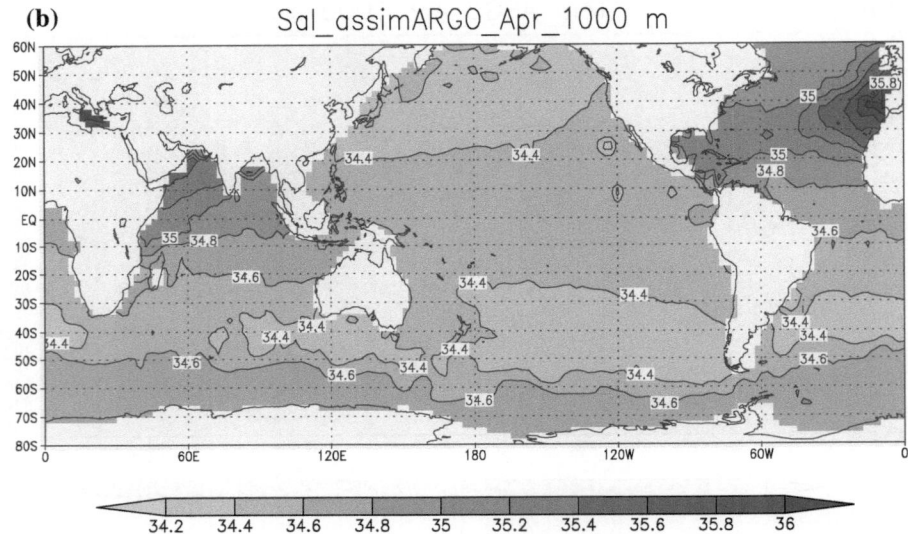

Figure 6
a Salinity at 100 m depth. Model results for the variational initialization–prediction run, 15 April 2008. **b** Salinity at 1,000 m depth. Model results for the variational initialization–prediction run, 15 April 2008

- It is a primitive equation σ-coordinate global ocean model;
- Model equations are written in a special symmetrized form that makes it possible to decompose the model operator into a sum of simpler, nonnegative suboperators;
- The numerical technique is based on the method of splitting by physical processes and space coordinates.

Numerical experiments showed that, due to the splitting method, a stable implicit scheme can be

efficiently implemented and a sufficiently large time step could be used for simulation (ZALESNY *et al.*, 2010). This procedure allows construction of a set of nonnegative split subsystems of adjoint operators of the problem which is also solved with an implicit time scheme (see Appendix). The splitting method and special symmetrized form of the governing equations allow the use of the same procedures for solving the adjoint equations as for the forward model and saving good accuracy of gradient descent algorithm (WENZEL and ZALESNY, 1996; ZALESNY and GUSEV, 2009).

We presented the numerical technique and results for the solution of the 4D-variational ocean data assimilation problem. The numerical procedure is intended for a σ-coordinate free surface primitive equation model. The process of solving the forward model is split into a number of stages. Multicomponent splitting is realized with respect to the physical processes and space coordinates. As a result of this splitting, a rather simple subsystem of equations is solved at each separate stage. We construct an adjoint subsystem for each separate splitting stage, and the set of subsystems yields a full adjoint model. The method is the constructive basis for the modular computing system of simulation and initialization of the ocean hydrographic fields.

Numerical experiments were performed for the Indian Ocean and the World Ocean model as examples. The numerical experiments confirmed the obtained theoretical conclusions and demonstrated the expedience of using a model with a block of assimilation of observational data.

Sea level data assimilation experiments support the necessity of including unknown external forces as additional unknown/control parameters into modeling and assimilation tools.

The results of numerical experiments with the variational data initialization procedure confirm that even a partial solution to this problem, i.e., only finding the initial conditions for temperature and salinity, makes it possible to construct dynamically consistent hydrographic fields and improve the quality of numerical simulation of the World Ocean circulation.

Acknowledgments

We are grateful to N.A. Diansky, E.I. Parmuzin, and A.V. Gusev for discussing this paper and making valuable comments which refined the description of the model results presented, and to S.A. Lebedev and N.B. Zakharova for the presented results of observed data.This work was supported by the Russian Foundation for Basic Research (Projects 09-05-00421, 10-01-00806), the program of the Presidium of the Russian Academy of Sciences no. 17, and the Federal Target Program "Scientific and Scientific–Pedagogical Specialists of an Innovative Russia."

Appendix: splitting algorithm for numerical solution of the 4DVAR data initialization problem

Let us consider equation with self-adjoint operator $B = B^* \geq 0$ and complicated nonnegative operator $A(\varphi) = \sum_1^I A_i(\varphi)$ which can be expressed in the following form:

$$B \frac{\partial \varphi}{\partial t} + \left(\sum_{i=1}^I A_i(\varphi) \right) \varphi = f, \ 0 \leq t \leq \bar{t}, \qquad (41)$$

where $A_i \geq 0$ are operators for splitting subproblems, and φ and φ^0 are problem solution and unknown initial condition: $\varphi \in \mathbf{\Omega} \times (0, \bar{t})$, $\varphi^0 \in \mathbf{\Omega}$.

Divide the time interval $(0, \bar{t})$ into N subintervals $t_{j-1} \leq t \leq t_j$, $\tau = t_j - t_{j-1}$, linearize operators $A_i(\varphi)$, and approximate equation (41) in each subinterval by the implicit scheme

$$
\begin{aligned}
&B \frac{\varphi_1 - \varphi^{j-1}}{\tau} + A_1\left(\varphi^{j-1}\right)\varphi_1 = f^{j-1} \\
&B \frac{\varphi_2 - \varphi_1}{\tau} + A_2\left(\varphi^{j-1}\right)\varphi_2 = 0 \\
&\cdots\cdots \\
&B \frac{\varphi^j - \varphi_{I-1}}{\tau} + A_I\left(\varphi^{j-1}\right)\varphi^j = 0, \ t_{j-1} \leq t \leq t_j.
\end{aligned}
\qquad (42)
$$

Excluding functions at intermediate splitting steps $\varphi_1, \ldots, \varphi_{I-1}$, we have an unconditionally stable first-order approximation of Eq. (41) with respect to time (MARCHUK, 1990)

$$
\begin{aligned}
&\left(B + \tau A_1\left(\varphi^{j-1}\right)\right)\left(B + \tau A_2\left(\varphi^{j-1}\right)\right) \ldots \left(B + \tau A_I\left(\varphi^{j-1}\right)\right) \\
&\varphi^j = \varphi^{j-1} + \tau f^{j-1}, \\
&j = 1, \ldots, N.
\end{aligned}
\qquad (43)
$$

Now, instead of the original problem (41), we consider an initialization problem for (43) assuming for simplicity that observed data exist everywhere (characteristic function equal to unity).

Suppose that $\varphi^0_{obs} \in \mathbf{\Omega}$, $\varphi_{obs} \in \mathbf{\Omega} \times (0, \bar{t})$ are observed functions and introduce the cost function

$$J = \frac{1}{2 \cdot mes(\Omega)} \left[\int_{\Omega} \alpha^0 (\varphi^0 - \varphi^0_{obs})^2 d\Omega \right.$$

$$\left. + \frac{1}{N} \sum_{j=1}^{N} \int_{\Omega} \alpha^j (\varphi^j - \varphi^j_{obs})^2 d\Omega \right], \qquad (44)$$

where α^0, α are corresponding regularization parameters and coefficients. Applying the traditional procedure (MARCHUK, ZALESNY, 1993; BLUM et al., 2008), we can write an optimality system which consists of two parts: forward and adjoint (backward) equations. The forward equations are

$$(B + \tau A_{1,0})(B + \tau A_{2,0})\ldots(B + \tau A_{I,0})\varphi^1 = \varphi^0 + \tau f^0$$

$$(B + \tau A_{1,1})(B + \tau A_{2,1})\ldots(B + \tau A_{I,1})\varphi^2 = \varphi^1 + \tau f^1$$

$$\ldots$$

$$(B + \tau A_{1,j-1})(B + \tau A_{2,j-1})\ldots(B + \tau A_{I,j-1})\varphi^j$$
$$= \varphi^{j-1} + \tau f^{j-1}$$

$$\ldots$$

$$(B + \tau A_{1,N-2})(B + \tau A_{2,N-2})\ldots(B + \tau A_{I,N-2})\varphi^{N-1}$$
$$= \varphi^{N-2} + \tau f^{N-2}$$

$$(B + \tau A_{1,N-1})(B + \tau A_{2,N-1})\ldots(B + \tau A_{I,N-1})\varphi^{N}$$
$$= \varphi^{N-1} + \tau f^{N-1}, \qquad (45)$$

where $A_{i,n} = A_i(\varphi^n)$. Taking into account (44) the adjoint equations solving back in time are

$$\varphi^{*N+1} = 0,$$

$$(B + \tau A^*_{I,N-1})(B + \tau A^*_{I-1,N-1})\ldots(B + \tau A^*_{1,N-1})\varphi^{*N}$$

$$- \varphi^{*N+1} + \frac{1}{N}\alpha^N(\varphi^N - \varphi^N_{obs}) = 0,$$

$$\ldots$$

$$(B + \tau A^*_{I,j-1})(B + \tau A^*_{I-1,j-1})\ldots(B + \tau A^*_{1,j-1})\varphi^{*j}$$

$$- \varphi^{*j+1} + \frac{1}{N}\alpha^j(\varphi^j - \varphi^j_{obs}) = g(\varphi^j, \varphi^{j+1}, \varphi^{*j+1})$$

$$\ldots$$

$$(B + \tau A^*_{I,0})(B + \tau A^*_{I-1,0})\ldots(B + \tau A^*_{1,0})\varphi^{*1}$$

$$- \varphi^{*2} + \frac{1}{N}\alpha^1(\varphi^1 - \varphi^1_{obs}) = g(\varphi^1, \varphi^2, \varphi^{*2}),$$

$$- \varphi^{*1} + \alpha^0(\varphi^0 - \varphi^0_{obs}) = g(\varphi^0, \varphi^1, \varphi^{*1}). \qquad (46)$$

Here $A_{i,j} \equiv A_i(\varphi^j)$, $A^*_{i,j}$ is its adjoint, and $g(\varphi^{j-1}, \varphi^j, \varphi^{*j})$ is a function depending on the nonlinearity of

operator $A_{i,j} \equiv A_i(\varphi^j)$. For the linear case, $g(\varphi^{j-1}, \varphi^j, \varphi^{*j}) = 0$.

To solve the optimality system (45–46) we can use an iterative method (AGOSHKOV, 2003) or one of the gradient descent procedures (BLUM et al., 2008).

In the second case we formulate an iterative procedure for initial condition

$$(\varphi^0)^{k+1} = (\varphi^0)^k - \mu \cdot \operatorname{grad} J. \qquad (47)$$

The expression for the cost function gradient results from the last equation of (46)

$$\operatorname{grad} J = -\varphi^{*1} + \alpha^0(\varphi^0 - \varphi^0_{obs}) - g(\varphi^0, \varphi^1, \varphi^{*1}), \qquad (48)$$

and the iterative parameter μ can be found by standard procedures, for example, M1QN3 (Gilbert and Lemarechal 1989).

We can see by looking at (46) that, using the splitting method, we do not need to construct an adjoint operator for the whole complicated operator $A(\varphi)$. Rather, we construct the adjoint of the suboperators $A_i(\varphi)$, each of which has a simpler structure. This is quite convenient for writing and testing algorithms and codes.

Note that the procedure for the adjoint system solution (46) has a similar split structure to the process for the forward problem solution (42); for example, at time interval $t_{j-1} \leq t \leq t_j$, we have

$$-B\frac{\varphi^{*j} - \varphi^*_1}{\tau} + A^*_I(\varphi^j)\varphi^*_1 = -\frac{1}{\tau N}\alpha^{j-1}(\varphi^{j-1} - \varphi^{j-1}_{obs})$$

$$+ \frac{1}{\tau}g(\varphi^j, \varphi^{*j})$$

$$-B\frac{\varphi^*_1 - \varphi^*_2}{\tau} + A^*_{I-1}(\varphi^j)\varphi^*_2 = 0$$

$$\ldots$$

$$-B\frac{\varphi^*_I - \varphi^{*j-1}}{\tau} + A^*_1(\varphi^j)\varphi^{*j-1} = 0. \qquad (49)$$

Because $A^*_i \geq 0$, this is also an unconditionally stable implicit scheme solving backward in time.

REFERENCES

AGOSHKOV, V. I. (2003), *Methods of Optimal Control and Adjoint Equations in Problems of Mathematical Physics* (IVM RAN, Moscow, 2003, 256 p.) [in Russian].

AGOSHKOV, V.I. (2005), *Inverse problems of the mathematical theory of tides: boundary-function problem*. Russ. J. Numer. Anal. Math. Modelling. V. 20. N. 1. P. 1-18.

AGOSHKOV, V.I., GUSEV A.V., DIANSKI N.A., OLEINIKOV R.V. (2007), *An algorithm for the solution of the ocean hydrothermodynamics problem with variational assimilation of the sea level function data*. Russ. J. Numer. Anal. Math. Modelling. V. 22. N. 2. P. 133-161.

AGOSHKOV, V.I., V. M. IPATOVA (2007), *Solvability of the Observational Data Assimilation Problem for a 3D Ocean Dynamics Model*. Differential Equations N. 8. P. 1064-1075.

AGOSHKOV, V.I., V. M. IPATOVA (2007), *Existence Theorems for a 3D Ocean Dynamics Model and Data Assimila tion Problems*. Dokl. Akad. Nauk. V. 412. N. 2. P. 151-153.

AGOSHKOV, V.I., V. M. IPATOVA, V.B. ZALESNY, E.I. PARMUZIN, and V.P. SHUTYAEV (2010), *Problems of Variational Assimilation of Observational Data into Ocean General Circulation Models and Methods for Their Solution*, Izvestiya, Atmospheric and Oceanic Physics. V. 46. N. 6. P. 677-712.

AGOSHKOV, V.I., S. A. LEBEDEV, and E. I. PARMUZIN (2009), *Numerical Solution to the Problem of Variational Assimilation of Operational Observational Data on the Ocean Surface Temperature*. Izv. Akad. Nauk, Fiz. Atmos. Okeana V. 45. N. 1. P. 76-107 [Izv., Atmos. Ocean. Phys. V. 45. N. 1. P. 69-101.

AGOSHKOV V.I., MARCHUK G.I. (1993), *On solvability and numerical solution of data assimilation problems*. Russ. J. Numer. Anal. Math. Modelling. V. 8. N 1. P. 1-16.

AGOSHKOV, V.I., MINYUK F.P., RUSAKOV A.S., ZALESNY V.B. (2005), *Study and solution of identification problems for nonstationary 2D- and 3D-convection-diffusion equation*. Russ. J. Numer. Anal. Math. Modelling. V. 20. N. 1. P. 19-43.

AGOSHKOV, V.I., E. I. PARMUZIN, and V. P. SHUTYAEV (2008), *Numerical Algorythm of Variational Assimilation of Observational Data on Oceanic Surface Temperature*. Zh. Vysch. Matem. Matem. Fiz. V. 48. N. 8. P. 1371-1391.

AGOSHKOV, V.I., PARMUZIN E.I., SHUTYAEV V.P. (2008), *A numerical algorithm of variational data assimilation for reconstruction of salinity fluxes on the ocean surface*. Russ. J. Numer. Anal. Math. Modelling. V. 23. N. 2. P. 135-161.

BELLMAN R.E. (1957), Dynamic Programming. Princeton University Press, Princeton, NJ 1957. Republished 2003: Dover, ISBN 0486428095.

BENNETT A. F. (2002), *Inverse modeling of the ocean and atmosphere*. Cambridge: Cambridge University Press, 234 p.

BLUM J., LE DIMET F.-X., NAVON I.M. (2008), Data assimilation for geophysical fluids, in Giarlet (Ed.): Computational Methods for the Atmosphere and the Oceans. Special volume. Handbook of Numerical analysis. V. XIV. P. 377-434.

CAYA A., SUN J., SNYDER C. (2005), *A Comparison between the 4DVAR and the ensemble Kalman filter techniques for radar data assimilation*. Monthly Weather Review. V. 133. N. 11. P. 3081-3094.

COURTIER P., ANDERSSON E., HECKLEY W., PAILLEUX J., VASILJEVIC D., HAMRUD M., HOLLINGSWORTH A., RABIER F., FISHER M. (1998), *The ECMWF implementation of three-dimensional variational assimilation (3D-Var). I: Formulation*. Quart. J. R. Meteorol. Soc. V. 124. P. 1783-1807.

CHASSIGNET E.P. AND VERRON J. (eds., 2005). *Ocean Weather Forecasting. An Integrated View of Oceanography*, Springer, Dordrecht, the Netherlands, 577 p.

DIANSKY, N.A., A. V. BAGNO, AND V. B. ZALESNY (2002), *Sigma Model of Global Ocean Circulation and Its Sensitivity to Variations in Wind Stress*. Izv. Akad. Nauk, Fiz. Atm. Okeana V. 38. N. 4. P. 537-556 (2002) [Izv., Atmos. Ocean. Phys. V. 38. N. 4. P. 477-494.

DIANSKY, N.A., V. B. ZALESNY, S. N. MOSHONKIN, A.S. RUSAKOV (2006), *High Resolution Modeling of the Monsoon Circulation in the Indian Ocean*. Okeanologiya V. 46. N. 4. P. 11-12 [Oceanology V. 46. N. 5. P. 608-628 (2006)].

EVENSEN G. (2007), *Data assimilation: The ensemble Kalman filter*. Berlin: Springer, 307 pp.

LE DIMET F.-X., SHUTYAEV V.P. (2005), *On deterministic error analysis in variational data assimilation*. Nonlinear Processes in Geophysics. V. 12. P. 481-490.

LE DIMET F.-X., TALAGRAND O. (1986), *Variational algorithms for analysis and assimilation of meteorological observations: Theoretical Aspects*. Tellus. V. 38A. P. 97-110.

FERTIG E.J., HARLIM J., and HUNT B.R. (2007), *A Comparative Study of 4D-VAR and a 4D Ensemble Kalman Filter: Perfect Model Simulations with Lorenz-96*. Tellus. V. 59A. P. 96-100.

GANDIN L. (1963), *Impartial Analysis of Hydrometeorological Fields* (Gidrometizdat, Leningrad) [in Russian].

GILBERT J.-C., LEMARECHAL C. (1989), *Some numerical experiment with variable storage quasi-Newton algorithms*. Math. Program. B25. P. 408-435.

IPATOVA V. M. (2008), *Solvability of the ocean hydrothermodynamics problem under a nonlinear state equation*. Russ. J. Numer. Anal. Math. Modelling. V. 23. N. 2. P. 185-196.

IVCHENKO, V. O., S. D. DANILOV, D. V. SIDORENKO, J. SCHROTER, M. WENZEL, and D. L. ALEYNIK, (2007), *Comparing the steric height in the Northern Atlantic with satellite altimetry*. Ocean Science, 3. P. 485-490.

IVCHENKO, V. O., N. C. WELLS, and D. L. ALEYNIK, (2006), *Anomaly of heat content in the northern Atlantic in the last 7 years: Is the ocean warming or cooling?* Geophysical Research Letters, 33, 6.

JAZWINSKI A.H. (1970), Stochastic Processes and Filtering Theory. Academic Press, London, 376 p.

KALMAN R.E. (1961), *A new approach to linear filtering and prediction problems*. Trans. ASME J. Basic Eng. V. 82D. P. 35-45.

KALMAN R.E., BUCY R.S. (1961), *New results in linear filtering and prediction theory*. Trans. ASME J. Basic Eng. V. 83D. P. 95-108.

KALNAY E. (2003), Atmospheric Modeling. Data Assimilation and Predictibility. Cambridge University Press, 457 p.

KALNAY E., LI H., MIYOSHI T., YANG S.-C. and BALLABRERA-POY J. (2007), *4D-Var or Ensemble Kalman Filter?* Tellus. V. A 59. P. 758-773.

KOROTAEV, G.K. and V. N. EREMEEV (2006), *Introduction to Operative Oceanography of the Black Sea* (NPTs EKOSI Gidrofizika, Sevastopol) [in Russian].

KUZIN, V.I. (1985), Finite Element Method in Simulation of Oceanic Processes (CC SO AN SSSR, Novosibirsk) [in Russian].

LORENC A.C. (1981), *A global three-dimensional multivariate statistical analysis scheme*. Mon. Wea. Rev. V. 109. P. 701-721.

LORENC A.C. (1986), *Analysis methods for numerical weather prediction*. Quart. J. R. Meteorol. Soc. V. 112. P. 1177-1194.

LIONS J.-L. (1968), *Control Optimal des Systemes Gouvernes par des Equations aux Derivees Partielles*. Dunod, Paris, 426 p.

MARCHUK G.I. (1995), Adjoint Equations and Analysis of Complex Systems. Kluwer, Dordrecht, 466 p.

MARCHUK G.I. (1975), *Formulation of the theory of perturbations for complicated models*. Appl. Math. Optimiz., V.2. N. 1. P. 1-33.

MARCHUK G.I. (1982), Methods of Computational Mathematics. Springer-Verlag, New York.

MARCHUK, G.I. (1990), *Splitting and alternating direction methods*, in: P.G. CIARLET, J.L. LIONS (Eds.), Handbook of Numerical Analysis, V. 1, North-Holland, Amsterdam. P. 197-462.

MARCHUK G.I., AGOSHKOV V.I., SHUTYAEV V.P. (1996), *Adjoint Equations and Perturbation Algorithms in Nonlinear Problems*, CRC Press Inc., New York, 275 p.

MARCHUK G.I., PENENKO V.V. (1978), *Application of optimization methods to the problem of mathematical simulation of atmospheric processes and environment*, in G.I.Marchuk (ed.), Modelling and Optimization of Complex Systems: Proc. Of the IFIP-TC7 Working conf., Springer, New York. P. 240-252.

MARCHUK G.I., SHUTYAEV V.P. (1994), *Iteration methods for solving a data assimilation problem*. Russ. J. Numer. Anal. Math. Modelling. V. 9. N. 3. P. 265-279.

MARCHUK, G.I., V. P. DYMNIKOV, AND V. B. ZALESNY (1987), *Mathematical Models in Geophysical Hydrodynamics and Numerical Methods of Their Realization* Gidrometeoizdat, Leningrad) [in Russian].

MARCHUK G.I. and KUZIN V.I. (1982), *On the combination of the finite element and splitting-up methods in the solution of the parabolic equations*. J. Comp. Phys. V. 52. N 2. P. 237-272.

MARCHUK G.I., RUSAKOV A.S., ZALESNY V.B., DIANSKY N.A. (2005), *Splitting Numerical Technique with Application to the High Resolution Simulation of the Indian Ocean Circulation*. Pure and Applied Geophysics, Birkhauser Verlag, Basel. N. 162. P. 1407-1429.

MARCHUK G., SHUTYAEV V., ZALESNY V. (2001), *Approaches to the solution of data assimilation problems*, in MENALDI J.L., ROFMAN E., SULEM A. (eds.) Optimal Control and Partial Differential Equations, IOS Press, Amsterdam. P. 489-497.

MARCHUK G.I., ZALESNY V.B. (1993), *A numerical technique for geophysical data assimilation problem using Pontryagin's principle and splitting-up method*. Russ. J. Numer. Anal. Math. Modelling. V. 8, N. 4. P. 311-326.

NAVON I.M. (2009), *Data assimilation for numerical weather prediction: a review*, in Park S.K., Xu L. Data Assimilation for Atmospheric, Oceanic and Hydrologic Applications. Springer-Verlag, Berlin Heidelberg. P. 21-65.

NECHAEV, D, SCROETER J., YAREMCHUK M, (2003), *A diagnostic stabilized finite-element ocean circulation model*. Ocean Modelling. V. 5. P. 37-63.

NELEPO, B.A., V. V. KNYSH, A. S. SARKISYAN, TIMCHENKO I.E., (1979), *Study of Synoptic Variability of the Ocean Based on Dynamic–Stochastic Approach*. Dokl. Akad. Nauk V. 246. N. 4. P. 974–978.

NERGER, L., J. SCHRÖTER, S. DANILOV, W. HILLER. (2006) *Using sea-level data to constrain a finite-element primitive-equation ocean model with a local SEIK filter*. Ocean Dynamics 56 634-649, doi: 10.1007/s10236-006-0083-0.

PENENKO, V.V. and N. V. OBRAZTSOV, (1976), *Variational Method for Fields of Meteorological Elements*. Meteorol. Gidrol., N. 11. P. 1–11.

PONTRYAGIN, L.S. et al. (1962), *The Mathematical Theory of Optimal Processes*, V. 4. Interscience. Translation of a Russian book. ISBN 2881240771 and ISBN 978-2881240775.

LE PROVOST C., SALMON R. (1986), *A variational method for inverting hydrographic data*. J. Mar. Res. V. 44. P. 1-34.

SARKISYAN A.S. and SUENDERMANN J. (2009), *Modelling Ocean Climate Variability*. Springer, 374 p.

SARKISYAN, A.S., V. B. ZALESNY, N. A. DIANSKY, et al. (2005), *Mathematical Models of Ocean and Sea Circulation* in *Modern Problems in Computational Mathematics and Mathematical Simulation*, (Nauka, Moscow), V. 2. P. 175–278 [in Russian].

SASAKI Y. (1970), *Some basic formalisms in numerical variational analysis*. Mon. Wea. Rev. V. 98. P. 875-883.

SHUTYAEV, V.P. (2001), *Control Operators and Iterative Algorithms in Variational Data Assimilation Problems* (Nauka, Moscow) [in Russian].

Shutyaev V.P., Parmuzin E.I. (2009), *Some algorithms for studying solution sensitivity in the problem of variational assimilation of observation data for a model of ocean thermodynamics*. Russ. J. Numer. Anal. Math. Modelling. V. 24. N. 2. P. 145-160.

TIAN X., XIE J., AND DAI A. (2008), *An ensemble-based explicit 4D-Var assimilation method*. Journal of Geophysical Research. V. 113 (D21124). 13 p.

WENZEL, M and V. B. ZALESNY, (1996), *Data Assimilation in One Dimensional Model of Heat Convection–Diffusion in the Ocean*. Izv. Akad. Nauk, Fiz. Atm. Okeana. V. 32. N. 5. P. 613–629.

WENZEL M., J. SCHRÖTER, D. OLBERS, (2001), *The annual cycle of the global ocean circulation as determined by 4D VAR data assimilation*. Prog. in Oceanogr., 48. P. 73-119.

ZALESNY V.B., GUSEV A.V. (2009), *Mathematical model of the World ocean dynamics with algorithms of variational assimilation of temperature and salinity fields*. Russ. J. Numer. Anal. Math. Modelling. V. 24. No. 2. P. 171-190.

ZALESNY V.B., RUSAKOV A.S. (2007), *Numerical algorithm of data assimilation based on splitting and adjoint equation methods*. Russ. J. Numer. Anal. Math. Modelling. V. 22. N. 2. P. 199-219.

ZALESNY V.B., MARCHUK G.I., AGOSHKOV V.I., BAGNO A.V., GUSEV A.V., DIANSKY N.A., MOSHONKIN S.N., TAMSALU R (2010), *Numerical simulation of large-scale ocean circulation based on the multicomponent splitting method*. Russ. J. Numer. Anal. Math. Modelling. V. 25. N. 6. P. 581-609.

ZHANG F. Q., ZHANG M., AND HANSEN J. A. (2009), *Coupling ensemble Kalman filter with four dimensional variational data assimilation*. Adv. Atmos. Sci. V. 26. N. 1. P. 1–8.

(Received November 11, 2010, accepted March 16, 2011, Published online September 11, 2011)

Pure Appl. Geophys. 169 (2012), 579–594
© 2011 Springer Basel AG
DOI 10.1007/s00024-011-0387-y

Barrier Layer and Relevant Variability of the Salinity Field in the Equatorial Pacific Estimated in an Ocean Reanalysis Experiment

Y. Fujii,[1] M. Kamachi,[1] S. Matsumoto,[2] and S. Ishizaki[2]

Abstract—This paper investigates the feasibility of an ocean data assimilation system to analyze the salinity variability associated with the barrier layer in the equatorial Pacific. In order to validate reproducibility of the temperature and salinity fields, we perform an assimilation run where some temperature and salinity observations by TRITON buoys and Argo floats are withheld. The assimilation run reproduces interannual variability of salinity in the equatorial Pacific exhibited in the data that are withheld. Statistics shows that salinity values and variations in the assimilation run are closer to the data than the climatology and in the model free run. We also confirm that zonal currents in the equatorial Pacific in the reanalysis, where all available temperature and salinity data are assimilated, are consistent with an observation-based mapping and the data of the Acoustic Doppler Current Profiler mounted on TAO buoys. Variability of the barrier layer and relevant salinity field in the reanalysis is consistent with former studies. A thick barrier layer area generally exists west of the equatorial salinity front and is displaced zonally with the migration of the front in the response to El Niño-Southern Oscillation, although the area moved to the east over the front in the 1997 El Niño. It is confirmed that the barrier layer thickness is closely correlated with the near-surface temperature in the equatorial Pacific.

1. Introduction

The salinity field in the equatorial Pacific has great zonal contrast. A fresh water pool associated with large precipitation occupies the near-surface layer in the western equatorial Pacific. In contrast, high salinity water, known as the South Pacific Tropical Water (SPTW), is generated by vigorous evaporation in the stable atmospheric anticyclone in the wide area of the tropical South Pacific. SPTW subducts into the subsurface layer, and is advected by the equatorial shallow overturn (McPhaden and Zhang, 2002) to the equator. It then surfaces through equatorial upwelling, resulting in an area of high salinity in the near-surface layer in the central equatorial Pacific.

A sharp equatorial salinity front exits between the fresh water pool and the central high salinity area in the near-surface layer. Observational studies indicate that the salinity front migrates further than 1000 km zonally (e.g., Hénin et al., 1998). Other studies (e.g., Picaut et al., 1996; Delcroix and Picaut, 1998) have also suggested that the migration is closely correlated with El Niño-Southern Oscillation (ENSO). Zonal displacement of the warm water pool, which may be instrumental in generating the ENSO cycle, is represented by the migration of the salinity front (Picaut et al., 1997). In addition, the horizontal gradient of the near surface salinity field across the front enhances an eastward current and advection of the warm water (Roemmich et al., 1994).

It should be noted that salinity stratification is essential for the formation of the barrier layer (Lukas and Lindstrom, 1991). Former studies (e.g., Vialard and Delecluse, 1998; Cronin and McPhaden, 2002) suggested that a thick barrier layer is generated west of the equatorial salinity front by the high salinity water subducting under the fresh water pool. This barrier layer can affect surface currents by thinning the mixed layer and concentrating the effect of wind stress (e.g., Vialard et al., 2002; Maes et al., 2002). It is also likely to induce a temperature rise in the mixed layer because it prevents warm water near the surface from mixing with cold water in the thermocline (e.g., Ando and McPhaden, 1997; Maes et al., 2006; Bosc et al. 2009). Several studies (Maes et al.

---
[1] Oceanographic Research Department, Meteorological Research Institute, Nagamine 1-1, Tsukuba, Ibaraki 305-0052, Japan. E-mail: yfujii@mri-jma.go.jp
[2] Global Environment and Marine Department, Japan Meteorological Agency, Ohtemachi 1-3-4, Chiyoda, Tokyo 100-8122, Japan.

2002, 2005) using coupled ocean-atmosphere general circulation models further suggested the impact of the barrier layer on onsets of El Niño events.

While the variability of near-surface salinity in the equatorial Pacific is the focus of many studies, that of subsurface salinity has been examined in only a few studies. SPTW makes a salinity maximum in the subsurface at the equator. KESSLER (1999) reported that the salinity maximum in the equatorial Pacific increases in La Niña periods. Intrusion of SPTW into the northern hemisphere in the early stages of El Niño periods has been pointed out in several studies (e.g., DELCROIX et al., 1992; JOHNSON and McPHADEN, 2000).

Former studies on the equatorial Pacific salinity field mainly analyzed observational data directly. The sparseness of salinity observations, however, has made it difficult to grasp the overall picture of the salinity variation. Although Ocean General Circulation Models (OGCMs) have been used in a few studies (e.g., VIALARD et al., 2002), they suffer from serious errors in salinity and may not be able to reproduce part of its variation. Other studies (e.g., MAES et al., 2000; FUJII and KAMACHI, 2003b) also took advantage of the temperature–salinity (T–S) relation to supplement salinity data. In particular, FUJII and KAMACHI (2003b) (hereafter, FK03b) applied a Three-Dimensional Variational (3DVAR) analysis scheme of the T–S relation to the mapping of the salinity field from temperature and salinity observations (without OGCMs) and reconstructed the migration of the salinity front and the relationship between the barrier layer and near-surface temperature. However, it is apparent that the T–S relation is insufficient for reconstructing the full variation of the salinity field because part of it is independent of temperature change.

Ocean data assimilation systems are powerful tools for analyzing the real fluctuation of the ocean, because they synthesize a variety of observation data under the constraints of OGCMs, and reproduce the realistic ocean state and its variations, including the salinity field (LEE et al. 2009; STAMMER et al., 2010). Salinity has not been estimated carefully in data assimilation systems until recently mainly because of the shortage of observations. However, recent improvements (KAMACHI et al., 2002; FUJII et al.,

2010), as well as a rapid increase of salinity observations by Argo floats (e.g., BOUTIN and MARTIN, 2006; GOULD and TURTON, 2006), enable us to provide time series of realistic salinity fields. We, therefore, aim to confirm the feasibility of using data assimilation to increase our understanding of the salinity variability associated with the barrier layer in the equatorial Pacific.

In this study, we use a global ocean data assimilation system that adopts the variational scheme of FK03b to create analysis fields reflecting both model prediction and observation data. In order to confirm the feasibility, we first validate the reproducibility of salinity and other ocean fields by the data assimilation system. We then examine the variability of the barrier layer and relevant salinity field in the assimilation result. The ocean data assimilation system is introduced in Sect. 2. The indices used in this study are described in Sect. 3. The ocean data assimilation result is validated in Sect. 4. The variability of the barrier layer and its connection to the temperature field are examined in Sects. 5 and 6, respectively. This study is summarized in Sect. 7.

2. Ocean Reanalysis Experiments

2.1. Outline of the Data Assimilation System

In this study, we use the results of an ocean reanalysis experiment using the ocean data assimilation system developed at the Meteorological Research Institute (MRI), introduced in USUI et al. (2006). The system is composed of an OGCM based on the MRI Community Ocean Model (MRI.COM; TSUJINO et al., 2010), and the Multivariate Ocean Variational Estimation System (MOVE), a 3DVAR analysis scheme based on FK03b. This experiment is named MOVE-Global Version Reanalysis 2007 (MOVE-G RA07; FUJII et al., 2009).

The domain of the ocean model is nearly global, extending from 75°S to 75°N. The grid spacing in the zonal direction is 1°, and that in the meridional direction changes from 0.3° (within 5.7°S–5.7°N) to 1° (poleward of 16°S and 16°N). There are 50 levels in the vertical direction. The first seven levels are 1,

3, 6, 10, 16, 22, and 30 m in depth, and the interval between 30 and 200 m depth is 10 m. For momentum advection, a generalized enstrophy-preserving scheme and a scheme that involves the concept of diagonally upward/downward mass momentum fluxes along the sloping bottom are applied (ISHIZAKI and MOTOI, 1999). Isopycnal diffusion (REDI, 1982), isopycnal thickness diffusion (GENT and MCWILLIAMS, 1990), and the vertical mixing scheme of NOH and KIM (1999) are also adopted in the model.

2.2. Analysis and Assimilation Scheme

The analysis scheme, MOVE, adopts the 3DVAR method using vertical coupled T–S Empirical Orthogonal Function (EOF) modal decomposition for the background error covariance matrix. The model domain is partitioned into 40 horizontal subdomains, and the T–S EOF modes are calculated for each subdomain from historical profile data in the World Ocean Database 2001 (WOD01; CONKRIGHT *et al.*, 2002) and the Global Temperature–Salinity Profile Program (GTSPP) database (HAMILTON, 1994). The subdomains overlap in the boundary areas.

Analysis of the temperature and salinity profiles at the pth horizontal grid, $\mathbf{x}_p$, are as follows:

$$\mathbf{x}_p = \mathbf{x}_p^f + \sum_m \sum_l w_{m,l,p}\left(a_{m,p}\lambda_{m,l,p}\mathbf{S}_p\mathbf{u}_{m,l}\right) \quad (1)$$

where $w_{m,l,p}$ is the amplitude of the lth EOF mode in the mth subdomain, $a_{m,p}$ is the weight coefficient of the mth subdomain, $\lambda_{m,l,p}$ is the singular value for the lth mode, $\mathbf{S}_p^f$ is the diagonal matrix composed of the respective standard deviations of temperature and salinity from their first guess at model levels, $\mathbf{u}_{m,l}$ is the vector representing the lth mode, and $\mathbf{x}_p^f$ is the first guess of the profiles. Thus, correction of temperature and salinity fields to the first guess is represented by a linear combination of the EOF modes. The weight coefficients $a_{m,p}$ satisfy $\sum_m a_{m,p}^2 = 1$, avoiding the loss or gain of total variances for each control variable by the area partition (FUKUMORI, 2002).

The mode amplitudes, $w_{m,l,p}$, are estimated so that the cost function is minimized. The cost function, $J(\mathbf{w})$, is defined as follows:

$$J(\mathbf{w}) = \frac{1}{2}\sum_m \sum_l \mathbf{w}_{m,l}^T \mathbf{B}_m^{-1} \mathbf{w}_{m,l}$$
$$+ \frac{1}{2}(\mathbf{Hx} - \mathbf{y})^T \mathbf{R}^{-1}((\mathbf{Hx} - \mathbf{y}))$$
$$+ \frac{1}{2}[\mathcal{H}_h(\mathbf{x}) - \mathbf{y}_h]^T \mathbf{R}_h^{-1}[\mathcal{H}_h(\mathbf{x}) - \mathbf{y}_h]$$
$$+ \frac{1}{2}[\mathcal{C}(\mathbf{x})]^2. \quad (2)$$

where $\mathbf{w}$ is the vector composed of $w_{m,l,p}$, and $\mathbf{w}_{m,l}$ is the partial vector of $\mathbf{w}$ whose elements are amplitudes of the lth mode in the mth region. The matrix $\mathbf{B}_m$ represents the horizontal correlation of background (first-guess) errors for the mth subdomain modeled by the Gaussian function. The vector $\mathbf{y}$ is composed of temperature and salinity observations, and $\mathbf{y}_h$ is satellite Sea Surface Height (SSH) data. The vector $\mathbf{x} = \mathbf{x}^f + \mathbf{Gw}$ is the state vector of temperature and salinity analysis fields composed of $\mathbf{x}_p$, where $\mathbf{x}^f$ is the first guess and $\mathbf{G}$ denotes the transformation represented by (1). The matrix $\mathbf{H}$ represents spatial interpolation for acquiring the values equivalent to the temperature and salinity observation, and $\mathcal{H}_h$ is the nonlinear operator that includes calculation of the Sea Surface Dynamic Height (SDH) from gridded temperature and salinity data and interpolation. The matrix $\mathbf{R}$ ($\mathbf{R}_h$) is the observation error covariance matrix for the temperature and salinity observation (the satellite SSH data). The fourth term on the right-hand side represents constraints for avoiding density inversion and $\mathcal{C}$ is defined as follows:

$$\mathcal{C}(\mathbf{x}) = (1/b) \sum_p \sum_k D(\rho_{p,k} - \rho_{p,k+1}) \quad (3)$$

where $\rho_{p,k}$ denotes the density at the kth level at the pth grid point ($k = 1$: surface), and b are coefficients for the strength of the constraints and set to 0.02 kg m^{-3} Here, D is the nonlinear function: $D(x) = 0$ when $x < 0$, and $D(x) = x$ when $x > 0$.

The calculation of SDH in the nonlinear operator $\mathcal{H}_h$ is implemented as follows:

$$h = -\frac{1}{\rho_s} \int_0^{z_m} \rho'(T, S, p)\mathrm{d}z \quad (4)$$

where h is the difference of SDH from a reference state, z denotes the vertical coordinate, T is temperature, S is salinity, p is pressure, ρ_s is surface density,

z_m is the reference depth (1500 m), and ρ' is the difference of the density from the reference state ($T = 0°C$ and $S = 35$).

The gradient of the cost function is written as follows:

$$\mathbf{g} = \sum_m \sum_l \mathbf{B}_m^{-1}\mathbf{w}_{m,l} + \mathbf{G}^T\left[\mathbf{H}^T\mathbf{R}^{-1}(\mathbf{Hx} - \mathbf{y})\right.$$
$$\left. +\mathbf{H}_h^*\mathbf{R}_h^{-1}\{\mathcal{H}_h(\mathbf{x}) - \mathbf{y}_h\} + \mathbf{C}^*\mathcal{C}(\mathbf{x})\right] \quad (5)$$

where $\mathbf{H}_h^*(\mathbf{C}^*)$ is the adjoint code of the nonlinear operator $\mathcal{H}_h(\mathcal{C})$. We also apply the variational quality control procedure introduced in Fujii et al. (2005). It should be noted that the cost function is non-quadratic for $\mathbf{w}$, and that the calculation of $\mathbf{g}$ includes inversion of the non-diagonal matrix $\mathbf{B}_m$. In MOVE, we adopt the preconditioned quasi-Newton method introduced in Fujii and Kamachi (2003c) and Fujii (2005) to minimize this non-quadratic function without implementing the inversion directly.

The strong point of this analysis scheme is that variation of salinity coupled with that of temperature can be estimated through the coupled T–S EOF modes, even if little salinity data is available (Fujii and Kamachi, 2003a). For example, vertical displacement of water mass associated with oceanic baroclinic modes causes a coupled T–S change. MOVE can adequately estimate this sort of changes. In contrast, salinity variation in the North Pacific Intermediate Water is difficult to estimate because it is independent of changes due to vertical displacement. Estimation through the T–S coupled modes improves the reliability of the salinity in assimilation fields (e.g., Maes et al., 2000).

The system adopts the Incremental Analysis Updates (IAU) technique (Bloom et al., 1997) coupled with an online model-bias estimation using the one-step bias-correction algorithm (Balmaseda et al., 2007) to reflect the analysis result in the model fields. Variational analysis is performed once a month using all available observations each month. The first guess for the analysis is given as $\mathbf{x}_f = (1 - w)(\mathbf{x}_s + \mathbf{b}) + w\mathbf{x}_c$, where $\mathbf{x}_s$ is the half-month prediction from the assimilation result at the end of the previous month, $\mathbf{b}$ is the three-dimensional bias estimate, $\mathbf{x}_c$ is the climatology, and w is the weight coefficient and is set to 0.04. The difference between the analysis and the model-predicted fields (i.e., analysis increment) is applied to correct the temperature and salinity fields in the model. Current fields are adjusted to the corrected temperature and salinity fields through the model dynamics, and thus establish the geostrophic balance in most areas. The bias estimate is also updated by taking a weighted mean of its original and analysis increment. The weight coefficient for analysis increment is set to 1/60.

2.3. Observation and Forcing

The atmospheric reanalysis dataset produced by the National Center for Environmental Prediction and the National Center for Atmospheric Research (NCEP-R1; Kalnay et al., 1996) is employed as the external forcing in the oceanic reanalysis experiment. The fresh water flux is corrected using a constant correction term (Vialard et al., 2001) and a term restoring sea surface salinity to its climatology (Vialard et al., 2002) with a restoring time of 1 year. The zonal wind stress near the equator is adjusted by multiplying by 1.4 in order to be balanced with the zonal pressure gradient generated by temperature and salinity analysis fields (Ishizaki et al., 2006).

In the reanalysis, in situ temperature and salinity profiles and satellite SSH anomaly data are assimilated into the model. The temperature and salinity profiles are collected from WOD01, GTSPP, and the data of the TAO/TRITON array (Hayes et al., 1991; McPhaden et al., 1998; Kuroda et al., 2002). Profiles of Argo floats are included in GTSPP. The SSH data is the along-track data from TOPEX/Poseidon, Jason-1, European Remote Sensing satellite (ERS)-1/2, and Environmental Satellite multimission altimeters (ENVISAT), extracted from Ssalto/Duacs delayed-time multimission altimeter products (CLS, 2004).

In this study, we also use the results of model free simulation and another assimilation run to validate the temperature and salinity fields reproduced by the data assimilation system. Free simulation is performed using the same ocean model that was used in the data assimilation system in the same setting as for MOVE-G RA07. The assimilation run is performed using the same system in the same setting as for MOVE-G RA07, although some of the temperature and salinity profiles are excluded from the data assimilated in order to use them as independent reference data. The excluded profiles are those observed by TRITON buoys positioned at 5°N–156°E, 0°–156°E, and 5°S–156°E; the profiles of Argo floats whose last digit of the World Meteorology Organization (WMO) ID is "4"; and the profiles whose position and date are similar to those of one of the above excluded profiles (within 0.1° in longitude and latitude, and on the same date). It should be noted that this exclusion reduces the number of assimilated salinity profiles by approximately 10%. Hereafter, this experiment is called MOVE-G VAL.

All analyses in this study are performed with the monthly averaged fields of the reanalysis, assimilation run, and free simulation described above.

3. Indices

We examine the distribution of the barrier layer thickness (BLT) and its relationship to the temperature and current fields. In this study, BLT is calculated following the method of de Boyer MONTÉGUT et al. (2004) and BOSC et al. (2009). Their method is based on SPRINTALL and TOMCZAK (1992). BLT is defined as the difference between isothermal layer depth (ILD) and mixed layer depth (MLD) when ILD is deeper than MLD. ILD (MLD) is estimated as the depth where temperature (density) is ΔT lower($\Delta \rho$ higher) than that at 10 m depth, where $\Delta T = 0.2°C$ and $\Delta \rho = -(\partial \rho / \partial T)\Delta T$. BLT is regarded as equal to 0 when MLD is deeper than ILD.

Temperature averaged over 0–300 m is generally employed as the ocean heat content (OHC), particularly in studies of ENSO. We also define the heat content of the water exceeding 28°C in a water column per unit area as the warm water heat content (WWHC). It is defined by analogy with tropical cyclone heat potential (TCHP; LEIPPER and VOLGENAU, 1972). TCHP is defined as the heat content of water exceeding 26°C and is used as an indicator of the oceanic potential for intensifying a tropical cyclone (e.g., WADA and CHAN, 2008; YABLONSKY and GINIS, 2008). WWHC is expected to indicate the oceanic potential to promote atmospheric convection in the warm water region.

We also define the SPTW Salt Content (SPTWSC) as an indicator of the influence of SPTW in the equatorial Pacific. SPTWSC is calculated as the salinity content in water whose salinity exceeds 35.2 in a water column above 300 m depth per unit area.

4. Validation

4.1. Salinity

In this subsection, we validate the reproducibility of the salinity field in the equatorial Pacific by the data assimilation system. Here we compare the result of MOVE-G VAL with the result of the free simulation and the independent reference profiles that are withheld in the experiment (MOVE-G VAL). It should be noted that the accuracy of MOVE-G RA07 is generally expected to be similar to or better than that of MOVE-G VAL, because all data assimilated in MOVE-G VAL, as well as the data withheld, are assimilated in MOVE-G RA07.

Comparison of time-depth sections of the salinity field in MOVE-G VAL with those observed by TRITON buoys demonstrates that the data assimilation system successfully recovers the variability of the high salinity water in the subsurface layer (Fig. 1a, b). For example, the relatively high salinity (>34.8) period between 2002–2004 in the subsurface (90–180 m) at 5°N is recovered; the existence of high salinity water (>35.4) at the equator in 2000, 2001, 2006, and 2008 is estimated; and the variability of the depth of the subsurface salinity maximum at 5°S is reproduced in the assimilation run. Interannual variations of fresh water in the near-surface layer at these three points are also estimated in the assimilation run. For example, fresh water with salinity of less than 34.4 occupies the near-surface layer between the second half of 2001 and the end of 2005 at the equator. Although salinity increases in the first half of 2006, it decreases again in the second half and extremely fresh water (<34.0) reaches deeper than in the former low salinity period. This variation is successfully estimated in the assimilation run.

In contrast, the salinity field of the free simulation has high salinity bias in the subsurface layer mainly occupied by SPTW (Fig. 1c). This bias seems to be caused by an inaccurate surface fresh water flux. This flux is generally considered one of the main error

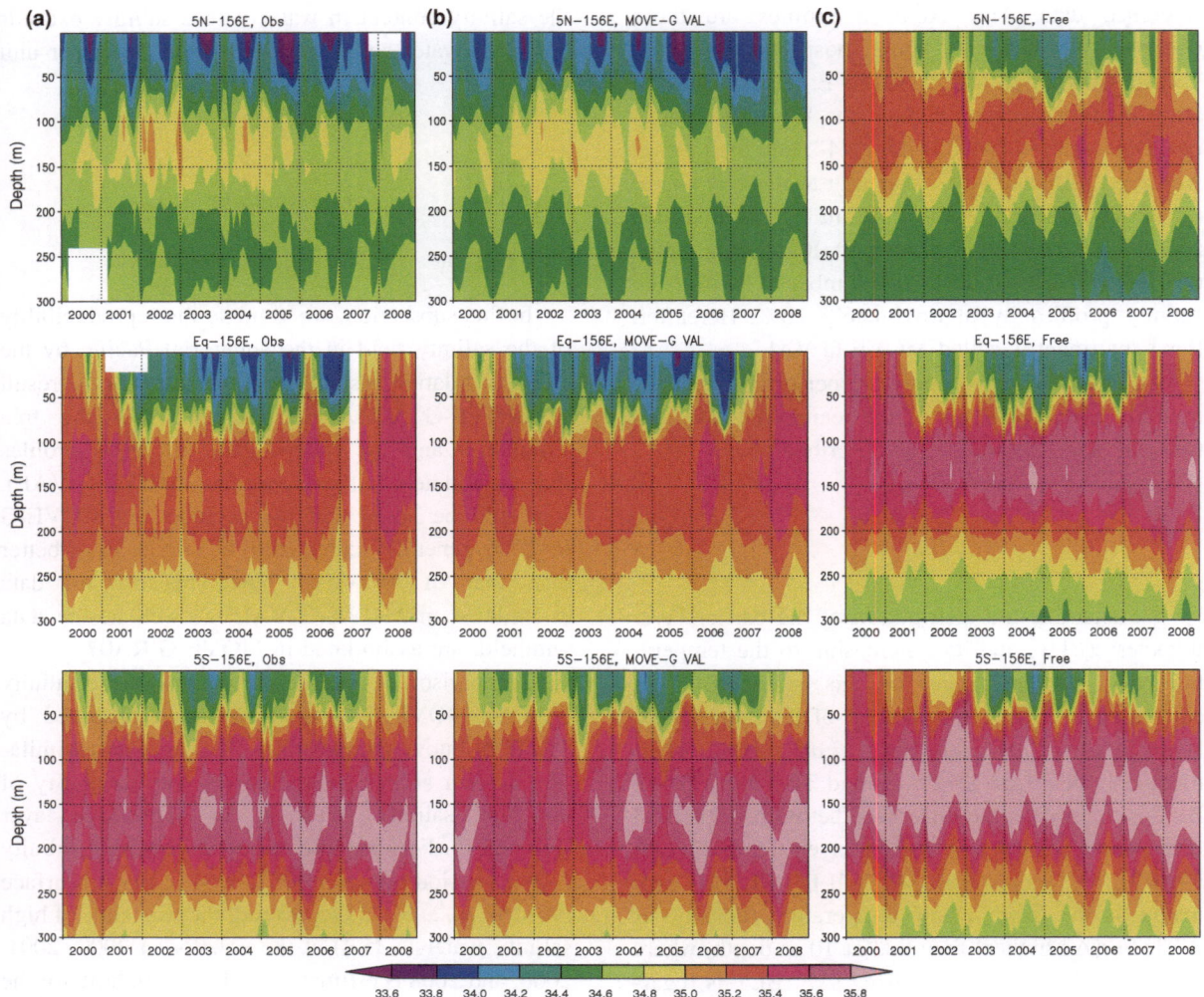

Figure 1
Time-depth sections of the salinity field during 2000–2008 observed by (**a**) TRITON buoys (monthly average), together with the sections
estimated in (**b**) MOVE-G VAL and (**c**) free simulation. *Top* 5°N-156°E. *Middle* 0°–156°E. *Bottom* 5°S–156°E

sources (e.g., ZHANG *et al.*, 2010), and our elaborate
adjustment of the flux, described in Sect. 3, may still
be insufficient for the simulation. The smaller
meridional contrast of salinity in the subsurface layer
between 5°S and the equator, as well as the larger
vertical contrast between the subsurface and the
lower-salinity layer under it, indicates another possibility that dissipation of SPTW by vertical mixing is
underestimated. This bias is, however, removed
almost completely in MOVE-G VAL.

The salinity field is also compared with the time
series of salinity profiles (not averaged monthly) for
two Argo floats (WMO ID: 5900384 and 5900304)

that are withheld in the experiment in Fig. 2.
Although the high frequency variations in the observation are smoothed out in the assimilation run partly
due to the use of monthly data, changes in the value
and position of the salinity maximum in the subsurface layer are reproduced. The variations of near-
surface salinity are also estimated well, particularly
for float 5900304. MOVE-G VAL recovers the
extremely low salinity from the boreal summer to
the winter of 2003–04, and the low salinity in 2006 in
the near-surface layer for the float.

Table 1 presents statistics for the accuracy of
salinity at fixed depths in MOVE-G VAL with those

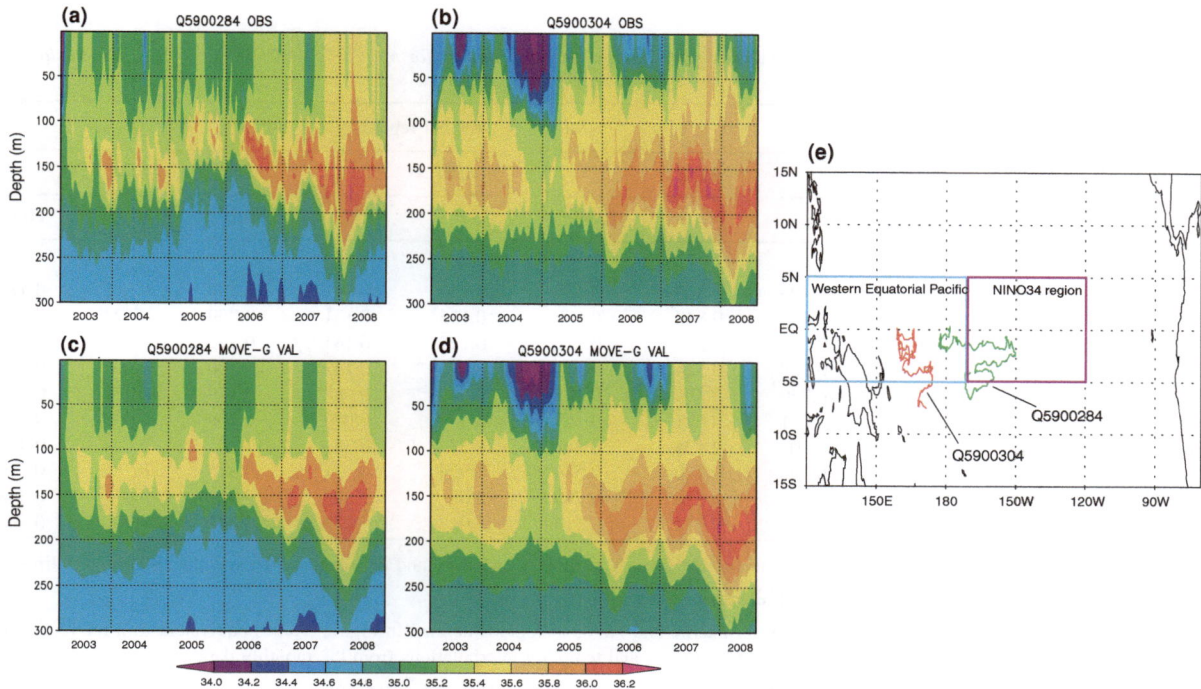

Figure 2

Time-depth sections of the salinity field observed by Argo floats whose WMO ID are (**a**) 5900284, and (**b**) 5900304, and the equivalent salinity sections for MOVE-G VAL (**c** and **d**). The tracks of those Argo floats are depicted in (**e**) together with the areas for the calculation of statistics in Table 1

for the free simulation. Here, the data of the Argo floats that are withheld in the assimilation run is employed as the reference. Monthly data of World Ocean Atlas 2009 (WOA09; LOCARNINI *et al.*, 2010; ANTONOV *et al.*, 2010) is used for calculating anomalies. In the western equatorial Pacific, salinity at 10 and 50 m depths in MOVE-G VAL has excellent accuracy; the root mean square deviations (RMSDs) are less than 60% of the RMSDs for the climatology, and the anomaly correlation coefficients (ACCs) are more than 0.8. The accuracy is still acceptable for salinity at 100–300 m depths in the western equatorial Pacific, and 10–300 m depths in the NINO34 region; the RMSDs are still smaller than those for the climatology, and the ACCs are usually above 0.5. It should be noted that the RMSDs (ACCs) for MOVE-G VAL are much smaller (higher) than those for the free simulation. These statistics indicate that the data assimilation not only reduces the bias in the surface–subsurface salinity field but also improves its variability.

Table 1 also presents statistics for BLT, OHC, WWHC, and SPTWSC. The ACCs for BLT are around 0.5, which is higher than those for the free simulation. The RMSDs for BLT are slightly reduced by the data assimilation in MOVE-G VAL. Influence of high frequency variations in the atmosphere probably decreases the correlation, and it is expected that the variability of BLT in MOVE-G VAL is more reliable for seasonal-to-interannual time scales. The vertically integrated indices (OHC, WWHC and SPTWSC) have good accuracy because the integration can cancel out some of the small-scale noise. In particular, the ACC of SPTWSC in the western equatorial Pacific is 0.2 higher than that for salinity in the subsurface (100–200 m). The ACC is also high (over 0.7) for the NINO34 region. Data assimilation has increased the ACCs by over 0.1. SPTWSC is thus calculated reliably from the result of the assimilation run.

4.2. Current Field

In this subsection, we attempt to validate the current field in MOVE-G RA07. First, we compare

Table 1

Statistics associated with the accuracy of salinity values at the fixed depths and indices for MOVE-G VAL (VAL) and the free simulation (Free)

Element	Western equatorial Pacific					NINO34 region				
	RMSD			ACC		RMSD			ACC	
	Clim	VAL	Free	VAL	Free	Clim	VAL	Free	VAL	Free
Salinity										
10 m	0.371	0.218	0.406	0.825	0.674	0.167	0.142	0.290	0.512	0.411
50 m	0.353	0.192	0.435	0.842	0.698	0.155	0.131	0.259	0.523	0.403
100 m	0.185	0.156	0.285	0.525	0.118	0.177	0.141	0.221	0.629	0.574
150 m	0.181	0.153	0.328	0.524	−0.074	0.176	0.142	0.200	0.603	0.342
200 m	0.182	0.153	0.249	0.565	0.225	0.099	0.088	0.189	0.478	0.343
300 m	0.064	0.051	0.221	0.621	0.159	0.037	0.029	0.209	0.633	0.121
BLT	21.84	20.14	21.10	0.494	0.448	18.24	16.38	17.34	0.524	0.484
OHC	0.786	0.562	0.796	0.710	0.586	0.924	0.571	0.875	0.785	0.719
WWHC	34.55	10.92	17.14	0.896	0.789	13.08	5.806	8.227	0.875	0.748
SPTWSC	20.53	10.60	35.43	0.774	0.514	13.02	9.079	21.06	0.720	0.590

RMSD RMSD between VAL/Free and reference values. RMSD between the climatology and reference values are also presented (Clim)

ACC ACC between anomalies of VAL/Free and reference values

The statistics are estimated for the western equatorial Pacific and NINO34 regions defined in Fig. 2. Values equivalent to a reference for MOVE-G VAL and the free simulation are calculated by spatial and temporal interpolations from the monthly data

Units of RMSDs for BLT, OHC, WWHC, and SPTWSC are m, °C, 10^7 J m^{-2}, and kg m^{-2}, respectively

the variation of the near-surface zonal current with the equivalent data from Ocean Surface Current Analysis-Real Time (OSCAR; JOHNSON et al., 2007) in Fig. 3. The two current fields correspond well in the area east of 150°E, although the spatially noisy features in the OSCAR field are smoothed out in MOVE-G RA07. The eastward extensions of the eastward current that occurred in the early stage of the EL Niños in 1997, 2003, and 2006 are estimated both in OSCAR and MOVE-G RA07. The substantial reversals of the westward South Equatorial Current (SEC) that occurred in 1997, 1999–2001, 2006, 2008 are found in both datasets. These zonal current variations also seem to be consistent with the displacement of the salinity front (35.0 salinity contours). In contrast, the prevailing eastward currents west of 150°E in OSCAR are not estimated in MOVE-G RA07 although the weakening of the currents in 1995–1996, and 1998–2001 in OSCAR is expressed as intensification of the westward current in MOVE-G RA07. This probably stems from excess easterly trade winds in the atmospheric forcing. The low resolution of the model may also cause an inaccurate current field.

Vertical current profiles at the equator are also compared with the Acoustic Doppler Current Profiler (ADCP) mounted on TAO buoys. Figure 4 illustrates that seasonal-to-interannual variations of the zonal currents at 0°–165°E are reasonably recovered in MOVE-G RA07, although the maximum speed of the subsurface Equatorial Undercurrent (EUC) is underestimated. MOVE-G RA07 estimates the near surface eastward currents that occurred before El Niños in 1997, 2002, and 2006. The shallowing of EUC in 1998, and the upward extension of the eastward current in the first half of 2000 are also recovered. In contrast, the substantial southward currents in 150–250 m depths are not reproduced, and the strong southward current around 80 m depth in 1998 is underestimated. This poor expression of meridional currents stems from insufficient model physics, inaccurate forcing of wind stress, and the absence of a close connection like the geostrophic balance between the current and density (temperature and salinity) fields. Similar features are also found for the current profiles at 0°–170°W.

These results confirm that the zonal current field in the equatorial Pacific in MOVE-G RA07 is reliable

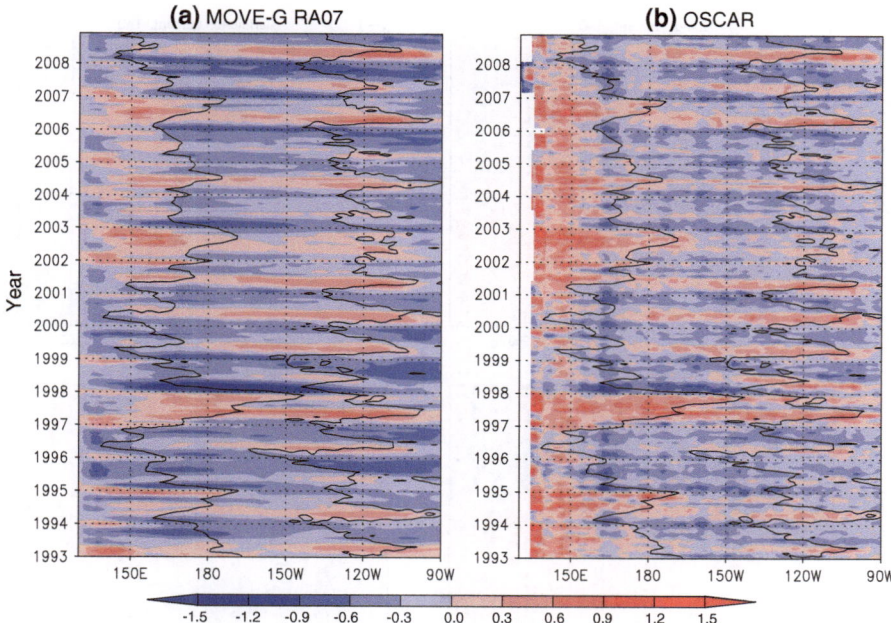

Figure 3
Longitude-time sections of **a** eastward current averaged over 0–30 m depth in MOVE-G RA07, and **b** near-surface eastward current calculated from OSCAR (monthly average) averaged over 1°S–1°N during 1993–2008. Units in m s^{-1}. *Thick black lines* denote 35.0 contours of SSS, which represent the position of the salinity fronts

enough for studies of interannual variability. However, the reproducibility of the meridional currents is insufficient and we must be careful to use the current field in analyses.

5. Distribution and Variability of the Barrier Layer

In this section, we examine the climatological features and variability of the barrier layer in the equatorial Pacific in MOVE-G RA07. First, we present the BLT climatology fields for December–May (D–M) and June–November (J–N) in Fig. 5.

The area of the thick barrier layer extends from 0°–160°E to 12°S–170°W in both periods. This area corresponds to the salinity front where the high salinity water associated with SPTW advected by SEC confronts the fresh water in the western equatorial Pacific. The existence of this area is consistent with Ando and McPhaden (1997) (hereafter AM97), Mignot *et al.* (2007) (hereafter, MEA07), and FK03b. This area has two peaks of BLT. One is

located at 160°E near the equator, and is estimated in AM97. It shifts southward and intensified in J–N.

Another peak is located at 10°S–180°. It thickens during the development of the salinity front in D–M and shifts southward in J–N. Another area of the thick barrier layer exists north of the sea surface salinity (SSS) maximum in the south Pacific and the barrier layer is thicker in J–N in this area. MEA07 also indicated that this maximum exists in May–October.

In the North Pacific, a band of the thick barrier layer extends from the Philippine Sea along the 34.4 isohaline in D–M. This band is also seen in MEA07 and FK03b. Sato *et al.* (2004) indicated that the thick barrier layer is often observed by Argo floats in this area. This barrier layer tends to form by capping the thick isothermal layer generated by oceanic deep convection in winter with a thin low salinity layer. In contrast, a band of the thick barrier layer emerges along 5°N in J–N, probably accompanying the intensification of the Intertropical Convergence Zone (ITCZ) in this season. This band is also found in AM97 and MEA07.

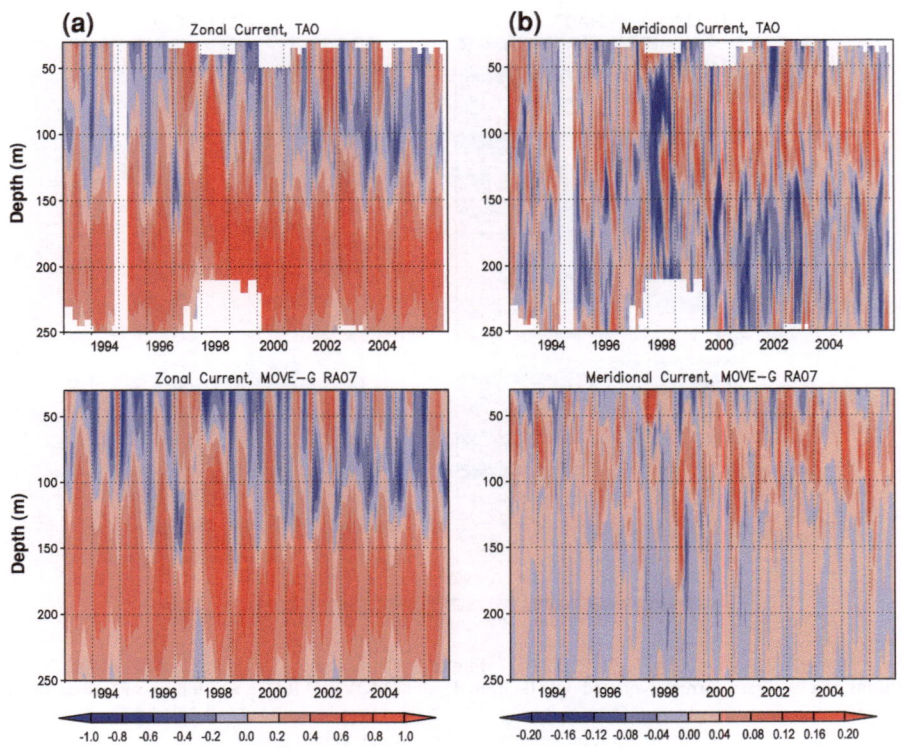

Figure 4
Time-depth sections of the current fields during 1993–2006 observed by ADCP mounted on the TAO buoy at 0°–165°E (monthly average, *top*), together with the sections estimated in MOVE-G RA07 (*bottom*). Units in m s^{-1}. **a** Eastward current. **b** Northward current

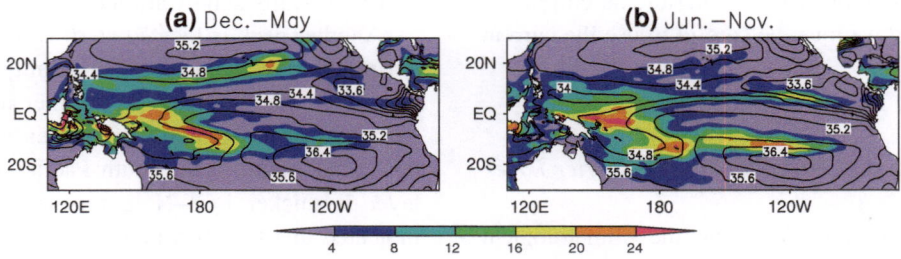

Figure 5
Climatological maps of BLT (m; *shading*) and SSS (*contour*) for **a** December–May and **b** June–November in 1993–2008

AM97 indicated that the area of the thick barrier layer in the equatorial Pacific shifts to the east in El Niño periods. FK03b also suggested a zonal shift of the thick barrier layer in response to ENSO, based on objective mapping results. This shift can also be confirmed in MOVE-G RA07 (Fig. 6a). The thick barrier layer is generally positioned at 165°E, which is consistent with the climatology of BLT (Fig. 5). It then shifts eastward over the date line in the boreal winters of 1995–96, 1997–98, 2002–03, and 2006–07

when El Niño-like events occur. In particular, an unusual eastward shift of the thick barrier to 90–150°W occurs in the strong El Niño of 1997.

The shift of the thick barrier layer is associated with the migration of the equatorial near-surface salinity front. The variation of SSS in the equatorial Pacific is illustrated in Fig. 6b. In this figure, the front is recognized as the zone with a large zonal gradient existing generally at 170°E, and roughly represented by the 35.0 contour of SSS. The salinity front

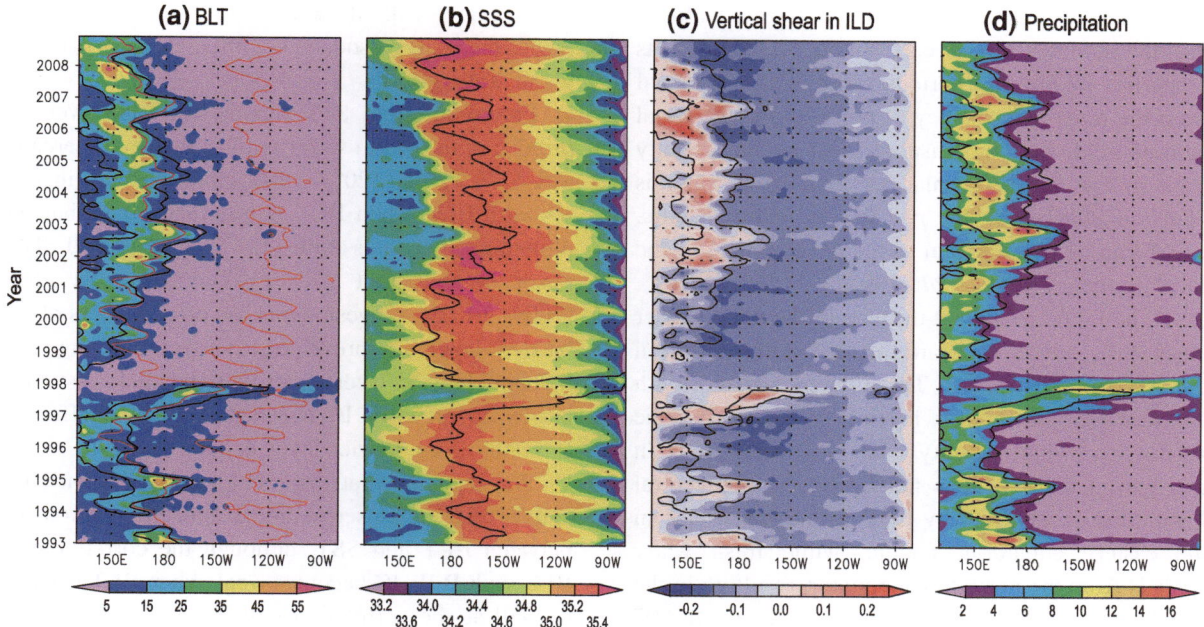

Figure 6
Longitude-time sections of **a** BLT (m), **b** SSS, **c** vertical shear of zonal currents in the isothermal layer (m s^{-1}) in MOVE-G RA07, and **d** precipitation (monthly average, mm day^{-1}) from CMAP during 1993–2008 averaged over 2°S–2°N. *Black lines* denote 40×10^7 J m^{-2} contours of WWHC in **a** and **d**, 28°C contours of SST in **b**, and 15 m contours of BLT in **a**. *Red lines* in **b** denote 35.0 contours of SSS

migrates zonally in response to ENSO, as indicated by former studies (e.g., PICAUT *et al.*, 1996; DELECROIX and PICAUT, 1998). The salinity front travels to the east, along with the eastward extension of the fresh water in mature phases of El Niños (e.g., at the end of 1994, 1997, 2002, and 2006). In particular, the front reaches 135°W in December 1997. This variation is consistent with FK03b.

The 35.0 contour of SSS, representing the salinity front, is also superimposed in Fig. 6a. Consistent with FK03b, the thick barrier layer usually exists west of the salinity front and shifts along with the fronts. Climatological fields also indicate the existence of the thick barrier layer near the equatorial salinity front at the eastern edge of the fresh water (Fig. 5). The variability of the barrier layer in MOVE-G RA07 is thus consistent with the idea that the thick barrier layer is generated west of the front by the high salinity water in the central equatorial Pacific subducting under the fresh water pool (e.g., VIALARD and DELECLUSE, 1998; CRONIN and MCPHADEN, 2002).

Figure 6c indicates a close relationship between BLT and the vertical shear of the zonal currents in the

isothermal layer in the equatorial Pacific. Here, the vertical shear is calculated as the difference between the zonal currents averaged over the upper and lower halves of the isothermal layer. A thick barrier layer tends to develop when the vertical shear is positive (i.e., the current in the lower isothermal layer is westward relative to the above) or slightly negative at the equator. Actually, a small negative value for the vertical shear does not necessarily mean that there is no substantial layer in which the current is westward relative to the near-surface layer because a strong eastward EUC occasionally penetrates into the lower part of the isothermal layer and decreases the value. This feature also suggests that the barrier layer is developed when high salinity water is advected by the relatively westward current subducting under surface fresh water.

Another important feature is that the 28°C contour of SST, which represents the eastern edge of the warm water pool, roughly moves with the salinity front (Fig. 6b) as in FK03b. For this reason, the salinity front is considered to be an indicator of the eastern edge of the warm water pool, and implies that the warm and fresh water in the western equatorial Pacific advects to the east without changing its properties in El Niño periods. This is also assumed from the rough consistency of the near-surface zonal current field with the displacement of the salinity front in Fig. 3. The displacement of fresh water is likely to induce an SST rise in the central Pacific, which can be instrumental in the development of the El Niño (e.g., PICAUT et al., 1997).

It should be noted that the area of the thick barrier layer shifts further eastward than the salinity front in the second half of 1997. This is not consistent with the idea that the barrier layer is generated by the subduction of high salinity water east of the front. In this period, SPTW does not upwell in the central Pacific but is advected by EUC to the subsurface in the eastern equatorial Pacific, while fresh water spreads from the north to the equator, resulting in the development of a thick barrier layer there. MEA07 also indicated that a thick barrier layer can exist in the eastern equatorial Pacific in El Niño years. It should be also noted that SST unusually rises over 28°C around the barrier layer. This rise is not explained by the advection of warm and fresh water from the western equatorial Pacific. Although the fresh water from the north already has a temperature of about 28°C, the SST rise may be enhanced by the effect of the barrier layer on SST discussed in the next section.

6. Relationship Between the Barrier Layer and Near-Surface Temperature

In this section, we confirm the relationship between the barrier layer and near-surface temperature. Former studies suggested that the barrier layer induces an SST rise (e.g., LUKAS AND LINDSTROM, 1992), and indicated that thick barrier layers tend to coexist with high SSTs (e.g., ANDO and MCPHADEN, 1997; FUJII and KAMACHI, 2003; MAES et al., 2006; BOSC et al., 2009). Figure 7 demonstrates that the frequency distribution of SST in the equatorial Pacific warm water pool (SST >28°C) is affected by BLT. The frequency monotonically decreases (increases) with the increase of SST when BLT is less than 5 m (greater than 25 m). The peak of occurrence moves to

higher SST for the thicker barrier layer. Thus, larger BLT is closely associated with higher SST in the warm water pool.

Figure 8a plots SST against BLT for vertical profiles in the warm water pool in the square area of 5°S–5°N, 120°E–120°W. SST monotonically increases with BLT when BLT is less than 20 m, but the plotted line becomes almost flat for greater BLT. This result is consistent with FK03b and Bosc et al. (2009). FK03b suggested that the effectiveness of the barrier layer in preventing vertical mixing and inducing high SST correlates with the thickness of the barrier layer when BLT is small, but the effect is saturated if the barrier layer is thick enough.

We also find that the relationship between BLT and WWHC is closer to linear than the relationship between BLT and SST, although the correlation of BLT with WWHC becomes weak when BLT is large, as well as the correlation with SST (Fig. 8b). Thus, the thick barrier layer tends to be associated with a high heat content in the warm water. The relationship between the thick barrier layer and large WWHC is also demonstrated in Fig. 6a, in which the area of the thick barrier layer corresponds well with the area of large WWHC. In particular, the large WWHC area moves to the east along with the large BLT area. The large WWHC area corresponds well to the area of heavy precipitation (Fig. 6d). Thus, we assume that the thick barrier layer supports the build-up of WWHC, and induces atmospheric convection.

Good correspondence of a large precipitation area (Fig. 6d) with a large BLT area (Fig. 6a) can lead to the assumption that precipitation generates a barrier layer. Actually, precipitation may generate a barrier layer if the subsurface layer is originally occupied by warm salty water such as SPTW. In that case, the barrier layer probably increases WWHC and further enhances precipitation. Thus, development of a barrier layer seems to take place simultaneously with the increase of WWHC and the enhancement of precipitation.

7. Summary

In this paper, we examined the feasibility of the state-of-the-art ocean data assimilation system

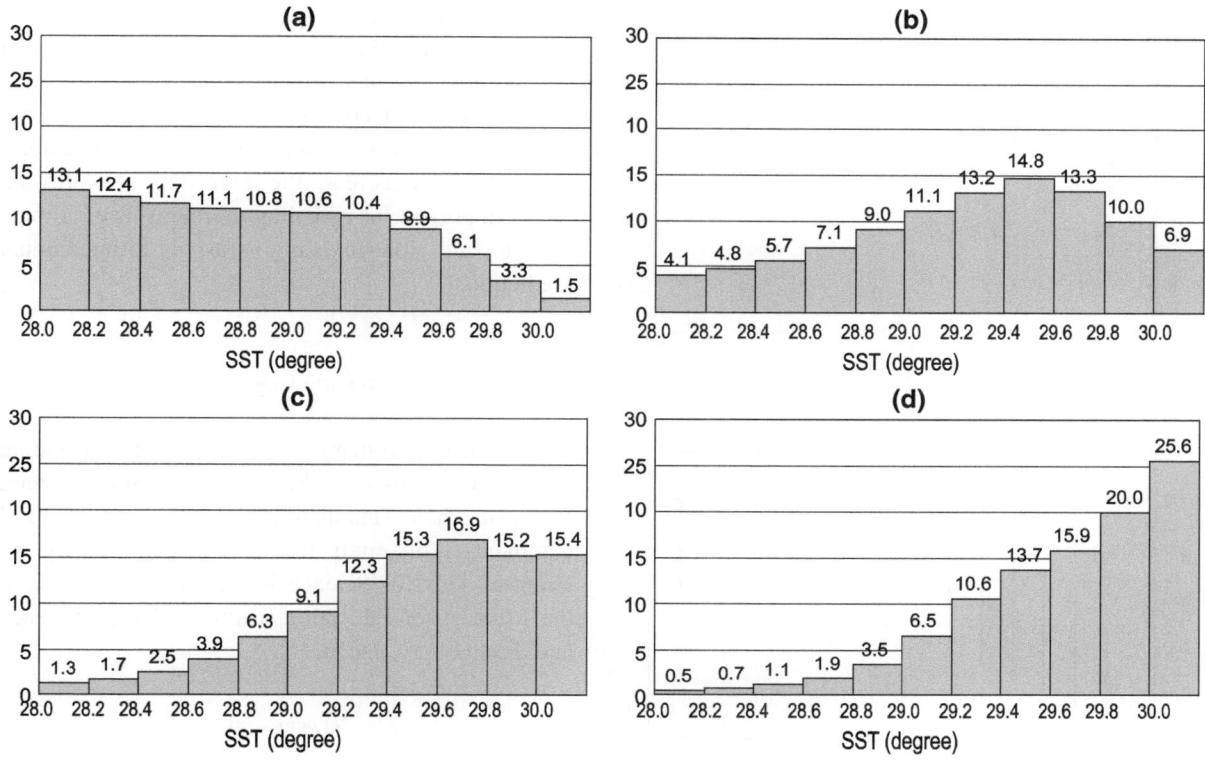

Figure 7
Histograms of SST (°C) for BLTs of **a** 0–5 m, **b** 5–15 m, **c** 15–25 m, and **d** greater than 25 m based on the vertical profiles at the model grid points with SST exceeding 28°C in 5°S–5°N, 120°E–120°W in 1993–2008

developed in MRI to increase our understanding of the salinity variability associated with the barrier layer in the equatorial Pacific. First, we validated the reproducibility of salinity and temperature fields in the equatorial Pacific by the data assimilation system. We performed an assimilation run where some temperature and salinity observations by TRITON buoys and Argo floats are withheld. The assimilation run reproduces interannual variability of near-surface low salinity and subsurface high salinity observed by the TRITON buoys and Argo floats that are withheld, although a relatively large high salinity bias exists in the subsurface layer in the free run. RMSDs of salinity at 10–300 m depth between the assimilation run and the independent data are smaller than those between the climatology and the data. Compared with the free run, RMSDs are reduced, and ACCs are increased in the assimilation run. Statistics also indicated that BLT and indices associated with heat and salinity contents are also estimated better in the

assimilation run than the climatology and in the free run.

We also validated the current fields in the reanalysis where all available temperature and salinity data are assimilated. The near-surface zonal current field east of 150°E is consistent with the equivalent data from OSCAR, although eastward bias is found west of the longitude. Comparison with ADCP mounted on TAO buoys also showed the reliability of the zonal currents although the meridional currents are not reconstructed sufficiently.

We then analyzed the barrier layer and relevant variability of the salinity field in the reanalysis. The thick barrier layer area is generally found to the west of the salinity front, and is displaced zonally with the migration of the front responding to ENSO, although the area went to east over the front in the 1997 El Niño. It is shown that the vertical shear of the zonal current in the isothermal layer correlates with BLT. These results remind us that the barrier layer is

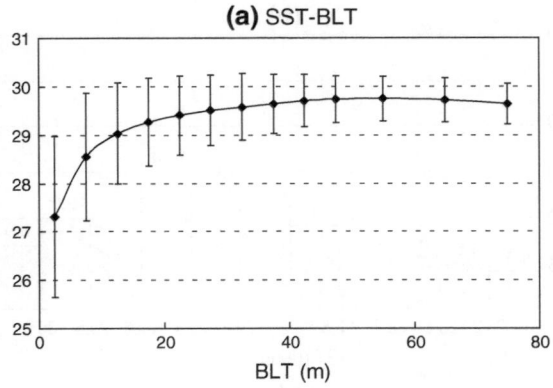

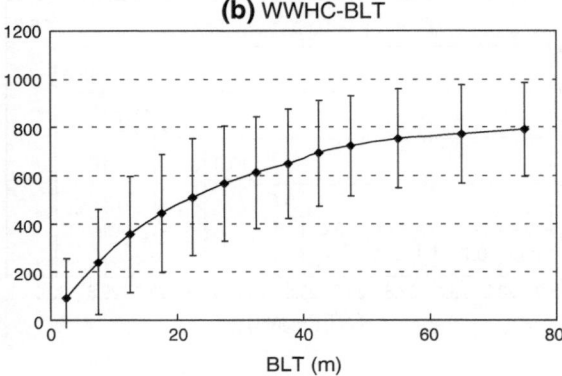

Figure 8

Plots of **a** SST (°C) and **b** WWHC (10^7 J m^{-2}) against BLT (m) for vertical profiles at the model grid points in 5°S–5°N, 120°E–120°W in 1993–2008. Profiles with BLT smaller (larger) than 50 m are binned every 5 m (10 m) of BLT. The averages and the standard deviations in each bins are denoted by *plots* and *error bars*

when we examine the seasonal-to-interannual variability of the climate. Improvement of the data assimilation systems will provide a more precise salinity field and contribute to the progress in the understanding on it. In addition, sustainable salinity observing systems (e.g., the Argo float network) are surely important for revealing mechanisms of climate change and for realizing seasonal-to-interannual prediction.

Acknowledgments

We would like to thank Dr. T. Yasuda for providing us the OGCM used in the data assimilation system. We also thank anomalous reviewers for helpful comments. This study was partly supported by the Grant-in-Aids for Science Research 19540469 from the Ministry of Education, Culture, Sports, Science and Technology, Japan.

formed by subducting high salinity water in the central equatorial Pacific under the fresh water. It is also confirmed that BLT is closely correlated with the near-surface temperature in the equatorial Pacific.

Although the salinity effect seems to be secondary, the salinity variability may not be negligible for the detailed description of ENSO. This study reconfirmed the relationship between high SST and the thick barrier layer. MAES *et al.* (2002, 2005) indicated the importance of the barrier layer for the preconditioning of El Niños. ZHANG *et al.* (2010) also suggested that the control of the MLD by salinity associated with the surface fresh water flux can affect the amplitude of ENSO. We should therefore pay more attention to salinity variability in the ocean

REFERENCES

ANDO, K., and MCPHADEN, M. J. (1997), *Variability of surface layer hydrography in the tropical Pacific Ocean*, J Geophys Res *102*, 23063–23078.

ANTONOV, J. I., SEIDOV, D., BOYER, T. P., LOCARNINI, R. A., MISHONOV, A. V., GARCIA, H. E., BARANOVA, O. K., ZWENG, M. M., and JOHNSON, D. R. (2010), *World Ocean Atlas 2009, Volume 2: Salinity* (ed. S. Levitus). NOAA Atlas NESDIS *69* (U.S. Government Printing Office, Washington, D.C., 2010), 184 pp.

BALMASEDA, M. A., DEE, D., VIALARD, A., and ANDERSON, D. L. T. (2007), *A multivariate treatment of bias for sequential data assimilation: application to the tropical oceans*, Q J R Meteorol Soc *133*, 167–179.

BLOOM, S. C., TAKACS, L. L., DASILVA, A. M., and LEDVINA, D. (1996), *Data assimilation using incremental analysis updates*, Mon Wea Rev *124*, 1256–1271.

BOSC, C., DELCROIX, T., and MAES, C. (2009), *Barrier layer variability in the western Pacific warm pool from 2000 to 2007*, J Geophys Res *114*, C06023, doi:10.1029/2008JC005187.

BOUTIN, J., and MARTIN, N. (2006), *ARGO upper salinity measurements: perspectives for L-band radiometers calibration and retrieved sea surface salinity validation*, IEEE Geosci Remote Sens Lett *3*, 202–206, doi:10.1029/LGRS.2005.861930.

CLS (2004), *SSALTO/DUACS user handbook (M)SLA and (M)ADT near-real time and delayed time products*, CLS-DOS-NT-04, *103* (Collecte Localisation, Satellites (CLS), Toulouse, France, 2004), 42 pp.

CONKRIGHT, M. E., ANTONOV, J. I., BARANOVA, O., BOYER, T. P., GARCIA, H. E., GELFELD, R., JOHNSON, D., LOCARNINI, R. A., O'BRIEN, T. D., SMOLYAR, I., and STEPHENS, C. (2002), *World Ocean Database 2001, volume 1, introduction* (eds. S. Levitus)

NOAA Atlas NESDIS, *42* (U. S. Government Printing Office, Washington D. C., 2002), 167 pp.

CRONIN, M. F., and MCPHADEN, M. J. (2002), *Barrier layer formation during westerly wind bursts*, J Geophys Res *107*, 8020, doi:10.1029/2001JC001171.

DE BOYER MONTÉGUT, C., MADEC, G., FISCHER, A. S., LAZAR, A., and LUDICONE, D. (2004), *Mixed layer depth over the global ocean: an examination of profile data and a profile-based climatology*, J Geophys Res *109*, C12003, doi:10.1029/2004JC002378.

DELCROIX, T., ELDIN, G., RADENAC, M. H., TOOLE, J., and FIRING, E. (1992), *Variations of the western equatorial Pacific Ocean, 1986–1988*, J Geophys Res *97*, 5423–5445.

DELCROIX, T., and PICAUT J. (1998), *Zonal displacement of the western equatorial Pacific "fresh pool"*, J Geophys Res *103*, 1087–1098.

FUJII, Y. (2005), *Preconditioned Optimizing Utility for Large-dimensional analyses (POpULar)*, J Oceanogr *61*, 167–181.

FUJII, Y., ISHIZAKI, S., and KAMACHI, M. (2005), *Application of nonlinear constraints in a three-dimensional variational ocean analysis*, J Oceanogr *61*, 665–662.

FUJII, Y., and KAMACHI, M. (2003a), *A reconstruction of observed profiles in the sea east of Japan using vertical coupled temperature-salinity EOF modes*, J Oceanogr *59*, 173–186.

FUJII, Y., and KAMACHI, M. (2003b), *Three-dimensional analysis of temperature and salinity in the equatorial Pacific using a variational method with vertical coupled temperature-salinity empirical orthogonal function modes*, J Geophys Res *108*, 3297, doi:10.1029/2002JC001745.

FUJII, Y., and KAMACHI, M. (2003c), *A nonlinear preconditioned quasi-Newton method without inversion of a first-guess covariance matrix in variational analyses*, Tellus *55A*, 450-454.

FUJII Y., MATSUMOTO, S., KAMACHI, M., and ISHIZAKI, S. (2010), *Estimation of the equatorial Pacific salinity field using ocean data assimilation systems*, Adv Geosci *18*, 197–212.

FUJII Y., NAKAEGAWA, T., MATSUMOTO, S., YASUDA, T., YAMANAKA, G., and KAMACHI M. (2009), *Coupled climate simulation by constraining ocean fields in a coupled model with ocean data*, J Climate *22*, 5541–5557.

FUKUMORI, I. (2002), *A partitioned Kalman filter and smoother*, Mon Wea Rev *130*, 1370–1383.

GENT, P. R., and MCWILLIAMS, J. C. (1990), *Isopycnal mixing in ocean circulation models*, J Phys Oceanogr *20*, 150–155.

GOULD, W. J., and TURTON, J. (2006), *Argo-sounding the oceans*, Weather, *61*, 17–21, doi:10.1256/wea.56.05.

HAMILTON, D. (1994), *GTSPP builds an ocean temperature-salinity database*, Earth System Monitor, *4(4)*, 4–5.

HAYES, S. P., MANGUM, L. J., PICAUT, J., SUMI, A., TAKEUCHI, K. (1991), *TOGA-TAO, a moored array for real-time measurements in tropical Pacific Ocean*, Bull Amer Meteor Soc *72*, 339–347.

HÉNIN, C., du PENHOAT, Y., and IOUALALEN, M. (1998), *Observations of sea surface salinity in the western Pacific fresh pool, large-scale changes in 1992–1995*, J Geophys Res *103*, 7523–7536.

ISHIZAKI, H., and MOTOI, T. (1999), *Reevaluation of the Takano-Oonishi scheme for momentum advection on bottom relief in ocean models*, J Atmos Oceanic Tech *16*, 1994–2010.

ISHIZAKI, S., MATSUMOTO, S., FUJII, Y., YASUDA, T., and KAMACHI, M. (2006), *Correction of the zonal wind stress in the tropical Pacific using the ocean data assimilation system in Meteorological Research Institute* (in Japanese), Proc The Oceanographic

Society of Japan 2006 Spring Meeting (Yokohama, Japan, 2006) p. 23.

JOHNSON, E. S., BONJEAN, F., LAGERLOEF, G. S. E., GUNN, J. T., and MITCHUM, G. T. (2007), *Validation and Error Analysis of OSCAR Sea Surface Currents*, J Atmos Oceanic Technol *24*, 688–701

JOHNSON, G. C., and MCPHADEN, M. J. (2000), *Upper equatorial Pacific Ocean current and salinity variability during the 1996–1998 El Niño-La Niña cycle*, J Geophys Res *105*, 1037–1053.

KALNAY, E., KANAMITSU, M., KISTLER, R., COLLINS, W., DEAVEN, D., GANDIN, L., IREDELL, M., SAHA, S., WHITE, G., and WOOLLEN, J. (1996), *The NCEP/NCAR 40-year reanalysis project*, Bull Am Meteorol Soc *77*, 437–471.

KAMACHI, M., FUJII, Y., and ZHOU, X. (2002), *Ocean data assimilation in the tropical Pacific: A short survey*, J Oceanogr *58*, 45–55.

KESSLER, W. S. (1999), *Interannual variability of the subsurface high salinity tongue south of the equator at 165°E*, J Phys Oceanogr *29*, 2038–2049.

KURODA, Y. (2002), *TRITON, present status and future plan, Report for the international workshop for review of the tropical moored buoy network* (JAMSTEC, Yokosuka, Japan, 2002) 77 pp.

LEE, T., AWAJI, T., BALMASEDA, M. A., E. GREINER, and STAMMER, D. (2009), *Ocean state estimation for climate research*, Oceanogr *22*, 160–167.

LEIPPER, D. F., and VOLGENAU, D. (1972), *Hurricane heat potential of the Gulf of Mexico*, J Phys Oceanogr *2*, 218–224.

LOCARNINI, R. A., MISHONOV, A. V., ANTONOV, J. I., BOYER, T. P., GARCIA, H. E., BARANOVA, O. K., ZWENG, M. M., and JOHNSON, D. R. (2010), *World Ocean Atlas 2009, Volume 1: Temperature* (ed. S. Levitus) NOAA Atlas NESDIS *68* (U.S. Government Printing Office, Washington, D.C., 2010), 184 pp.

LUKAS, R., and LINDSTROM, E. (1991), *The mixed layer of the western equatorial Pacific ocean*, J Geophys Res *96*, 3343–3357.

MAES, C., ANDO, K., DELCROIX, T., KESSLER, W. S., MCPHADEN, M. J., and ROEMMICH, D. (2006), *Observed correlation of surface salinity, temperature and barrier layer at the eastern edge of the western Pacific warm pool*, Geophys Res Lett *33*, L06601, doi: 10.1029/2005GL024772.

MAES, C., BEHRINGER, D., REYNOLDS, R. W., and JI, M. (2000), *Retrospective analysis of the salinity variability in the western tropical Pacific Ocean using an indirect minimization approach*, J Atmos Oceanic Technol *17*, 512–524.

MAES, C., PICAUT, J., and BELAMARI, S. (2002), *Salinity barrier layer and onset of El Niño in a Pacific coupled model*, Geophys Res Lett *29*, 2206, doi:10.1029/2002GL016029.

MAES, C., PICAUT, J., and BELAMARI, S. (2005), *Importance of salinity barrier layer for the buildup of El Niño*, J Climate *18*, 104–118.

MCPHADEN, M. J., BUSALACCHI, A. J., CHENEY, R., DONGUY, J., GAGE, K. S., HALPERN, D., JI, M., JULIAN, P., MEYERS, G., MITCHUM, T. G., NIILER, P. P., PICAUT, J., REYNOLDS, R. W., SMITH, N., and TAKEUCHI, K. (1998), *The tropical Ocean-Global Atmosphere observing system, a decade of progress*, J Geophys Res *103*, 14169–14240.

MCPHADEN, M. J., and ZHANG, D. (2002), *Slowdown of the meridional overturning circulation in the upper Pacific Ocean*, Nature, *415*, 603–608.

MIGNOT, J., DE BOYER MONTÉGUT, C., LAZAR, A., and CRAVATTE, S. (2007): *Control of salinity on the mixed layer depth in the world* ocean: 2. Tropical areas, J Geophys Res 112, C10010, doi: 10.1029/2006JC003954.

NOH, Y., and KIM, H. J. (1999), Simulations of temperature and turbulence structure of the oceanic boundary layer with the improved near-surface process, J Geophys Res 104, 15621–15634.

PICAUT, J., IOUALALEN, M., MENKES, C., DELCROIX, T., and McPHADEN, M. I. (1996), Mechanism of the zonal displacements of the Pacific warm water pool, implication for ENSO, Science, 274, 1486–1489.

PICAUT, J., MASIA, F., and DU PENHOAT, Y. (1997), An advective-reflective conceptual model for the oscillatory nature of the ENSO, Science, 277, 663–666.

REDI, M. H. (1982), Oceanic isopycnal mixing by coordinate rotation, J Phys Oceanogr 12, 1154–1158.

ROEMMICH, D., MORRIS, M., YOUNG, W. R., and DONGUY, J.-R. (1994), Fresh equatorial jet, J Phys Oceanogr 24, 540–558.

SATO, K., SUGA, T., and HANAWA, K. (2004), Barrier layer in the North Pacific subtropical gyre, Geophys Res Lett 31, L05301, doi:10.1029/2003GL018590.

SPRINTALL, J., and TOMCZAK, M. (1992), Evidence of the barrier layer in the surface layer of the Tropics, J Geophys Res 97, 7305–7316.

STAMMER, D., and Co-Authors (2010), Ocean Information Provided through Ensemble Ocean Syntheses, In Proceedings of OceanObs'09: Sustained Ocean Observations and Information for Society, 2 (Venice, Italy, 21–25 September 2009) (eds. J. Hall, D. E. Harrison, D. E. Stammer) (ESA Publication WPP-306) (in press).

TSUJINO, H., MOTOI, T., ISHIKAWA, I., HIRABARA, M., NAKANO, H., YAMANAKA, G., and YASUDA, T. (2010), Meteorological Research

Institute Community Ocean Model Version 3 (MRI.COM3) Manual, Technical Report of Meteorological Research Institute 59 (Meteorological Research Institite, Tsukuba, Japan, 2009), 241 pp.

USUI N., ISHIZAKI, S., FUJII, Y., TSUJINO, H., YASUDA, T., and KAMACHI M. (2006), Meteorological Research Institute Multivariate Ocean Variational Estimation (MOVE) System: Some Early Results, Adv Spa Res 37, 806–822.

VIALARD, J., and DELECLUSE, P. (1998), A OGCM study for the TOGA decade, II, Barrier-layer formation and variability, J Phys Oceanogr 28, 1089–1106.

VIALARD, J., DELECLUSE, P., and MENKES, C. (2002), A modeling study of salinity variability and its effects in the tropical pacific Ocean during the 1993–1999 period, J Geophys Res 107, 8005, doi:10.1029/2000JC000758.

VIALARD, J., MENKES, C., BOULANGER, J.-P., DELECLUSE, P., and GUILYARDI, E. (2001), A model study of oceanic mechanisms affecting equatorial Pacific sea surface temperature during the 1997-98 El Niño, J Phys Oceanogr 31, 1649–1675.

WADA, A., and CHAN, J. C. L. (2008), Relationship between typhoon activity and upper ocean heat content, Geophys Res Lett 35, L17603, doi:10.1029/2008GL035129.

YABLONSKY, R. M., and GINIS, I. (2008), Improving the ocean initialization of coupled hurricane-ocean models using feature-based data assimilation, Mon Wea Rev 136, 2592–2607.

ZHANG, R.-H., WANG, G., CHEN, D., BUSALACCHI, A. J., and HACKERT, E. C. (2010), Interannual Biases induced by freshwater flux and coupled feedback in the tropical Pacific, Mon Wea Rev 138, 1715–1737.

(Received November 8, 2010, accepted May 9, 2011, Published online July 15, 2011)